A Feasible Approach for Natural Products to Treatment of Diseases

A Feasible Approach for Natural Products to Treatment of Diseases

Editors

Sokcheon Pak
Soo Liang Ooi

MDPI • Basel • Beijing • Wuhan • Barcelona • Belgrade • Manchester • Tokyo • Cluj • Tianjin

Editors
Sokcheon Pak
School of Dentistry and
Medical Sciences
Charles Sturt University
Bathurst
Australia

Soo Liang Ooi
School of Dentisry and
Medical Sciences
Charles Sturt University
Bathurst
Australia

Editorial Office
MDPI
St. Alban-Anlage 66
4052 Basel, Switzerland

This is a reprint of articles from the Special Issue published online in the open access journal *Molecules* (ISSN 1420-3049) (available at: www.mdpi.com/journal/molecules/special_issues/diseases_naturalproducts).

For citation purposes, cite each article independently as indicated on the article page online and as indicated below:

LastName, A.A.; LastName, B.B.; LastName, C.C. Article Title. *Journal Name* **Year**, *Volume Number*, Page Range.

ISBN 978-3-0365-7873-6 (Hbk)
ISBN 978-3-0365-7872-9 (PDF)

Contents

About the Editors

Sokcheon Pak

Sok Cheon Pak has taught and researched at Charles Sturt University in Australia since 2007. His expertise has been in the field of complementary medicine and has become recognised nationally and internationally through ongoing external research collaborations. His area of interest relates specifically to introducing evidence-based practice to complementary medicine research and practice. This has been based on laboratory experiments incorporating modern medical technologies to identify and evidence the underlying rationale for prescribing therapeutic substances for treatment. Over the years, his principal focus has been on the experimental/clinical application of bee venom to human diseases. The following are two examples of where his research related to bee venom has been acknowledged: (a) His research has been recognised by an invitation from the publishing company Springer to write two chapters on "Chemical Composition of Bee Venom" and "Health Benefits and Uses in Medicine of Bee Venom" for the publication of a book entitled "Bee products - chemical and biological properties". This invitation is directly related to his research on honeybee venom, focusing on health benefits and uses in medicine. (b) He was appointed as the Guest Editor of a Special Issue on "Bee and wasp venoms: biological characteristics and therapeutic application" for a reputable journal. He currently leads and guides research into nutraceuticals to provide relevant and impactful clinical applications.

Soo Liang Ooi

Soo Liang Ooi is a practising nutritionist and naturopath with clinical experience in Australia and Singapore. He holds a Bachelor of Health Science degree from Charles Sturt University, a Master of Business Administration degree from the National University of Singapore, and a Master of Mathematics degree from the University of Waterloo in Canada. Soo Liang is completing his Doctor of Philosophy (PhD) in investigating the immunomodulating properties of rice bran arabinoxylan compound and its effects on the quality of life of cancer patients at Charles Sturt University. His research interests include evidence-based complementary medicine, nutritional medicine, naturopathy, herbal medicine, meditation, and integrative cancer therapies. As an avid academic writer, Soo Liang has published extensively in major peer-reviewed academic journals. He was nominated for the prestigious Elsevier Atlas award in 2017 for his work on "Transcendental meditation for lowering blood pressure: An overview of systematic reviews and meta-analyses". He is also a co-editor of several academic books, including "Health Benefits of Meditation", "Nutraceuticals in Immune Function", and "Modified Rice Bran Arabinoxylan–Therapeutic Applications in Cancer and Other Diseases".

Preface to "A Feasible Approach for Natural Products to Treatment of Diseases"

In a world where the pursuit of improved health and well-being is a universal goal, the significance of natural products in the treatment of diseases cannot be underestimated. Nature, with its vast repertoire of plants, herbs, and other organic sources, has long provided humanity with a remarkable array of therapeutic compounds. From ancient civilizations to modern medical advancements, the potential of natural products to positively impact our health has remained an enduring subject of fascination and exploration.

This reprint, titled "A Feasible Approach for Natural Products to Treatment of Diseases", delves into the fascinating realm of natural products, unravelling their inherent properties, mechanisms of action, and their potential role in the pursuit of effective disease management. Through comprehensive experiments of scientific research and reviews, this book aims to shed light on the immense potential of natural products as a viable approach to enhancing health and combating diseases.

We thank all the authors for their invaluable contributions to this reprint. Their collective expertise, extensive research, and dedication have shaped a comprehensive resource that bridges the gap between traditional wisdom and contemporary scientific understanding. Their insights, experimental findings, and critical analyses have enriched the contents of this reprint, fostering a deeper understanding of the therapeutic potential locked within nature's abundant offerings. It is through their tireless efforts that we are able to present a diverse and multidimensional exploration of natural products, offering new avenues for healthcare professionals, researchers, and enthusiasts to harness the power of these fascinating resources from nature.

Sokcheon Pak and Soo Liang Ooi
Editors

molecules

Editorial

Editorial: A Feasible Approach for Natural Products to Treatment of Diseases

Soo Liang Ooi and Sok Cheon Pak *

School of Dentistry and Medical Sciences, Charles Sturt University, Bathurst, NSW 2795, Australia; sooi@csu.edu.au
* Correspondence: spak@csu.edu.au

Citation: Ooi, S.L.; Pak, S.C. Editorial: A Feasible Approach for Natural Products to Treatment of Diseases. *Molecules* **2023**, *28*, 3791. https://doi.org/10.3390/molecules28093791

Received: 24 April 2023
Accepted: 26 April 2023
Published: 28 April 2023

The potential of natural products from both plant and animal sources to treat diseases remains enormous, as our understating forms just the tip of the iceberg. There are also other issues to address regarding extraction, isolation, and standardization of any naturally derived compounds and their effective mode of therapeutic delivery. Hence, much research is needed before any natural products can be feasibly applied in treatments. The aim of this Special Issue was to first bring together academics and researchers whose work focuses on natural products to explore and debate different therapeutic perspectives, methodological approaches, and analytical findings. The second aim was to explore key developments and highlight new areas of research and ideas emerging on natural products. Finally, the third aim was to highlight and debate the potential translation of research from the bench to the bedside, which may directly benefit humans.

The opening article by Zhang et al. [1] explores a novel class of antifungal antibiotics from *Icacina trichantha* Oliv., a potential plant source of various bioactive endophytes. The authors document the endophytic fungus *Penicillium oxalicum* derived from the *Icacina* species and its related bioactivity. A novel macrolactam named oxalactam A is detailed, and dipeptides and alkaloids were isolated from this fungus. The article describes the isolation, structural elucidation, and anti-fungal activities of these isolates. These findings indicate an alternative natural source of new fungicides.

Somwong et al. [2] next explore the underlying mechanisms of the fresh stem bark extract of *Holoptelea integrifolia* (Roxb.) Planch. in the context of treating human cutaneous diseases. They used thin layer and gas chromatography to screen phytochemicals from this sample. Two bioactive compounds, friedelin and lupeol, were identified, and their activity in wound healing was investigated in keratinocytes. Both compounds exhibited wound healing activity by increasing keratinocyte migration and matrix metalloproteinase-9 production. In addition, gene expressions for wound healing and pro-inflammation were increased and reduced, respectively, after treatment with *n*-hexane extracts of *H. integrifolia* and its bioactive compounds. Thus, the wound healing and anti-inflammation properties are mediated by regulating the gene expression involved in skin re-epithelialization.

As major active compounds from *Angelica sinensis* and *Astragalus membranaceus*, *Angelica sinensis* polysaccharide (ASP) and *Astragalus membranaceus* polysaccharide (AMP) are known to exert anti-fibrosis and hepatoprotective effects. Wen et al. [3] studied their roles and related mechanisms in liver regeneration. The study demonstrated that ASP and AMP promoted hepatocyte proliferation at various concentrations in vitro while inducing hepatoprotection of liver injuries, demonstrating an enhanced liver/body weight ratio and reduced serum transaminase and total bilirubin levels after partial hepatectomy in mice. Further analyses confirmed the involvement of the JAK2/STAT3/HK2 pathway in ASP- and AMP-accelerated liver regeneration. These results indicate ASP and AMP as potential hepatoprotective agents.

In another combined in vivo and in vitro experiment, Moon et al. [4] showed that walnuts (*Juglans regia*) could be a potential functional food to improve diabetic cognitive deficits and neuronal impairments. The study found evidence that walnuts inhibit reactive oxygen species production in high-glucose-induced neuronal and hippocampal cells and ameliorate behavioral and memory dysfunction in mice with cognitive impairment induced by a high fat diet. Further amelioration of cerebral damage with walnut treatment was observed via protein expression of the JNK signaling and apoptosis pathways. Hence, walnuts could protect against cerebral disorders, insulin resistance, oxidative stress, and inflammation.

Molineria recurvata (MR) is used to manage diabetes mellitus. Dey et al. [5] investigated the protective effects of MR extracts versus nephropathy in streptozotocin-induced diabetic rats. The MR extracts showed anti-fibrotic properties, demonstrated by downregulated expressions of fibronection, collagen, and α-smooth muscle actin. Increased oxidative stress caused by induced hyperglycemia was improved after administration of MR extracts, with the marked restoration of antioxidant enzymes. Diabetic-induced renal injuries with accompanying inflammation were also ameliorated by MR extracts, with increased levels of SIRT1 and SIRT3 and a reduced claudin-1 level in the kidneys. Thus, MR could promote renal repair by repressing inflammation and oxidative stress.

Idriss et al. [6] investigated inhibitors from *Saussurea costus* against the main protease, as it is an ideal target for COVID-19 treatment. The authors found that eight of the active inhibitors were carbohydrates, five were fatty acids, three were terpenoids, two were carboxylic acids, one was a tannin, one was a phenolic compound, and one was a sterol. In addition, the *Saussurea costus* aqueous extract had no virucidal effect and inhibited the virus after cell entry. As such, the findings justify using this plant, especially in rural communities, for preventing and treating COVID-19.

As a flagellated parasitic protozoan, *Trypanosoma brucei* causes African trypanosomiasis, a neglected tropical disease. According to Chaudhuri et al. [7], two natural antimetabolites, cholesta-5,7,22,24-tetraenol (CHT) and ergosta-5,7,22,24(28)-tetraenol (ERGT), possess antiprotozoal properties as a trypanocidal agent against *T. brucei*. In addition, CHT/ERGT protected mice infected with *T. brucei* by increasing their survival time. Hence, CHT and ERGT are promising candidates for antitrypanosomal drugs.

When food products containing arginine, glucose, or maltose undergo heat processing, arginyl-fructose (AF) and arginyl-fructosyl-glucose (AFG) are formed. It is known that AF and AFG can modulate the glucose uptake from dietary carbohydrates by exhibiting an inhibitory effect on α-glucosidases. Lee et al. [8] investigated AFs anti-hyperglycemic effect in barley using AF-enriched barley extract (BEE) in mice. Mice who were fed a high fat diet and treated with BEE showed a suppressed body weight gain and increased serum adiponectin levels. The accumulation of intracellular lipids and the expressions of adipogenic genes in the preadipocytes were significantly decreased after treatment with BEE. Thus, BEE could be used as a weight management strategy by inhibiting adipogenesis and increasing adiponectin levels.

Hordyjewska et al. [9] critically examined whether electric cell-substrate impedance sensing (ECIS) could be a valuable tool for live monitoring of changes in the morphology and physiology of cancer cells. Betulin and betulinic acid are naturally occurring pentacyclic lupane-type triterpenoids, possessing a broad spectrum of biological activities, including antitumor activities. The ECIS results confirmed the great potential effects of betulin and betulinic acid's antitumor properties on lung and breast cancer cell lines. Moreover, both substances showed a negligible toxic effect on healthy epithelial cells. The ECIS method is an appropriate alternative to the currently used assay for testing the in vitro anticancer activity of compounds.

Lemieszek et al. [10] bring our attention to the significance of young barley on the human body in terms of colon cancer prevention. The water extracts of young green barley (*Hordeum vulgare*) were evaluated for their immunoenhancement properties. Polysaccharide-rich young green barley extracts may have immunomodulatory properties associated with enhancing the natural killer cells' ability to recognize and eliminate human colon cancer cells without any side effects on normal colon epithelial cells. The findings indicate the beneficial effect of the consumption of young barley regarding colon cancer.

This Special Issue includes three review articles. Yao and Liu [11] offer a comprehensive review of terpenoids, focusing on non-alcoholic fatty liver disease (NAFLD). Despite a lack of clinical research, the authors found that terpenoids could have a therapeutic role in NAFLD by regulating lipid metabolism, insulin resistance, oxidative stress, and inflammation.

Song et al. [12] introduce Citri Reticulatae Pericarpium (CRP), derived from the ripe peel of the Rutaceae plant *Citrus reticulate Blanco* and its cultivars. It is often used to treat diseases with coughs, expectoration, nausea, and vomiting as the main symptoms. The authors review the pharmacology of CRP and the mechanism of flavonoids as the key components of CRP against cancers with a high diagnosis rate. Although CRP does not prevent certain aspects of cancer, it reverses or suppresses the development of cancer through various pathways.

Another review article by Phillips et al. [13] explores the current research on medicinal mushrooms in the context of COVID-19. The authors identify the key properties claimed to confer health benefits. Also considered are the barriers or limitations that may impact the general recommendations of medicinal mushrooms as a therapy. Notably, the authors include mushrooms commonly available for culinary use and obtainable as a dietary supplement for medicinal purposes.

In the last few decades, there has been a wealth of research into natural products. This Special Issue has brought together active researchers who have explored a diverse range of products and provided rich insights into their underlying mechanisms. We believe that further preclinical and clinical research is still required to uncover the active constituents, mechanisms of action, pharmacokinetics, safety, and toxicity of natural products for therapeutic use. The diverse and critical perspectives within this Special Issue provide a springboard to enable the continual development and enhancement of natural products in therapeutic applications.

Conflicts of Interest: The authors declare no conflict of interest.

References

1. Zhang, R.; Ma, Y.; Xu, M.-M.; Wei, X.; Yang, C.-B.; Zeng, F.; Duan, J.-A.; Che, C.-T.; Zhou, J.; Zhao, M. Oxalactam A, a Novel Macrolactam with Potent Anti-Rhizoctonia solani Activity from the Endophytic Fungus Penicillium oxalicum. *Molecules* **2022**, *27*, 8811. [CrossRef] [PubMed]
2. Somwong, K.; Lertpatipanpong, P.; Nimlamool, W.; Panya, A.; Tragoolpua, Y.; Yongsawas, R.; Gritsanapan, W.; Pandith, H.; Baek, S.J. Effect of *Holoptelea integrifolia* (Roxb.) Planch. n-Hexane Extract and Its Bioactive Compounds on Wound Healing and Anti-Inflammatory Activity. *Molecules* **2022**, *27*, 8540. [CrossRef] [PubMed]
3. Wen, X.-D.; Zhang, Y.-L.; Yang, L.; Ye, Z.; Fu, G.-C.; Hu, Y.-H.; Pan, T.; Ye, Q.-B. Angelica sinensis Polysaccharide and Astragalus membranaceus Polysaccharide Accelerate Liver Regeneration by Enhanced Glycolysis via Activation of JAK2/STAT3/HK2 Pathway. *Molecules* **2022**, *27*, 7890. [CrossRef] [PubMed]
4. Moon, J.H.; Kim, J.M.; Lee, U.; Kang, J.Y.; Kim, M.J.; Lee, H.L.; Jeong, H.R.; Go, M.J.; Kim, H.-J.; Park, H.W.; et al. Walnut Prevents Cognitive Impairment by Regulating the Synaptic and Mitochondrial Dysfunction via JNK Signaling and Apoptosis Pathway in High-Fat Diet-Induced C57BL/6 Mice. *Molecules* **2022**, *27*, 5316. [CrossRef] [PubMed]
5. Dey, P.; Kundu, A.; Lee, H.E.; Kar, B.; Vishal, V.; Dash, S.; Kim, I.S.; Bhakta, T.; Kim, H.S. Molineria recurvata Ameliorates Streptozotocin-Induced Diabetic Nephropathy through Antioxidant and Anti-Inflammatory Pathways. *Molecules* **2022**, *27*, 4985. [CrossRef] [PubMed]
6. Idriss, H.; Siddig, B.; Maldonado, P.G.; Elkhair, H.M.; Alakhras, A.I.; Abdallah, E.M.; Torres, P.H.; Elzupir, A.O. Phytochemical Discrimination, Biological Activity and Molecular Docking of Water-Soluble Inhibitors from Saussurea costus Herb against Main Protease of SARS-CoV-2. *Molecules* **2022**, *27*, 4908. [CrossRef] [PubMed]
7. Chaudhuri, M.; Singha, U.K.; Vanderloop, B.H.; Tripathi, A.; Nes, W.D. Steroidal Antimetabolites Protect Mice against Trypanosoma brucei. *Molecules* **2022**, *27*, 4088. [CrossRef] [PubMed]
8. Lee, S.-Y.; Kim, T.-Y.; Hong, J.-Y.; Kim, G.-J.; Oh, J.-B.; Kim, M.-J.; Apostolidis, E.; Lee, J.-Y.; Kwon, Y.-I. Anti-Obesity and Anti-Adipogenic Effects of Administration of Arginyl-Fructose-Enriched Jeju Barley (*Hordeum vulgare* L.) Extract in C57BL/6 Mice and in 3T3-L1 Preadipocytes Models. *Molecules* **2022**, *27*, 3248. [CrossRef] [PubMed]

9. Hordyjewska, A.; Prendecka-Wróbel, M.; Kurach, Ł.; Horecka, A.; Olszewska, A.; Pigoń-Zając, D.; Małecka-Massalska, T.; Kurzepa, J. Antiproliferative Properties of Triterpenoids by ECIS Method—A New Promising Approach in Anticancer Studies? *Molecules* **2022**, *27*, 3150. [CrossRef] [PubMed]
10. Lemieszek, M.K.; Komaniecka, I.; Chojnacki, M.; Choma, A.; Rzeski, W. Immunomodulatory Properties of Polysaccharide-Rich Young Green Barley (*Hordeum vulgare*) Extract and Its Structural Characterization. *Molecules* **2022**, *27*, 1742. [CrossRef] [PubMed]
11. Yao, P.; Liu, Y. Terpenoids: Natural Compounds for Non-Alcoholic Fatty Liver Disease (NAFLD) Therapy. *Molecules* **2023**, *28*, 272.
12. Song, L.; Xiong, P.; Zhang, W.; Hu, H.; Tang, S.; Jia, B.; Huang, W. Mechanism of Citri Reticulatae Pericarpium as an Anticancer Agent from the Perspective of Flavonoids: A Review. *Molecules* **2022**, *27*, 5622. [CrossRef] [PubMed]
13. Phillips, J.M.; Ooi, S.L.; Pak, S.C. Health-Promoting Properties of Medicinal Mushrooms and Their Bioactive Compounds for the COVID-19 Era—An Appraisal: Do the Pro-Health Claims Measure Up? *Molecules* **2022**, *27*, 2302. [CrossRef] [PubMed]

molecules MDPI

Review

Terpenoids: Natural Compounds for Non-Alcoholic Fatty Liver Disease (NAFLD) Therapy

Pengyu Yao [1] and Yajuan Liu [2,*]

1 Shandong Laboratory of Engineering Technology, Suzhou Institution of Biomedical Engineering and Technology, Chinese Academy of Sciences, Jinan 250101, China
2 Department of Febrile Diseases, Shandong University of Traditional Chinese Medicine, Jinan 250355, China
* Correspondence: tgqolyj@163.com

Abstract: Natural products have been the most productive source for the development of drugs. Terpenoids are a class of natural active products with a wide range of pharmacological activities and therapeutic effects, which can be used to treat a variety of diseases. Non-alcoholic fatty liver disease (NAFLD), a common metabolic disorder worldwide, results in a health burden and economic problems. A literature search was conducted to obtain information relevant to the treatment of NAFLD with terpenoids using electronic databases, namely PubMed, Web of Science, Science Direct, and Springer, for the period 2011–2021. In total, we found 43 terpenoids used in the treatment of NAFLD. Over a dozen terpenoid compounds of natural origin were classified into five categories according to their structure: monoterpenoids, sesquiterpenoids, diterpenoids, triterpenoids, and tetraterpenoids. We found that terpenoids play a therapeutic role in NAFLD, mainly by regulating lipid metabolism disorder, insulin resistance, oxidative stress, and inflammation. The AMPK, PPARs, Nrf-2, and SIRT 1 pathways are the main targets for terpenoid treatment. Terpenoids are promising drugs and will potentially create more opportunities for the treatment of NAFLD. However, current studies are restricted to animal and cell experiments, with a lack of clinical research and systematic structure–activity relationship (SAR) studies. In the future, we should further enrich the research on the mechanism of terpenoids, and carry out SAR studies and clinical research, which will increase the likelihood of breakthrough insights in the field.

Keywords: non-alcoholic fatty liver disease (NAFLD); natural products; terpenoids; mechanisms; treatment

Citation: Yao, P.; Liu, Y. Terpenoids: Natural Compounds for Non-Alcoholic Fatty Liver Disease (NAFLD) Therapy. *Molecules* **2023**, *28*, 272. https://doi.org/10.3390/molecules28010272

Academic Editors: Sokcheon Pak and Soo Liang Ooi

Received: 11 November 2022
Revised: 15 December 2022
Accepted: 26 December 2022
Published: 29 December 2022

1. Introduction

In the epidemiology of liver disease, there has been a gradual transition in focus from infectious diseases to metabolic diseases. Non-alcoholic fatty liver disease (NAFLD) has become a serious public health issue, affecting the health of approximately one-quarter of adults worldwide, causing wide-ranging social and economic implications [1]. The prevalence of NAFLD is 25% globally, and it has become the most rapidly increasing cause of liver-related mortality [2]. The prevalence of NAFLD is increasing rapidly worldwide and is predicted to become more prevalent in the future as the obese and diabetic populations increase [3].

NAFLD is a clinicopathological syndrome characterized by parenchymal cell steatosis and fat storage without history of excessive alcohol consumption. Its disease pathophysiological development ranges from non-alcoholic hepatic steatosis to non-alcoholic steatohepatitis (NASH) and hepatic fibrosis, potentially evolving into hepatic cirrhosis, hepatocellular carcinoma (HCC), and liver failure. The prevalence of NASH is approximately 30% for patients with NAFLD, and approximately 20% of NASH patients with fibrosis progress to cirrhosis [4]. However, NAFLD is a diagnosis of exclusion rather than one of inclusion, and so the new nomenclature "metabolic-associated fatty liver disease (MAFLD)" has been proposed [5]. The standardization of the nomenclature for NAFLD still needs to be explored. So far, the above two terms have been accepted. NAFLD is a complex disease in which disease severity is influenced by genetic, environmental, and behavioral factors (e.g., diet, physical activity, and socioeconomic influences) [6]. NAFLD is considered the hepatic manifestation of metabolic syndrome and involves pathological changes in multiple systems, including metabolic disease (such as metabolic syndrome, type 2 diabetes mellitus, obesity, and hypertension) [7,8], cardiovascular disease (such as coronary heart disease, atherosclerosis, cardiomyopathy, and arrhythmia) [9], extrahepatic carcinomas (such as incident gastric and colorectal cancer) [10], and chronic kidney disease [11].

The underlying pathophysiological mechanism of NAFLD remains unclear, although it is typically characterized by an accumulation of lipids in the liver that stems from multiple factors. Furthermore, it coexists with inflammation, hepatic cell injury, and the deposition of collagen fibers. The "two-hit hypothesis" explains how simple fatty liver or steatosis progresses to severe NASH and liver fibrosis [12], in the pathogenesis of early NAFLD. With the in-depth study of the pathological mechanism, this view seems unable to fully summarize the complexity of NAFLD, and the "multiple impact model" has been proposed and widely used. Factors such as lipid accumulation, insulin resistance [13], oxidative stress [14], the gut microbiome [15], and inflammation [16] have been implicated in the pathogenesis of NAFLD, but the mechanisms that drive disease progression are not fully understood. NAFLD is a disease with a favorable prognosis, but when NAFLD progresses to NASH, it can be fatal, as inflammation and fibrosis progress to end-stage liver disease. Therefore, the prevention and early treatment of NAFLD are very important. Despite this high prevalence of NAFLD, there are currently no approved treatments.

The identification of drug candidates that can successfully treat NAFLD, with few or no side effects, is a huge challenge. Natural products have played an important role in drug discovery and were the basis of most early medicines [17], especially in the treatment of chronic and metabolic diseases. Some of the most famous drugs worldwide, including aspirin, morphine, artemisinin, berberine, and paclitaxel, are derived from natural sources [18]. With the advances in natural medicinal chemistry technology, it is possible to determine the chemical composition of plants and their application in drug discovery. Over the nearly four decades from 1981 to 2019, natural products, as sources of new drugs, still exist and account for a large share of new drug discovery [19]. Terpenoids are a class of active natural products with a wide range of pharmacological effects and that represent a rich reservoir of candidate compounds for drug discovery [20]. Based on their chemical structure, terpenoids are composed of several subclasses, including hemiterpenoids, monoterpenoids, sesquiterpenoids, diterpenoids, sesterterpenoids, triterpenoids, and tetraterpenoids. Terpenoids have been widely used for the treatment of many diseases due to their broad range of biological activities, such as anti-microbial, anti-cancer, hypotensive, anti-hyperlipidemic, anti-hyperglycemic, anti-inflammatory, anti-oxidant, anti-parasitic, immunomodulatory, and anti-cholinesterasic activities [21]. The therapeutic effects of terpenoids in NAFLD have been increasingly discussed, suggesting their great potential in the treatment of NAFLD, and potentially representing a new direction for breakthroughs in drug development. There are many studies on terpenoids in NAFLD, but there is still a lack of a systematic analysis of these therapeutic applications. The purpose of this article is to review the recent research progress concerning terpenoids in the treatment of NAFLD and the underlying mechanism of action, and to provide a comprehensive introduction of this class of compounds and their potential in the treatment of NAFLD.

2. Methods

We searched the electronic databases PubMed, Excerpt Medica, Web of Science, Science Direct, and Springer for the period 2011–2021 regarding the use of terpenoids to treat NAFLD, using the following search terms: ("terpenoid" OR "monoterpenoids" OR "sesquiterpenoids" OR "diterpenoids" OR "triterpenoids" OR "tetraterpenoids") AND ("non-alcoholic fatty liver disease" OR "metabolically associated fatty liver disease" OR "non-alcoholic fatty liver" OR "non-alcoholic steatohepatitis" OR "NAFLD" OR "NAFL" OR "MAFLD" OR "NASH").

Studies were excluded from this review if they were found to harbor significant methodological errors or to lack scientific value. In order to aid in classification efforts, studies regarding mixtures of different compounds or crude extracts were also excluded from this study, in addition to those focused on terpenoids with poorly defined chemical structures. Not all terpenoids have been described in detail in the literature; for example, studies of the carotenoid family often ignore their presence as terpenoids. Therefore, according to the results of the literature search, we conducted a secondary literature search for the obtained terpenoids. The limitations of the first search were supplemented and our research system was greatly enriched. Finally, 43 terpenoids were obtained.

In total, 43 natural compounds were classified into five categories according to their structure, namely monoterpenoids, sesquiterpenoids, diterpenoids, triterpenoids, and tetraterpenoids. Figure 1 shows the numbers of the different types of terpenoids.

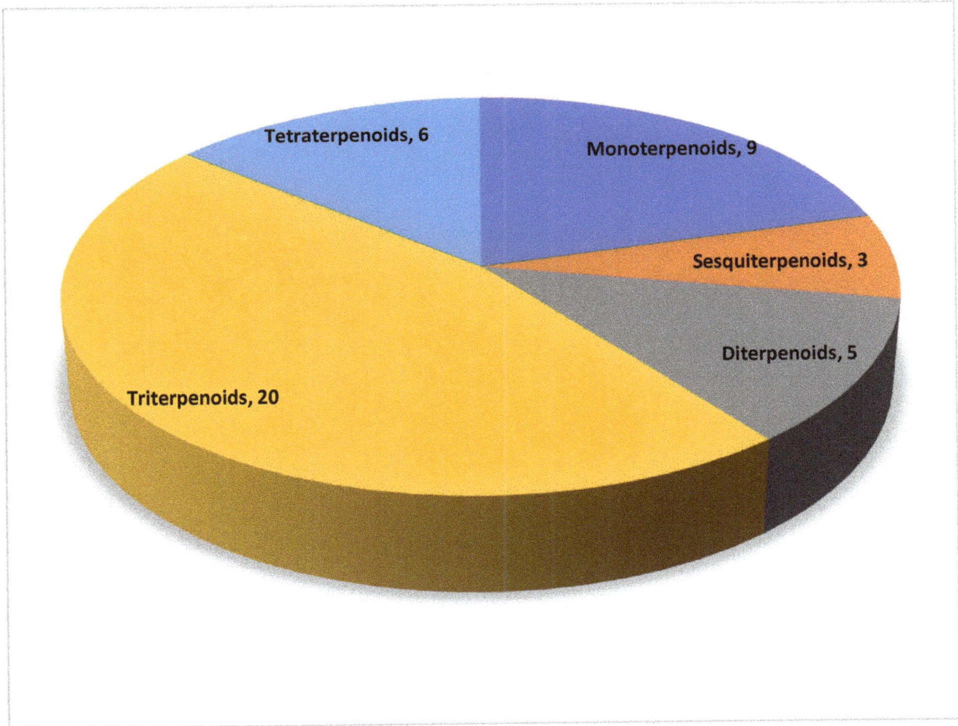

Figure 1. Distribution of different subclasses of terpenoids.

3. Terpenoids and Their Mechanism of Action in the Treatment of Non-Alcoholic Fatty Liver Disease

Increasing evidence indicates that terpenoids can effectively inhibit the progression of FALD, play a therapeutic role in different stages of the pathological process of disease, and effectively prevent and treat FALD through a variety of approaches, including improving lipid metabolism, inhibiting oxidative stress, inhibiting inflammation, and preventing fibrosis. Table 1 provides and introduction to the basic information of terpenoids, including the subclass, compounds, molecular formula, weight, and sources. Table 2 summarizes the effects and the mechanisms of action of 43 terpenoids, including the animal/cell model, dosage, target, mechanism, and effect. Figure 2 shows the chemical structure of

43 terpenoids. We will classify these terpenoids according to their chemical structure and introduce the results of research into their use in the treatment of NAFLD.

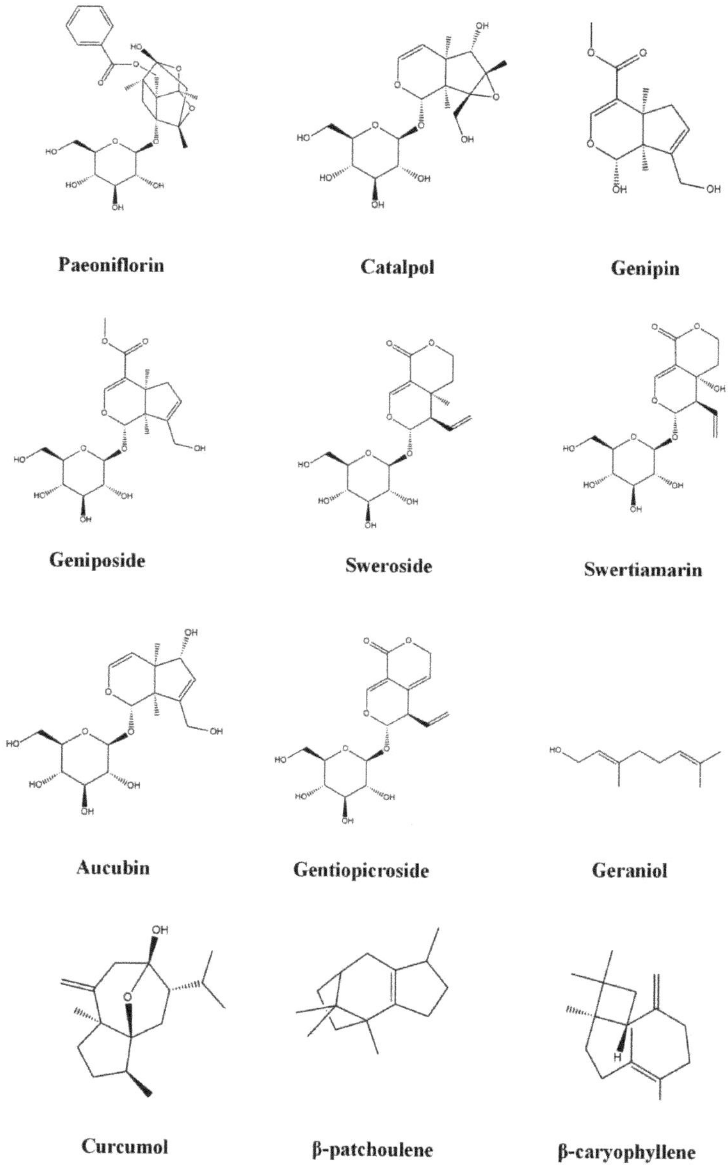

Figure 2. *Cont.*

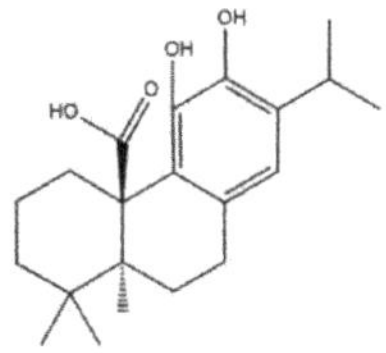

Carnosic acid

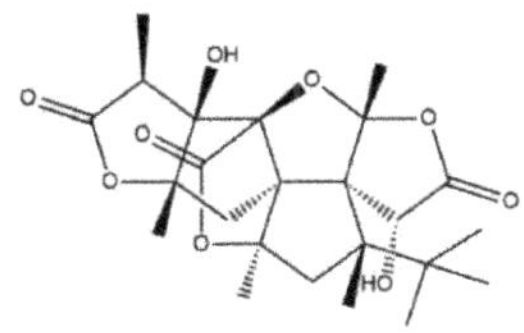

Ginkgolide A

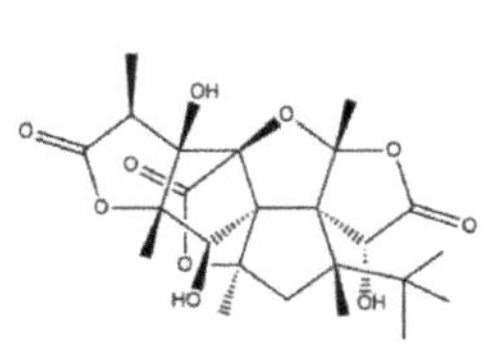

Ginkgolide B

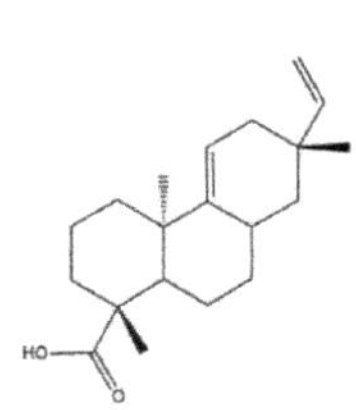

Acanthoic acid

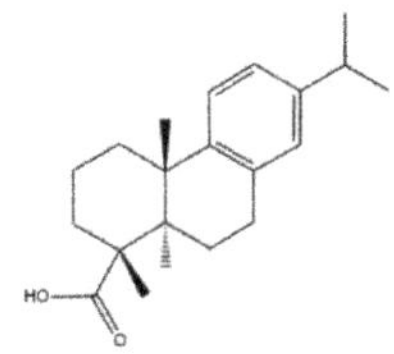

Dehydroabietic acid

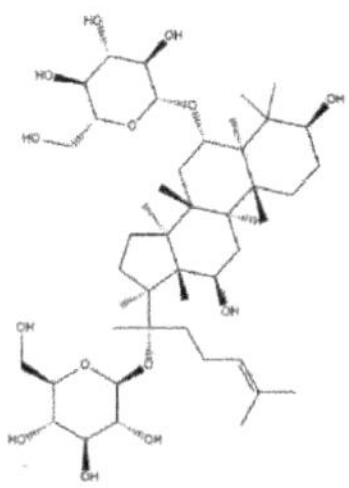

Ginsenoside Rg1

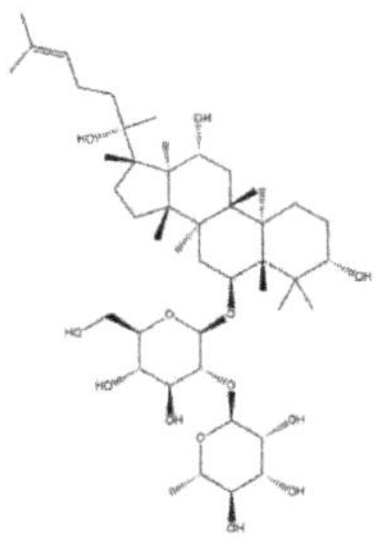

Ginsenoside Rg2

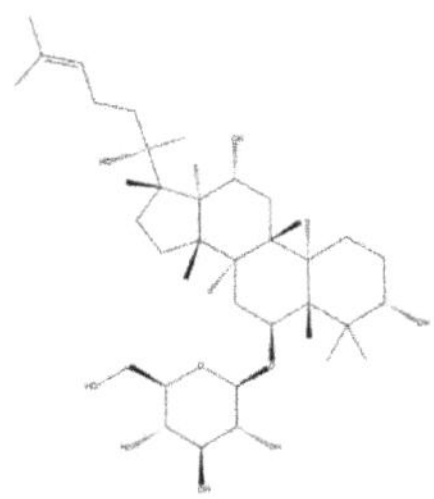

Ginsenoside Rh1

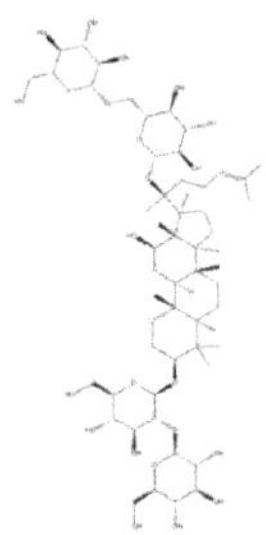

Ginsenoside Rb1

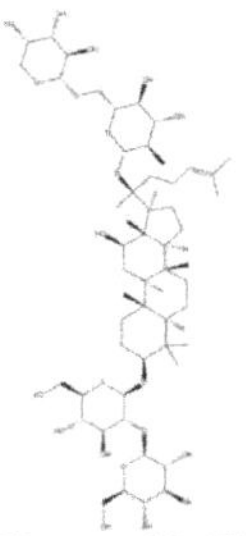

Ginsenoside Rb2

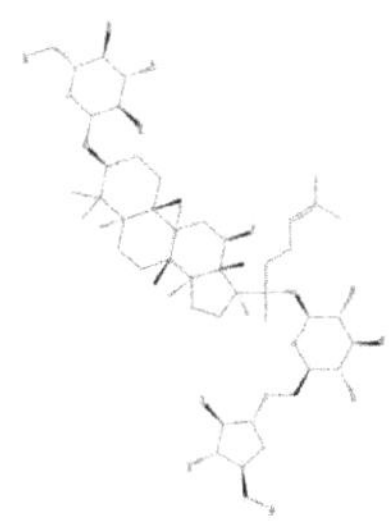

Ginsenoside Mc1

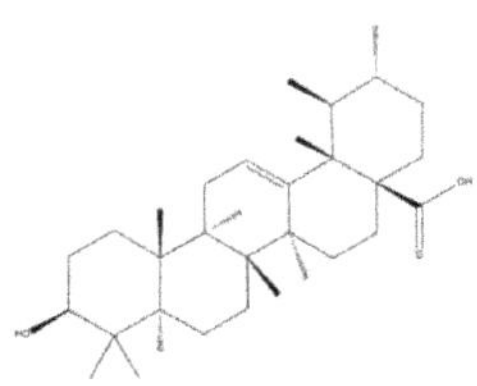

Ursolic acid

Figure 2. *Cont.*

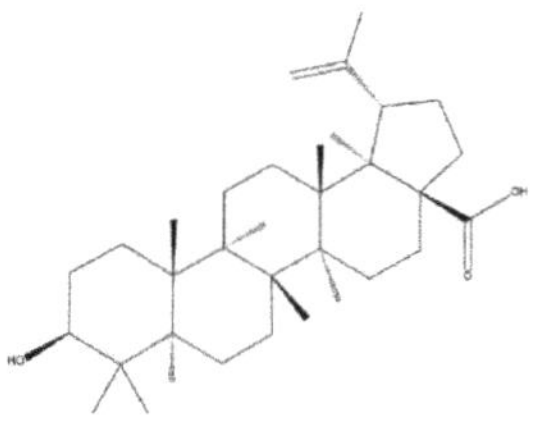

Betulinic acid

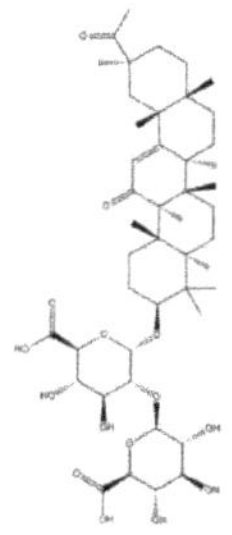

Glycyrrhizic acid

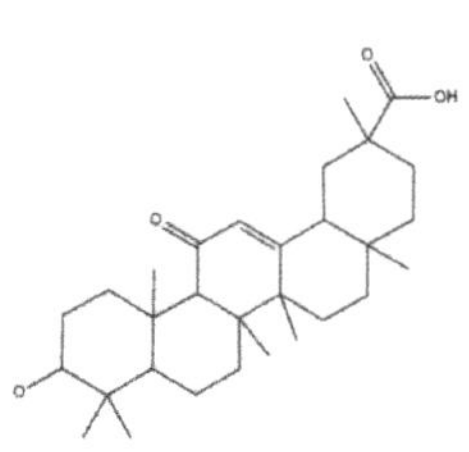

Glycyrrhetinic acid

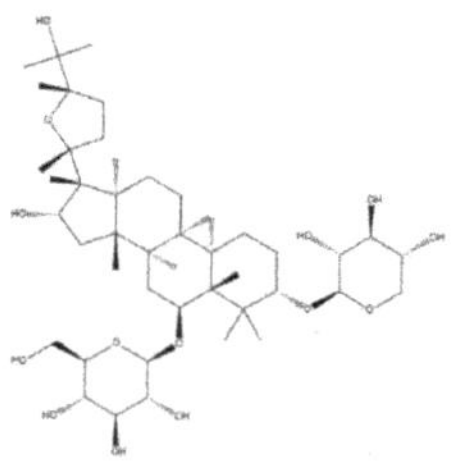

Astragaloside IV

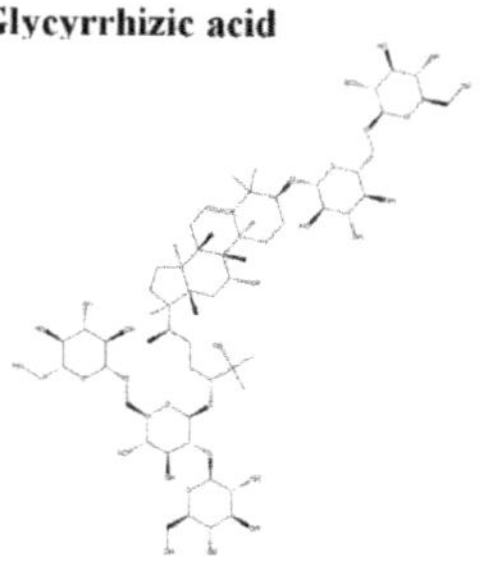

Mogroside V

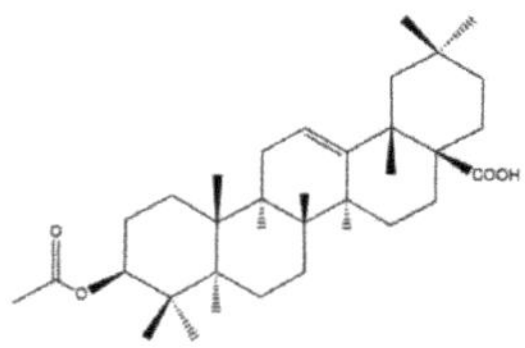

3-Acetyl-oleanolic acid

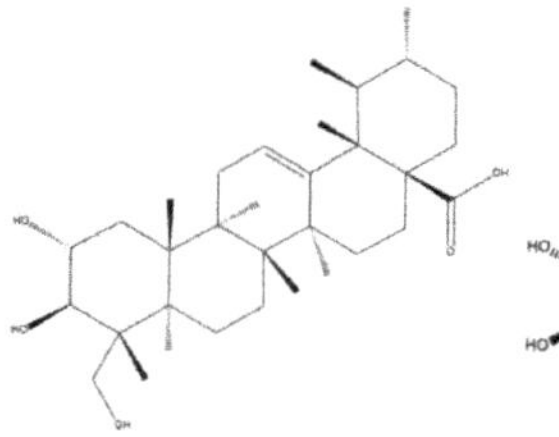

Asiatic acid

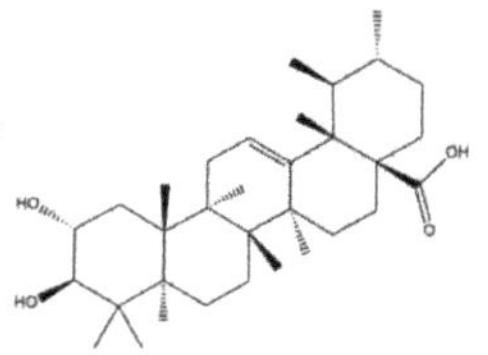

Corosolic acid

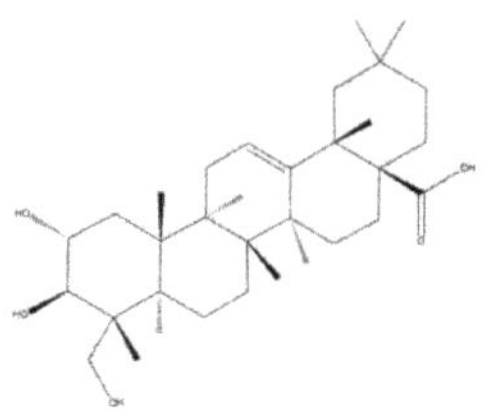

Arjunolic acid

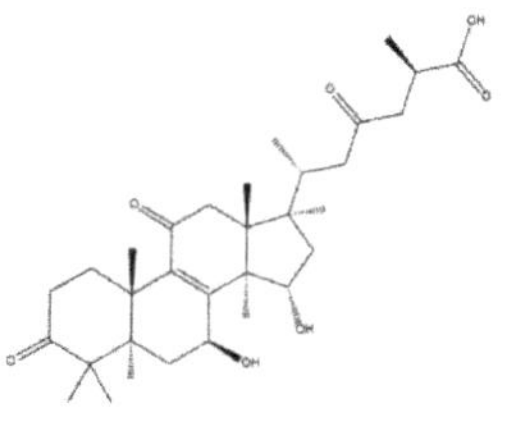

Ganoderic acid A

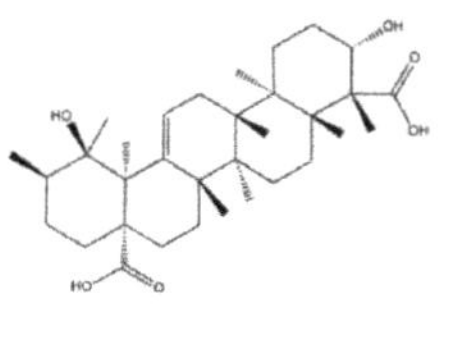

Ilexgenin A

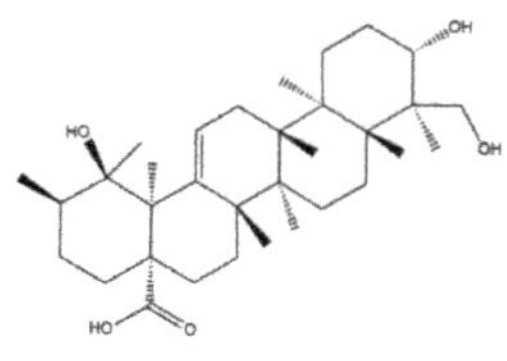

Rotundic acid

Figure 2. *Cont.*

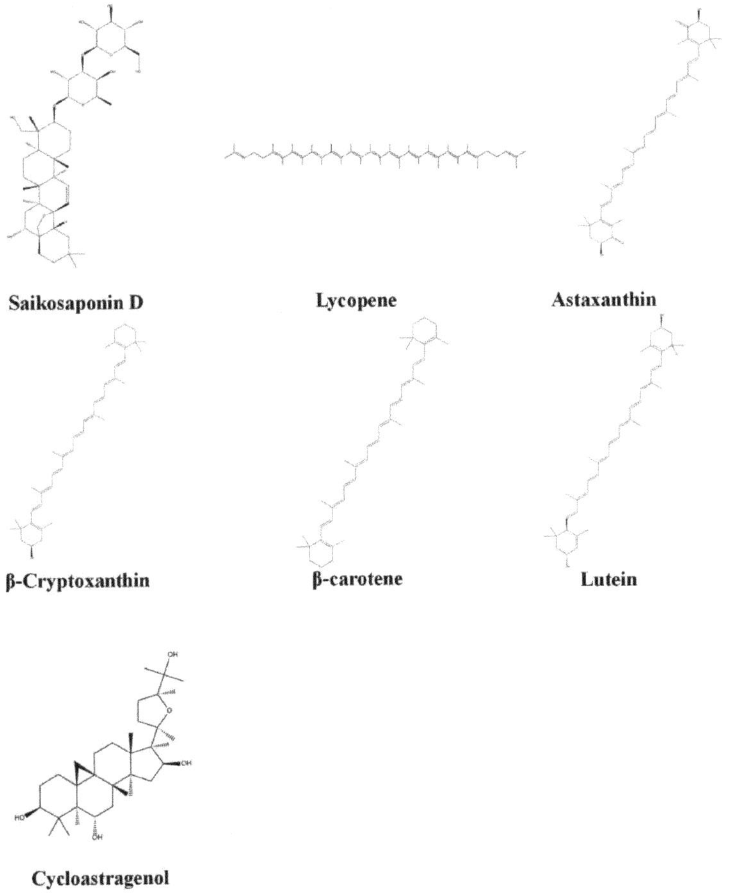

Figure 2. Chemical structure of terpenoids.

3.1. Monoterpenoids

Monoterpenoids are compounds with a basic carbon frame that consists of two isoprene and C10, constituting a large family of natural products. Iridoids belong to the monoterpenoids. In total, our study identified nine monoterpenoids capable of treating NAFLD.

3.1.1. Paeoniflorin

Paeoniflorin is a monoterpenoid glycoside extracted from *Paeonia albiflora* Pallas, which has been extensively studied due to its remarkable pharmacological effects, low toxicity, and few side effects [22]. Paeoniflorin has been reported to possess anti-inflammatory, anti-oxidative stress, anti-hyperlipidemic, and hepatoprotective activities [23]. The anti-NAFLD effect of paeoniflorin is involved in many aspects. It could be used to improve lipid metabolism, improve glucose metabolism, and inhibit inflammation for the protection of liver function [24]. Insulin is a hormone with multiple functions, among which is to inhibit the lipolysis of adipose tissue. The first hit of the "two-hit hypothesis" is that insulin resistance causes hepatic steatosis. The insulin-sensitizing effect of paeoniflorin is among the reasons behind its efficacy in preventing NAFLD. Paeoniflorin could exert an insulin-sensitizing effect by regulating the insulin signaling pathway IRS/Akt/GSK3β [25].

This is also why paeoniflorin can be of benefit for a variety of metabolic diseases, such as diabetes, hypertension, hyperlipidemia, and NAFLD. The activation of the LKB1/AMPK signaling pathway is another key mechanism by which paeoniflorin exerts its effects. The protective effects of paeoniflorin might be involved in the activation of LKB1/AMPK signaling pathways, which results in the inhibition of lipogenesis, insulin resistance, and hepatic steatosis [26]. NAFLD is a chronic inflammatory disease, and anti-inflammatory therapy is beneficial throughout its pathological process. The significant anti-inflammatory effect of paeoniflorin is key in preventing NAFLD from transforming into NASH, and the inhibition of the ROCK/NF-κB signaling pathway in the NASH liver is associated with the mechanism of its anti-inflammatory action [27].

3.1.2. Catalpol

Catalpol is an iridoid compound widely distributed in many plant families and is mainly extracted from the roots of *Rehmannia glutinosa* (Gaertn.) Libosch. ex Fisch. and C. A. Mey. Catalpol possesses extensive pharmacological activities, such as anti-inflammatory, anti-oxidative stress, anti-diabetic, anti-stroke, and anti-osteoporosis activity [28]. Catalpol's role in metabolic diseases is widely recognized. Activated AMPK can enhance lipocatabolism and produce ATP to regulate lipid metabolism. Catalpol regulates lipid metabolism through the activation of AMPK, thereby reducing hepatic steatosis [29]. AMPK is a potential target for treating diseases, which is often associated with changes in metabolism [30]. Catalpol induces autophagy and attenuates lipotoxicity, the mechanism of action of which is also through AMPK activation [31].

The activation of the p66shc/cytochrome C cascade is responsible for causing ROS metabolism, hepatic steatosis, and apoptosis in NAFLD. The anti-oxidative stress and liver-protective effects of catalpol are dependent on the miR-96-5p/p66shc/cytochrome C cascade [32]. In addition, anti-inflammatory activity has been shown to be a key effect of catalpaol in the treatment of NAFLD [33]. Due to the complexity of the mechanism of action, the underlying mechanism of catalpol's anti-inflammatory effect has not been fully elucidated.

3.1.3. Geniposide

Geniposide, as a kind of iridoid glycoside extracted from *Gardenia jasminoides* J. Ellis, has many biological effects, such as anti-inflammatory, hepatoprotective, and cholagogic effects [34]. Insulin resistance leads to fat accumulation in the liver, which is the first hit in the onset of NAFLD. Excess fat accumulation in the liver leads to a vicious cycle of aggravated insulin resistance and induced oxidative stress, which increases the pathological complexity of NAFLD. Geniposide inhibits lipid accumulation via enhancing anti-oxidative stress and anti-inflammation activity, which mostly depend on upregulating the Nrf2/HO-1 and AMPK signaling pathways [35]. Geniposide exerts protective effects against hepatic steatosis, the underlying mechanism of which may be associated with its regulation of adipocytokine release and expression of PPARa [36].

3.1.4. Genipin

Genipin is a metabolite derived from genipioside and has been recognized as a beneficial compound against metabolic disorders. For decades, genipin has been extensively studied and used in the field of liver disease, including in the treatment of acute liver injury, fulminant hepatitis, NAFLD, and other non-cancer liver diseases [37]. Genipin potentially serves as an effective therapeutic intervention against NAFLD, effectively antagonizing hyperlipidemia and hepatic lipid accumulation, by regulating the miR-142a-5p/SREBP-1c axis [38]. Pyroptosis not only causes liver cell death, but also aggravates the inflammatory response and process of fibrosis. Pyroptosis plays an important role in the development of NAFLD; thus, it is important to stop pyroptosis to prevent disease. Genipin reverses liver damage and inhibits UCP2-mediated pyroptosis [39].

3.1.5. Sweroside

Sweroside is a natural product found in *Swertia bimaculata* (Siebold and Zucc.) Hook. f. and Thomson ex C. B. Clarke. Sweroside is a typical iridoid that exhibits diverse biological activities, such as hepatoprotective, anti-diabetic, and anti-inflammatory effects [40]. The peroxisome proliferator-activated receptors (PPAR-α, PPAR-β/δ, and PPAR-γ) are members of the nuclear receptor super-family and play crucial roles in glucose and lipid metabolism [41]. The activation of PPARα induces the oxidation of fatty acids, and regulates liver fat metabolism. Sweroside may ameliorate obesity and the fatty liver via the regulation of lipid metabolism and its anti-inflammatory activity; these effects are closely associated with the regulation of PPAR-α [42]. The activation of the NLRP3 inflammasome increases NASH, induces intense inflammatory responses, causes pyrodeath, increases lipid accumulation, and promotes fibrosis. Sweroside can inhibit these pathological changes by blocking the activation of the NLRP3 inflammasome in macrophages and liver tissues [43].

3.1.6. Swertiamarin

Swertiamarin is a typical natural iridoid found in *Swertia bimaculata* (Siebold and Zucc.) Hook. f. and Thomson ex C. B. Clarke, which has been reported to cure many metabolic diseases, such as diabetes and hyperlipemia [44]. Hepatic steatosis is a common liver pathological lesion, defined as the presence of large and small vesicles of fat, predominantly triglycerides, accumulating within hepatocytes. Swertiamarin ameliorates obesity and improves insulin resistance by improving dyslipidemia and attenuating inflammation, which further improves hepatic steatosis. The p38 MAPK and NF-κB pathways participate in some of its effects, but its mechanism of action is still not fully understood [45].

3.1.7. Aucubin

Aucubin is widely distributed in plants, such as Eucommia ulmoides Oliv., Plantago asiatica L., and Scrophularia ningpoensis Hemsl. It has shown many positive effects, such as anti-oxidant, anti-aging, anti-inflammatory, anti-fibrotic, anti-cancer, hepatoprotective, neuroprotective, and osteoprotective properties [46]. Nrf2 is the main regulator of the cellular defense system against oxidative stress. Nrf2, similar to AMPK, is often concerned in the treatment of NAFLD. Aucubin inhibits lipid accumulation, the inflammatory response, and oxidative stress via the Nrf2/HO-1 and AMPK signaling pathways [47].

3.1.8. Gentiopicroside

Gentiopicroside, the main active ingredient of *Gentiana scabra* Bunge, has anti-inflammatory, anti-fibrosis, anti-oxidative stress, and anti-apoptosis activity [48]. These effects make gentiopicroside a significant contributor to the treatment of NAFLD. Gentiopicroside may be a useful therapeutic strategy for NAFLD through the alleviation of oxidative damage and lipid accumulation in the liver. The upregulation of the Nrf2 anti-oxidant pathway is thought to be a key mechanism behind its function [49].

3.1.9. Geraniol

Geraniol is an acyclic isoprenoid monoterpene isolated from the essential oils of aromatic plants, including *Elsholtzia ciliata* (Thunb.) Hyl., *Murraya exotica* L., *Rosa rugosa* Thunb., and several other plants. Treatment with geraniol reduced blood lipids, attenuated hepatic fibrosis and apoptosis, and suppressed inflammation in NASH [50]. The mechanism of action of geraniol in NAFLD has not been thoroughly studied, and its pathway and target in this disease have not been mentioned in the literature.

3.2. *Sesquiterpenoids*

Sesquiterpenoids are a class of enormously diverse natural products derived from a 15-carbon precursor, contains three isoprene units. Sesquiterpenoids have a variety of skeletal structures due to the diversity of sesquiterpene hydrocarbon backbones. In total, our study identified three sesquiterpenoids capable of treating NAFLD.

3.2.1. Curcumol

Curcumol, as an important component of *Curcuma phaeocaulis* Valeton, structurally belongs to the guaiacane sesquiterpene natural products. While curcuminoids have been extensively studied for their anti-microbial, anti-oxidant, anti-inflammatory, and other effects, the therapeutic efficacy of curcumol is still emerging [51]. The treatment of NAFLD using curcumol is a research hotspot. Cellular senescence has attracted much interest from researchers due to its involvement in NAFLD. Hepatocyte senescence can be found in the liver of NAFLD patients, which aggravates steatosis in the liver. When the liver is injured, the ferritinophagy signaling pathway can be activated, inducing the decomposition of ferritin storage and iron-rich mitochondrial proteins, releasing excess iron into the cytoplasm, leading to disordered iron metabolism in the liver, and eventually pathological iron overload. Excess iron can cause cell senescence, and further promote the pathological progression of NAFLD. Previous research has clarified the mechanism of the curcumol inhibition of hepatocyte senescence through the YAP/NCOA4 regulation of ferritinophagy in NAFLD [52]. In addition, curcumerol has a significant protective effect on liver function and liver fibrosis, which may be related to the regulation of the TLR4, TAK1, and NF-κB/p65 signaling pathways to reduce inflammatory factors and increase anti-inflammatory factors [53].

3.2.2. β-Patchoulene

Patchoulene is among the natural compounds derived from *Pogostemon cablin* (Blanco) Benth. In the volatile oil of patchouli there exist four isoforms, of which β-patchoulene is one. β-patchoulene is well known for its anti-inflammatory and anti-oxidative functions in various diseases. β-patchoulene exerts a positive effect against NASH by interrupting the vicious cycle between oxidative stress, histanoxia, and lipid accumulation. The activation of the CD36/AMPK signaling pathway to balance lipid metabolism disorders is its internal mechanism [54,55].

3.2.3. β-Caryophyllene

β-caryophyllene is an odoriferous bicyclic sesquiterpene found in various herbs and spices, such as *Rosmarinus officinalis* L., *Cinnamomum cassia* (L.) D. Don, *Ocimum basilicum* L., and *Lavandula angustifolia* Mill. β-caryophyllene has potential efficacy in preventing and ameliorating non-alcoholic fatty liver disease and its associated metabolic disorders. AMPK is considered to be an effective target for regulating insulin synergism in the treatment of metabolic diseases, and has been highly studied in NAFLD. β-caryophyllene could attenuate lipogenesis and lipid accumulation by upregulating the AMPK signaling pathway. Further mechanistic studies revealed that the β-caryophyllene-induced activation of AMPK could be mediated by the CB2 receptor-dependent $Ca2^+$/CaMKK signaling pathway [56].

3.3. *Diterpenoids*

Diterpenoids are terpenoids composed of four isoprene structural units. More than 126 different diterpenoid carbon skeletons have been identified, which can give rise to more than 18,000 compounds [57]. Diterpenoids are widely exploited in the clinic and in research. In total, our study identified five diterpenoids capable of treating NAFLD.

3.3.1. Carnosic Acid

Carnosic acid is a phenolic diterpene isolated from *Rosmarinus officinalis* Linnaeus and *Salvia japonica* Thunb., which possesses anti-tumor, anti-inflammatory, neuroprotection, anti-oxidative, and anti-microbial properties [58,59]. The effectiveness of carnosic acid in preventing fat accumulation, especially in fatty livers, and in relieving glucose intolerance has been proven [60]. Studies have shown that carnosic acid is an effective anti-obesity agent that regulates fatty acid metabolism in C57BL/6J-ob/ob mice. It can regulate the expression of hepatic lipogenesis-related genes (L-FABP, SCD1, and FAS), which decreased, whereas lipolysis-related gene (CPT1) expression increased [61]. However, other studies have confirmed that carnosic acid has a certain level of hepatotoxicity in a dose-dependent manner, so a proper safety assessment is required before use [62]. How to avoid the toxic damage of drugs and give full play to their effects is a problem worth discussing.

Carnosic acid may be used as a potential therapeutic agent in the treatment of NAFLD-related metabolic diseases. Related research has revealed that carnosic acid possesses the ability to improve high-fat-diet-induced NAFLD in mice through reducing the lipogenesis and inflammation in the liver [63]. Studies have found a link between liver cell apoptosis and the miR-34a/SIRT1/p66shc pathway, which can be regulated by CA in NAFLD [64]. Brain damage in non-alcoholic fatty liver disease is clearly present and has been demonstrated in many studies. Data in the literature suggest a role of NAFLD in promoting early cerebral alterations with cognitive impairment, subclinical ischemic lesions, and cerebrovascular accidents [65]. HFD-induced NAFLD can exacerbate dopaminergic neuron injury, while carnosic acid is able to prevent such impairment [66]. The development of this compound may provide a new path for the prevention and treatment of NAFLD complications.

3.3.2. Ginkgolide (A,B)

Ginkgolide, a natural substance extracted from *Ginkgo biloba* L., is a terpenoid compound composed of sesquiterpene lactone and diterpene lactone. Eleven terpene lactones were isolated from ginkgo biloba leaves. Ginkgolides A, B, C, J, K, L, M, N, P and Q belong to the diterpene lactones, while ginkgolides belong to the sesquiterpene lactones [67]. Ginkgolide has been widely used in clinical practice because of its rich pharmacological effects and weak side effects. It has anti-oxidation, anti-inflammation, anti-aging, anti-platelet aggregation, and anti-apoptosis effects, and lowers blood pressure, promotes blood circulation, and protects the central nervous system.

Ginkgolide A may be a natural compound with great therapeutic promise, especially for the treatment of cardiovascular, hepatological, and neurological diseases [68]. Ginkgolide A is non-toxic at high concentrations, and may be feasible as a therapeutic agent for NAFLD patients. Studies have confirmed that ginkgolide A's effects on NAFLD mainly include anti-oxidative stress and anti-inflammation activity. Ginkgolide A showed hepatoprotective efficacy by inducing cellular lipoapoptosis and by inhibiting inflammation [69].

Ginkgolide B is a diterpenoid compound isolated from ginkgo biloba leaves, which is the most significant active component in active lactones. Pregnane X receptor (PXR, NR1I2) is a ligand-activated nuclear hormone receptor and its value in metabolic diseases has been emphasized in recent years. Targeting PXR may be a strategy for the therapy of metabolic diseases. Ginkgolide B may have beneficial effects on metabolic disorders, possibly through the activation of PXR, such as blocking body weight gain, attenuating hypertriglyceridemia and hepatic steatosis, and improving bile acid homeostasis in DIO mice [70].

The ferroptosis of hepatocytes and intrahepatic macrophages may lead to the progression of simple fatty liver degeneration to NASH, and the inhibition of ferroptosis is gradually becoming a new treatment strategy for NAFLD. Ginkgolide B treatment has a specific effect on lipid accumulation and oxidative-stress-induced ferroptosis in NAFLD, the mechanism of action of which is through the regulation of the Nrf2 signaling pathway [71].

3.3.3. Acanthoic Acid

Acanthoic acid is a pimaradiene diterpene isolated from the root of *Eleutherococcus senticosus* (Rupr. and Maxim.) Maxim. The treatment of liver disease is an important aspect of acanthoic acid applications, and the value of acanthoic acid in liver disease has been widely explored, having been studied in alcoholic liver disease [72], drug-induced hepatotoxicity [73], and fulminant liver failure [74]. Acanthoic acid may be proved to be an attractive candidate for the treatment of NAFLD, as it can attenuate liver steatosis and fibrosis in NAFLD. The farnesoid X receptor (FXR) and liver X receptor (LXR) are involved in lipid metabolism, glucose metabolism, and inflammatory activation, and maintain the nutrient/energy balance of the liver. Acanthoic acid can regulate fat metabolism, and especially prevent lipid accumulation and fatty acid synthesis, the mechanism of which is to activate the FXR and LXR signaling pathways, contributing to the increased expression of the AMPK-SIRT1 signaling pathway [75].

3.3.4. Dehydroabietic Acid

Dehydroabietic acid is a diterpene found in tree pine, derived from Pinaceae plants such as *Pinus massoniana* Lamb. and *Picea asperata* Mast. Various bioactive effects of dehydroabietic acid have been studied, including anti-bacterial, anti-fungal, and anti-cancer activities [76]. The accumulation of iron-dependent lipid peroxides is among the important causes of NAFLD. Dehydroabietic acid can improve NAFLD by regulating lipid metabolism and inhibiting ferroptosis, and its ability to reduce triglyceride (TG), total cholesterol (TC), and lipid peroxidation levels is significant [77].

3.4. *Triterpenoids*

Triterpenoids, composed of four isoprene structural units, are derived from (E, E, E) -geranylgeranyl diphosphate (GGPP). More than 126 different triterpenoid carbon skeletons have been identified, which give rise to more than 18,000 compounds [78]. The complexity and diversity of the structure of triterpenoids results in different biological activities, and are widely exploited in the clinic and in research. In total, our study identified 20 triterpenoids capable of treating NAFLD.

3.4.1. Ginsenoside (Rg1, Rg2, Rh1, Rb1, Rb2 and Mc1)

Panax ginseng C. A. Mey. is among the most widely used natural medicinal plants worldwide. Ginsenosides, the major bioactive constituents of *Panax ginseng* C. A. Mey., are a series of glycosylated triterpenoids which belong to protopanaxadiol (PPD)-, protopanaxatriol (PPT)-, ocotillol (OCT)-, and oleanane (OA)-type saponins. Approximately 300 ginsenosides have been isolated and identified from different Panax species [79]. The known ginsenoside monomers can be divided into three categories: ginsenodiols (such as Rb1, Rb2, Rc, Rd, and F2), ginsenotriols (such as Re, Rg1, Rg2, Rf, and Rh1), and pentacyclic triterpenoid saponins (such as RO). Ginsenoside Rb1, Rb2, Rg1, Rg2, Rh1 and Mc1 and other components have been proven to have liver-protective effects and can be used to treat NAFLD.

Ginsenoside Rg1, a bioactive phytochemical, is the most reported ginsenoside in the field of NAFLD therapeutic research. Improving fat synthesis and metabolism is an important measure of its function. Ginsenoside Rg1 significantly improves fat metabolism and synthesis, inhibits lipid synthesis, decreases lipid uptake, enhances lipid oxidation and reduces hepatic steatosis, by regulating PPAR α and PPAR γ expression [80]. FOXO1 may be involved in many aspects of liver pathology, such as hepatic aging, steatosis, and glucose and lipid metabolism dysregulation. Ginsenoside Rg1 exerts the pharmaceutic effect of maintaining FOXO1 activity in the liver, thus protecting livers from senescence- and metabolic-abnormality-induced fatty liver disease [81]. Ginsenoside Rg1 may protect NAFLD through inflammation; the anti-inflammatory activity of ginsenoside Rg1 includes the inhibition of endoplasmic reticulum(ER) stress and inflammasome activation [82]. The changes in the transcriptome also suggest that the efficiency of ginsenoside Rg1 treatment on NAFLD may be associated with two hub genes, Atf3 and Acox2 [83]. This has been demonstrated in its ability to improve liver function and alleviate pathological processes in animal models of NAFLD.

Ginsenosides Rg2 and Rh1 belong to the category of ginsenotriols. Saponin extract contains amounts of ginsenosides Rg2 and Rh1, which can inhibit inflammation-mediated pathological inflammasome activation in macrophages, thereby preventing NAFLD development [84]. In addition to its anti-inflammatory effects, the administration of ginsenosides Rg2 significantly ameliorated HFD-induced hepatic oxidative stress and apoptosis in a SIRT1-dependent manner [85]. The effects of these compounds are not singular, but complex and multifaceted. Ginsenoside Rh1 has a positive effect on NAFLD via its anti-fibrotic and hepatoprotective activity [86].

Ginsenosides Rb1 and Rb2 belong to the category of ginsenodiols. Ginsenoside Rb1 can reduce liver cell apoptosis, and the activation of PPAR-γ may be the internal mechanism [87]. Ginsenoside Rb2 has a significant positive effect on NAFLD and glucose intolerance, and the underlying molecular mechanism consists of alleviating hepatic lipid accumulation and restoring hepatic autophagy via sirt1 induction and AMPK activation [88].

Ginsenoside Mc1, a newly identified deglycosylated ginsenoside, is converted from the major ginsenoside Rc. Ginsenoside Mc1 exerts protective effects against apoptotic damage, insulin resistance and lipogenesis in the liver [89].

Therefore, ginsenoside supplementation could be a potential therapeutic strategy to prevent NAFLD in patients. Of course, the therapeutic value of many different ginsenosides in NAFLD remains to be explored.

3.4.2. Ursolic Acid

Ursolic acid is a natural triterpene compound and is widely distributed in nature. It is known to be found in at least hundreds of plants, many of which are aromatic plants and fruit trees used in traditional Chinese medicine. Ursolic acid exists in various forms in plants, some in the free state, some in the binding state of ester or glycoside, and some as other polysubstituted derivatives. Ursolic acid has widespread pharmacologic activity, including anti-tumor, anti-inflammatory, anti-oxidant, anti-apoptotic, anti-allergy, and anti-carcinogenic effects [90]. Ursolic acid effectively ameliorated HFD-induced hepatic steatosis through a PPAR-an involved pathway, via improving key enzymes in the control of lipid metabolism [91]. Ursolic acid treatment significantly prevents the development of NAFLD and liver injury in db/db mice, most likely through increasing lipid β-oxidation and inhibiting hepatic ER stress [92]. The LXRα is a multifunctional nuclear receptor that controls lipid homeostasis. Ursolic acid, a novel LXRα antagonist, can inhibit adipogenesis through this mechanism in NAFLD [93]. Ursolic acid administration not only has therapeutic effects, but can also prevent the occurrence of disease when ingested at an early stage. The administration of ursolic acid during periods of developmental plasticity shows prophylactic potential against dietary-fructose-induced NAFLD [94].

3.4.3. Betulinic Acid

Betulinic acid is a pentacyclic triterpene distributed in a variety of plants, such as *Betula platyphylla* Sukaczev. It shows a wide spectrum of biological and pharmacological properties, such as anti-inflammatory, anti-diabetic, and anti-hyperlipidemic effects. The FXR plays an important role in hepatic homeostasis. The activation of FXR can reduce lipotoxicity and increase cholesterol excretion, thus playing a role in improving insulin resistance, with anti-inflammatory and anti-fibrosis effects. Betulinic acid is a novel FXR agonist, and can alleviate endoplasmic reticulum stress-mediated NAFLD through the activation of FXR. The effect of BA is via the FXR-mediated inhibition of PERK/EIF2α/ATF4/CHOP signaling [95]. Betulinic acid can regulate fat metabolism by delaying lipid accumulation, reducing fat synthesis, inhibiting hepatic steatosis, and promoting fatty acid oxidation, among other effects. Betulinic acid effectively ameliorates intracellular lipid accumulation in liver cells by modulating the AMPK-SREBP signaling pathway [96]. Betulinic acid protects hepatocytes from abnormal lipid deposition in NAFLD through the YY1/FAS pathway [97].

3.4.4. Glycyrrhizic Acid

Glycyrrhizic acid is a triterpene glycoside isolated from *Glycyrrhiza uralensis* Fisch., which is an edible and medicinal plant. Glycyrrhizin has anti-inflammatory, anti-allergy, and anti-oxidative stress effects. As a sweetener, glycyrrhizin is widely used in all kinds of food. Glycyrrhizin can reduce liver lipogenesis, increase lipid metabolism, reduce liver inflammation, and prevent anti-liver fibrosis by intervening in the pathological process of NAFLD [98]. Glycyrrhizic acid protects against NAFLD through the suppression of lipid synthesis, promoting fatty acid β-oxidation and triglyceride-rich lipoproteins lipolysis, inhibiting gluconeogenesis and the promotion of glycogen synthesis, and reversing insulin resistance [99]. However, the internal mechanism of glycyrrhizin intervention in the pathology of NAFLD remains to be further clarified. The bile acids (BAs) modulate lipid and glucose metabolism, inflammation, and fibrosis, and are closely related to the pathological process of NAFLD [100]. Glycyrrhizic acid alleviates non-alcoholic steatohepatitis via modulating bile acids [101].

3.4.5. Glycyrrhetinic Acid

Glycyrrhetinic acid, a pentacyclic triterpenoid saponin metabolite of glycyrrhizin, is extracted from *Glycyrrhiza uralensis* Fisch. 18β glycyrrhetinic acid, uralenic acid, and enoxolone are the aliases of glycyrrhetinic acid. Vitamin A (VA) plays critical roles in various physiological functions, including the regulation of glucose and lipid homeostasis in the liver [102]. Glycyrrhetinic acid, as a novel AKR1B10 inhibitor, could promote retinoic acid synthesis. Glycyrrhizic acid restored the balance of VA metabolism in NAFLD/NASH mice by metabolizing to glycyrrhetinic acid [103].

Hepatocyte nuclear factor 4-alpha (HNF4α) is best known for its role as a master regulator of liver-specific gene expression. Glycyrrhetinic acid was reported to act as a partial HNF4α antagonist to modulate hepatic very low-density lipoprotein (VLDL) secretion and gluconeogenesis [104]. Free fatty acid (FFA)-induced lipotoxicity plays a crucial role in the pathogenesis of NAFLD. Glycyrrhetinic acid reduces hepatic lipotoxicity by stabilizing the integrity of lysosomes and mitochondria and inhibiting cathepsin B expression and enzyme activity [105].

3.4.6. Oleanolic Acid

Oleanolic acid is a pentacyclic triterpenoid found in many plants, which has been isolated from more than 1620 plant species [106]. It has attracted the attention of the scientific community because of its wide range of biological activities. The regulation of metabolism by oleanolic acid is a combination of its multiple mechanisms of action in NAFLD. The interplay between the gut and liver, the so-called "gut–liver axis" (GLA), has been widely considered as a potential therapeutic target for NAFLD. Oleanolic acid has a protective influence on imbalances in GLA homeostasis, has anti-oxidative stress and anti-inflammatory effects, and improves glucose tolerance and insulin resistance by regulating GLA in NAFLD [107]. The expression of LXRs is correlated with the degree of hepatic fat deposition, as well as with hepatic inflammation and fibrosis in NAFLD patients. Oleanolic acid may be a novel antagonist of LXRα- and PXR-mediated lipogenesis, and plays a role in the treatment of NAFLD by inhibiting liver X receptor alpha and pregnane X receptor to attenuate ligand-induced lipogenesis [108]. In addition, it has been reported that dietary supplementation with oleanolic acid in the neonatal phase of development potentially encourages hepatoprotection against the development of NAFLD in adult life [109].This suggests that oleanolic acid can be used as an early preventative measure. As a derivative of oleanolic acid, 3-acetyl-oleanolic acid shows good potential against NAFLD, including the amelioration of lipid accumulation, anti-steatotic effects, and the inhibition of apoptosis. The therapeutic effect of 3-acetyl-oleanolic may lie in two aspects: altering the secretion of multiple adipokines and activating AMPK-related pathways [110]. Ha-20 is a newly discovered oleanolic acid derivative, which effectively suppresses fat accumulation in the liver. FABP4/aP2 is among the key carrier proteins in fat accumulation, and HA-20 inhibits adipogenesis in a manner involving the PPARγ-FABP4/aP2 pathway [111].

3.4.7. Astragaloside IV

Astragaloside IV, among the major compounds from the extract of Astragalus membranaceus, is a cycloartane-type triterpene glycoside chemical, which has a series of pharmacological effects including anti-inflammatory and anti-oxidant activity and the regulation of energy metabolism [112]. AS-IV attenuates free fatty acid (FFA)-induced ER stress and lipid accumulation via the AMPK signaling pathway, which further supports its use as a promising therapeutic for the treatment of NAFLD [113]. Insulin resistance participates in liver fat accumulation, hepatocyte steatosis, and steatohepatitis, and is the central link between the occurrence and development of NAFLD. AS-IV inhibits PTP1B and effectively improves insulin resistance, and also has an effect on the prevention and treatment of NAFLD [114].

3.4.8. Mogroside V

Mogroside, the main sweet component of *Siraitia grosvenorii* (Swingle) C. Jeffrey ex A. M. Lu and Zhi Y. Zhang, is a triterpene glucoside, and more than 20 triterpene saponins have been isolated and identified. It has the functions of lowering blood glucose, lowering blood lipids, and scavenging free radicals, alongside anti-inflammatory and anti-oxidant effects [115]. Mogrosides protect against the development of NAFLD, and the mechanisms of action underlying this inhibitory effect may be associated with the promotion of AMPK phosphorylation and the enhancement of anti-oxidative defenses through the upregulation of p62 expression [116]. Mogroside V is the representative active ingredient in mogroside, and it has been confirmed to play an important role in NAFLD after in-depth study. Mogroside V exerts a pronounced effect, improving hepatic steatosis and alleviating lipid accumulation through the regulation of the disequilibrium of lipid metabolism in the liver via an AMPK-dependent pathway [117].

3.4.9. Asiatic Acid

Asiatic acid, an active substance of *Centella asiatica* (L.) Urb., belongs to the category of triterpenoids. It possesses a wide spectrum of biological activities, notably anti-inflammatory, anti-diabetic, anti-oxidant, hepatoprotective, and anti-viral effects (specifically, hepatitis C virus) [118]. The overexpression of NF-KB plays an important role in the occurrence and development of liver inflammation, hepatocyte apoptosis, and hepatic fibrosis. Asiatic acid can relieve oxidative stress and inflammation by inhibiting the NF-kB signaling pathway, effectively alleviating hepatic steatosis, hepatocyte apoptosis, and hepatocyte injury [119].

3.4.10. Corosolic Acid

Corosolic acid is a natural pentacyclic triterpenoid, mainly derived from plants such as *Eriobotrya japonica* (Thunb.) Lindl., *Eriobotrya japonica* (Thunb.) Lindl., Actinidia chinensis Planch., and *Lagerstroemia speciosa* (L.) Pers. It exerts anti-diabetic, anti-obesity, anti-inflammatory, anti-hyperlipidemic, and anti-viral effects [120]. Among the many cytokines, TGF-β1 plays the most important role in liver fibrosis. The Smad protein family is involved in TGF-β signa'ling, of which Smad2 is a member of the receptor-activated Smad class. Corosolic acid prevents non-alcoholic liver injury and fibrosis in the progression of NASH through regulating the TGF-β1/Smad2, NF-κB, and AMPK signaling pathways [121].

3.4.11. Arjunolic Acid

Arjunolic acid is a pentacyclic triterpenoid found in *Cymbidium goeringii* (Rchb. f.) Rchb. f. and *Chrysanthemum morifolium* (Ramat.) Hemsl. The multifunctional therapeutic application of arjunolic acid has already been documented in terms of its various biological functions, including anti-oxidant, anticholinesterase, anti-tumor, and anti-asthmatic effects [122]. Arjunolic acid could ameliorate lipid accumulation, inhibit steatosis, and reduce blood fat. The upregulation of PPARγ expression may be an important molecular mechanism of action [123].

3.4.12. Ganoderic Acid A

Ganoderic acid A is a triterpenoid compound, extracted from *Ganoderma lucidum* (Curtis) P. Karst. Ganoderic acid A shows a variety of pharmacological activities, such as reducing blood lipid levels, lowering blood pressure, protecting the liver, and regulating liver function [124]. Ganoderic acid A could improve NAFLD by regulating the levels of signaling events involved in free fatty acid production, lipid oxidation, and liver inflammation. Previous research confirms that the alteration of the expression of signaling events in the AMPK-mediated pathway is its key mechanism [125].

3.4.13. Ilexgenin A

Ilexgenin A, a pentacyclic triterpenoid, is among the main bioactive compounds in *Quercus aliena* Blume and Ilex brachyphylla (Hand.-Mazz.) S. Y. Hu. Ilexgenin A showed beneficial effects on rats with NAFLD by lowering their liver weight, regulating lipid metabolism, and ameliorating steatosis in the liver. In addition, Ilexgenin A synergistic effect on other drugs in NAFLD has been further explored. Ilexgenin A enhances the effects of simvastatin and combinations of drugs on NAFLD without changes in the pharmacokinetics of simvastin [126].

3.4.14. Rotundic Acid

Rotundic acid is a natural pentacyclic triterpene compound in *Ilex rotunda* Thunb. with many pharmacological activities. As a single pure compound, rotundic acid's therapeutic effect and mechanism have been reported in relation to liver diseases such as hepatocellular carcinoma [127]. The value of rotundic acid in NASH treatment is gradually being recognized. The sterol regulatory element binding proteins-1c (SREBP-1c) and sterol-coA desaturase-1 (SCD-1) play an important role in the process of fatty acid anabolism. Rotundic acid could attenuate the symptoms of NASH, and the mechanism of action was through downregulating the expression of the SREBP-1c/SCD1 signaling pathway [128].

3.4.15. Saikosaponin D

Saikosaponin D, the epoxy ether-type pentacyclic triterpenoid compound extracted from Radix bupleuri, exerts various pharmacological properties, with its anti-inflammatory activity being among the key benefits [129]. Saikosaponin D has a hepatoprotective effect in NAFLD via its anti-inflammatory action and ability to act as an anti-oxidant [130]. However, the specific mechanism of saikosaponin D is still unclear and needs to be further studied.

3.5. Tetraterpenoids

Tetraterpenoids are composed of eight isoprene units and C40 polyene. All carotenoids are derivatives of tetraterpenes, and are thus produced from eight isoprene molecules (four terpene units) [131]. In total, our study identified six tetraterpenoids capable of treating NAFLD.

3.5.1. Lycopene

Lycopene belongs to the tetraterpene carotenoid family and is found in red fruits and vegetables, such as *Solanum lycopersicum* L., *Citrullus lanatus* (Thunb.) Matsum. and Nakai, *Citrus aurantium* (Lour.) Engl., and *Daucus carota* var. sativa Hoffm. Eleven conjugated double bonds predetermine the anti-oxidant properties of lycopene and its ability to scavenge lipid peroxyl radicals, reactive oxygen species, and nitric oxide [132]. Lycopene has numerous biological activities, such as anti-cancer, anti-oxidant, cardioprotective, anti-inflammatory, anti-platelet aggregative, and anti-hypertensive action [133]. Carotenoid metabolism in animals is mainly dependent on two enzymes—BCMO1 and β-carotene-9′,10′-dioxygenase (BCO2)—and lycopene is the same. Lycopene feeding has widespread effects on hepatic metabolism, stress, nuclear receptors, and nuclear coregulator gene expression, of which some are substantially dependent upon the BCO2 genotype. A total of 19 genes were affected by lycopene feeding [134]. NAFLD is strongly associated with mesenteric adipose tissue (MAT), and represents an immune dialogue between the gut and liver. The study of the enterohepatic immune axis provides evidence that MAT inflammation contributes to the pathology of NAFLD. The lycopene-mediated ability to prevent hepatic steatosis was associated with modulations in MAT lipid metabolism. Lycopene reduced steatosis by increasing MAT fatty acid utilization, and upregulating PPARa-inducible genes may be among its functions [135].

Lycopene prevents the progression of lipotoxicity-induced non-alcoholic steatohepatitis by decreasing oxidative stress in mice. The intrinsic mechanism of action is lycopene decreasing LPS-/IFN-γ-/TNFα-induced M1 marker mRNA levels in peritoneal macrophages, as well as the TGF-β1-induced expression of fibrogenic genes in a stellate cell line [136]. The hepatoprotective and anti-oxidant effects of lycopene on NAFLD, downregulating the expression of TNF-α and CYP2E1, may be among the mechanisms of action [137]. Lycopene reduced hepatic steatosis in mice fed a high-fat diet by upregulating miR-21, identifying FABP7 as a novel target of miR-21 [138]. In addition, lycopene can reduce multiple risk factors of non-alcoholic fatty liver disease. Studies have confirmed that lycopene can reduce the damage to the liver caused by obesity [139], smoking [140], hypertension [141], diabetes [142], hyperlipidemia [143], and other risk factors, which is conducive to curbing the occurrence and development of NAFLD. Of course, many diseases could benefit from the effects described above.

The synergistic effect of lycopene in combination with other drugs is a new development for lycopene. Luteolin and lycopene in combination can effectively ameliorate NAFLD in a "two-hit" manner through the activation of the Sirt1/AMPK pathway. The luteolin (20 μM) + lycopene (10 μM) therapeutic combination was found to be the best, and significantly improved cell viability and lipid accumulation in PA-induced HepG2 cells and primary hepatocytes [144].

3.5.2. Astaxanthin

Astaxanthin is a kind of ketone carotenoid belonging to the tetraterpenoids [145]. Astaxanthin is widely found in nature (especially in shrimp, fish, crab, algae, etc.), and shows a wide range of pharmacological activities, include anti-oxidant, anti-inflammatory, anti-lipid peroxidation, and immune enhancement effects. Studies have shown that astaxanthin has important preventive and therapeutic effects on liver fibrosis, non-alcoholic fatty liver, liver cancer, and drug- and ischemia-induced liver injury, and has potential as a therapeutic agent in both healthy and diseased livers [146]. One animal study found that the expression of 8848 genes was associated with NASH in mice. Among these genes, 1137 were significantly up- or downregulated by astaxanthin [147]. Previous research suggests that astaxanthin might be a novel and promising treatment for NAFLD, and can effectively prevent and treat the disease at multiple stages. The increased hepatic expression of endogenous anti-oxidant genes is an effective way for astaxanthin to intervene in NAFLD. In an astaxanthin-supplemented liver, NRF-2 mRNA expression doubled, and the expression of its target endogenous anti-oxidant genes increased [148]. Fibrosis results from the dysregulation of fibrogenesis in hepatic stellate cells (HSCs). Astaxanthin prevents TGFβ1-induced pro-fibrogenic gene expression by inhibiting Smad3 activation in hepatic stellate cells [149]. Vitamin E has become a standard treatment for NASH. Astaxanthin displays multiple functions in the inhibition of NASH progression via modulating intrahepatic immunity comparison with vitamin E [150]. Liver macrophages play a central role in inflammatory cell infiltration and immune response in NASH. The beneficial effects of astaxanthin were attributable in part to both the decreased hepatic recruitment of T cells and macrophages, as well as an M2-dominant polarization of macrophages/Kupffer cells to attenuate hepatic inflammation and fibrosis [151]. Mitochondria regulate hepatic lipid metabolism, cell death, and oxidative stress. Astaxanthin attenuated hepatocyte damage and mitochondrial dysfunction in NAFLD by upregulating the FGF21/PGC-1α pathway [152].

3.5.3. β-Cryptoxanthin

β-Cryptoxanthin, a xanthophyll carotenoid, has been isolated from a variety of sources, including orange, papaya, egg yolk, butter, and apples. β-cryptoxanthin has been shown to contribute to the improvement of NAFLD through a multifaceted approach, including improved insulin resistance, the suppression of inflammation and oxidative stress, a reduction in macrophages and a shift of their subsets, and the control of lipid metabolism by PPAR family activation [153]. A study recruited 92 NAFLD outpatients for a

12-week, single-center, parallel-group, double-blind RCT, and confirmed that a hypocaloric high-protein diet supplemented with β-cryptoxanthin safely and efficaciously improves NAFLD [154]. This clinical study also revealed that an energy-restricted HPD supplemented with BCX more efficaciously alleviates oxidative stress and inflammation in NAFLD as compared with a standard energy-restricted diet [155]. β-Carotene oxygenases 1 and 2 (BCO1 and BCO2) are the enzymes that metabolize carotenoids. β-cryptoxanthin feeding mitigates high-refined-carbohydrate diet (HRCD)-induced NAFLD in both wild-type (WT) and $BCO1^{-/-}/BCO2^{-/-}$ double-knockout (DKO) mice through different mechanisms in the liver–mesenteric adipose tissues axis, depending on the presence or absence of BCO1/BCO2 [156]. The consumption of β-cryptoxanthin possibly prevents NASH, which contributes to the anti-oxidative stress, anti-inflammation, immunoregulatory, and fibrosis-suppressing effects of β-cryptoxanthin [157]. The expression of lipopolysaccharide-inducible and/or TNF-α-inducible genes was suppressed by cryptoxanthin, probably via the inhibition of macrophage activation. NAFLD can be defined as lipotoxic liver injury and progression to NASH. β-cryptoxanthin prevents and reverses insulin resistance and steatohepatitis, at least in part, through an M2-dominant shift in macrophages/Kupffer cells in a lipotoxic model of NASH [158].

3.5.4. β-Carotene

β-carotene is a common carotenoid that is important for human health. Major sources of β-carotene in the human diet are primarily green leafy vegetables, carrots, apricots, sweet potatoes, red palm oil, mature squashes, pumpkins, and mangoes [159]. β-carotene might be a promising preventive and protective nutrient for fatty liver disease, and has been reported to alleviate hepatic steatosis (SS), inflammation, and fibrosis in in vivo and in vitro studies [160]. In the dietary carotenoids and NAFLD among US Adults across 2003–2014 study, it was confirmed that higher intake and serum levels of most carotenoids were associated with lower odds of having NAFLD [161]. A community-based cross-sectional study involving total of 2935 participants aged 40–75 years found that higher levels of α-carotene, β-carotene, lutein + zeaxanthin, and total carotenoids were significantly associated with a decrease in the degree of NAFLD [162]. β-carotene is beneficial for the treatment and prevention of NAFLD, and can enhance the therapeutic effect of other drugs on NAFLD. The combination of rosuvastatin with β-carotene is more effective than rosuvastatin alone [163]. α-carotene was recognized as having therapeutic value for NAFLD, but α-carotene has consistently been included in broader carotenoid studies, with a lack of more specific work.

3.5.5. Lutein

Lutein and its isomers, zeaxanthin and meso-zeaxanthin, are xanthophyll carotenoids found commonly in green leafy vegetables, avocados, and eggs, which play significant roles in human health [164]. Lutein is involved in the treatment of NAFLD by regulating the expression of the key factors related to insulin signaling and lipid metabolism in the liver. Lutein supplementation could ameliorate hepatic lipid accumulation and insulin resistance induced by a HFD, possibly via the activation of the expression of SIRT1/PPAR-α and other key factors in insulin signaling pathways [165].

3.5.6. Cycloastragenol

Cycloastragenol, a tetracyclic triterpenoid compound, is a secondary metabolite isolated from *Astragalus membranaceus var. mongholicus* (Bunge) P. K. Hsiao, and has a wide spectrum of pharmacological functions, which are attracting attention in the research community. It has been shown to have anti-aging, anti-oxidation, anti-inflammation, anti-cancer, and cardiovascular-protective effects [166]. Nuclear receptor FXR, although known as a bile acid receptor, is also related to liver lipid metabolism and glucose metabolism. Cycloastragenol alleviates hepatic steatosis, reduces lipid accumulation, and lowers blood glucose in NAFLD via the stimulation and enhancement of the FXR signaling pathway [167].

Table 1. Introduction to the basic information of terpenoids.

Subclass	Compound	Molecular Formula	Weight (g/mol)	Sources
Monoterpenoid	Paeoniflorin	$C_{23}H_{28}O_{11}$	480.5	*Paeonia albiflora* Pallas
Iridoid	Catalpol	$C_{15}H_{22}O_{10}$	362.3	*Rehmannia glutinosa* (Gaertn.) Libosch. ex Fisch. and C. A. Mey.
Iridoid	Geniposide	$C_{17}H_{24}O_{10}$	388.4	*Gardenia jasminoides* J. Ellis
Iridoid	Genipin	$C_{11}H_{14}O_5$	226.2	*Gardenia jasminoides* J. Ellis
Iridoid	Sweroside	$C_{16}H_{22}O_9$	358.3	*Swertia bimaculata* (Siebold and Zucc.) Hook. f. and Thomson ex C. B. Clarke
Iridoid	Swertiamarin	$C_{16}H_{22}O_{10}$	374.3	*Swertia bimaculata* (Siebold and Zucc.) Hook. f. and Thomson ex C. B. Clarke
Iridoid	Aucubin	$C_{15}H_{22}O_9$	346.3	*Eucommia ulmoides* Oliv., *Plantago asiatica* L., *Scrophularia ningpoensis* Hemsl.
Iridoid	Gentiopicroside	$C_{16}H_{20}O_9$	356.3	*Gentiana scabra* Bunge
Monoterpenoid	Geraniol	$C_{10}H_{18}O$	154.3	*Elsholtzia ciliata (Thunb.) Hyl., Murraya exotica L., Rosa rugosa Thunb.*
Sesquiterpenoid	Curcumol	$C_{15}H_{24}O_2$	236.4	*Curcuma phaeocaulis* Valeton
Sesquiterpenoid	β-patchoulene	$C_{15}H_{24}$	204.4	*Pogostemon cablin* (Blanco) Benth.
Sesquiterpenoid	β-caryophyllene	$C_{15}H_{24}$	204.4	*Rosmarinus officinalis* L., *Cinnamomum cassia* (L.) D. Don, *Ocimum basilicum* L., *Lavandula angustifolia* Mill.
Diterpenoid	Carnosic acid	$C_{20}H_{28}O_4$	332.4	*Rosmarinus officinalis* Linnaeus, *Salvia japonica* Thunb.
Diterpenoid	Ginkgolide A	$C_{20}H_{24}O_9$	408.4	*Ginkgo biloba* L.
Diterpenoid	Ginkgolide B	$C_{20}H_{24}O_{10}$	424.4	*Ginkgo biloba* L.
Diterpenoid	Acanthoic acid	$C_{20}H_{30}O_2$	302.5	*Eleutherococcus senticosus* (Rupr. and Maxim.) Maxim.
Diterpenoid	Dehydroabietic acid	$C_{20}H_{28}O_2$	300.4	*Pinus massoniana* Lamb., *Picea asperata* Mast.
Triterpenoid	Ginsenoside Rg1	$C_{42}H_{72}O_{14}$	801	*Panax ginseng* C. A. Mey.
Triterpenoid	Ginsenoside Rg2	$C_{42}H_{72}O_{13}$	785	*Panax ginseng* C. A. Mey.
Triterpenoid	Ginsenoside Rh1	$C_{36}H_{62}O_9$	638.9	*Panax ginseng* C. A. Mey.
Triterpenoid	Ginsenoside Rb1	$C_{54}H_{92}O_{23}$	1109.3	*Panax ginseng* C. A. Mey.
Triterpenoid	Ginsenoside Rb2	$C_{53}H_{90}O_{22}$	1079.3	*Panax ginseng* C. A. Mey.
Triterpenoid	Ginsenoside Mc1	$C_{47}H_{80}O_{17}$	917.1	*Panax ginseng* C. A. Mey.
Triterpenoid	Ursolic acid	$C_{30}H_{48}O_3$	456.7	*Gentiana scabra* Bunge, *Pseudocydonia sinensis* (Thouin) C. K. Schneid., *Eriobotrya japonica* (Thunb.) Lindl., *Pyrrosia lingua* (Thunb.) Farw. and *Plantago asiatica* L.
Triterpenoid	Betulinic acid	$C_{30}H_{48}O_3$	456.7	*Betula platyphylla* Sukaczev
Triterpenoid	Glycyrrhizic acid	$C_{42}H_{62}O_{16}$	822.9	*Glycyrrhiza uralensis* Fisch.
Triterpenoid	Glycyrrhetinic acid	$C_{30}H_{46}O_4$	470.7	*Glycyrrhiza uralensis* Fisch.
Triterpenoid	Oleanolic acid	$C_{30}H_{48}O_3$	456.7	*Ophiopogon japonicus* (L. f.) Ker Gawl., *Swertia leducii* Franch., *Ligustrum lucidum* W. T. Aiton
Triterpenoid	Astragaloside IV	$C_{41}H_{68}O_{14}$	785	*Astragalus membranaceus var. mongholicus* (Bunge) P. K. Hsiao
Triterpenoid	Mogroside V	$C_{60}H_{102}O_{29}$	1287.4	*Siraitia grosvenorii* (Swingle) C. Jeffrey ex A. M. Lu and Zhi Y. Zhang
Triterpenoid	Asiatic acid	$C_{30}H_{48}O_5$	488.7	*Centella asiatica* (L.) Urb.

Table 1. *Cont.*

Subclass	Compound	Molecular Formula	Weight (g/mol)	Sources
Triterpenoid	Corosolic acid	$C_{30}H_{48}O_4$	472.7	*Eriobotrya japonica* (Thunb.) Lindl., *Actinidia chinensis* Planch., *Hippophae rhamnoides* L. *and Lagerstroemia speciosa* (L.) Pers
Triterpenoid	Arjunolic acid	$C_{30}H_{48}O_5$	488.7	*Cymbidium goeringii* (Rchb. f.) Rchb. f., Chrysanthemum morifolium (Ramat.) Hemsl.
Triterpenoid	Ganoderic acid A	$C_{30}H_{44}O_7$	516.7	*Ganoderma lucidum* (Curtis) P. Karst.
Triterpenoid	Ilexgenin A	$C_{30}H_{46}O_6$	502.7	*Quercus aliena* Blume, *Ilex brachyphylla* (Hand.-Mazz.) S. Y. Hu
Triterpenoid	Rotundic acid	$C_{30}H_{48}O_5$	488.7	*Ilex rotunda* Thunb.
Triterpenoid	Saikosaponin D	$C_{42}H_{68}O_{13}$	781	*Radix bupleuri*
Tetraterpenoid	Lycopene	$C_{40}H_{56}$	536.9	*Solanum lycopersicum* L., *Citrullus lanatus* (Thunb.) Matsum. and Nakai, *Citrus aurantium* (Lour.) Engl., *Daucus carota var.* sativa Hoffm.
Tetraterpenoid	Astaxanthin	$C_{40}H_{52}O_4$	596.8	Shrimp, fish, crab, algae, etc.
Tetraterpenoid	β-cryptoxanthin	$C_{40}H_{56}O$	552.9	Orange, papaya, egg yolk, butter, apples, etc.
Tetraterpenoid	β-carotene	$C_{40}H_{56}$	536.9	Carrots, apricots, sweet potatoes, mature squashes, pumpkins, mangoes, etc.
Tetraterpenoid	Lutein	$C_{40}H_{56}O_2$	568.9	Spinach, kale, yellow carrots, etc.
Tetraterpenoid	Cycloastragenol	$C_{30}H_{50}O_5$	490.7	*Astragalus membranaceus var. mongholicus* (Bunge) P. K. Hsiao

Table 2. The effects and mechanisms of 43 terpenoids on NAFLD, from recent studies.

Compound	Animal/Cell Model	Dosage (mg/kg/d; μM/24 h)	Target/Pathways/Mechanism	Effects	Reference
Paeoniflorin	HFD-induced NAFLD mice	0.05% in diet	Activation of the CD36/AMPK signaling pathway	Reduced body weight, improved insulin resistance, anti-inflammatory, inhibition of lipid accumulation, attenuated hepatic adipose infiltration	[24]
	HFD-induced NAFLD rats	20	Regulation of the IRS/Akt/GSK3β signaling pathway	Inhibition of lipid accumulation, improved insulin resistance, anti-oxidative stress, liver protection	[25]
	Fructose-induced metabolic syndrome rats	10, 20, 40	Activation of the AMPK signaling pathway	Inhibition of hepatic lipid accumulation, improved insulin resistance, inhibition of hepatic steatosis, inhibition of hepatic lipogenesis, promotion of fatty acid oxidation	[26]
	HFD-induced NAFLD rats	20, 60, 100	Inhibition of the ROCK/NF-κB signaling pathway	Anti-inflammatory, ameliorated hepatic steatosis, reduced lipids	[27]

Table 2. *Cont.*

Compound	Animal/Cell Model	Dosage (mg/kg/d; μM/24 h)	Target/Pathways/Mechanism	Effects	Reference
Catalpol	HFD-induced NAFLD mice; PA-induced HepG2 cells	100, 200, 400; 100, 200, 400	Activation of the AMPK signaling pathway	Ameliorated hepatic steatosis, reduced body weight, inhibition of lipid accumulation	[29]
	HFD-induced NAFLD mice; PA-induced HepG2 cells	100; 10 μg/mL/24h	/	Inhibition of autophagy, ameliorated hepatic steatosis, reduced liver weight, reduced liver fat	[31]
	($LDLr^{-/-}$)+ HFD-induced NAFLD mice	100	Regulation of the p66shc/cytochrome C signaling pathway	Attenuated liver injury, anti-oxidative stress, inhibition of hepatic steatosis, anti-apoptosis	[32]
	HFD-induced NAFLD mice	100, 200, 400	/	Ameliorated hepatic steatosis, reduced body weight, improvement of lipid metabolism disorders, inhibition of lipid accumulation, anti-inflammatory, anti-apoptosis	[33]
Geniposide	(PA + OA)-induced HepG2 cells	0, 65, 130, 260, 390, 520 μmol/L/24 h	Upregulation of the Nrf2/AMPK/mTOR signaling pathways	Inhibition of lipid accumulation, anti-oxidative stress, anti-inflammatory	[35]
	HFD-induced NAFLD rats	25, 50, 100	Increased expression of PPARa gene	Ameliorated hepatic steatosis, anti-oxidative stress	[36]
Genipin	HFD-induced NAFLD mice; (PA + OA)-induced cells primary hepatocytes of mice	5, 20; 5, 20	Regulation of the miR-142a-5p/SREBP-1c axis	Reduced body weight gain, increased locomotor activity, improved insulin resistance, alleviated hyperlipidemia, inhibition of lipid accumulation	[38]
	HFD-induced NAFLD mice	5, 20	Suppressed UCP2	Reversed liver damage, anti-pyroptosis	[39]
Sweroside	HFD-induced NAFLD mice	60, 120, 240	Increases expression of PPARa gene	Reduced body weight, improved insulin resistance, inhibited hepatic steatosis, anti-inflammatory	[42]
	MCD diet-induced NAFLD mice	5, 30	Suppressed activation of the NLRP3 inflammasome	Anti-inflammatory, inhibition of hepatic lipid accumulation, anti-fibrosis	[43]
Swertiamarin	HFD-induced NAFLD mice; LPSO-induced murine monocytic cells	10, 100; 1, 10, 50	Suppressed activation of the p38 MAPK and NF-κB signaling pathways	Ameliorated hepatic steatosis, anti-inflammatory, reduced body weight, improved insulin resistance	[45]
Aucubin	Tyloxapol-induced NAFLD mice	10, 20, 40	Activation of the Nrf2/HO-1 and AMPK signaling pathways	Inhibition of lipid accumulation, anti-oxidative stress, anti-inflammatory	[47]
Gentiopicroside	Tyloxapol-induced NAFLD mice; (PA + OA)-induced HepG2 cells	20, 40, 80; 0, 4, 20, 100, 200, 500	Upregulation of the Nrf2 signaling pathway	Inhibition of hepatic lipid accumulation, anti-oxidative stress	[49]
Geraniol	MCD-induced NAFLD rats	25, 100, 200	/	Inhibition of hepatic lipid accumulation, anti-fibrosis, anti-apoptosis, anti-inflammatory, anti-oxidative stress	[50]
Curcumol	HFD-induced NAFLD mice	15, 30, 60	Regulation of the YAP/NCOA4 signaling pathway	Inhibition of hepatocyte senescence, suppressed ferritinophagy	[52]

Table 2. *Cont.*

Compound	Animal/Cell Model	Dosage (mg/kg/d; μM/24 h)	Target/Pathways/Mechanism	Effects	Reference
β-patchoulene	HFD-induced NAFLD rats	25, 50, 100	Regulation of the TLR4, TAK1, and NF-κB/P65 signaling pathways	Anti-inflammatory, improved liver function, anti-fibrosis, anti-apoptosis	[53]
	HFD-induced NAFLD rats; (FFA + PA + OA)-induced HepG2 cells	10, 20, 40; 40	Activation of the AMPK signaling pathway	Inhibition of hepatic lipid accumulation, improved insulin resistance, ameliorated hepatic steatosis, inhibition of hepatic lipogenesis, promotion of fatty acid oxidation	[54]
	HFD-induced NAFLD rats	10, 20, 40	Activation of the CD36/AMPK signaling pathway	Reduced body weight, reversed liver damage, ameliorated hepatic steatosis, anti-oxidative stress, anti-inflammatory	[55]
β-caryophyllene	PA-induced HepG2 cells	40	Activation o the f AMPK signaling pathway	Inhibition of hepatic lipid accumulation	[56]
Carnosic acid	Obese leptin-deficient (ob/ob) mice	0.05% in diet	/	Reduced body weight, inhibition of lipid accumulation, recovered glucose tolerance	[60]
	Obese leptin-deficient (ob/ob) mice	0.01, 0.02% in diet	/	Improved glucose tolerance, inhibition of lipid accumulation, reduced body weight	[61]
	HFD-induced NAFLD mice	15	Upregulation of MARCKS expression/impairment of the PI3K/AKT and NLRP3 inflammasome signaling pathways	Inhibition of lipid accumulation, anti-inflammatory, improved insulin resistance	[63]
	HFD-induced NAFLD mice, PA-induced human L0246 hepatic cell	30, 60; 10	Inhibition of the miR-34a/SIRT1/p66shc signaling pathway	Inhibition of lipid accumulation, anti-apoptosis	[64]
Ginkgolide A	HFD-induced NAFLD mice; NEFA-induced HepG2 cell	5; 0, 10, 50, 100		Inhibition of lipid accumulation, induced cellular lipoapoptosis, anti-inflammatory, reduced body weight	[69]
Ginkgolide B	HFD-induced NAFLD mice	0.1 % in diet	Activation of pregnane X receptor	Reduced body weight, ameliorated hepatic steatosis	[70]
	HFD-induced NAFLD mice; (PA + OA)-induced HepG2 cells	20, 30; 4, 8, 16	Increased Nrf2 expression	Anti-oxidative stress, reduced body weight, inhibition of lipid accumulation, anti-inflammatory	[71]
Acanthoic acid	Modified Lieber–DeCarli diet-induced mice	20, 40	Via FXR–LXR axis	Inhibition of hepatic lipid accumulation, anti-fibrosis, regulation of fatty acid synthesis	[75]
Dehydroabietic acid	HFD-induced NAFLD mice; OA-induced HL7702 cells	10, 20; 2.5, 5, 10	Activation of the Keap1/Nrf2-ARE signaling pathway	Reduced blood lipid, inhibition of ferroptosis	[77]

Table 2. *Cont.*

Compound	Animal/Cell Model	Dosage (mg/kg/d; μM/24 h)	Target/Pathways/Mechanism	Effects	Reference
	(PA + OA)-induced HepG2 cells	25, 50	Regulation of PPAR α and PPAR γ expression	Inhibition of lipid accumulation, ameliorated hepatic steatosis	[80]
Ginsenoside Rg1	D-galactose-induced fatty liver disease mice	40	Upregulation of FOXO1 gene	Anti-inflammatory, inhibition of lipid accumulation	[81]
	HFD-induced NAFLD mice	20, 40	/	Anti-inflammatory, reduced body weight, alleviated ER stress	[82]
	HSD-induced NAFLD rats	100	Regulation of Atf3 and Acox2 gene	Reduced body weight, reduced blood lipid, alleviated hepatic steatosis	[83]
Ginsenoside Rg2	HFD-induced NAFLD mice	2.5, 5, 10	Regulation of the SIRT1 signaling pathways	Improvement of lipid and glucose disorders, anti-oxidative stress, anti-apoptosis, inhibition of lipid accumulation	[85]
Ginsenoside Rh1	HFD-induced NAFLD rats	3	/	Anti-fibrotic	[86]
Ginsenoside Rb1	HFD-induced NAFLD mice	10	Activation of PPAR-γ expression	Reduced body weight, improved glucose metabolism, inhibition of lipid accumulation, anti-apoptosis	[87]
Ginsenoside Rb2	db/db mice, OA-induced HepG2 cells	10; pretreated with 0.1, 1, 10, 50, 100 μmol/L/4h	Regulation of the SIRT1 and AMPK signaling pathways	Alleviated hepatic steatosis, improved glucose tolerance, regulation of hepatic autophagy, inhibition of lipid accumulation	[88]
Ginsenoside Mc1	HFD-induced NAFLD mice	10	/	Alleviated ER stress, anti-apoptosis, improved insulin resistance, alleviated hepatic steatosis	[89]
	HFD-induced NAFLD rats; human normally hepatic immortal cell line HL-7702	0.125, 0.25, 0.5% in diet; 0, 25, 50, 100	Regulation of PPAR α expression	Reduced body weight, alleviated hepatic steatosis, improved metabolic disorders, improved insulin resistance, anti-inflammatory, anti-oxidative stress	[91]
Ursolic acid	db/db mice (type 2 diabetic mouse model); PA-induced HepG2 cells	0.14% in diets; 10–30	/	Inhibition of lipid accumulation, alleviated ER stress, reduced liver weight and reduced liver injury, alleviated hepatic steatosis	[92]
	DMSO-induced human hepatocellular carcinoma cell	5, 10	Regulation of LXRα	Inhibition of lipid accumulation, alleviated hepatic steatosis, reduced blood lipids	[93]
	Fructose induced NAFLD newborn rats	10	/	Inhibition of lipid accumulation	[94]
Betulinic acid	HFD-induced NAFLD rats	0.1% in diet	Regulation of the PERK/EIF2α/ATF4/CHOP signaling pathway	Enhanced energy expenditure, modulation of bile acids, alleviated hepatic steatosis, anti-inflammatory, alleviated ER stress	[95]
	HFD-induced NAFLD rats; insulin-resistant HepG2 cells	5, 10; 10–40	Regulation of the AMPK–mTOR SREBP signaling pathway	Inhibition of lipid accumulation	[96]

Table 2. *Cont.*

Compound	Animal/Cell Model	Dosage (mg/kg/d; μM/24 h)	Target/Pathways/Mechanism	Effects	Reference
Glycyrrhizic acid	HFD-induced NAFLD mice; (PA + OA)-induced mice primary hepatocytes	150; 10	Inhibition of the YY1/FAS signaling pathway	Inhibition of lipid accumulation, alleviated fatty acid synthesis, anti-fibrosis, anti-inflammatory, inhibition of excessive lipogenesis	[97]
	MCD diet-induced NAFLD mice	12.5, 25, 50	/	Inhibition of lipid accumulation, anti-inflammatory, anti-fibrosis, improved lipid metabolism	[98]
	HFD-induced NAFLD mice	15, 30, 60	/	Inhibition of lipid accumulation, reduced hepatic lipogenesis, reduced body weight, ameliorated hepatic steatosis, reduced serum glucose, improved glucose tolerance and insulin sensitivity	[99]
	MCD diet-induced NAFLD mice	30, 50	/	Inhibition of lipid accumulation, modulation of bile acids, anti-inflammatory, anti-fibrosis	[101]
Glycyrrhetinic acid	HFD-induced NAFLD mice	15, 30, 60	/	Regulation of vitamin A metabolism, protection against hepatic injury	[103]
	HFD-induced NAFLD mice	60	Suppression of HNF4α	Reduced blood glucose, improved insulin resistance	[104]
Oleanolic acid	HFD-induced NAFLD rats	25, 50, 100	Inhibition of LXRs, activation of the AMPK pathways	Alleviated hepatic steatosis, anti-inflammatory, anti-oxidative stress, improved insulin resistance	[107]
	HFHCD-induced NAFLD rats	80	/	Decreased blood lipids, anti-oxidative stress, reversed hepatic degeneration	[109]
Astragaloside IV	(PA + OA)-induced HepG2 cells and primary murine hepatocytes	50–200	Activation of the AMPK signaling pathway	Inhibition of lipid accumulation, inhibition of lipogenesis, alleviated ER stress	[113]
Mogroside V	High-concentration insulin or OA-induced HepG2 cells	25.6, 51.2, 102.4	Inhibition of protein tyrosine phosphatase 1B	Improved insulin resistance, inhibition of lipid accumulation	[114]
	HFD-induced NAFLD mice	400, 800	Upregulation of pAMPK expression	Inhibition of lipid accumulation, anti-inflammatory, anti-oxidative stress	[116]
	HFD-induced NAFLD mice; (PA + OA)-induced human LO2 cells	25, 50, 100; 15, 30, 60,120	Activation of the AMPK signaling pathway	Inhibition of lipid accumulation, ameliorated hepatic steatosis	[117]
Asiatic acid	HFD-induced NAFLD rats	4, 8	Inhibition of the ERS signaling pathway	Inhibition of lipid accumulation, anti-inflammatory, anti-oxidative stress	[119]
Corosolic acid	HFD + CCl4-induced NAFLD mice; FFA + OA + PA-induced HepG2 cells	10, 20; 5, 10, 20	Regulation of the TGF-β1/Smad2, NF-κB, and AMPK signaling pathways	Inhibition of lipid accumulation, anti-inflammatory, anti-fibrosis	[121]
Arjunolic acid	HFD-induced NAFLD rats; (PA + OA)-induced HepG2 cells	100, 200; 12.5, 50	Upregulation of PPARγ expression	Inhibition of lipid accumulation, ameliorated hepatic steatosis, reduced blood lipids	[123]

Table 2. *Cont.*

Compound	Animal/Cell Model	Dosage (mg/kg/d; μM/24 h)	Target/Pathways/Mechanism	Effects	Reference
Ganoderic acid A	HFD-induced NAFLD rats	20, 40	Activation of the AMPK signaling pathway	Inhibition of lipid accumulation, anti-inflammatory, reduced live weight	[125]
Ilexgenin A	HFD-induced NAFLD rats	80	/	Ameliorated hepatic steatosis, hypolipidemic, anti-inflammatory, enhanced effects of simvastatin	[126]
Rotundic acid	HFD-induced NAFLD rats; insulin-induced primary hepatocytes	10, 30, 100; 6.25–200	Downregulation of the SREBP-1c/SCD1 signaling pathway	Inhibition of lipid accumulation, improved dyslipidemia, protection against hepatic injury, anti-inflammatory, inhibition of excessive lipogenesis	[128]
Saikosaponin D	TAA-induced liver injury mice; HFD-induced NAFLD mice	2	/	Reduced blood lipids, anti-oxidative stress, anti-inflammatory	[130]
Lycopene	HFD-induced NAFLD mice	100, 1000	Upregulation of PPARa-inducible genes	Ameliorated hepatic steatosis	[135]
	HFD-induced NAFLD mice	0.004, 0.012% in diet	/	Inhibition of lipid accumulation, improved insulin resistance, anti-fibrosis, anti-inflammatory, anti-oxidative stress	[136]
	HFD-induced NAFLD rats	5, 10, 20	Downregulated expression of TNF-α and CYP2E1	Improved lipid profiles, reduced lipid peroxides, reduced blood lipids	[137]
	HFD-induced NAFLD mice	0.05% in diet	microRNA-21-induced downregulation of fatty-acid-binding protein 7	Ameliorated hepatic steatosis, inhibition of hepatic lipid accumulation	[138]
Astaxanthin	HFD-induced NAFLD rats	20	/	Ameliorated hepatic steatosis, reduced liver weight, reduced blood lipids	[140]
	HFD-induced NAFLD mice	0.02% in diet	Inhibition o the f eIF-2 signaling pathway	Inhibition of lipid accumulation, anti-inflammatory, anti-fibrosis, anti-oxidative stress	[147]
	HFD-induced NAFLD mice	0.003, 0.01, 0.03% in diet	/	Alleviated hepatic steatosis, anti-inflammatory, anti-oxidative stress	[148]
	Hepatic stellate cells from humans and mice	0–200	Inhibition of the TGFβ1–Smad3 signaling pathway	Anti-oxidative stress, anti-fibrosis	[149]
	HFD-induced NAFLD mice	80	/	Alleviated hepatic steatosis, anti-inflammatory, anti-oxidative stress	[150]
	HFD-induced NAFLD mice	0.0067, 0.02% in diet	/	Inhibition of lipid accumulation, alleviated hepatic steatosis, improved glucose intolerance, improved insulin resistance, anti-inflammatory, anti-fibrosis	[151]
	HFD-induced NAFLD mice; human liver cell line	10, 30, 60; 30, 60, 90	Upregulating the FGF21/PGC-1α signaling pathway	Inhibition of lipid accumulation, anti-oxidative stress, anti-apoptosis, anti-inflammatory, anti-fibrosis, attenuated mitochondrial dysfunction	[152]

Table 2. *Cont.*

Compound	Animal/Cell Model	Dosage (mg/kg/d; μM/24 h)	Target/Pathways/Mechanism	Effects	Reference
β-cryptoxanthin	HRCD + DKO-induced NAFLD mice	10	Activation of the SIRT1/AMPK signaling pathway	Inhibition of lipid accumulation, alleviated hepatic steatosis, increased cholesterol efflux	[156]
	HFD-induced NASH mice	0.003% in diet	/	Anti-inflammatory, anti-oxidative stress, anti-fibrosis, alleviated hepatic steatosis, inhibition of lipid accumulation	[157]
	HFD-induced NASH mice	0.003% in diet	/	Anti-inflammatory, anti-oxidative stress, anti-fibrosis, alleviated hepatic steatosis, inhibition of lipid accumulation, improved liver dysfunction	[158]
β-carotene	HFD-induced NAFLD rats	70	/	Alleviated hepatic steatosis, anti-inflammatory	[163]
Lutein	HFD-induced NAFLD rats	0, 12.5, 25, 50	Activation of the SIRT1/PPAR-α signaling pathway	Reduced body weight, alleviated hepatic steatosis, improved insulin resistance	[165]
Cycloastragenol	HFD-induced NAFLD mice; FXR deletion HepG2 cells	0.1% in diet; 25	Regulation of the FXR signaling pathway	Alleviated hepatic steatosis, inhibition of lipid accumulation, reduced blood glucose, anti-oxidative stress	[167]

4. Conclusions

The data reported in this review highlight that, despite the very large number of studies investigating the health effects of terpenoids in humans, triterpenoids have been the main focus, with 20 triterpenoids proving effective in the treatment of NAFLD. Different classes of terpenoids have been shown to be present in different amounts in NAFLD treatment, and can be ranked as follows: triterpenoids > monoterpenoids > tetraterpenoids > diterpenoids > sesquiterpenoids. This is because compared with other terpenoids, triterpenoids are more diverse, widely distributed, and abundant. In particular in the 21st century, the diversity and importance of the biological activities of triterpenes have attracted much attention and have become a popular area of natural medicinal chemistry research. In addition, their effects may be related to the unique side chains of these triterpenes, which may facilitate the formation of links with the target proteins and thus exhibit better activity. It is necessary to conduct some research on molecular docking technology to further evaluate the affinity of ligand binding to the target protein receptor, so as to better explain the systematic structure–activity relationship.

The pathogenesis of NAFLD is complex and implicates a cross talk between various metabolically active sites. Experimental models play a crucial role in elucidating the pathophysiology of diseases and the pharmacological effects of the drug. In this review, animal models and cell models are the main methods of study. In order to study the pathogenesis of non-alcoholic fatty liver disease and the effect of drug treatment, transgenic animals, chemical-induced animals, and three high-fat-diet-induced experimental animal models are often used as research models for NAFLD. The animal model of fatty liver induced by a high-fat diet has similar pathological characteristics to those in humans, with a high success rate and low mortality. The method is simple, easy to repeat, and has become the main choice for drug research. Mice and rats have been used most frequently in NAFLD modeling and therefore constitute the main focus of animal model research. Although it is easy to establish NAFLD models in rats and mice, long-term drug observation cannot be achieved, which has also led to the lack of studies on the development of NALFD to end-stage cirrhosis. The cell lines used to establish cell models include primary hepatocytes from animals and humans, immortalized cell lines, and hepatoid cells derived from induced pluripotent stem cells. HepG2 is a popular hepatic cell line, used in a wide range of studies. As a class of nontumorigenic cells, with high proliferation rates and an epithelial-like morphology, it performs many differentiated hepatic functions. The HepG2 cell manufacturing model induced by palmitate and oleic acid is a common choice of NAFLD cell model. The induced pluripotent hepatocyte technology still has some problems, such as the low success rate of inducing cell transformation and easy genetic mutation in the process of induction, which limit its wide application. This review found that animal and cell experiments are often used to study the effects of terpenoids on NAFLD, and their complementarity can better reveal the mechanism of action and efficacy characteristics of terpenoids.

The different doses and methods of administration in these studies are an issue worth exploring. In animal studies, there are generally two ways to administer a drug: one is to administer a certain amount of the drug, and the other is to add it to the food in a certain percentage. Of the 76 animal modeling studies, 15 were conducted with a diet supplemented with terpenoids. Most of them use the method of administration according to a certain dose, which is more convenient for calculation and statistics, and can be easily controlled. The dosage of terpenoids in these studies was different, with even the dosage of one terpene compound being different in different experiments. In fact, different doses of terpenoids play a different role in the effective treatment of NAFLD, but there is only limited information on the effects of these different doses of terpenoids on hepatic cells. Drug administration in cell experiments mainly takes two forms: one is the direct addition into the medium, and the other is the pretreatment of the drug; the latter is less used. The cells were usually treated with different doses of terpenoids for 24 h to discover their effects on the cells. Unnecessarily high doses can increase the risk of complications and adverse effects associated with the drug. The current research is inadequate, so more studies are required to define the best dose.

Based on the above reports, there is a high probability that multiple edible terpenoids are potential candidates for managing NAFLD. Although the efficacy of terpenoids in treating NAFLD has been demonstrated in both animal and cell studies, the efficacy of most terpenoids has not been supported by clinical evidence. Some drugs have already been put to use in clinical practice, but the small number of cases in which they have been applied to date precludes an evaluation of their long-term durability. In China, oleanolic acid has been used as an over-the-counter (OTC) hepatoprotective drug for decades [168]. As early as 1993, it was used in clinical controlled trials as a classic anti-hepatitis drug [169], but there are not yet any clinical studies on NAFLD, although it is a drug with a long clinical history. Dietary carotenoids are thought to provide health benefits in terms of decreasing the risk of NAFLD and have been confirmed both in vitro and in vivo. Lycopene, astaxanthin, β-cryptoxanthin, β-carotene, and lutein are popular research topics in the field of carotenoid therapy for NAFLD. Clinical studies on these compounds are richer than those on other terpenoids. This is because these terpenoids are more commonly found in the daily diet. Ginkgo biloba extract is a popular topic in natural compound research and has a wide clinical research base. The Ginkgo biloba special extract EGb 761® is currently listed in local clinical guidelines in Germany, Switzerland, and some Asian countries, including Indonesia, the Philippines, and China [170]. Although ginkgolides have been proven to be effective for NAFLD, there is no clinical research support. Glycyrrhizic acid has been developed in Japan and China as a hepatoprotective drug in cases of chronic hepatitis, such as Stronger Neo-Minophagen C and compound clycyrrhizin tablets. Licorice and its derivatives are recognized as safe by the US Food and Drug Administration (FDA). The long medicinal history and rich clinical studies of glycyrrhizic acid provide possibilities for its development and application in NAFLD. However, glycyrrhizic acid can cause a metabolic syndrome presenting as pseudohyperaldosteronism, which limits its clinical application [171]. There exists the idea that "The best way to discover a new drug is to start with an old one"—these terpenoids seem to be able to achieve clinical breakthroughs more quickly than other terpenoids.

NAFLD is not a single disease, but encompasses a range of liver diseases, from hepatic steatosis to NASH, and can eventually lead to cirrhosis and even death. Lipid metabolism disorder, insulin resistance, oxidative injury, inflammatory reaction, apoptosis, cytosis, and autophagy interact with each other to form a complex regulatory network, resulting in a series of pathological cascades of NAFLD. Terpenoids can protect the injured liver tissue by inhibiting and regulating the above factors. NAFLD is the result of a serious systemic disorder of lipid metabolism. Regulating lipid metabolism is a common effect of all terpenoids included in this review, included the inhibition of hepatic lipid accumulation, the regulation of steatosis, the inhibition of hepatic lipogenesis, and the promotion of fatty acid oxidation. Inflammation is among the features necessary for a histologic diagnosis of NASH, involved in the whole pathological process of NAFLD. A total of 28 terpenoids have been shown to play an anti-inflammatory role in NAFLD, thereby inhibiting the progression of the disease. The anti-inflammatory effects of these terpenoids also inhibit programmed cell death, liver fibrosis, and liver tissue damage. The increased production of ROS in the liver leads to lipid peroxidation/oxidative stress, which then leads to the functional dysfunction of the liver cytoplasmic body, as well as apoptosis and deterioration. A total of 19 terpenoids play a therapeutic role in NAFLD via their anti-oxidative stress activity. Insulin resistance plays a key role in the pathological process of NAFLD, with the prevalence of NAFLD being 5-fold higher in patients with diabetes compared to those without [172]. Insulin resistance is involved in the metabolic basis of NAFLD. In this review, 13 terpenoids are reported to improve insulin resistance in the study of NAFLD. In fact, a large number of terpenoids are reported to improve insulin resistance in the study of diabetes, but they have not been mentioned in the study of NAFLD. Liver fibrosis is the most important predictor of mortality in NAFLD. Untreated NASH may lead to liver fibrosis and eventually HCC. A total of 10 terpenoids have been proved to have anti-fibrotic effects, which is of great significance for the long-term treatment of NAFLD. Pyroptosis, as a novel form of pro-inflammatory programmed cell death, has become a popular research topic in the life sciences. Pyroptosis is considered to play an important role in low-grade chronic inflammation, which is thought to be the core of the pathological mechanism of NAFLD [173]. Although the involvement of pyroptosis in NAFLD has been well established, little has been written about the role of pyroptosis through this natural compound. Only one terpenoid has been reported to cause pyroptosis in NAFLD. The maintenance of iron distribution is crucial to the metabolic response of hepatocytes. The ectopic storage of iron between hepatocytes and astrocytes is a key factor in the development of non-alcoholic fatty liver disease [174]. However, only one terpenoid has been revealed to regulate iron homeostasis and protect the liver. To accelerate the application of the latest pathological mechanisms in the field of drug research, and provide new potential targets for the further drug development of terpenoids, in the future, we need to enrich the research on the effects of terpenoids on NAFLD, encouraging new ideas for drug use and disease treatment.

We found that the signaling pathways and targets of these terpenoids were consistent to a certain extent. The evolution of AMPK and its homologs enabled excellent responsivity and control of cellular energetic homeostasis [175]. The central role of AMPK in maintaining cellular energy homeostasis has made it a promising target for drugs aimed at preventing and/or treating NAFLD. The effects of 14 terpenoids on NAFLD through the AMPK signaling pathway were reported in the literature. The peroxisome proliferator-activated receptors (PPAR-α, PPAR-β/δ, and PPAR-γ) are members of the nuclear receptor superfamily and play crucial roles in lipid metabolism [41]. Eight terpenoids were confirmed to exert an intervention effect through this receptor, and PPAR-α and PPAR-γ were the key to this effect, of which PPAR-α was the most frequently mentioned. PPAR-α and PPAR-γ are the master regulators of adipogenesis and adipose tissue development and both of them have specific expression in the liver. The gene expression program induced by Nrf2 transcription factor plays a critical role in cell defense responses against a broad variety of cellular stresses, including oxidative stress, metabolism, and inflammation [176]. Five terpenoids exert anti-oxidative stress, anti-inflammatory, and liver-protective effects by regulating Nrf2. In recent years, accumulating evidence has indicated that sirtuins play important roles in regulating the metabolic processes related to fatty liver diseases [177]. SIRT 1 is the most extensively studied sirtuin and is involved in fatty liver diseases. Five terpenoids regulate lipid metabolism, oxidative stress, and inflammation in the liver via SIRT 1.

The existence of these common pathways and targets provides a reference for the further improvement of the study of terpenoids in the treatment of NAFLD and promotes the study of the mechanism of action of these terpenoids.

Some terpenoids reviewed in this paper show potent activity in the treatment of NAFLD. However, current studies are restricted to animal and cell experiments, with a lack of clinical research and systematic SAR studies. Further clinical research and SAR studies with terpenoids could provide more insight into the effectiveness of these complicated pharmacological properties, enabling terpenoids to be used safely and efficiently.

Author Contributions: Concept, collection of material, methodology and writing, P.Y.; revision, supervision and funding acquisition, Y.L. All authors have read and agreed to the published version of the manuscript.

Funding: This research received no external funding.

Institutional Review Board Statement: Not applicable.

Informed Consent Statement: Not applicable.

Data Availability Statement: Not applicable.

Conflicts of Interest: The authors declare no conflict of interest.

Sample Availability: Not applicable.

Abbreviations

non-alcoholic fatty liver disease	NAFLD
non-alcoholic fatty liver	NAFL
non-alcoholic steatohepatitis	NASH
metabolic-associated fatty liver disease	MAFLD
hepatic fibrosis	HF
palmitate	PA
methionine-choline deficient	MCD
low-density lipoprotein receptor knockout	$LDLr^{-/-}$
lipopolysaccharide	LPSO
non-esterified fatty acid	NEFA
Yin Yang 1	YY1
fatty acid synthase	FAS
endoplasmic reticulum	ER
free fatty acids	FFA
carbon tetrachloride	CCl4
thioacetamide	TAA
high-fat + high-carbohydrate diet	HFHCD
uncoupling protein-2	UCP2
liver X receptors	LXRs
anti-oxidant redux elements	ARE
over the counter	OTC
Food and Drug Administration	FDA
high-refined carbohydrate diet	HRCD
double knockout	DKO
wild-type	WT
β-carotene oxygenase 1	BCO1
β-carotene oxygenase 2	BCO2
sirtuin 1	SIRT 1
farnesoid X receptor	FXR
liver X receptor	LXR
sterol regulatory element binding proteins-1c	SREBP-1c
steroyl-coA desaturase-1	SCD-1

References

1. Lazarus, J.V.; Mark, H.E.; Anstee, Q.M.; Arab, J.P.; Batterham, R.L.; Castera, L.; Cortez-Pinto, H.; Crespo, J.; Cusi, K.; Dirac, M.A.; et al. Advancing the global public health agenda for NAFLD: A consensus statement. *Nat. Rev. Gastroenterol. Hepatol.* **2022**, *19*, 60–78. [CrossRef] [PubMed]
2. Powell, E.E.; Wong, V.W.; Rinella, M. Non-alcoholic fatty liver disease. *Lancet* **2021**, *397*, 2212–2224. [CrossRef] [PubMed]
3. Kang, S.H.; Lee, H.W.; Yoo, J.J.; Cho, Y.; Kim, S.U.; Lee, T.H.; Jang, B.K.; Kim, S.G.; Ahn, S.B.; Kim, H.; et al. KASL clinical practice guidelines: Management of nonalcoholic fatty liver disease. *Clin. Mol. Hepatol.* **2021**, *27*, 363–401. [CrossRef] [PubMed]
4. Kim, H.; Lee, D.S.; An, T.H.; Park, H.J.; Kim, W.K.; Bae, K.H.; Oh, K.J. Metabolic Spectrum of Liver Failure in Type 2 Diabetes and Obesity: From NAFLD to NASH to HCC. *Int. J. Mol. Sci.* **2021**, *22*, 4495. [CrossRef] [PubMed]
5. Eslam, M.; Sarin, S.K.; Wong, V.W.; Fan, J.G.; Kawaguchi, T.; Ahn, S.H.; Zheng, M.H.; Shiha, G.; Yilmaz, Y.; Gani, R.; et al. The Asian Pacific Association for the Study of the Liver clinical practice guidelines for the diagnosis and management of metabolic associated fatty liver disease. *Hepatol. Int.* **2020**, *14*, 889–919. [CrossRef]
6. Rinella, M.E.; Tacke, F.; Sanyal, A.J.; Anstee, Q.M. Report on the AASLD/EASL Joint Workshop on Clinical Trial Endpoints in NAFLD. *Hepatology* **2019**, *70*, 1424–1436. [CrossRef]
7. Marchisello, S.; Di Pino, A.; Scicali, R.; Urbano, F.; Piro, S.; Purrello, F.; Rabuazzo, A.M. Pathophysiological, Molecular and Therapeutic Issues of Nonalcoholic Fatty Liver Disease: An Overview. *Int. J. Mol. Sci.* **2019**, *20*, 1948. [CrossRef]
8. Rhee, E.J. Nonalcoholic Fatty Liver Disease and Diabetes: An Epidemiological Perspective. *Endocrinol. Metab.* **2019**, *34*, 226–233. [CrossRef]
9. Stahl, E.P.; Dhindsa, D.S.; Lee, S.K.; Sandesara, P.B.; Chalasani, N.P.; Sperling, L.S. Nonalcoholic Fatty Liver Disease and the Heart: JACC State-of-the-Art Review. *J. Am. Coll. Cardiol.* **2019**, *73*, 948–963. [CrossRef]
10. Kasper, P.; Martin, A.; Lang, S.; Kütting, F.; Goeser, T.; Demir, M.; Steffen, H.M. NAFLD and cardiovascular diseases: A clinical review. *Clin. Res. Cardiol.* **2021**, *110*, 921–937. [CrossRef]

11. Hamaguchi, M.; Hashimoto, Y.; Obora, A.; Kojima, T.; Fukui, M. Non-alcoholic fatty liver disease with obesity as an independent predictor for incident gastric and colorectal cancer: A population-based longitudinal study. *BMJ Open Gastroenterol.* **2019**, *6*, e000295. [CrossRef] [PubMed]
12. Fernando, D.H.; Forbes, J.M.; Angus, P.W.; Herath, C.B. Development and Progression of Non-Alcoholic Fatty Liver Disease: The Role of Advanced Glycation End Products. *Int. J. Mol. Sci.* **2019**, *20*, 5037. [CrossRef] [PubMed]
13. Fougerat, A.; Montagner, A.; Loiseau, N.; Guillou, H.; Wahli, W. Peroxisome Proliferator-Activated Receptors and Their Novel Ligands as Candidates for the Treatment of Non-Alcoholic Fatty Liver Disease. *Cells* **2020**, *9*, 1638. [CrossRef] [PubMed]
14. Delli Bovi, A.P.; Marciano, F.; Mandato, C.; Siano, M.A.; Savoia, M.; Vajro, P. Oxidative Stress in Non-alcoholic Fatty Liver Disease. An Updated Mini Review. *Front. Med.* **2021**, *8*, 595371. [CrossRef] [PubMed]
15. Di Ciaula, A.; Passarella, S.; Shanmugam, H.; Noviello, M.; Bonfrate, L.; Wang, D.Q.; Portincasa, P. Nonalcoholic Fatty Liver Disease (NAFLD). Mitochondria as Players and Targets of Therapies? *Int. J. Mol. Sci.* **2021**, *22*, 5375. [CrossRef]
16. Fujii, H.; Kawada, N.; Japan Study Group Of Nafld Jsg-Nafld. The Role of Insulin Resistance and Diabetes in Nonalcoholic Fatty Liver Disease. *Int. J. Mol. Sci.* **2020**, *21*, 3863. [CrossRef]
17. Chopra, B.; Dhingra, A.K. Natural products: A lead for drug discovery and development. *Phytother. Res.* **2021**, *35*, 4660–4702. [CrossRef]
18. Cheng, C.; Zhuo, S.; Zhang, B.; Zhao, X.; Liu, Y.; Liao, C.; Quan, J.; Li, Z.; Bode, A.M.; Cao, Y.; et al. Treatment implications of natural compounds targeting lipid metabolism in nonalcoholic fatty liver disease, obesity and cancer. *Int. J. Biol. Sci.* **2019**, *15*, 1654–1663. [CrossRef]
19. Newman, D.J.; Cragg, G.M. Natural Products as Sources of New Drugs over the Nearly Four Decades from 01/1981 to 09/2019. *J. Nat. Prod.* **2020**, *83*, 770–803. [CrossRef]
20. Huang, M.; Lu, J.J.; Huang, M.Q.; Bao, J.L.; Chen, X.P.; Wang, Y.T. Terpenoids: Natural products for cancer therapy. *Expert Opin. Investig. Drugs* **2012**, *21*, 1801–1818. [CrossRef]
21. Lai Shi Min, S.; Liew, S.Y.; Chear, N.J.Y.; Goh, B.H.; Tan, W.N.; Khaw, K.Y. Plant Terpenoids as the Promising Source of Cholinesterase Inhibitors for Anti-AD Therapy. *Biology* **2022**, *11*, 307. [CrossRef] [PubMed]
22. Wang, Y.; Zhu, Y.; Lu, S.; Hu, C.; Zhong, W.; Chai, Y. Beneficial effects of paeoniflorin on osteoporosis induced by high-carbohydrate, high-fat diet-associated hyperlipidemia in vivo. *Biochem. Biophys. Res. Commun.* **2018**, *498*, 981–987. [CrossRef] [PubMed]
23. Zhang, Y.; Liang, Y.; Liu, H.; Huang, Y.; Li, H.; Chen, B. Paeoniflorin attenuates gestational diabetes via Akt/mTOR pathway in a rat model. *Food. Nutr. Res.* **2020**, *64*. [CrossRef]
24. Zhang, L.; Yang, B.; Yu, B. Paeoniflorin Protects against Nonalcoholic Fatty Liver Disease Induced by a High-Fat Diet in Mice. *Biol. Pharm. Bull.* **2015**, *38*, 1005–1011. [CrossRef] [PubMed]
25. Ma, Z.; Chu, L.; Liu, H.; Wang, W.; Li, J.; Yao, W.; Yi, J.; Gao, Y. Beneficial effects of paeoniflorin on non-alcoholic fatty liver disease induced by high-fat diet in rats. *Sci. Rep.* **2017**, *7*, 44819. [CrossRef] [PubMed]
26. Li, Y.C.; Qiao, J.Y.; Wang, B.Y.; Bai, M.; Shen, J.D.; Cheng, Y.X. Paeoniflorin Ameliorates Fructose-Induced Insulin Resistance and Hepatic Steatosis by Activating LKB1/AMPK and AKT Pathways. *Nutrients* **2018**, *10*, 1024. [CrossRef]
27. Ma, Z.; Chu, L.; Liu, H.; Li, J.; Zhang, Y.; Liu, W.; Dai, J.; Yi, J.; Gao, Y. Paeoniflorin alleviates non-alcoholic steatohepatitis in rats: Involvement with the ROCK/NF-κB pathway. *Int. Immunopharmacol.* **2016**, *38*, 377–384. [CrossRef]
28. Chen, D.; Guo, J.; Li, L. Catalpol promotes mitochondrial biogenesis in chondrocytes. *Arch. Physiol. Biochem.* **2022**, *128*, 802–808. [CrossRef]
29. Tian, X.; Ru, Q.; Xiong, Q.; Wen, R.; Chen, Y. Catalpol Attenuates Hepatic Steatosis by Regulating Lipid Metabolism via AMP-Activated Protein Kinase Activation. *Biomed. Res. Int.* **2020**, *2020*, 6708061. [CrossRef]
30. Herzig, S.; Shaw, R.J. AMPK: Guardian of metabolism and mitochondrial homeostasis. *Nat. Rev. Mol. Cell Biol.* **2018**, *19*, 121–135. [CrossRef]
31. Ren, H.; Wang, D.; Zhang, L.; Kang, X.; Li, Y.; Zhou, X.; Yuan, G. Catalpol induces autophagy and attenuates liver steatosis in ob/ob and high-fat diet-induced obese mice. *Aging* **2019**, *11*, 9461–9477. [CrossRef] [PubMed]
32. Zhang, Y.; Wang, C.; Lu, J.; Jin, Y.; Xu, C.; Meng, Q.; Liu, Q.; Dong, D.; Ma, X.; Liu, K.; et al. Targeting of miR-96-5p by catalpol ameliorates oxidative stress and hepatic steatosis in LDLr-/- mice via p66shc/cytochrome C cascade. *Aging* **2020**, *12*, 2049–2069. [CrossRef] [PubMed]
33. Tian, X.; Xiong, Q.; Chen, L.; Wen, J.R.; Ru, Q. Intervention of Catalpol on High-fat Diet Induced Nonalcoholic Fatty Liver Disease in Mice. *Zhongguo Yi Xue Ke Xue Yuan Xue Bao* **2019**, *41*, 746–755. (In Chinese) [CrossRef]
34. Wen, M.; Liu, Y.; Chen, R.; He, P.; Wu, F.; Li, R.; Lin, Y. Geniposide suppresses liver injury in a mouse model of DDC-induced sclerosing cholangitis. *Phytother. Res.* **2021**, *35*, 3799–3811. [CrossRef] [PubMed]
35. Shen, B.; Feng, H.; Cheng, J.; Li, Z.; Jin, M.; Zhao, L.; Wang, Q.; Qin, H.; Liu, G. Geniposide alleviates non-alcohol fatty liver disease via regulating Nrf2/AMPK/mTOR signalling pathways. *J. Cell Mol. Med.* **2020**, *24*, 5097–5108. [CrossRef]
36. Ma, T.; Huang, C.; Zong, G.; Zha, D.; Meng, X.; Li, J.; Tang, W. Hepatoprotective effects of geniposide in a rat model of nonalcoholic steatohepatitis. *J. Pharm. Pharmacol.* **2011**, *63*, 587–593. [CrossRef]
37. Fan, X.; Lin, L.; Cui, B.; Zhao, T.; Mao, L.; Song, Y.; Wang, X.; Feng, H.; Qingxiang, Y.; Zhang, J.; et al. Therapeutic potential of genipin in various acute liver injury, fulminant hepatitis, NAFLD and other non-cancer liver diseases: More friend than foe. *Pharmacol. Res.* **2020**, *159*, 104945. [CrossRef]

38. Zhong, H.; Chen, K.; Feng, M.; Shao, W.; Wu, J.; Chen, K.; Liang, T.; Liu, C. Genipin alleviates high-fat diet-induced hyperlipidemia and hepatic lipid accumulation in mice via miR-142a-5p/SREBP-1c axis. *FEBS J.* **2018**, *285*, 501–517. [CrossRef]
39. Zhong, H.; Liu, M.; Ji, Y.; Ma, M.; Chen, K.; Liang, T.; Liu, C. Genipin Reverses HFD-Induced Liver Damage and Inhibits UCP2-Mediated Pyroptosis in Mice. *Cell Physiol. Biochem.* **2018**, *49*, 1885–1897. [CrossRef]
40. Wang, R.; Dong, Z.; Lan, X.; Liao, Z.; Chen, M. Sweroside Alleviated LPS-Induced Inflammation via SIRT1 Mediating NF-κB and FOXO1 Signaling Pathways in RAW264.7 Cells. *Molecules* **2019**, *24*, 872. [CrossRef]
41. Mirza, A.Z.; Althagafi, I.I.; Shamshad, H. Role of PPAR receptor in different diseases and their ligands: Physiological importance and clinical implications. *Eur. J. Med. Chem.* **2019**, *166*, 502–513. [CrossRef] [PubMed]
42. Yang, Q.; Shu, F.; Gong, J.; Ding, P.; Cheng, R.; Li, J.; Tong, R.; Ding, L.; Sun, H.; Huang, W.; et al. Sweroside ameliorates NAFLD in high-fat diet induced obese mice through the regulation of lipid metabolism and inflammatory response. *J. Ethnopharmacol.* **2020**, *255*, 112556. [CrossRef] [PubMed]
43. Yang, G.; Jang, J.H.; Kim, S.W.; Han, S.H.; Ma, K.H.; Jang, J.K.; Kang, H.C.; Cho, Y.Y.; Lee, H.S.; Lee, J.Y. Sweroside Prevents Non-Alcoholic Steatohepatitis by Suppressing Activation of the NLRP3 Inflammasome. *Int. J. Mol. Sci.* **2020**, *21*, 2790. [CrossRef] [PubMed]
44. Jaishree, V.; Narsimha, S. Swertiamarin and quercetin combination ameliorates hyperglycemia, hyperlipidemia and oxidative stress in streptozotocin-induced type 2 diabetes mellitus in wistar rats. *Biomed. Pharmacother.* **2020**, *130*, 110561. [CrossRef] [PubMed]
45. Xu, L.; Li, D.; Zhu, Y.; Cai, S.; Liang, X.; Tang, Y.; Jin, S.; Ding, C. Swertiamarin supplementation prevents obesity-related chronic inflammation and insulin resistance in mice fed a high-fat diet. *Adipocyte* **2021**, *10*, 160–173. [CrossRef] [PubMed]
46. Zeng, X.; Guo, F.; Ouyang, D. A review of the pharmacology and toxicology of aucubin. *Fitoterapia* **2020**, *140*, 104443. [CrossRef] [PubMed]
47. Shen, B.; Zhao, C.; Wang, Y.; Peng, Y.; Cheng, J.; Li, Z.; Wu, L.; Jin, M.; Feng, H. Aucubin inhibited lipid accumulation and oxidative stress via Nrf2/HO-1 and AMPK signalling pathways. *J. Cell. Mol. Med.* **2019**, *23*, 4063–4075. [CrossRef]
48. Chang, Y.; Tian, Y.; Zhou, D.; Yang, L.; Liu, T.M.; Liu, Z.G.; Wang, S.W. Gentiopicroside ameliorates ethanol-induced gastritis via regulating MMP-10 and pERK1/2 signaling. *Int. Immunopharmacol.* **2021**, *90*, 107213. [CrossRef]
49. Jin, M.; Feng, H.; Wang, Y.; Yan, S.; Shen, B.; Li, Z.; Qin, H.; Wang, Q.; Li, J.; Liu, G. Gentiopicroside Ameliorates Oxidative Stress and Lipid Accumulation through Nuclear Factor Erythroid 2-Related Factor 2 Activation. *Oxid. Med. Cell Longev.* **2020**, *2020*, 2940746. [CrossRef]
50. Chen, J.; Fan, X.; Zhou, L.; Gao, X. Treatment with geraniol ameliorates methionine-choline-deficient diet-induced non-alcoholic steatohepatitis in rats. *J. Gastroenterol. Hepatol.* **2016**, *31*, 1357–1365. [CrossRef]
51. Hashem, S.; Nisar, S.; Sageena, G.; Macha, M.A.; Yadav, S.K.; Krishnankutty, R.; Uddin, S.; Haris, M.; Bhat, A.A. Therapeutic Effects of Curcumol in Several Diseases; An Overview. *Nutr. Cancer* **2021**, *73*, 181–195. [CrossRef] [PubMed]
52. Qi, X.; Song, A.; Ma, M.; Wang, P.; Zhang, X.; Lu, C.; Zhang, J.; Zheng, S.; Jin, H. Curcumol inhibits ferritinophagy to restrain hepatocyte senescence through YAP/NCOA4 in non-alcoholic fatty liver disease. *Cell Prolif.* **2021**, *54*, e13107. [CrossRef] [PubMed]
53. Qi, S.Y.; Huang, H.; Li, Y.K.; Pei, L.G.; Zhang, W.J. Effects of curcumol on liver function and fibrosis in rats of nonalcoholic fatty liver disease and its mechanism. *Zhongguo Ying Yong Sheng Li Xue Za Zhi* **2021**, *37*, 611–615. (In Chinese) [CrossRef] [PubMed]
54. Xu, N.; Luo, H.; Li, M.; Wu, J.; Wu, X.; Chen, L.; Gan, Y.; Guan, F.; Li, M.; Su, Z.; et al. β-patchoulene improves lipid metabolism to alleviate non-alcoholic fatty liver disease via activating AMPK signaling pathway. *Biomed. Pharmacother.* **2021**, *134*, 111104. [CrossRef]
55. Luo, H.; Xu, N.; Wu, J.; Gan, Y.; Chen, L.; Guan, F.; Li, M.; Li, Y.; Chen, J.; Su, Z.; et al. β-patchoulene protects against non-alcoholic steatohepatitis via interrupting the vicious circle among oxidative stress, histanoxia and lipid accumulation in rats. *Int. Immunopharmacol.* **2021**, *98*, 107915. [CrossRef]
56. Kamikubo, R.; Kai, K.; Tsuji-Naito, K.; Akagawa, M. β-Caryophyllene attenuates palmitate-induced lipid accumulation through AMPK signaling by activating CB2 receptor in human HepG2 hepatocytes. *Mol. Nutr. Food. Res.* **2016**, *60*, 2228–2242. [CrossRef]
57. Zi, J.; Mafu, S.; Peters, R.J. To gibberellins and beyond! Surveying the evolution of (di)terpenoid metabolism. *Annu. Rev. Plant. Biol.* **2014**, *65*, 259–286. [CrossRef]
58. Lou, T.Y.; Ma, B.B.; Liang, Y.Y.; Wang, C.X.; Liu, J.H.; Li, R.J.; Yu, S.Y.; Wang, Z.B. Analysis of carnosic acid metabolites in rats by UHPLC-Q-Exactive M.S. *Zhongguo Zhong Yao Za Zhi* **2020**, *45*, 3952–3960. (In Chinese) [CrossRef]
59. Birtić, S.; Dussort, P.; Pierre, F.X.; Bily, A.C.; Roller, M. Carnosic acid. *Phytochemistry* **2015**, *115*, 9–19. [CrossRef]
60. Wang, T.; Takikawa, Y.; Satoh, T.; Yoshioka, Y.; Kosaka, K.; Tatemichi, Y.; Suzuki, K. Carnosic acid prevents obesity and hepatic steatosis in ob/ob mice. *Hepatol. Res.* **2011**, *41*, 87–92. [CrossRef]
61. Park, M.Y.; Sung, M.K. Carnosic acid attenuates obesity-induced glucose intolerance and hepatic fat accumulation by modulating genes of lipid metabolism in C57BL/6J-ob/ob mice. *J. Sci. Food. Agric.* **2015**, *95*, 828–835. [CrossRef] [PubMed]
62. Dickmann, L.J.; VandenBrink, B.M.; Lin, Y.S. In vitro hepatotoxicity and cytochrome P450 induction and inhibition characteristics of carnosic acid, a dietary supplement with antiadipogenic properties. *Drug. Metab. Dispos.* **2012**, *40*, 1263–1267. [CrossRef] [PubMed]
63. Song, H.M.; Li, X.; Liu, Y.Y.; Lu, W.P.; Cui, Z.H.; Zhou, L.; Yao, D.; Zhang, H.M. Carnosic acid protects mice from high-fat diet-induced NAFLD by regulating MARCKS. *Int. J. Mol. Med.* **2018**, *42*, 193–207. [CrossRef] [PubMed]

64. Shan, W.; Gao, L.; Zeng, W.; Hu, Y.; Wang, G.; Li, M.; Zhou, J.; Ma, X.; Tian, X.; Yao, J. Activation of the SIRT1/p66shc antiapoptosis pathway via carnosic acid-induced inhibition of miR-34a protects rats against nonalcoholic fatty liver disease. *Cell Death. Dis.* **2015**, *6*, e1833. [CrossRef]
65. Lombardi, R.; Fargion, S.; Fracanzani, A.L. Brain involvement in non-alcoholic fatty liver disease (NAFLD): A systematic review. *Dig. Liver Dis.* **2019**, *51*, 1214–1222. [CrossRef] [PubMed]
66. Xu, T.; Zhou, J.; Zhu, J.; Zhang, S.; Zhang, N.; Zhao, Y.; Ding, C.; Shi, X.; Yao, J. Carnosic acid protects non-alcoholic fatty liver-induced dopaminergic neuron injury in rats. *Metab. Brain Dis.* **2017**, *32*, 483–491. [CrossRef] [PubMed]
67. Du, J.F.; Li, P.; Lu, X. Advance in biosynthesis and metabolic regulation of ginkgolides. *Zhongguo Zhong Yao Za Zhi* **2021**, *46*, 3288–3297. (In Chinese) [CrossRef]
68. Sarkar, C.; Quispe, C.; Jamaddar, S.; Hossain, R.; Ray, P.; Mondal, M.; Abdulwanis Mohamed, Z.; Sani Jaafaru, M.; Salehi, B.; Islam, M.T.; et al. Therapeutic promises of ginkgolide A: A literature-based review. *Biomed. Pharmacother.* **2020**, *132*, 110908. [CrossRef]
69. Jeong, H.S.; Kim, K.H.; Lee, I.S.; Park, J.Y.; Kim, Y.; Kim, K.S.; Jang, H.J. Ginkgolide A ameliorates non-alcoholic fatty liver diseases on high fat diet mice. *Biomed. Pharmacother.* **2017**, *88*, 625–634. [CrossRef]
70. Luo, L.; Yin, L.; Dongshan, W.; Zhao, Y.; Wang, Y.; Li, F.; Fang, J.; Chen, H.; Fan, S.; Huang, C. Ginkgolide B lowers body weight and ameliorates hepatic steatosis in high-fat diet-induced obese mice correlated with pregnane X receptor activation. *RSC Advances* **2017**, *7*, 37858–37866. [CrossRef]
71. Yang, Y.; Chen, J.; Gao, Q.; Shan, X.; Wang, J.; Lv, Z. Study on the attenuated effect of Ginkgolide B on ferroptosis in high fat diet induced nonalcoholic fatty liver disease. *Toxicology* **2020**, *445*, 152599. [CrossRef] [PubMed]
72. Song, J.; Han, X.; Yao, Y.L.; Li, Y.M.; Zhang, J.; Shao, D.Y.; Hou, L.S.; Fan, Y.; Song, S.Z.; Lian, L.H.; et al. Acanthoic acid suppresses lipin1/2 via TLR4 and IRAK4 signalling pathways in EtOH- and lipopolysaccharide-induced hepatic lipogenesis. *J. Pharm. Pharmacol.* **2018**, *70*, 393–403. [CrossRef] [PubMed]
73. Wu, Y.L.; Jiang, Y.Z.; Jin, X.J.; Lian, L.H.; Piao, J.Y.; Wan, Y.; Jin, H.R.; Joon Lee, J.; Nan, J.X. Acanthoic acid, a diterpene in Acanthopanax koreanum, protects acetaminophen-induced hepatic toxicity in mice. *Phytomedicine* **2010**, *17*, 475–479. [CrossRef] [PubMed]
74. Nan, J.X.; Jin, X.J.; Lian, L.H.; Cai, X.F.; Jiang, Y.Z.; Jin, H.R.; Lee, J.J. A diterpenoid acanthoic acid from Acanthopanax koreanum protects against D-galactosamine/lipopolysaccharide-induced fulminant hepatic failure in mice. *Biol. Pharm. Bull.* **2008**, *31*, 738–742. [CrossRef]
75. Han, X.; Cui, Z.Y.; Song, J.; Piao, H.Q.; Lian, L.H.; Hou, L.S.; Wang, G.; Zheng, S.; Dong, X.X.; Nan, J.X.; et al. Acanthoic acid modulates lipogenesis in nonalcoholic fatty liver disease via FXR/LXRs-dependent manner. *Chem. Biol. Interact.* **2019**, *311*, 108794. [CrossRef]
76. Kim, E.; Kang, Y.G.; Kim, Y.J.; Lee, T.R.; Yoo, B.C.; Jo, M.; Kim, J.H.; Kim, J.H.; Kim, D.; Cho, J.Y. Dehydroabietic Acid Suppresses Inflammatory Response Via Suppression of Src-, Syk-, and TAK1-Mediated Pathways. *Int. J. Mol. Sci.* **2019**, *20*, 1593. [CrossRef]
77. Gao, G.; Xie, Z.; Li, E.W.; Yuan, Y.; Fu, Y.; Wang, P.; Zhang, X.; Qiao, Y.; Xu, J.; Hölscher, C.; et al. Dehydroabietic acid improves nonalcoholic fatty liver disease through activating the Keap1/Nrf2-ARE signaling pathway to reduce ferroptosis. *J. Nat. Med.* **2021**, *75*, 540–552. [CrossRef]
78. Abboud, R.; Charcosset, C.; Greige-Gerges, H. Interaction of triterpenoids with human serum albumin: A review. *Chem. Phys. Lipids* **2017**, *207*, 260–270. [CrossRef]
79. Hou, M.; Wang, R.; Zhao, S.; Wang, Z. Ginsenosides in Panax genus and their biosynthesis. *Acta Pharm. Sin. B.* **2021**, *11*, 1813–1834. [CrossRef]
80. Gao, Y.; Zhang, S.; Li, J.; Zhao, J.; Xiao, Q.; Zhu, Y.; Zhang, J.; Huang, W. Effect and mechanism of ginsenoside Rg1-regulating hepatic steatosis in HepG2 cells induced by free fatty acid. *Biosci. Biotechnol. Biochem.* **2020**, *84*, 2228–2240. [CrossRef]
81. Qi, R.; Jiang, R.; Xiao, H.; Wang, Z.; He, S.; Wang, L.; Wang, Y. Ginsenoside Rg1 protects against d-galactose induced fatty liver disease in a mouse model via FOXO1 transcriptional factor. *Life Sci.* **2020**, *254*, 117776. [CrossRef] [PubMed]
82. Xu, Y.; Yang, C.; Zhang, S.; Li, J.; Xiao, Q.; Huang, W. Ginsenoside Rg1 Protects against Non-alcoholic Fatty Liver Disease by Ameliorating Lipid Peroxidation, Endoplasmic Reticulum Stress, and Inflammasome Activation. *Biol. Pharm. Bull.* **2018**, *41*, 1638–1644. [CrossRef] [PubMed]
83. Gu, D.; Yi, H.; Jiang, K.; Fakhar, S.H.; Shi, J.; He, Y.; Liu, B.; Guo, Y.; Fan, X.; Li, S. Transcriptome analysis reveals the efficacy of ginsenoside-Rg1 in the treatment of nonalcoholic fatty liver disease. *Life Sci.* **2021**, *267*, 118986. [CrossRef]
84. Wang, F.; Park, J.S.; Ma, Y.; Ma, H.; Lee, Y.J.; Lee, G.R.; Yoo, H.S.; Hong, J.T.; Roh, Y.S. Ginseng Saponin Enriched in Rh1 and Rg2 Ameliorates Nonalcoholic Fatty Liver Disease by Inhibiting Inflammasome Activation. *Nutrients* **2021**, *13*, 856. [CrossRef] [PubMed]
85. Cheng, B.; Gao, W.; Wu, X.; Zheng, M.; Yu, Y.; Song, C.; Miao, W.; Yang, Z.; He, Y.; Liu, C.; et al. Ginsenoside Rg2 Ameliorates High-Fat Diet-Induced Metabolic Disease through SIRT1. *J. Agric. Food. Chem.* **2020**, *68*, 4215–4226. [CrossRef]
86. Chen, X.J.; Liu, W.J.; Wen, M.L.; Liang, H.; Wu, S.M.; Zhu, Y.Z.; Zhao, J.Y.; Dong, X.Q.; Li, M.G.; Bian, L.; et al. Ameliorative effects of Compound K and ginsenoside Rh1 on non-alcoholic fatty liver disease in rats. *Sci. Rep.* **2017**, *7*, 41144. [CrossRef]
87. Song, B.; Sun, Y.; Chu, Y.; Wang, J.; Zheng, H.; Liu, L.; Cai, W.; Zhang, H. Ginsenoside Rb1 Alleviated High-Fat-Diet-Induced Hepatocytic Apoptosis via Peroxisome Proliferator-Activated Receptor γ. *Biomed. Res. Int.* **2020**, *2020*, 2315230. [CrossRef]

88. Huang, Q.; Wang, T.; Yang, L.; Wang, H.Y. Ginsenoside Rb2 Alleviates Hepatic Lipid Accumulation by Restoring Autophagy via Induction of Sirt1 and Activation of AMPK. *Int. J. Mol. Sci.* **2017**, *18*, 1063. [CrossRef]
89. Roh, E.; Hwang, H.J.; Kim, J.W.; Hong, S.H.; Kim, J.A.; Lee, Y.B.; Choi, K.M.; Baik, S.H.; Yoo, H.J. Ginsenoside Mc1 improves liver steatosis and insulin resistance by attenuating ER stress. *J. Ethnopharmacol.* **2020**, *259*, 112927. [CrossRef]
90. Jinhua, W. Ursolic acid: Pharmacokinetics process in vitro and in vivo, a mini review. *Arch. Pharm.* **2019**, *352*, e1800222. [CrossRef]
91. Li, S.; Liao, X.; Meng, F.; Wang, Y.; Sun, Z.; Guo, F.; Li, X.; Meng, M.; Li, Y.; Sun, C. Therapeutic role of ursolic acid on ameliorating hepatic steatosis and improving metabolic disorders in high-fat diet-induced non-alcoholic fatty liver disease rats. *PLoS ONE* **2014**, *9*, e86724, Erratum in *PLoS ONE* **2014**, *9*, e92364. [CrossRef] [PubMed]
92. Li, J.S.; Wang, W.J.; Sun, Y.; Zhang, Y.H.; Zheng, L. Ursolic acid inhibits the development of nonalcoholic fatty liver disease by attenuating endoplasmic reticulum stress. *Food Funct.* **2015**, *6*, 1643–1651. [CrossRef] [PubMed]
93. Lin, Y.N.; Wang, C.C.N.; Chang, H.Y.; Chu, F.Y.; Hsu, Y.A.; Cheng, W.K.; Ma, W.C.; Chen, C.J.; Wan, L.; Lim, Y.P. Ursolic Acid, a Novel Liver X Receptor α (LXRα) Antagonist Inhibiting Ligand-Induced Nonalcoholic Fatty Liver and Drug-Induced Lipogenesis. *J. Agric. Food. Chem.* **2018**, *66*, 11647–11662. [CrossRef] [PubMed]
94. Mukonowenzou, N.C.; Dangarembizi, R.; Chivandi, E.; Nkomozepi, P.; Erlwanger, K.H. Administration of ursolic acid to new-born pups prevents dietary fructose-induced non-alcoholic fatty liver disease in Sprague Dawley rats. *J. Dev. Orig. Health Dis.* **2021**, *12*, 101–112. [CrossRef]
95. Gu, M.; Zhao, P.; Zhang, S.; Fan, S.; Yang, L.; Tong, Q.; Ji, G.; Huang, C. Betulinic acid alleviates endoplasmic reticulum stress-mediated nonalcoholic fatty liver disease through activation of farnesoid X receptors in mice. *Br. J. Pharmacol.* **2019**, *176*, 847–863. [CrossRef]
96. Quan, H.Y.; Kim, D.Y.; Kim, S.J.; Jo, H.K.; Kim, G.W.; Chung, S.H. Betulinic acid alleviates non-alcoholic fatty liver by inhibiting SREBP1 activity via the AMPK-mTOR-SREBP signaling pathway. *Biochem. Pharmacol.* **2013**, *85*, 1330–1340. [CrossRef]
97. Mu, Q.; Wang, H.; Tong, L.; Fang, Q.; Xiang, M.; Han, L.; Jin, L.; Yang, J.; Qian, Z.; Ning, G.; et al. Betulinic acid improves nonalcoholic fatty liver disease through YY1/FAS signaling pathway. *FASEB J.* **2020**, *34*, 13033–13048. [CrossRef]
98. Wang, C.; Duan, X.; Sun, X.; Liu, Z.; Sun, P.; Yang, X.; Sun, H.; Liu, K.; Meng, Q. Protective effects of glycyrrhizic acid from edible botanical glycyrrhiza glabra against non-alcoholic steatohepatitis in mice. *Food Funct.* **2016**, *7*, 3716–3723. [CrossRef]
99. Sun, X.; Duan, X.; Wang, C.; Liu, Z.; Sun, P.; Huo, X.; Ma, X.; Sun, H.; Liu, K.; Meng, Q. Protective effects of glycyrrhizic acid against non-alcoholic fatty liver disease in mice. *Eur. J. Pharmacol.* **2017**, *806*, 75–82, *Erratum in Eur. J. Pharmacol.* **2022**, *928*, 175106. [CrossRef]
100. Gillard, J.; Clerbaux, L.A.; Nachit, M.; Sempoux, C.; Staels, B.; Bindels, L.B.; Tailleux, A.; Leclercq, I.A. Bile acids contribute to the development of non-alcoholic steatohepatitis in mice. *JHEP Rep.* **2021**, *4*, 100387. [CrossRef]
101. Yan, T.; Wang, H.; Cao, L.; Wang, Q.; Takahashi, S.; Yagai, T.; Li, G.; Krausz, K.W.; Wang, G.; Gonzalez, F.J.; et al. Glycyrrhizin Alleviates Nonalcoholic Steatohepatitis via Modulating Bile Acids and Meta-Inflammation. *Drug Metab. Dispos.* **2018**, *46*, 1310–1319. [CrossRef] [PubMed]
102. Chen, G. The link between Hepatic Vitamin A Metabolism and Nonalcoholic Fatty Liver Disease. *Curr. Drug Targets* **2015**, *16*, 1281–1292. [CrossRef]
103. Shi, L.; Guo, S.; Zhang, S.; Gao, X.; Liu, A.; Wang, Q.; Zhang, T.; Zhang, Y.; Wen, A. Glycyrrhetinic acid attenuates disturbed vitamin a metabolism in non-alcoholic fatty liver disease through AKR1B10. *Eur. J. Pharmacol.* **2020**, *883*, 173167. [CrossRef] [PubMed]
104. Yang, M.; Zhang, M.; Liu, Q.; Xu, T.; Huang, T.; Yao, D.; Wong, C.W.; Liu, J.; Guan, M. 18β-Glycyrrhetinic acid acts through hepatocyte nuclear factor 4 alpha to modulate lipid and carbohydrate metabolism. *Pharmacol. Res.* **2020**, *157*, 104840. [CrossRef] [PubMed]
105. Wu, X.; Zhang, L.; Gurley, E.; Studer, E.; Shang, J.; Wang, T.; Wang, C.; Yan, M.; Jiang, Z.; Hylemon, P.B.; et al. Prevention of free fatty acid-induced hepatic lipotoxicity by 18beta-glycyrrhetinic acid through lysosomal and mitochondrial pathways. *Hepatology* **2008**, *47*, 1905–1915. [CrossRef] [PubMed]
106. Pollier, J.; Goossens, A. Oleanolic acid. *Phytochemistry* **2012**, *77*, 10–15. [CrossRef]
107. Xue, C.; Li, Y.; Lv, H.; Zhang, L.; Bi, C.; Dong, N.; Shan, A.; Wang, J. Oleanolic Acid Targets the Gut-Liver Axis to Alleviate Metabolic Disorders and Hepatic Steatosis. *J. Agric. Food Chem.* **2021**, *69*, 7884–7897. [CrossRef]
108. Lin, Y.N.; Chang, H.Y.; Wang, C.C.N.; Chu, F.Y.; Shen, H.Y.; Chen, C.J.; Lim, Y.P. Oleanolic Acid Inhibits Liver X Receptor Alpha and Pregnane X Receptor to Attenuate Ligand-Induced Lipogenesis. *J. Agric. Food Chem.* **2018**, *66*, 10964–10976. [CrossRef]
109. Gamede, M.; Mabuza, L.; Ngubane, P.; Khathi, A. Plant-derived oleanolic acid ameliorates markers associated with non-alcoholic fatty liver disease in a diet-induced pre-diabetes rat model. *Diabetes Metab. Syndr. Obes.* **2019**, *12*, 1953–1962. [CrossRef]
110. Ou-Yang, Q.; Xuan, C.X.; Wang, X.; Luo, H.Q.; Liu, J.E.; Wang, L.L.; Li, T.T.; Chen, Y.P.; Liu, J. 3-Acetyl-oleanolic acid ameliorates non-alcoholic fatty liver disease in high fat diet-treated rats by activating AMPK-related pathways. *Acta Pharmacol. Sin.* **2018**, *39*, 1284–1293. [CrossRef]
111. Wang, J.; Zhang, Y.; Shen, Q.; Wu, J.; Li, J.X. Oleanolic acid derivative HA-20 inhibits adipogenesis in a manner involving PPARγ-FABP4/aP2 pathway. *J. Mol. Endocrinol.* **2021**, *66*, 245–258. [CrossRef] [PubMed]
112. You, L.Z.; Lin, Y.X.; Fang, Z.H.; Shen, G.M.; Zhao, J.D.; Wang, T.T. Research advances on astragaloside-IV in treatment of diabetes mellitus and its complications pharmacological effects. *Zhongguo Zhong Yao Za Zhi* **2017**, *42*, 4700–4706. (In Chinese) [CrossRef] [PubMed]

113. Zhou, B.; Zhou, D.L.; Wei, X.H.; Zhong, R.Y.; Xu, J.; Sun, L. Astragaloside I.V attenuates free fatty acid-induced ER stress and lipid accumulation in hepatocytes via AMPK activation. *Acta Pharmacol. Sin.* **2017**, *38*, 998–1008. [CrossRef] [PubMed]
114. Zhou, X.; Wang, L.L.; Tang, W.J.; Tang, B. Astragaloside I.V inhibits protein tyrosine phosphatase 1B and improves insulin resistance in insulin-resistant HepG2 cells and triglyceride accumulation in oleic acid (OA)-treated HepG2 cells. *J. Ethnopharmacol.* **2021**, *268*, 113556. [CrossRef] [PubMed]
115. Liu, C.; Dai, L.; Liu, Y.; Dou, D.; Sun, Y.; Ma, L. Pharmacological activities of mogrosides. *Future Med. Chem.* **2018**, *10*, 845–850. [CrossRef]
116. Zhang, X.; Song, Y.; Ding, Y.; Wang, W.; Liao, L.; Zhong, J.; Sun, P.; Lei, F.; Zhang, Y.; Xie, W. Effects of Mogrosides on High-Fat-Diet-Induced Obesity and Nonalcoholic Fatty Liver Disease in Mice. *Molecules* **2018**, *23*, 1894. [CrossRef]
117. Li, L.; Zheng, W.; Wang, C.; Qi, J.; Li, H. Mogroside V Protects against Hepatic Steatosis in Mice on a High-Fat Diet and LO2 Cells Treated with Free Fatty Acids via AMPK Activation. *Evid. Based Complement. Alternat. Med.* **2020**, *2020*, 7826874. [CrossRef]
118. Lv, J.; Sharma, A.; Zhang, T.; Wu, Y.; Ding, X. Pharmacological Review on Asiatic Acid and Its Derivatives: A Potential Compound. *SLAS Technol.* **2018**, *23*, 111–127. [CrossRef]
119. Wang, D.; Lao, L.; Pang, X.; Qiao, Q.; Pang, L.; Feng, Z.; Bai, F.; Sun, X.; Lin, X.; Wei, J. Asiatic acid from Potentilla chinensis alleviates non-alcoholic fatty liver by regulating endoplasmic reticulum stress and lipid metabolism. *Int. Immunopharmacol.* **2018**, *65*, 256–267, *Erratum in Int. Immunopharmacol.* **2020**, *84*, 106291. [CrossRef]
120. Zhao, J.; Zhou, H.; An, Y.; Shen, K.; Yu, L. Biological effects of corosolic acid as an anti-inflammatory, anti-metabolic syndrome and anti-neoplasic natural compound. *Oncol. Lett.* **2021**, *21*, 84. [CrossRef]
121. Liu, G.; Cui, Z.; Gao, X.; Liu, H.; Wang, L.; Gong, J.; Wang, A.; Zhang, J.; Ma, Q.; Huang, Y.; et al. Corosolic acid ameliorates non-alcoholic steatohepatitis induced by high-fat diet and carbon tetrachloride by regulating TGF-β1/Smad2, N.F-κB, and AMPK signaling pathways. *Phytother. Res.* **2021**, *35*, 5214–5226. [CrossRef] [PubMed]
122. Ghosh, J.; Sil, P.C. Arjunolic acid: A new multifunctional therapeutic promise of alternative medicine. *Biochimie* **2013**, *95*, 1098–1109. [CrossRef] [PubMed]
123. Toppo, E.; Sylvester Darvin, S.; Esakkimuthu, S.; Buvanesvaragurunathan, K.; Ajeesh Krishna, T.P.; Antony Caesar, S.; Stalin, A.; Balakrishna, K.; Pandikumar, P.; Ignacimuthu, S.; et al. Curative effect of arjunolic acid from Terminalia arjuna in non-alcoholic fatty liver disease models. *Biomed. Pharmacother.* **2018**, *107*, 979–988. [CrossRef] [PubMed]
124. Lu, X.; Xu, C.; Yang, R.; Zhang, G. Ganoderic Acid A Alleviates OVA-Induced Asthma in Mice. *Inflammation* **2021**, *44*, 1908–1915. [CrossRef] [PubMed]
125. Liu, F.; Shi, K.; Dong, J.; Jin, Z.; Wu, Y.; Cai, Y.; Lin, T.; Cai, Q.; Liu, L.; Zhang, Y. Ganoderic acid A attenuates high-fat-diet-induced liver injury in rats by regulating the lipid oxidation and liver inflammation. *Arch. Pharm. Res.* **2020**, *43*, 744–754. [CrossRef] [PubMed]
126. Lu, Y.W.; Zhu, Y.C.; Zhang, L.; Li, P.; Yang, J.; Wen, X.D. Ilexgenin A enhances the effects of simvastatin on non-alcoholic fatty liver disease without changes in simvastatin pharmacokinetics. *Chin. J. Nat. Med.* **2018**, *16*, 436–445. [CrossRef] [PubMed]
127. Roy, G.; Guan, S.; Liu, H.; Zhang, L. Rotundic Acid Induces DNA Damage and Cell Death in Hepatocellular Carcinoma Through AKT/mTOR and MAPK Pathways. *Front. Oncol.* **2019**, *9*, 545. [CrossRef] [PubMed]
128. Liu, H.J.; Cao, S.T.; Wen, B.Y.; Han, X.; Li, Y.; Li, S.; Li, J.; Zhang, L. Rotundic acid ameliorates non-alcoholic steatohepatitis via SREBP-1c/ SCD1 signaling pathway and modulating gut microbiota. *Int. Immunopharmacol.* **2021**, *99*, 108065. [CrossRef] [PubMed]
129. Zhao, L.; Li, J.; Sun, Z.B.; Sun, C.; Yu, Z.H.; Guo, X. Saikosaponin D inhibits proliferation of human osteosarcoma cells via the p53 signaling pathway. *Exp. Ther. Med.* **2019**, *17*, 488–494. [CrossRef] [PubMed]
130. Chang, G.R.; Lin, W.L.; Lin, T.C.; Liao, H.J.; Lu, Y.W. The Ameliorative Effects of Saikosaponin in Thioacetamide-Induced Liver Injury and Non-Alcoholic Fatty Liver Disease in Mice. *Int. J. Mol. Sci.* **2021**, *22*, 11383. [CrossRef]
131. Avila, C. Terpenoids in Marine Heterobranch Molluscs. *Mar. Drugs* **2020**, *18*, 162. [CrossRef] [PubMed]
132. Petyaev, I.M. Lycopene Deficiency in Ageing and Cardiovascular Disease. *Oxid. Med. Cell Longev.* **2016**, *2016*, 3218605. [CrossRef] [PubMed]
133. Khan, U.M.; Sevindik, M.; Zarrabi, A.; Nami, M.; Ozdemir, B.; Kaplan, D.N.; Selamoglu, Z.; Hasan, M.; Kumar, M.; Alshehri, M.M.; et al. Lycopene: Food Sources, Biological Activities, and Human Health Benefits. *Oxid. Med. Cell Longev.* **2021**, *2021*, 2713511. [CrossRef]
134. Tan, H.L.; Moran, N.E.; Cichon, M.J.; Riedl, K.M.; Schwartz, S.J.; Erdman, J.W., Jr.; Pearl, D.K.; Thomas-Ahner, J.M.; Clinton, S.K. β-Carotene-9′,10′-oxygenase status modulates the impact of dietary tomato and lycopene on hepatic nuclear receptor-, stress-, and metabolism-related gene expression in mice. *J. Nutr.* **2014**, *144*, 431–439, *Erratum in J. Nutr.* **2014**, *144*, 1664. [CrossRef] [PubMed]
135. Ip, B.C.; Liu, C.; Lichtenstein, A.H.; Von Lintig, J.; Wang, X.D. Lycopene and apo-10′-lycopenoic acid have differential mechanisms of protection against hepatic steatosis in β-carotene-9′,10′-oxygenase knockout male mice. *J. Nutr.* **2015**, *145*, 268–276. [CrossRef]
136. Ni, Y.; Zhuge, F.; Nagashimada, M.; Nagata, N.; Xu, L.; Yamamoto, S.; Fuke, N.; Ushida, Y.; Suganuma, H.; Kaneko, S.; et al. Lycopene prevents the progression of lipotoxicity-induced nonalcoholic steatohepatitis by decreasing oxidative stress in mice. *Free Radic. Biol. Med.* **2020**, *152*, 571–582. [CrossRef]
137. Jiang, W.; Guo, M.H.; Hai, X. Hepatoprotective and antioxidant effects of lycopene on non-alcoholic fatty liver disease in rat. *World J. Gastroenterol.* **2016**, *22*, 10180–10188. [CrossRef]

138. Ahn, J.; Lee, H.; Jung, C.H.; Ha, T. Lycopene inhibits hepatic steatosis via microRNA-21-induced downregulation of fatty acid-binding protein 7 in mice fed a high-fat diet. *Mol. Nutr. Food Res.* **2012**, *56*, 1665–1674. [CrossRef]
139. Chen, G.; Ni, Y.; Nagata, N.; Zhuge, F.; Xu, L.; Nagashimada, M.; Yamamoto, S.; Ushida, Y.; Fuke, N.; Suganuma, H.; et al. Lycopene Alleviates Obesity-Induced Inflammation and Insulin Resistance by Regulating M1/M2 Status of Macrophages. *Mol. Nutr. Food Res.* **2019**, *63*, e1900602. [CrossRef]
140. Mustra Rakic, J.; Liu, C.; Veeramachaneni, S.; Wu, D.; Paul, L.; Ausman, L.M.; Wang, X.D. Dietary lycopene attenuates cigarette smoke-promoted nonalcoholic steatohepatitis by preventing suppression of antioxidant enzymes in ferrets. *J. Nutr. Biochem.* **2021**, *91*, 108596. [CrossRef]
141. Li, Z.; Chen, J.; Zhang, D. Association between dietary carotenoid intakes and hypertension in adults: National Health and Nutrition Examination Survey 2007–2014. *J. Hypertens.* **2019**, *37*, 2371–2379. [CrossRef] [PubMed]
142. Jha, P.; Kumari, S.; Jobby, R.; Desai, N.; Ali, A. Dietary Phytonutrients in the Prevention of Diabetes-related Complications. *Curr. Diabetes Rev.* **2020**, *16*, 657–673. [CrossRef] [PubMed]
143. Wang, Z.; Fan, J.; Wang, J.; Li, Y.; Xiao, L.; Duan, D.; Wang, Q. Protective effect of lycopene on high-fat diet-induced cognitive impairment in rats. *Neurosci. Lett.* **2016**, *627*, 185–191. [CrossRef] [PubMed]
144. Zhu, Y.; Liu, R.; Shen, Z.; Cai, G. Combination of luteolin and lycopene effectively protect against the "two-hit" in NAFLD through Sirt1/AMPK signal pathway. *Life Sci.* **2020**, *256*, 117990. [CrossRef] [PubMed]
145. Li, L.; Tang, X.; Luo, Y.; Hu, X.; Ren, L. Accumulation and conversion of β-carotene and astaxanthin induced by abiotic stresses in Schizochytrium sp. *Bioprocess Biosyst. Eng.* **2022**, *45*, 911–920. [CrossRef]
146. Li, J.; Guo, C.; Wu, J. Astaxanthin in Liver Health and Disease: A Potential Therapeutic Agent. *Drug Des. Dev. Ther.* **2020**, *14*, 2275–2285. [CrossRef]
147. Kobori, M.; Takahashi, Y.; Sakurai, M.; Ni, Y.; Chen, G.; Nagashimada, M.; Kaneko, S.; Ota, T. Hepatic Transcriptome Profiles of Mice with Diet-Induced Nonalcoholic Steatohepatitis Treated with Astaxanthin and Vitamin E. *Int. J. Mol. Sci.* **2017**, *18*, 593. [CrossRef]
148. Yang, Y.; Pham, T.X.; Wegner, C.J.; Kim, B.; Ku, C.S.; Park, Y.K.; Lee, J.Y. Astaxanthin lowers plasma TAG concentrations and increases hepatic antioxidant gene expression in diet-induced obesity mice. *Br. J. Nutr.* **2014**, *112*, 1797–1804. [CrossRef]
149. Yang, Y.; Kim, B.; Park, Y.K.; Koo, S.I.; Lee, J.Y. Astaxanthin prevents TGFβ1-induced pro-fibrogenic gene expression by inhibiting Smad3 activation in hepatic stellate cells. *Biochim. Biophys. Acta* **2015**, *1850*, 178–185. [CrossRef]
150. Yang, M.; Kimchi, E.T.; Staveley-O'Carroll, K.F.; Li, G. Astaxanthin Prevents Diet-Induced NASH Progression by Shaping Intrahepatic Immunity. *Int. J. Mol. Sci.* **2021**, *22*, 11037. [CrossRef]
151. Ni, Y.; Nagashimada, M.; Zhuge, F.; Zhan, L.; Nagata, N.; Tsutsui, A.; Nakanuma, Y.; Kaneko, S.; Ota, T. Astaxanthin prevents and reverses diet-induced insulin resistance and steatohepatitis in mice: A comparison with vitamin E. *Sci. Rep.* **2015**, *5*, 17192. [CrossRef] [PubMed]
152. Wu, L.; Mo, W.; Feng, J.; Li, J.; Yu, Q.; Li, S.; Zhang, J.; Chen, K.; Ji, J.; Dai, W.; et al. Astaxanthin attenuates hepatic damage and mitochondrial dysfunction in non-alcoholic fatty liver disease by up-regulating the FGF21/PGC-1βpathway. *Br. J. Pharmacol.* **2020**, *177*, 3760–3777. [CrossRef] [PubMed]
153. Nishino, A.; Maoka, T.; Yasui, H. Preventive Effects of β-Cryptoxanthin, a Potent Antioxidant and Provitamin A Carotenoid, on Lifestyle-Related Diseases-A Central Focus on Its Effects on Non-Alcoholic Fatty Liver Disease (NAFLD). *Antioxidants* **2021**, *11*, 43. [CrossRef] [PubMed]
154. Haidari, F.; Hojhabrimanesh, A.; Helli, B.; Seyedian, S.S.; Ahmadi-Angali, K.; Abiri, B. A hypocaloric high-protein diet supplemented with β-cryptoxanthin improves non-alcoholic fatty liver disease: A randomized controlled trial. *BMC Gastroenterol.* **2020**, *20*, 349. [CrossRef] [PubMed]
155. Haidari, F.; Hojhabrimanesh, A.; Helli, B.; Seyedian, S.S.; Ahmadi-Angali, K. An energy-restricted high-protein diet supplemented with β-cryptoxanthin alleviated oxidative stress and inflammation in nonalcoholic fatty liver disease: A randomized controlled trial. *Nutr. Res.* **2020**, *73*, 15–26. [CrossRef] [PubMed]
156. Lim, J.Y.; Liu, C.; Hu, K.Q.; Smith, D.E.; Wu, D.; Lamon-Fava, S.; Ausman, L.M.; Wang, X.D. Dietary β-Cryptoxanthin Inhibits High-Refined Carbohydrate Diet-Induced Fatty Liver via Differential Protective Mechanisms Depending on Carotenoid Cleavage Enzymes in Male Mice. *J. Nutr.* **2019**, *149*, 1553–1564. [CrossRef]
157. Kobori, M.; Ni, Y.; Takahashi, Y.; Watanabe, N.; Sugiura, M.; Ogawa, K.; Nagashimada, M.; Kaneko, S.; Naito, S.; Ota, T. β-Cryptoxanthin alleviates diet-induced nonalcoholic steatohepatitis by suppressing inflammatory gene expression in mice. *PLoS ONE* **2014**, *9*, e98294. [CrossRef]
158. Ni, Y.; Nagashimada, M.; Zhan, L.; Nagata, N.; Kobori, M.; Sugiura, M.; Ogawa, K.; Kaneko, S.; Ota, T. Prevention and reversal of lipotoxicity-induced hepatic insulin resistance and steatohepatitis in mice by an antioxidant carotenoid, β-cryptoxanthin. *Endocrinology* **2015**, *156*, 987–999. [CrossRef]
159. Yilmaz, B.; Sahin, K.; Bilen, H.; Bahcecioglu, I.H.; Bilir, B.; Ashraf, S.; Halazun, K.J.; Kucuk, O. Carotenoids and non-alcoholic fatty liver disease. *Hepatobiliary Surg. Nutr.* **2015**, *4*, 161–171. [CrossRef]
160. Wang, L.; Ding, C.; Zeng, F.; Zhu, H. Low Levels of Serum β-Carotene and β-Carotene/Retinol Ratio Are Associated with Histological Severity in Nonalcoholic Fatty Liver Disease Patients. *Ann. Nutr. Metab.* **2019**, *74*, 156–164. [CrossRef]
161. Christensen, K.; Lawler, T.; Mares, J. Dietary Carotenoids and Non-Alcoholic Fatty Liver Disease among U.S Adults, NHANES 2003–2014. *Nutrients* **2019**, *11*, 1101. [CrossRef] [PubMed]

162. Cao, Y.; Wang, C.; Liu, J.; Liu, Z.M.; Ling, W.H.; Chen, Y.M. Greater serum carotenoid levels associated with lower prevalence of nonalcoholic fatty liver disease in Chinese adults. *Sci. Rep.* **2015**, *5*, 12951. [CrossRef] [PubMed]
163. Seif El-Din, S.H.; El-Lakkany, N.M.; El-Naggar, A.A.; Hammam, O.A.; Abd El-Latif, H.A.; Ain-Shoka, A.A.; Ebeid, F.A. Effects of rosuvastatin and/or β-carotene on non-alcoholic fatty liver in rats. *Res. Pharm. Sci.* **2015**, *10*, 275–287. [PubMed]
164. Li, J.; Abdel-Aal, E.M. Dietary Lutein and Cognitive Function in Adults: A Meta-Analysis of Randomized Controlled Trials. *Molecules* **2021**, *26*, 5794. [CrossRef]
165. Qiu, X.; Gao, D.H.; Xiang, X.; Xiong, Y.F.; Zhu, T.S.; Liu, L.G.; Sun, X.F.; Hao, L.P. Ameliorative effects of lutein on non-alcoholic fatty liver disease in rats. *World J. Gastroenterol.* **2015**, *21*, 8061–8072. [CrossRef]
166. Chen, C.; Ni, Y.; Jiang, B.; Yan, S.; Xu, B.; Fan, B.; Huang, H.; Chen, G. Anti-aging derivatives of cycloastragenol produced by biotransformation. *Nat. Prod. Res.* **2021**, *35*, 2685–2690. [CrossRef]
167. Gu, M.; Zhang, S.; Zhao, Y.; Huang, J.; Wang, Y.; Li, Y.; Fan, S.; Yang, L.; Ji, G.; Tong, Q.; et al. Cycloastragenol improves hepatic steatosis by activating farnesoid X receptor signalling. *Pharmacol. Res.* **2017**, *121*, 22–32. [CrossRef]
168. Lin, C.; Wen, X.; Sun, H. Oleanolic acid derivatives for pharmaceutical use: A patent review. *Expert Opin. Ther. Pat.* **2016**, *26*, 643–655. [CrossRef]
169. Zhang, B.Z.; Ding, F.; Tan, L.W. Clinical and experimental study on yi-gan-ning granule in treating chronic hepatitis B. *Zhongguo Zhong Xi Yi Jie He Za Zhi* **1993**, *13*, 597–599, 580. (In Chinese)
170. Kandiah, N.; Ong, P.A.; Yuda, T.; Ng, L.L.; Mamun, K.; Merchant, R.A.; Chen, C.; Dominguez, J.; Marasigan, S.; Ampil, E.; et al. Treatment of dementia and mild cognitive impairment with or without cerebrovascular disease: Expert consensus on the use of Ginkgo biloba extract, EGb 761®. *CNS Neurosci. Ther.* **2019**, *25*, 288–298. [CrossRef]
171. Patel, P.; Aknouk, M.; Dawson, A.; Aya, A.; Kanukuntla, A.; Kata, P.; De Dona, A. How Much Is Too Much? Exploring Pseudohyperaldosteronism in Glycyrrhizic Acid Toxicity From Chronic Licorice Root Consumption. *Cureus* **2021**, *13*, e16454. [CrossRef] [PubMed]
172. Khan, R.S.; Bril, F.; Cusi, K.; Newsome, P.N. Modulation of Insulin Resistance in Nonalcoholic Fatty Liver Disease. *Hepatology* **2019**, *70*, 711–724. [CrossRef] [PubMed]
173. Wu, J.; Lin, S.; Wan, B.; Velani, B.; Zhu, Y. Pyroptosis in Liver Disease: New Insights into Disease Mechanisms. *Aging Dis.* **2019**, *10*, 1094–1108. [CrossRef] [PubMed]
174. Gao, H.; Jin, Z.; Bandyopadhyay, G.; Wang, G.; Zhang, D.; Rocha, K.C.E.; Liu, X.; Zhao, H.; Kisseleva, T.; Brenner, D.A.; et al. Aberrant iron distribution via hepatocyte-stellate cell axis drives liver lipogenesis and fibrosis. *Cell Metab.* **2022**, *34*, 1201–1213.e5. [CrossRef] [PubMed]
175. Carling, D. AMPK signalling in health and disease. *Curr. Opin. Cell Biol.* **2017**, *45*, 31–37. [CrossRef] [PubMed]
176. He, F.; Ru, X.; Wen, T. NRF2, a Transcription Factor for Stress Response and Beyond. *Int. J. Mol. Sci.* **2020**, *21*, 4777. [CrossRef]
177. Ding, R.B.; Bao, J.; Deng, C.X. Emerging roles of SIRT1 in fatty liver diseases. *Int. J. Biol. Sci.* **2017**, *13*, 852–867. [CrossRef]

molecules

Article

Oxalactam A, a Novel Macrolactam with Potent Anti-*Rhizoctonia solani* Activity from the Endophytic Fungus *Penicillium oxalicum*

Ruizhen Zhang [1,†], Yingrun Ma [1,†], Ming-Ming Xu [1], Xinyi Wei [1], Cheng-Bin Yang [1], Fei Zeng [1], Jin-Ao Duan [1], Chun-Tao Che [2,*], Junfei Zhou [1,*] and Ming Zhao [1,2,*]

1 National and Local Collaborative Engineering Center of Chinese Medicinal Resources Industrialization and Formulae Innovative Medicine, Jiangsu Collaborative Innovation Center of Chinese Medicinal Resources Industrialization, School of Pharmacy, Nanjing University of Chinese Medicine, Nanjing 210023, China

2 Department of Pharmaceutical Sciences, College of Pharmacy, The University of Illinois at Chicago, Chicago, IL 60612, USA

* Correspondence: chect@uic.edu (C.-T.C.); 300524@njucm.edu.cn (J.Z.); mingzhao@njucm.edu.cn (M.Z.); Tel.: +86-15895865208 (M.Z.)

† These authors contributed equally to this work.

Citation: Zhang, R.; Ma, Y.; Xu, M.-M.; Wei, X.; Yang, C.-B.; Zeng, F.; Duan, J.-A.; Che, C.-T.; Zhou, J.; Zhao, M. Oxalactam A, a Novel Macrolactam with Potent Anti-*Rhizoctonia solani* Activity from the Endophytic Fungus *Penicillium oxalicum*. *Molecules* **2022**, *27*, 8811. https://doi.org/10.3390/molecules27248811

Academic Editors: Sokcheon Pak and Soo Liang Ooi

Received: 27 September 2022
Accepted: 9 December 2022
Published: 12 December 2022

Abstract: A novel macrolactam named oxalactam A (**1**), three known dipeptides (**2–4**) as well as other known alkaloids (**5–7**) were obtained from the endophytic fungus *Penicillium oxalicum*, which was derived from the tuber of *Icacina trichantha* (Icacinaceae). All chemical structures were established based on spectroscopic data, chemical methods, ECD calculations, and ^{13}C-DP4+ analysis. Among them, oxalactam A (**1**) is a 16-membered polyenic macrolactam bearing a new skeleton of 2,9-dimethyl-azacyclohexadecane core and exhibited potent anti-*Rhizoctonia solani* activity with a MIC value of 10 μg/mL in vitro. The plausible biosynthetic pathway of **1** was also proposed via the alanyl protecting mechanism. Notably, three dipeptides (**2–4**) were first identified from the endophytic fungus *P. oxalicum* and the NMR data of cyclo(*L*-Trp-*L*-Glu) (**2**) was reported for the first time. In addition, the binding interactions between compound **1** and the sterol 14α-demethylase enzyme (CYP51) were studied by molecular docking and dynamics technologies, and the results revealed that the 16-membered polyenic macrolactam could be a promising CYP51 inhibitor to develop as a new anti-*Rhizoctonia solani* fungicide.

Keywords: oxalactam A; macrolactam; anti-*Rhizoctonia solani*; *Penicillium oxalicum*; molecular docking; molecular dynamics

1. Introduction

Rice sheath blight (RSB), also called "snakeskin disease", "mosaic footstalk", and "rotten footstalk", is one of the most devastating rice diseases caused by the necrotrophic pathogen *Rhizoctonia solani* [1,2]. The RSB incidence has witnessed a sharp increase owing to the unrestricted nitrogen fertilizers and semi-dwarf high-yield cultivars in recent decades [3]. In China, RSB was already the second most serious disease in rice, which could lead to a yield loss of 10~30% with 15~20 million hm^2 every year [4,5], even up to 50% in the Yangtze river valley in epidemic years [6,7]. Currently, chemical fungicides are major approaches to preventing RSB and related diseases with increased human health risks, environmental pollution, and resistant phytopathogens [8]. Therefore, natural anti-*R. solani* agents, such as strobilurins or QoI from *Strobilurus tenacellus* (wild mushroom) [8], have engaged worldwide attention owing to their environmental friendliness, high selectivity, and new mechanism of action.

Recently, macrolactam scaffolds are usually regarded as precursor components with high efficacy including antibiotics, antifungals, anticancer as well as immunosuppressants [9]. A typical macrocyclic glycopeptide antibiotic, vancomycin, has been applied to treat

Gram-positive bacterial infections including methicillin-resistant *Staphylococcus aureus* and penicillin-resistant *Streptococcus pneumonia* in the clinic for a long time [9]. Previous investigations indicated that 59% of approved small molecules bearing a macrolactam core are natural products or related derivatives [10]. Therefore, macrolactam components might be a promising lead scaffold for pharmaceutical purposes [11]. However, there is no report on their anti-*R. solani* activity in agricultural applications.

Icacina trichantha Oliv. (Icacinaceae), distributed in West and Central Africa [12], is an excellent plant resource that has the concomitant function of both medicine and foodstuff to relieve constipation, food poisoning, and malaria [13]. Due to the tropical rainforest climate near the equator in Nigeria, *I. trichantha* might be a potential plant source to seek various bioactive endophytes producing interesting secondary metabolites. Until now, no report on the endophytic fungus derived from *Icacina* species and related bioactivity has been documented. Previously, our group found that *Penicillium oxalicum* from the tuber of *I. trichantha* could significantly reduce the growth of *R. solani* strain with a 96.83% of inhibition rate (*Alternaria alternata*, 83.33%; *Fusarium solani*, 50.67%; *Colletotrichum gloeosporioides*, 76.04%) by the plate confrontation test (Figure S2, Table S1). Therefore, the further anti-*R. solani* activity-guided fractionation resulted in the isolation of one novel macrolactam (**1**), three dipeptides (**2**–**4**) as well as other alkaloids (**5**–**7**) (Figure 1). Notably, oxalactam A (**1**) represents the first example of a 16-membered polyenic macrolactam bearing a 2,9-dimethyl-azacyclohexadecane core, and this is the first report of the NMR data for cyclo(*L*-Trp-*L*-Glu) (**2**). Additionally, three dipeptides (**2**–**4**) were first identified from the endophytic fungus *P. oxalicum*. Among them, oxalactam A (**1**) exhibited potent anti-*R. solani* activity with a MIC value of 10 μg/mL by the mycelium growth rate method in vitro. Herein, the isolation, structural elucidation, anti-*R. solani* activity, molecular docking, and molecular dynamics simulation of these isolates were elaborated.

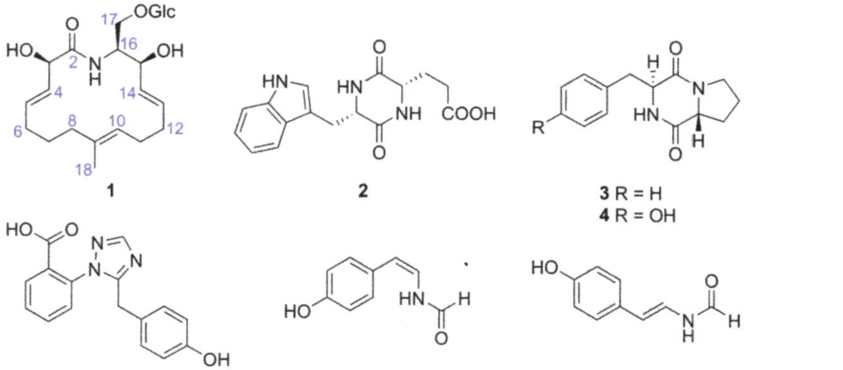

Figure 1. Chemical structures of compounds **1**–**7**.

2. Results and Discussion

2.1. Isolation and Structural Identification

Oxalactam A (**1**), $[\alpha]_D^{20}$ + 2.00 (*c* 0.1, MeOH), was isolated as a white amorphous powder. Its molecular formula could be verified as $C_{23}H_{37}NO_9$ in line with the HR-ESI-MS at *m/z* 518.2943 $[M + CH_3CH_2OH + H]^+$ (calcd for $C_{25}H_{44}NO_{10}$, 518.2965) and its ^{13}C NMR data, indicating six degrees of unsaturation. The presence of double bonds in **1** could be assigned based on a maximum absorption at 207 nm in the UV spectrum. The ^1H NMR spectrum (Table 1) showed resonances attributed to five olefinic protons at δ_H (5.84, 5.72, 5.49, 5.45, 5.15), two oxygenated methines at δ_H (4.44, 4.14), and one glucopyranosyl group in **1**. Additionally, the combination of ^{13}C NMR, DEPT135, and HSQC data suggested the presence of a methyl at δ_C 16.3, a carbonyl at δ_C 175.6, three double bonds at δ_C (136.9, 134.8, 134.7, 131.2, 129.2, 125.0), and a glucopyranosyl segment at δ_C (104.9, 78.1, 78.0, 75.1,

71.7, 62.8) in **1**. Three double bonds, a carbonyl, and a glucopyranosyl account for five degrees of unsaturation, the remaining one demonstrated the presence of a monocycle system in the aglycone of **1**.

Table 1. NMR Spectroscopic Data (500 MHz, CD_3OD) for Oxalactam A (**1**).

	Oxalactam A (1)	
Position	δ_H **(*J* in Hz)**	δ_C**, Type**
2		175.6, C
3	4.44 (d, *J* = 6.0 Hz)	74.2, CH
4	5.49 (dd, *J* = 16.0, 6.0 Hz)	129.2, CH
5	5.84 (dt, *J* = 16.0, 6.0 Hz)	134.8, CH
6	2.04, m	33.6, CH_2
7	1.40, m	29.3, CH_2
8	1.98 (t, *J* = 7.5 Hz)	41.0, CH_2
9		136.9, C
10	5.15 (t, *J* = 6.5 Hz)	125.0, CH
11	2.06, m	28.8, CH_2
12	2.04, m	33.9, CH_2
13	5.72 (dt, *J* = 15.1, 6.1 Hz)	134.7, CH
14	5.45 (dd, *J* = 15.1, 5.4 Hz)	131.2, CH
15	4.14 (dd, *J* = 10.1, 5.4 Hz)	73.0, CH
16	3.97, m	54.7, CH
17a	3.71 (dd, *J* = 10.3, 3.4 Hz);	69.8, CH_2
17b	4.13, overlap	
18	1.60, s	16.3, CH_3
1′	4.27 (d, *J* = 7.8 Hz)	104.9, CH
2′	3.19, m	75.1, CH
3′	3.28, m	78.1, CH
4′	3.27, m	71.7, CH
5′	3.36, m	78.0, CH
6a′	3.86, (d, *J* = 11.7 Hz)	62.8, CH_2
6b′	3.66, (dd, *J* = 11.7, 4.3 Hz)	

The $^1H-^1H$ COSY and HSQC data (Figure 2) suggested the presence of three spin systems in **1**, (a) H-3/H-4/H-5/H_2-6/H_2-7/H_2-8, (b) H-10/H_2-11/H_2-12/H-13/H-14/H-15/H-16/H_2-17, and (c) H-1′/H-2′/H-3′/H-4′/H-5′/H_2-6′. Additionally, HMBC correlations (Figure 2) from one methyl singlet H_3-18 to C-8/C-9/C-10 suggested the direct connections of C-8, C-10, and C-18 toward C-9. The connection between C-2 and C-3 was verified by the key HMBC correlation from H-3 to C-2. The HMBC correlation from the H-1′ to C-17 established the connections of one glucopyranosyl group to the oxygenated primary carbon C-17. Furthermore, H-16 (δ_H 3.97)/C-16 (δ_C 54.7) were different compared to a typical oxygenated secondary carbon, thus, a nitrogen bridge exists between C-2 and C-16 according to the molecular formula and resulting degrees of saturation of **1**. Notably, all three *trans* double bonds $\Delta^{4,5}$, $\Delta^{13,14}$, and $\Delta^{9,10}$ could be established by the coupling constants of $J_{H\text{-}4/H\text{-}5}$ = 16.0 Hz, $J_{H\text{-}13/H\text{-}14}$ = 15.1 Hz, and the key NOESY correlation between H-8 and H-10. Consequently, the planar structure of oxalactam A (**1**) was constructed as (4*E*,9*E*,13*E*)-3,15-dihydroxy-16-(glucopyranosyl-*O*-methyl)-9-methyl-azacyclo-hexadeca-4,9,13-trien-2-one.

However, due to the flexibility of the 16-membered ring, the relative configuration of oxalactam A (**1**) could not only be constructed from the NOESY data and require a special assignment. The stereochemistry of C-3 and C-15 were confirmed according to the DP4+ probability for theoretical ^{13}C NMR shifts of four possible isomers (3*R**,15*S**,16*S**)–**1**, (3*R**,15*R**,16*S**)–**1**, (3*S**,15*S**, 16*S**)–**1**, and (3*S**,15*R**,16*S**)–**1**. Finally, (3*R**,15*S**,16*S**) could be determined as the relative configuration of **1** with a 97.08% DP4+ probability (Figure 3) [14].

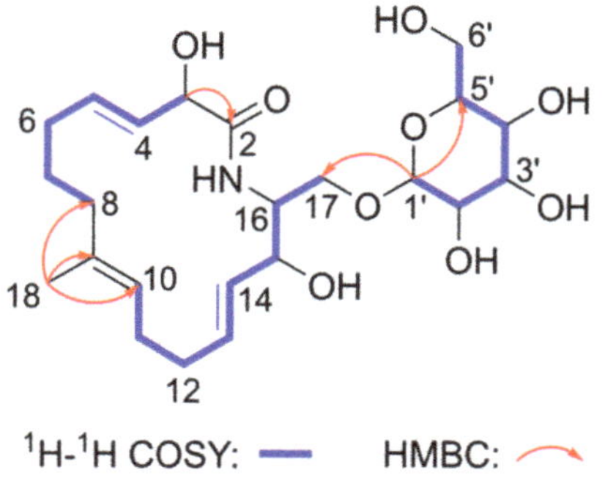

Figure 2. Key ^{1}H–^{1}H COSY and HMBC correlations of **1**.

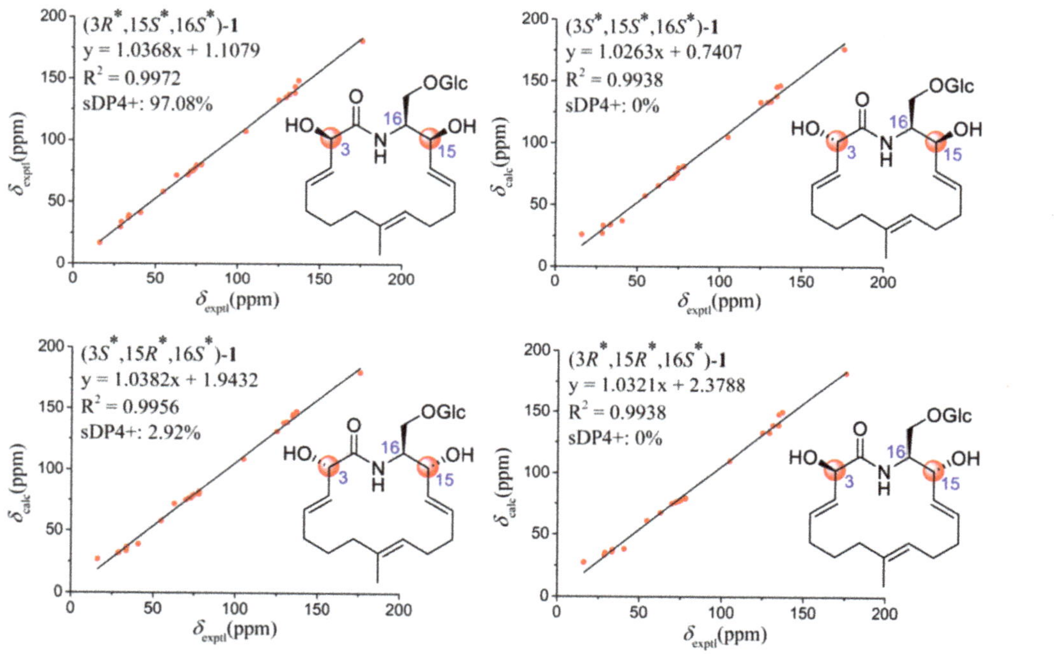

Figure 3. Linear correlation plots of calculated-experimental ^{13}C NMR chemical shift values for (3*R**,15*S**,16*S**)–**1**, (3*S**,15*S**,16*S**)–**1**, (3*S**,15*R**,16*S**)–**1**, and (3*R**,15*R**,16*S**)–**1**.

To confirm the assigned structure of **1**, (3*R**,15*S**,16*S**)–**1**/(3*R**,15*R**,16*S**)–**1**/(3*S**,15*S**,16*S**)–**1**/(3*S**,15*R**,16*S**)–**1** were further analyzed by the artificial neural networks (ANNs) method [15]. The NMR data of four possible isomers were recalculated at the OPT/HF/3-21G and GIAO/PCM/mPW1PW91/6-31G(d,p) levels in methanol, the ANNs calculation results classified four possible isomers of **1** into category 1 (correct) based on the analysis of 18 parameters (Figures S3–S6), and the ratio of category 1 (correct) to category 2 (incorrect) was calculated as 0.8573:0.1198 for (3*R**,15*S**,16*S**)–**1** [0.3963:0.5382 for (3*R**,15*R**,16*S**)–**1**, 0.4834:0.4027 for (3*S**,15*S**,16*S**)–**1**, and 0.6683:0.2713 for (3*S**,15*R**,16*S**)–**1**], confirming the relative configuration of **1** as (3*R**,15*S**,16*S**)–**1**.

Oxalactam A (**1**) could be proposed as a 3-aminobutyrate unit in the starter position based on the β-amino acid incorporation pathway in polyketide macrolactam biosynthesis reported previously [16]. Therefore, the absolute configuration of C-16 in **1** could be assigned as *S*. Moreover, the coupling constant of H-1′ (J = 7.8 Hz) assigned the *β*-glucopyranosyl linkage in **1**, and the *D*-glucose was proved by the comparison of the HPLC retention times of the monosaccharide derivative of hydrolysate of **1** (t_R = 21.924 min) with those of the standards *D*-glucose (t_R = 22.179 min) and *L*-glucose (t_R = 19.962 min) (Figures S28–S30). Further, the absolute configuration of **1** was verified by the electronic circular dichroism (ECD) method with time-dependent density functional theory [17]. The positive Cotton effect (222.4 nm) of **1** was consistent with that (224.2 nm) in the theoretically calculated ECD spectrum of (3*R*,15*S*,16*S*)–**1** (Figure 4). Thus, compound **1** was constructed as depicted and given a trivial name oxalactam A.

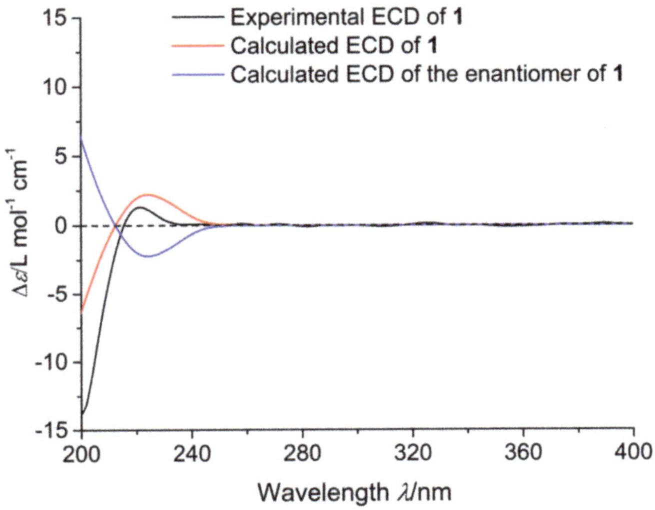

Figure 4. Experimental ECD for **1** and calculated ECD spectra for **1** and its enantiomer.

Oxalactam A (**1**) was the first example of a 16-membered polyenic macrolactam bearing a 2,9-dimethyl-azacyclohexadecane scaffold, and its biosynthetic pathway was also similar to those of vicenistatin from *Streptomyces halstedii* [18,19]. As shown in Figure 5, oxalactam A (**1**) biosynthesis originated from *L*-glutamate, which could convert into 3-aminobutyrate by VinH/VinI [20]. VinN, a adenylation enzyme, could recognize 3-aminobutyrate and transfer it to 3-aminobutyrate-VinL, which was aminoacylated with *L*-alanine under the catalysis of VinM to generate a dipeptidyl-VinL. The dipeptidyl group could be tied to the ACP domain of the modular PKS VinP1 by VinK. Then, modular PKSs VinP1–P4 elongated the polyketide chain with a terminal alanyl group, which could be removed by VinJ before intermediate **IV** generation by VinP4 thioesterase domain. Finally, the hydroxylase and glycosyl-transferase could successfully catalyze **IV** to yield oxalactam A (**1**). This alanyl protecting mechanism was also detected in the biosynthetic pathways of some other antibiotics including butirosin [21,22] and desertomycin [23].

Besides, three known dipeptides (**2**–**4**) and other alkaloids (**5**–**7**) were identified as cyclo(*L*-Trp-*L*-Glu) (**2**) [24], cyclo(*L*-Pro-*L*-Phe) (**3**) [25,26], cyclo(*L*-Pro-*L*-Tyr) (**4**) [27], penipanoid A (**5**) [28], (*Z*)-N-(4-Hydroxystyryl)-carboxamide (**6**) [28], and (*E*)-N-(4-Hydroxystyryl)-carboxamide (**7**) [28], by the specific rotation, HR-ESI-MS, and NMR data analyses, as well as the comparison with the literature data. Significantly, three dipeptides (**2**–**4**) were first identified from the fermentation liquid of *P. oxalicum*, and the NMR data of cyclo(*L*-Trp-*L*-Glu) (**2**) was reported for the first time. These isolated alkaloids in our study enriched the chemical diversity of secondary metabolites from *Penicillium* genus.

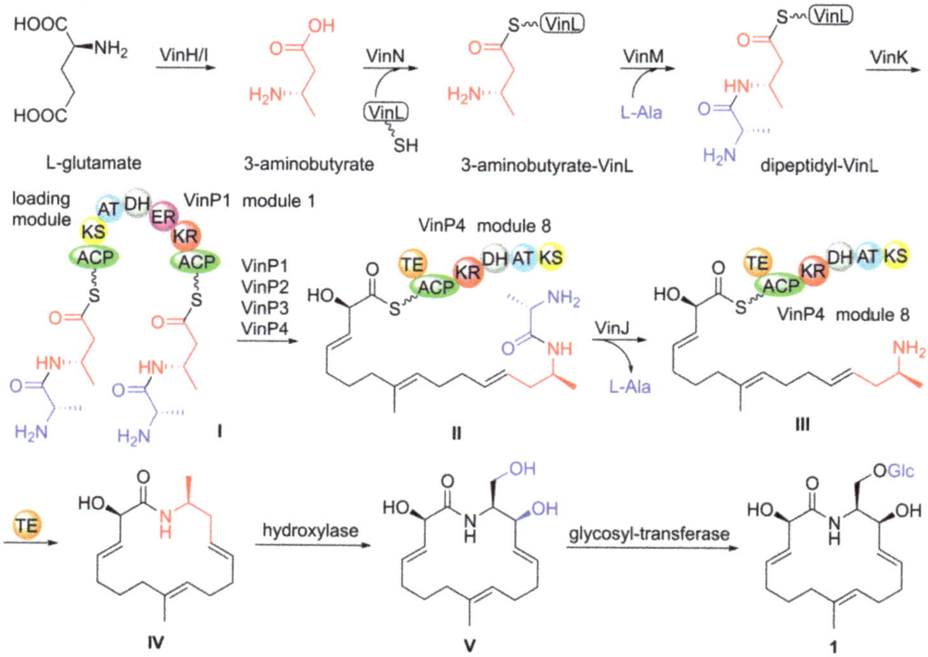

Figure 5. Proposed biosynthesis pathway of **1**.

2.2. *Anti-Rhizoctonia solani Activity*

The previous investigation indicated that *P. oxalicum* is a promising fungal agent to prevent plenty of plant diseases [29], including *Phytophthora* root rot of azalea [30], *Pythium* seed rot [31], and tomato *Fusarium* and *Verticillium* wilts [32]. These applications have aroused our interest to search for active secondary metabolites from the endophytic fungus *P. oxalicum* as anti-*R. solani* drugs to relieve rice sheath blight. Consequently, all isolates **1–7** were evaluated for their anti-*R. solani* activity using the mycelium growth rate method in vitro. Among these alkaloids, compound **1** was the most active component with a 29.30% of inhibition rate at the concentration of 100 μM (MIC = 10 μg/mL), whereas compounds **2–7** did not show potent anti-*R. solani* activity under the same condition (Table 2, Figures S27 and S28). Up to now, this is the first 16-membered polyenic macrolactam with anti-*R. solani* activity for agricultural tools.

Table 2. Anti-*Rhizoctonia solani* activity of compounds **1–7** in Vitro *.

No.	Inhibition Rate (%) **	MIC (μg/mL)	ED_{50} (μM)
1	29.30 ± 2.91	10	/
2	−7.82 ± 1.31	/	/
3	2.13 ± 2.29	/	/
4	0.83 ± 2.12	/	/
5	−3.34 ± 2.12	/	/
6	0.83 ± 2.12	/	/
7	2.11 ± 2.94	/	/
Carbendazim ***	82.39 ± 7.32	/	/
Hexaconazole ***	70.55 ± 5.8 [33]	/	2.44 [33]

* All measurements were carried out in triplicate. ** The test concentrations of compounds **1–7** and carbendazim are all 100 μM and hexaconazole is 10 μM. *** Standard anti-*Rhizoctonia solani* positive control substance.

2.3. Assessment of Binding Affinity of 1 and CYP51

The 14α-demethylase enzyme (CYP51, PDB ID: 3GW9) [34] is an essential component of the fungal cell membrane, which played an important role in the fungal-specific ergosterol biosynthesis pathway. Azole antifungals (hexaconazole for example) could compete with the CYP51 substrate by binding to the heme iron in the active site in a ligand–binding pocket, leading to ergosterol depletion, membrane fluidity reduction, and lipid layer destruction [35]. To analyze the binding mode of compound **1** to CYP51 enzyme, molecular docking and dynamics studies have been performed using Discovery Studio 2019. The result illustrated that 18-methyl of **1** showed favorable steric interactions with LUE356, MET460, and VAL461 residues (Figure 6), and MET460 might cause a misfolded fungus-specific loop that affects the binding efficiency of the cognate NADPH-cytochrome P450 reductase [36]. Moreover, the 1,2,4-triazole and 1,3-dichlorobenzene moieties of hexaconazole were observed to generate interactions with the ALA291, CYS422, and ILE423 (Figure 7), and the ALA291 substitution is vital for the microbial resistance to triazole drugs including voriconazole and fluconazole [37]. Notably, it showed that the higher negative binding energy value of **1** (−202.0196 kcal/mol) compared to that of hexaconazole (−105.4279 kcal/mol) indicated more favorable binding interactions between compound **1** and the CYP51 enzyme (Table 3) [33], which could explain the reason for the superiority of compound **1** over hexaconazole.

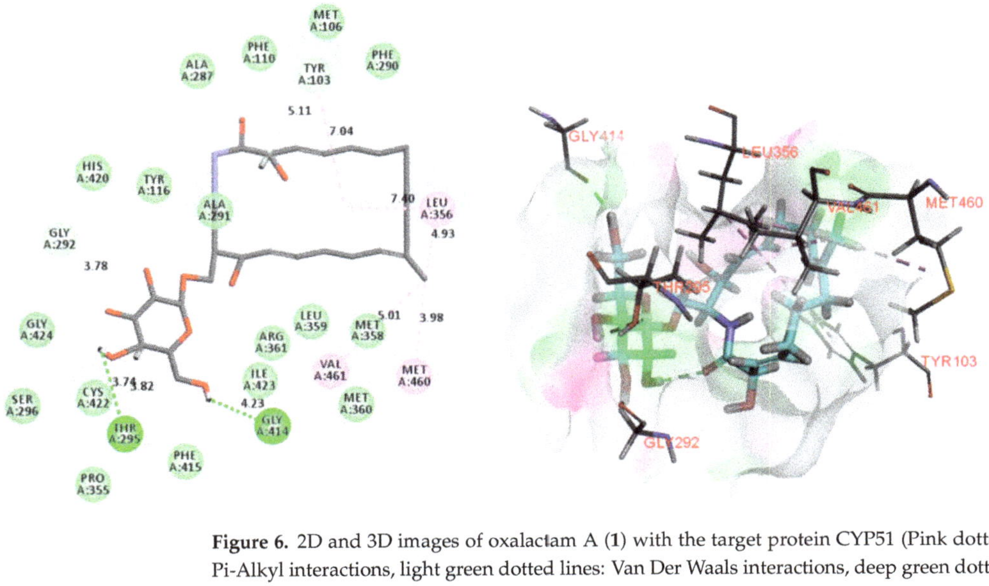

Figure 6. 2D and 3D images of oxalactam A (**1**) with the target protein CYP51 (Pink dotted lines: Pi-Alkyl interactions, light green dotted lines: Van Der Waals interactions, deep green dotted lines: Conventional hydrogen bond).

Table 3. The interaction analysis of the molecular docking study on 14α-demethylase CYP51 with oxalactam A (**1**) and hexaconazole.

Compound	Binding Energy (kcal/mol)	Interaction with Amino Acids
1	−202.0196	TYR103, THR295, LEU356, GLY414, MET460, VAL461
Hexaconazole *	−105.4279	ALA291, LEU356, LEU359, CYS422, ILE423

* Standard anti-*Rhizoctonia solani* positive control substance.

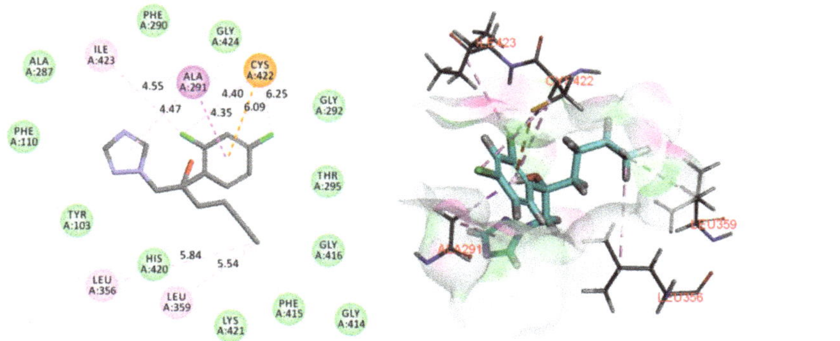

Figure 7. 2D and 3D images of hexaconazole with the target protein CYP51 (Orange dotted lines: Pi-Anion interactions, pink dotted lines: Pi-Alkyl interactions, purple dotted lines: Pi-Sigma interactions).

The steric interactions between CYP51 enzyme and oxalactam A (**1**)/hexaconazole were further estimated by molecular dynamics simulation according to RMSD (Root-mean-square deviation), RMSF (Root-mean-square fluctuation), and the total energy (Figure 8). Accordingly, RMSD values suggested that both complexes could reach an equilibrium state at 0.6–0.9 nm of the vibration amplitude within 200 ps. Additionally, the active and binding sites (THR295–LEU359/GLY414–VAL461) of CYP51–**1** have a slightly higher RMSF than those of the CYP51–hexaconazole, indicating the CYP51–hexaconazole complex possesses better thermodynamic stability. Notably, the CYP51–**1** complex showed lower total energy values than that of CYP51–hexaconazole with 55 kcal/mol per picosecond, which demonstrated that the binding affinity of CYP51–**1** was much stronger than that of CYP51–hexaconazole. These results indicated that compound **1** could directly bind to the high-affinity catalytic areas of the CYP51 enzyme and yield stable enzyme–ligand complexes in the saline condition.

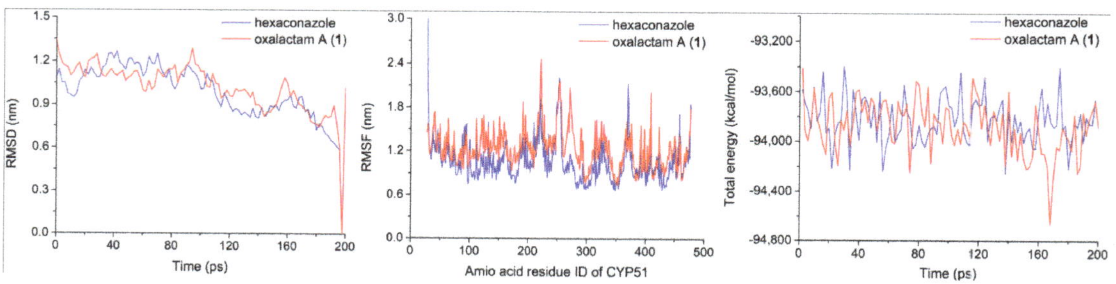

Figure 8. Molecular dynamics trajectory analysis of CYP51–**1** and CYP51–hexaconazole complexes, including the RMSD, RMSF, and the total energy.

3. Materials and Methods

3.1. General Experimental Procedures

A UV-2600 spectrophotometer (Shimadzu, Kyoto, Japan) was applied to measure the UV absorption spectra. An MCP 150 digital polarimeter (Anton Paar, Graz, Austria) was utilized to gain the optical rotations. A J-810 spectrometer (JASCO, Kyoto, Japan) could be exerted to acquire CD spectra in methanol at room temperature. The HR-ESI-MS data could be collected by a Waters SYNAPT G2-Si Q-TOF mass spectrometer (Waters, Milford, MA, USA). The NMR spectra were recorded on a Bruker AV-500 spectrometer (Bruker, Karlsruhe, Germany). The semi-preparative RP-HPLC included a dual-wavelength detector, a Shimadzu LC-20AR instrument, and an Ultimate XB-C_{18} column (10 × 250 mm, 5 μm, Welch).

3.2. Strain Material

The test endophytic fungus was obtained from the fresh tubers of the West African plant *I. trichantha*, the Orba village in Nsukka of the Enugu State, Nigeria, in June 2011. It was identified as *P. oxalicum* based on its morphological characteristics and 16S rRNA gene sequence by Shanghai Shenggong Bioengineering Co., Ltd. A voucher sample (ZRZ20210301) was stored in the Key Laboratory of prescriptions of Nanjing University of traditional Chinese medicine.

3.3. Fermentation and Isolation

The potato dextrose agar (PDA) was used to culture the *P. oxalicum* strain at room temperature for 72 h. Next, the PDA was cut into small pieces and then inoculated into a mass of sterilized Erlenmeyer flasks containing 10 g glucose, 4 g soluble starch, 2.5 g peptone, 1 g sodium chloride, 1 g calcium carbonate, 0.25 g magnesium sulfate, 1 g yeast extract, 0.25 g dipotassium hydrogen phosphate, and 500 mL distilled water. All flasks were incubated at room temperature for 72 h. The culture broths (82 L) and the smashed mycelium were extracted by ethanol three times, then combined and concentrated under reduced pressure to yield a crude extract, which was partitioned between ethyl acetate and water to obtain the EtOAc-soluble extract (42.58 g). This part was loaded on a 200~300 mesh silica gel eluted with CH_2Cl_2–MeOH (100:0 to 0:100, *v*/*v*) to furnish eight fractions (Fr.1~Fr.8).

Fr.6 (3.67 g) was separated by MPLC with an MCI column to yield seven subfractions Fr.6-1~Fr.6-7 (MeOH–H_2O, 10:90 to 100:0, *v*/*v*). Compound **2** (12.71 mg) was purified from Fr.6-6 (80.60 mg) by a semi-preparative HPLC with an Ultimate XB-C_{18} column (MeOH–H_2O, 45:55, *v*/*v*, 3 mL/min). Fr.6-8 (1.47 g) was treated with excess sodium carbonate, then loaded on a 200~300 mesh silica gel CC using CH_2Cl_2–MeOH (10:1, *v*/*v*) mixed solvent to yield compound **1** (140.59 mg). Fr.3 (4.58 g) was isolated as nine sub-fractions (Fr.3-1~Fr.3-9) by a Sephadex LH-20 column with pure methanol. A semi-preparative HPLC (MeOH–H_2O, 30:70, *v*/*v*, 3 mL/min) was applied to purify compound **3** (3.02 mg) from Fr.3-3 (0.78 g). Compounds **6** (10.88 mg) and **7** (5.84 mg) were also purified from Fr.3-5 (0.39 g) by a semi-preparative HPLC (MeOH–H_2O, 20:80, *v*/*v*, 3 mL/min). Fr.4 (11.18 g) was acquired by MPLC with an MCI column (MeOH–H_2O, 5:95 to 100:0, *v*/*v*) to afford nine fractions Fr.4-1~Fr.4-9. Compounds **4** (101.00 mg) and **5** (3.50 mg) were finally purified from Fr.4-7 (0.58 g, MeOH–H_2O, 17:83, *v*/*v*, 3 mL/min) and Fr.4-9 (0.21 g, MeOH–H_2O, 35:65, *v*/*v*, 3 mL/min) by a semi-preparative HPLC, respectively.

Oxalactam A (**1**): A white amorphous powder; $[\alpha]_D^{20}$+2.00 (*c* 0.1, MeOH); UV (MeOH) λ_{max} (log ε) 207 (3.24) nm; CD (MeOH): λ_{max} ($\Delta\varepsilon$) 222.0 (+1.30) nm; ^{1}H NMR (CD_3OD, 500 MHz) data, see Table 1; ^{13}C NMR (CD_3OD, 125 MHz), see Table 1; HR-ESI-MS *m/z* 518.2943 [M + CH_3CH_2OH + H]$^+$ (calcd for $C_{25}H_{44}NO_{10}$, 518.2965).

3.4. Enzymatic Hydrolysis

Oxalactam A (**1**) (2 mg) and β-cellulase (0.5 mg) were dissolved in an aqueous solution (2 mL) at 50 °C for 4 h. The reaction solution was extracted by EtOAc five times, 5 mL each. The EtOAc extract liquor was combined and evaporated, and analyzed by the TLC and HPLC methods. Finally, the aglycone of compound **1** was purified by a semipreparative HPLC with an Ultimate XB-C_{18} column.

3.5. Sugar Identification

The Enzymatic Hydrolysis mixtures were analyzed using HPLC with an Ultimate XB-C_{18} column ($CH_3CN/H_2O/CH_3COOH$ = 25:75:0.1, 0.8 mL/min). The sugar moiety of *D*-glucose in **1** (t_R = 21.924 min) was assigned by the comparison of the retention time of the monosaccharide derivative with those of *D*-glucose (t_R = 22.179 min) and *L*-glucose (t_R = 19.962 min) [38].

3.6. ECD and NMR Calculation Methods

HyperChem 8.0 program was carried out to search for the most stable stereoisomers of compound **1** with molecular mechanics MMFF94s. Gaussian 16 software was applied to optimize geometries to screen minimum-energy conformers at the OPT/B3LYP/6-31+G(d) (OPT/HF/3-21G for ANNs analysis) level. The optimized conformers were submitted to the ECD calculation program with the TD/B3LYP/6-311+G(d) level in methanol. The overall ECD curves were summed up by the Boltzmann distribution averaging of all conformers with SpecDis 1.71 software [39]. Additionally, the ^{13}C NMR shifts calculation of all stable conformers of **1** was conducted with the GIAO/PCM/mPW1PW91/6-311G(d,p) (GIAO/PCM/mPW1PW91/6-31G(d,p) for ANNs analysis) level in methanol. The DP4+ probability and linear regression were obtained by analyzing overall theoretical NMR data to verify the relative configurations of C-3/C-15 in **1** [14].

3.7. The Plate Confrontation Test

The fungal blocks of *Rhizoctonia solani, Alternaria alternata, Fusarium solani,* and *Colletotrichum gloeosporioides* (each diameter 5 mm) were, respectively inoculated in the center of the PDA plate (each diameter 150 mm), and purified endophytic fungus *P. oxalicum* was inoculated in the surroundings. All plates were incubated at room temperature for 72 h. Each treatment was repeated three times and the sterile water was used as a blank control. The antagonistic effect was determined by the width of the inhibition belt when colonies fully covered the whole Petri dish in blank control.

3.8. Anti-Rhizoctonia solani Assays In Vitro

The sensitivity of *R. solani* to compounds **1–7** from the endophytic fungus *P. oxalicum* was measured by the mycelium growth rate method in vitro. This phytopathogenic fungus, *R. solani*, was provided by Beijing Baioubowei Biotechnology Co., Ltd.

A mycelial plug (1 cm in diameter) was cut from the beforehand PDA medium fully covered by *R. solani* colonies, which was placed with their mycelia-side down on the PDA medium in the center of each plate containing 100 μM of pure compounds 1–7. The positive control is hexaconazole (100 μM), a commercial fungicide. An equal concentration of DMSO solution served as the solvent control. A thermostatic chamber was applied to inoculate the treatment plates, and the average diameter of each *R. solani* colony was recorded based on the previous literature [40]. The anti-*R. solani* activity was evaluated in terms of the inhibition rate, which was performed in triplicate for each treatment. The inhibition ratio could be acquired by the following formula: Inhibition (%) = (A − B)/A × 100% (A: average diameter of the control group, B: average diameter of the treatment group) [41].

3.9. Molecular Docking Study

The molecular docking simulation [42] of compound **1** and hexaconazole to the sterol 14α-demethylase enzyme (CYP51, PDB ID: 3GW9) was performed by the CDOCKER protocol of Discovery Studio 2019 (BIOVIA, San Diego, CaA). The CYP51 X-ray crystal structure was downloaded from RCSB Protein Data Bank (http://www.rcsb.org, accessed on 1 June 2022). All conformations of compound **1** were searched by ChemHyper 8.0 program with the density functional method (6-31G*). The CYP51 protein was protonated and deleted water at pH 7 using Prepare Protein tool under the CHARMm forcefield [43]. Additionally, the implicit solvent model was selected as the Generalized Born with Molecular Volume method. Additionally, the binding energy of CYP51–**1**/hexaconazole was calculated by the Spherical Cutoff mode. All other parameters were set as default.

3.10. Molecular Dynamics Simulation

Molecular dynamics simulation [42] was also performed for CYP51–**1**/hexaconazole based on molecular docking results using the Standard Dynamics Cascade protocol of Discovery Studio 2019 (BIOVIA, San Diego, CA, USA). This software brings together over 30 years of peer-reviewed research and world-class in silico techniques such as molecular mechanics, free energy calculations, and biotherapeutics developability and more into a common environment, and it could be available at the website: https://www.3ds.com/products-services/biovia/products/molecular-modeling-simulation/biovia-discovery-studio/ (accessed on 1 January 2022).

The CYP51–**1**/hexaconazole complex was in the aqueous environment with the Extended Simple Point Charge water model under the CHARMm force field [44]. The entire system was equilibrated under NVT (isothermal-isochoric) and NPT (isothermal-isobaric) ensembles to maintain the stabilized pressure [45]. All hydrogen bonds were constrained during equilibration using LINC algorithms [46]. Besides, the Particle Mesh Ewald module was applied for the long-range ionic interaction [47]. Finally, the entire trajectories were saved for analysis at a frequency of 1 ps.

4. Conclusions

In summary, a novel macrolactam (**1**), three dipeptides (**2**–**4**) as well as other alkaloids (**5**–**7**) were isolated from the fermentation liquid of *P. oxalicum* derived from the tuber of *I. trichantha*. Among them, only oxalactam A (**1**) displayed anti-*R. solani* activity at 100 μM with a 29.30% of inhibition rate in vitro (MIC = 10 μg/mL). Notably, the content of **1** is more than 1.70 mg/L in the fermentation liquid of the endophytic fungus *P. oxalicum*. These findings provided an alternative natural resource to obtain novel anti-*R. solani* macrolactam leads.

Supplementary Materials: The following supporting information can be downloaded at: https://www.mdpi.com/article/10.3390/molecules27248811/s1, Figure S1: Flow chart of extraction and isolation; Figure S2: *Penicillium oxalicum* resisted the growth of four plant pathogenic fungi strains; Figures S3–S9: 1D and 2D NMR of compound **1**; Figure S10: (+)-HR-ESI-MS of compound **1**; Figure S11: UV of compound **1**; Figures S12–S23: ^{1}H and ^{13}C NMR of compounds **2**–**7**; Figures S24–S26: HPLC chromatograms of the derivative of the standard *L*/*D*-Glucose and the hydrolysate of compound **1**; Figure S27: Compounds **1**–**7** and carbendazim (100 μM) inhibit the growth of *Rhizoctonia solani*; Figure S28: Growth inhibition of *Rhizoctonia solani* with different concentrations of compound **1**; Table S1: *Penicillium oxalicum* resisted the growth of four plant pathogenic fungi strains; Table S2: Cartesian coordinate for the lowest energy conformer of **1** calculated at TD/B3LYP/6-311G(d,p) level of theory in the gas phase; Table S3: Experimental and calculated chemical shifts of **1** for DP4+ probability statistical analysis.

Author Contributions: Conceptualization, R.Z., Y.M., J.Z., X.W. and M.Z.; methodology, R.Z., Y.M., J.Z. and M.Z.; software, J.Z.; validation, R.Z., Y.M., J.Z. and M.Z.; formal analysis, J.Z. and M.Z.; investigation, R.Z., Y.M., M.-M.X., X.W., C.-B.Y., F.Z., J.-A.D., C.-T.C., J.Z. and M.Z.; resources, J.-A.D., C.-T.C., J.Z. and M.Z.; data curation, R.Z. and J.Z.; writing—original draft preparation, J.Z.; writing—review and editing, J.Z. and M.Z.; visualization, M.Z.; supervision, J.Z. and M.Z.; project administration, J.Z. and M.Z.; funding acquisition, M.Z. All authors have read and agreed to the published version of the manuscript.

Funding: This research received no external funding.

Institutional Review Board Statement: Not applicable.

Informed Consent Statement: Not applicable.

Data Availability Statement: Not applicable.

Acknowledgments: This work was supported financially by the National Natural Science Foundation of China (No. 82204549), the Jiangsu Collaborative Innovation Center of Chinese Medicinal Resources Industrialization (ZDXM-2020-17), and the Jiangsu Province Specially Appointed Professor Program. We thank Er-Xin Shang in the Mass Spectrometry Center of Nanjing University of Chinese Medicine for assistance in collecting HRESIMS data.

Conflicts of Interest: The authors declare no conflict of interest.

Sample Availability: Compounds **2**–**7** are available from the authors.

References

1. Rao, T.B.; Chopperla, R.; Prathi, N.B.; Balakrishnan, M.; Prakasam, V.; Laha, G.S.; Balachandran, S.M.; Mangrauthia, S.K. A comprehensive gene expression profile of pectin degradation enzymes reveals the molecular events during cell wall degradation and pathogenesis of rice sheath blight pathogen Rhizoctonia solani AG1-IA. *J. Fungi.* **2020**, *6*, 71.
2. Molla, K.A.; Karmakar, S.; Molla, J.; Bajaj, P.; Varshney, R.K.; Datta, S.K.; Datta, K. Understanding sheath blight resistance in rice: The road behind and the road ahead. *Plant Biotechnol. J.* **2020**, *18*, 895–915. [CrossRef] [PubMed]
3. Senapati, M.; Tiwari, A.; Sharma, N.; Chandra, P.; Bashyal, B.M.; Ellur, R.K.; Bhowmick, P.K.; Bollinedi, H.; Vinod, K.K.; Singh, A.K.; et al. Rhizoctonia solani Kühn pathophysiology: Status and prospects of sheath blight disease management in rice. *Front. Plant Sci.* **2022**, *13*, 881116. [CrossRef] [PubMed]
4. Bernardes-de-Assis, J.; Storari, M.; Zala, M.; Wang, W.X.; Jiang, D.H.; Dong, L.S.; Jin, M.S.; McDonald, B.A.; Ceresini, P.C. Genetic structure of populations of the rice-infecting pathogen Rhizoctonia solani AG-1 IA from China. *Phytopathology* **2009**, *99*, 1090–1099. [CrossRef]
5. Shu, C.W.; Zhao, M.; Anderson, J.P.; Garg, G.; Singh, K.B.; Zheng, W.B.; Wang, C.J.; Yang, M.; Zhou, E.X. Transcriptome analysis reveals molecular mechanisms of sclerotial development in the rice sheath blight pathogen Rhizoctonia solani AG1-IA. *Funct. Integr. Genom.* **2019**, *19*, 743–758. [CrossRef]
6. Li, D.; Zhang, F.; Pinson, S.R.M.; Edwards, J.D.; Jackson, A.K.; Xia, X.; Eizenga, G.C. Assessment of rice sheath blight resistance including associations with plant architecture, as revealed by genome-wide association studies. *Rice* **2022**, *15*, 31. [CrossRef]
7. Zhu, G.; Liang, E.X.; Lan, X.; Li, Q.; Qian, J.J.; Tao, H.X.; Zhang, M.J.; Xiao, N.; Zuo, S.M.; Chen, J.M.; et al. ZmPGIP3 gene encodes a polygalacturonase-inhibiting protein that enhances resistance to sheath blight in rice. *Phytopathology* **2019**, *109*, 1732–1740. [CrossRef]
8. Singh, P.; Mazumdar, P.; Harikrishna, J.A.; Babu, S. Sheath blight of rice: A review and identification of priorities for future research. *Planta* **2019**, *250*, 1387–1407. [CrossRef]
9. Yu, X.; Sun, D. Macrocyclic drugs and synthetic methodologies toward macrocycles. *Molecules* **2013**, *18*, 6230–6268. [CrossRef]
10. Vitaku, E.; Smith, D.T.; Njardarson, J.T. Analysis of the structural diversity, substitution patterns, and frequency of nitrogen heterocycles among U.S. FDA approved pharmaceuticals. *J. Med. Chem.* **2014**, *57*, 10257–10274. [CrossRef]
11. Hügel, H.M.; Smith, A.T.; Rizzacasa, M.A. Macrolactam analogues of macrolide natural products. *Org. Biomol. Chem.* **2016**, *14*, 11301–11316. [CrossRef] [PubMed]
12. Che, C.T.; Zhao, M.; Guo, B.; Onakpa, M.M. Icacina trichantha, a tropical medicinal plant. *Nat. Prod. Commun.* **2016**, *11*, 1039–1042. [CrossRef] [PubMed]
13. Asuzu, I.U.; Ugwueze, E.E. Screening of Icacina trichantha extracts for pharmacological activity. *J. Ethnopharmacol.* **1990**, *28*, 151–156. [CrossRef] [PubMed]
14. Grimblat, N.; Zanardi, M.M.; Sarotti, A.M. Beyond DP4: An improved probability for the stereochemical assignment of isomeric compounds using quantum chemical calculations of NMR shifts. *J. Org. Chem.* **2015**, *80*, 12526–12534. [CrossRef] [PubMed]
15. Sarotti, A.M. Successful combination of computationally inexpensive GIAO ^{13}C NMR calculations and artificial neural network pattern recognition: A new strategy for simple and rapid detection of structural misassignments. *Org. Biomol. Chem.* **2013**, *11*, 4847–4859. [CrossRef]
16. Miyanaga, A.; Kudo, F.; Eguchi, T. Mechanisms of β-amino acid incorporation in polyketide macrolactam biosynthesis. *Curr. Opin. Chem. Biol.* **2016**, *35*, 58–64. [CrossRef]
17. Pescitelli, G.; Bari, L.D.; Berova, N. Application of electronic circular dichroism in the study of supramolecular systems. *Chem. Soc. Rev.* **2014**, *43*, 5211–5233. [CrossRef]
18. Shinohara, Y.; Kudo, F.; Eguchi, T. A natural protecting group strategy to carry an amino acid starter unit in the biosynthesis of macrolactam polyketide antibiotics. *J. Am. Chem. Soc.* **2011**, *133*, 18134–18137. [CrossRef]
19. Ogasawara, Y.; Katayama, K.; Minami, A.; Otsuka, M.; Eguchi, T.; Kakinuma, K. Cloning, sequencing, and functional analysis of the biosynthetic gene cluster of macrolactam antibiotic vicenistatin in Streptomyces halstedii. *Chem. Biol.* **2004**, *11*, 79–86. [CrossRef]
20. Ogasawara, Y.; Kakinuma, K.; Eguchi, T. Involvement of glutamate mutase in the biosynthesis of the unique starter unit of the macrolactam polyketide antibiotic vicenistatin. *J. Antibiot.* **2005**, *58*, 468–472. [CrossRef]
21. Li, Y.Y.; Llewellyn, N.M.; Giri, R.; Huang, F.L.; Spencer, J.B. Biosynthesis of the unique amino acid side chain of butirosin: Possible protective-group chemistry in an acyl carrier protein-mediated pathway. *Chem. Biol.* **2005**, *12*, 665–675. [CrossRef]
22. Llewellyn, N.M.; Li, Y.Y.; Spencer, J.B. Biosynthesis of butirosin: Transfer and deprotection of the unique amino acid side chain. *Chem. Biol.* **2007**, *14*, 379–386. [CrossRef] [PubMed]
23. Hong, H.; Samborskyy, M.; Lindner, F.; Leadlay, P.F. An amidinohydrolase provides the missing link in the biosynthesis of amino marginolactone antibiotics. *Angew. Chem. Int. Ed.* **2016**, *55*, 1118–1123. [CrossRef] [PubMed]
24. Bivin, D.B.; Kubota, S.; Pearlstein, R.; Morales, M.F. On how a myosin tryptophan may be perturbed. *Proc. Natl. Acad. Sci. USA* **1993**, *90*, 6791–6795. [CrossRef] [PubMed]
25. Gong, J.; Tang, H.; Geng, W.L.; Liu, B.S.; Zhang, W. Cyclic dipeptides in actinomycete Brevibacterium sp. associated with sea cucumber *Apostichopus japonicas* Selenka: Isolation and identification. *Acad. J. Second Military Med. Univ.* **2013**, *33*, 1284–1287. [CrossRef]

26. Adamczeski, M.; Reed, A.R.; Crews, P. New and known diketopiperazines from the Caribbean sponge, calyx cf. Podatypa. *J. Nat. Prod.* **1995**, *58*, 201–208. [CrossRef] [PubMed]
27. Wattana-Amorn, P.; Charoenwongsa, W.; Williams, C.; Crump, M.P.; Apichaisataienchote, B. Antibacterial activity of cyclo(L-Pro-L-Tyr) and cyclo(D-Pro-L-Tyr) from *Streptomyces* sp. strain 22-4 against phytopathogenic bacteria. *Nat. Prod. Res.* **2016**, *30*, 1980–1983. [CrossRef]
28. Li, X.; Li, X.M.; Zhang, P.; Wang, B.G. A new phenolic enamide and a new meroterpenoid from marine alga-derived endophytic fungus *Penicillium oxalicum* EN-290. *J. Asian Nat. Prod. Res.* **2015**, *17*, 1204–1212. [CrossRef]
29. Yang, L.P.; Xie, J.T.; Jiang, D.H.; Fu, Y.P.; Li, G.Q.; Lin, F.C. Antifungal substances produced by Penicillium oxalicum strain PY-1—Potential antibiotics against plant pathogenic fungi. *World J. Microbiol. Biotechnol.* **2008**, *24*, 909–915. [CrossRef]
30. Gintis, B.O.; Benson, D.M. Biological control of Phytophthora root rot of azalea with Penicillium oxalicum. *Phytopathology* **1987**, *77*, 1688.
31. Trapero-Casas, A.; Kaiser, W.J.; Ingram, D.M. Control of Pythium seed rot and preemergence damping-off of chickpea in the US Pacific Northwest and Spain. *Plant Dis.* **1990**, *74*, 563–569. [CrossRef]
32. Sabuquillo, P.; De Cal, A.; Melgarejo, P. Dispersal improvement of a powder formulation of Penicillium oxalicum, a biocontrol agent of tomato wilt. *Plant Dis.* **2005**, *89*, 1317–1323. [CrossRef] [PubMed]
33. Upadhyay, R.K.; Saini, K.K.; Deswal, N.; Singh, T.; Tripathi, K.P.; Kaushik, P.; Shakil, N.A.; Bhartid, A.C.; Kumar, R. Synthesis of benzothiazole-appended bis-triazolebased structural isomers with promising antifungal activity against Rhizoctonia solani. *RSC Adv.* **2022**, *12*, 24412. [CrossRef] [PubMed]
34. Lepesheva, G.I.; Park, H.W.; Hargrove, T.Y.; Vanhollebeke, B.; Wawrzak, Z.; Harp, J.M.; Sundaramoorthy, M.; Nes, W.D.; Pays, E.; Chaudhuri, M.; et al. Crystal structures of Trypanosoma brucei sterol 14alpha-demethylase and implications for selective treatment of human infections. *J. Biol. Chem.* **2010**, *285*, 1773–1780. [CrossRef] [PubMed]
35. Rosam, K.; Monk, B.C.; Lackner, M. Sterol 14-Demethylase Ligand-Binding Pocket-Mediated Acquired and Intrinsic Azole Resistance in Fungal Pathogens. *J. Fungi* **2021**, *7*, 1. [CrossRef]
36. Mullins, J.G.; Parker, J.E.; Cools, H.J.; Togawa, R.C.; Lucas, J.A.; Fraaije, B.A.; Kelly, D.E.; Kelly, S.L. Molecular modelling of the emergence of azole resistance in Mycosphaerella graminicola. *PLoS ONE* **2011**, *6*, e20973. [CrossRef]
37. Snelders, E.; Camps, S.M.; Karawajczyk, A.; Rijs, A.J.; Zoll, J.; Verweij, P.E.; Melchers, W.J. Genotype-phenotype complexity of the TR46/Y121F/T289A cyp51A azole resistance mechanism in Aspergillus fumigatus. *Fungal Genet. Biol.* **2015**, *82*, 129–135. [CrossRef]
38. Abdullah, F.O.; Hussain, F.H.S.; Sardar, A.S.; Gilardoni, G.; Tosi, S.; Vidari, G. Iridoids Isolation from a Phytochemical Study of the Medicinal Plant Teucrium parviflorum Collected in Iraqi Kurdistan. *Molecules* **2022**, *27*, 5963. [CrossRef]
39. Ma, Y.; Liu, X.; Liu, B.; Li, P.; Suo, X.; Zhu, T.; Ji, T.; Li, J.; Li, X. Hyperacmosin R, a New Decarbonyl Prenylphloroglucinol with Unusual Spiroketal Subunit from Hypericum acmosepalum. *Molecules* **2022**, *27*, 5932. [CrossRef]
40. Liang, H.J.; Di, Y.L.; Li, J.L.; Zhu, F.X. Baseline sensitivity and control efficacy of fluazinam against Sclerotinia sclerotiorum. *Eur. J. Plant Pathol.* **2015**, *142*, 691–699. [CrossRef]
41. Wang, Y.; Duan, Y.B.; Zhou, M.G. Baseline sensitivity and efficacy of fluazinam in controlling Sclerotinia stem rot of rapeseed. *Eur. J. Plant Pathol.* **2016**, *144*, 337–343. [CrossRef]
42. Karwasra, R.; Ahmad, S.; Bano, N.; Qazi, S.; Raza, K.; Singh, S.; Varma, S. Macrophage-Targeted Punicalagin Nanoengineering to Alleviate Methotrexate-Induced Neutropenia: A Molecular Docking, DFT, and MD Simulation Analysis. *Molecules* **2022**, *27*, 6034. [CrossRef] [PubMed]
43. Brooks, B.R.; Bruccoleri, R.E.; Olafson, B.D.; States, D.J.; Swaminathan, S.; Karplus, M. CHARMM: A program for macromolecular energy, minimization, and dynamics calculations. *J. Comput. Chem.* **1983**, *4*, 187–217. [CrossRef]
44. Armstrong, J.A.; Bresme, F. Water polarization induced by thermal gradients: The extended simple point charge model (SPC/E). *J. Chem. Phys.* **2013**, *139*, 014504. [CrossRef]
45. Zeng, Q.H.; Yu, A.B.; Lu, G.Q.; Standish, R.K. Molecular Dynamics Simulation of Organic-Inorganic Nanocomposites: Layering Behavior and Interlayer Structure of Organoclays. *Chem. Mater.* **2003**, *15*, 4732–4738. [CrossRef]
46. Hess, B.; Bekker, H.; Berendsen, H.J.C.; Fraaije, J.G.E.M. LINCS: A linear constraint solver for molecular simulations. *J. Comput. Chem.* **1997**, *18*, 1463–1472. [CrossRef]
47. Petersen, H.G. Accuracy and efficiency of the particle mesh Ewald method. *J. Chem. Phys.* **1995**, *103*, 3668–3679. [CrossRef]

molecules

Article

Effect of *Holoptelea integrifolia* (Roxb.) Planch. *n*-Hexane Extract and Its Bioactive Compounds on Wound Healing and Anti-Inflammatory Activity

Kanokwan Somwong [1,2], Pattawika Lertpatipanpong [2], Wutigri Nimlamool [3], Aussara Panya [1,4], Yingmanee Tragoolpua [1,4], Rujipas Yongsawas [1], Wandee Gritsanapan [5], Hataichanok Pandith [1,4,*] and Seung Joon Baek [2,*]

1 Department of Biology, Faculty of Science, Chiang Mai University, Chiang Mai 50200, Thailand
2 Laboratory of Signal Transduction, College of Veterinary Medicine and Research Institute for Veterinary Science, Seoul National University, Seoul 08826, Republic of Korea
3 Department of Pharmacology, Faculty of Medicine, Chiang Mai University, Chiang Mai 50200, Thailand
4 Research Center in Bioresources for Agriculture, Industry and Medicine, Chiang Mai 50200, Thailand
5 Department of Pharmacognosy, Faculty of Pharmacy, Mahidol University, Bangkok 10400, Thailand
* Correspondence: hataichanok064@gmail.com (H.P.); baeksj@snu.ac.kr (S.J.B.)

Abstract: The stem bark of *Holoptelea integrifolia* (Roxb.) Planch. has been applied for the treatment of human cutaneous diseases as well as canine demodicosis in several countries. However, no detailed mechanistic studies have been reported to support their use. In this study, thin-layer chromatography and gas chromatography were used to screen phytochemicals from the fresh stem bark extract of *H. integrifolia*. We found the two major bioactive compounds, friedelin and lupeol, and their activity on wound healing was further investigated in keratinocytes. Both bioactive compounds significantly reduced wound area and increased keratinocyte migration by increasing matrix metalloproteinases-9 production. Subsequently, we found that the mRNA gene expressions of cadherin 1 and desmoglobin 1 significantly decreased, whereas the gene expression involved in keratinocyte proliferation and homeostasis (keratin-17) increased in compound-treated human immortalized keratinocytes cells. The expression of inflammatory genes (cyclooxygenase-2 and inducible nitric oxide synthase) and pro-inflammatory cytokine genes (tumor necrosis factor-alpha and interleukin-6) was reduced by treatment with *n*-hexane extract of *H. integrifolia* and its bioactive compounds. Our results revealed that *H. integrifolia* extract and its bioactive compounds, friedelin and lupeol, exhibit wound-healing activity with anti-inflammatory properties, mediated by regulating the gene expression involved in skin re-epithelialization.

Keywords: *Holoptelea integrifolia*; MMP-9; anti-inflammation; wound healing; friedelin

Citation: Somwong, K.; Lertpatipanpong, P.; Nimlamool, W.; Panya, A.; Tragoolpua, Y.; Yongsawas, R.; Gritsanapan, W.; Pandith, H.; Baek, S.J. Effect of *Holoptelea integrifolia* (Roxb.) Planch. *n*-Hexane Extract and Its Bioactive Compounds on Wound Healing and Anti-Inflammatory Activity. *Molecules* **2022**, *27*, 8540. https://doi.org/10.3390/molecules27238540

Academic Editors: Sokcheon Pak and Soo Liang Ooi

Received: 10 November 2022
Accepted: 28 November 2022
Published: 4 December 2022

1. Introduction

Holoptelea integrifolia (Roxb.) Planch. is a medicinal plant that is traditionally used for the treatment and prevention of several diseases. The bark and leaves of this plant are the most investigated parts and have anti-inflammatory effects, and their extracts are externally applied to treat several skin diseases, such as leprosy, leukoderma, and chronic wounds [1,2]. The stem bark extract of this plant contains various phytochemicals, such as terpenoids, friedelin, β-amyrin, betulin, betulinic acid, and lupeol [3].

Cutaneous wound healing is a complex biological process that can be classified into three major phases: inflammation, proliferation, and maturation phase [4,5]. Re-epithelialization during the proliferative phase is important for successful wound closure. Matrix metalloproteinases (MMPs) play a crucial role in augmenting keratinocyte migration by degrading extracellular matrix proteins secreted by keratinocytes in the neo-epidermis. Among these, MMP-1 and MMP-9 have been reported to function as mediating enzymes for keratinocyte migration, and MMP-9 in particular promotes terminal cell differentiation [6]. Moreover, inflammation can occur immediately after skin injury, leading to the upregulation of several pro-inflammatory cytokines, including cyclooxygenase-2 (COX-2) and inducible nitric oxide synthase (iNOS) that play a critical role in inflammation. These enzymes are triggered to produce pro-inflammatory mediators, thereby enhancing the expression of pro-inflammatory cytokines, including tumor necrosis factor-alpha (TNF-α) and interleukin-6 (IL-6).

Various herbal drugs and formulations have been widely used in the treatment of wounds for years as several natural products, such as alkaloids, tannins, flavonoids, and terpenes promote the wound healing process. However, traditional herbal medicinal products for wound healing require elucidation of their molecular mechanisms. This will provide beneficial clues for accelerating wound healing and encourage fighting against skin infection.

Several studies have reported the wound-healing activity of *H. integrifolia*. For example, the wound-healing activity of *H. integrifolia* extracts was examined in excision wound model in Wister rats, the external application of these extracts on the wound sites show more than 90% wound healing was recorded in treated groups by 14 days of post-surgery, whereas only 62.99% was observed in the control group albino rats. In addition, in incision model, the higher collagen re-deposition than the control group was presented in treated group. Not only wound healing property, but also prevented microbes from invading the wound, protecting it against microbial infections [7,8].

In this study, we extracted the fresh stem bark of *H. integrifolia* using various solvents and investigated the bioactive components in these extracts for keratinocyte wound healing and anti-inflammatory activity. This is the first study suggesting that *H. integrifolia* stem bark extracts and their active compounds modulate the expression of re-epithelialization-related genes and MMP-9, thereby enhancing wound-healing activity.

2. Results

2.1. Effects of H. Integrifolia n-Hexane Extract on Cell Viability in Human and Canine Keratinocytes

According to the TLC result (Supplementary Figure S1), among those various solvent, *n*-hexane extract provides highest content of phytochemical components after extraction. Therefore, *n*-hexane extract was selected to use in this study. Human and canine keratinocytes were treated with various concentrations of *H. integrifolia n*-hexane extract (1–20 μg/mL), and cell viability was evaluated using the CellTiter 96® assay kit. After 24 h of treatment, the viability of human keratinocytes significantly decreased to 57.9% and 31.4% at 10 μg/mL and 20 μg/mL, respectively (Figure 1A), while it decreased to 70.8% and 3.87%, respectively, after 48 h of treatment. Similarly, the cell viability of canine keratinocytes significantly decreased to 27.0% and 18.1% at 10 and 20 μg/mL *n*-hexane extracts, respectively, after 24 h of treatment (Figure 1B), whereas it was completely abolished in both 10 and 20 μg/mL after 48 h of treatment. Subsequently, 1 μg/mL *n*-hexane extract was selected, and the scratch wound healing assay was conducted using human keratinocytes. Results showed that the wound area was marginally affected by the *n*-hexane extract; however, no significant difference in the percentage of wound area between the DMSO control and the *n*-hexane extract treatment was observed (Figure 1C).

2.2. Identification of Friedelin and Lupeol in n-Hexane Extract

H. integrifolia extract contains several phytochemicals, including friedelin and lupeol [3,9,10]. The *n*-hexane extract was analyzed using TLC, and the developed TLC indicated that two triterpenoids, friedelin (Rf = 0.9625, 0.9125) and lupeol (Rf = 0.8625), were the major components in the extract (Figure 2A). Subsequently, the chromatogram of GC-FID analysis showed the following compounds: lupeol (RT 97) and friedelin (RT 104) (Figure 2B), which had the standard specific peaks of each compound. Thus, friedelin and lupeol were the major phytochemical components found in the *n*-hexane extract of the fresh stem bark of *H. integrifolia*.

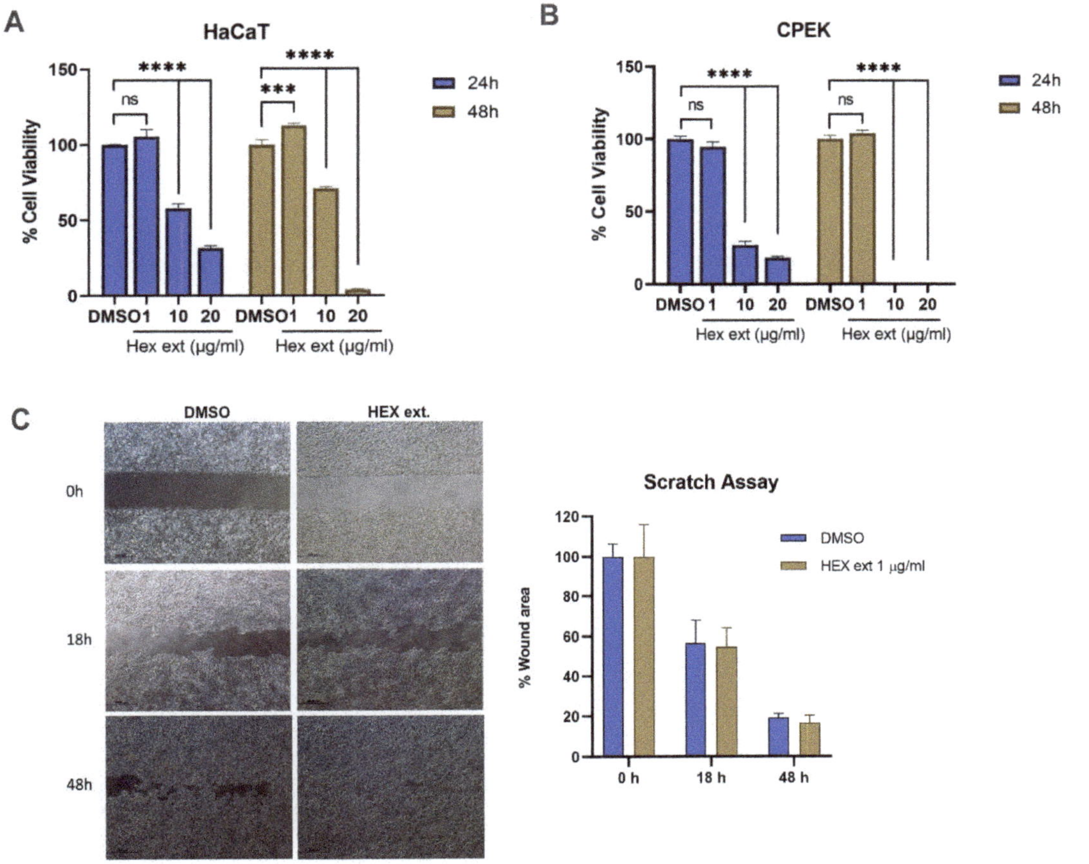

Figure 1. Cell growth inhibition by *H. integrifolia* in human and canine keratinocytes. (**A**) Human keratinocyte (HaCaT) cells and (**B**) canine keratinocyte (CPEK) cells treated with various concentrations of *H. integrifolia n*-hexane extract. The percentage of cell viability is presented as the mean of cell viability with standard error of the mean (SEM) compared to the DMSO treatment. Statistical significance is indicated as *** $p < 0.001$, **** $p < 0.0001$, ns = not significant. (**C**) The effect of 1 µg/mL of *n*-hexane extracts on wound healing using the scratch assay.

2.3. Effect of Friedelin and Lupeol on Cell Viability and Wound-Healing Activity in Keratinocytes

Human and canine keratinocytes were treated with various concentrations of friedelin (0.01 to 4 µM) and lupeol (0.2 to 20 µM). The cell viability results showed that the percentage of cell survival in 4 µM friedelin-treated HaCaT cells decreased to 57.7% and 65.3% at 24 and 48 h, respectively (Figure 3A, left); 20 µM lupeol-treated HaCaT cells viability decreased to 76.0% and 71.15% at 24 and 48 h, respectively. For the canine keratinocytes, 0.4 µM friedelin reduced canine keratinocyte viability to 49.3% and 59.2% at 24 and 48 h, respectively (Figure 3B, left). Lupeol at concentration 20 µM decreased canine keratinocytes cell viability to 63.1% and 54.7%, respectively, while lupeol at concentration 2 µM decreased it to 82.1% (Figure 3B, right). A scratch assay was conducted using friedelin and lupeol at non-toxic doses. Results revealed that friedelin (0.04 µM) and lupeol (0.2 µM) reduced the percentage of wound area by 78.84% ($p < 0.01$) and 11.17% ($p < 0.001$) at 48 h, respectively, compared to the DMSO treatment (Figure 3C). Thus, friedelin and lupeol treatment may facilitate wound closure after 48 h of treatment.

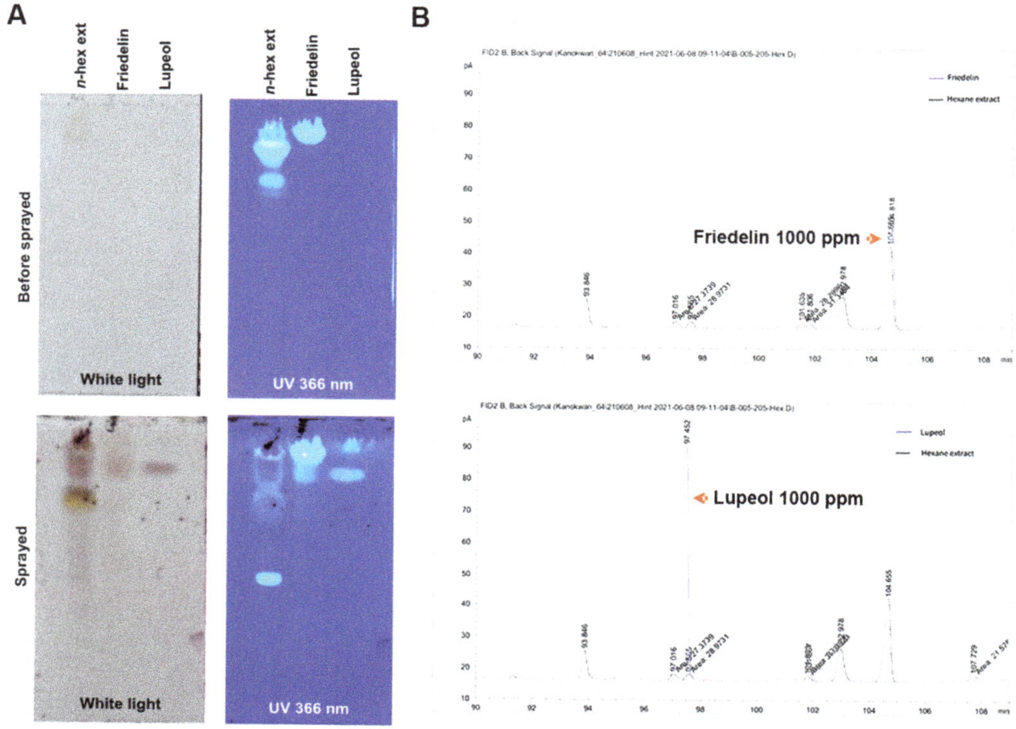

Figure 2. Chemical analysis of fresh stem bark *n*-hexane extract of *H. integrifolia*. (**A**) Thin-layer chromatography (TLC) shows two triterpenoids, friedelin and lupeol, in fresh stem bark *n*-hexane extraction of *H. integrifolia*. The upper panel indicates the TLC sheet before being sprayed with anisaldehyde-sulfuric acid reagent, whereas the lower panel presents the TLC sheet after being sprayed. (**B**) Gas chromatograph of friedelin and lupeol in the *n*-hexane crude extract of *H. integrifolia*. The blue line represents a peak of the standard compound.

2.4. MMPs Expression by Friedelin

The MMPs are well-known proteins that play important roles in wound-healing activity [11,12]. To investigate the effect of friedelin on proteins belonging to the MMP family involved in keratinocyte re-epithelialization, a human MMP array was performed. As shown in Figure 4A, friedelin treatment increased most MMPs and TIMPs, with the highest increase in the density of MMP-9 protein (2.36-fold increase) compared with the control (DMSO) treatment. The quantitative density of the protein was plotted as fold change compared to that of the control treatment (Figure 4B). To confirm the antibody array data, friedelin (0.04 μM) and lupeol (0.2 μM) were treated into HaCaT cells and the conditioned medium was collected to determine the activity of MMP-9 using gelatin zymography assay. TGF−β1 (0.5 μM) treatment was used as a positive control and the DMSO treatment was used as a negative control. As shown in Figure 4C, the activity of MMP-9 was significantly increased using lupeol (0.2 μM) treatment to 3.54-fold ($p < 0.01$) after 24 h and 4.66-fold ($p < 0.0001$) after 48 h, respectively, in a dose-dependent manner, when compared with the negative control. However, following friedelin treatment, a significant increase in MMP-9 activity (3.1-fold) was observed only after 48 h of treatment.

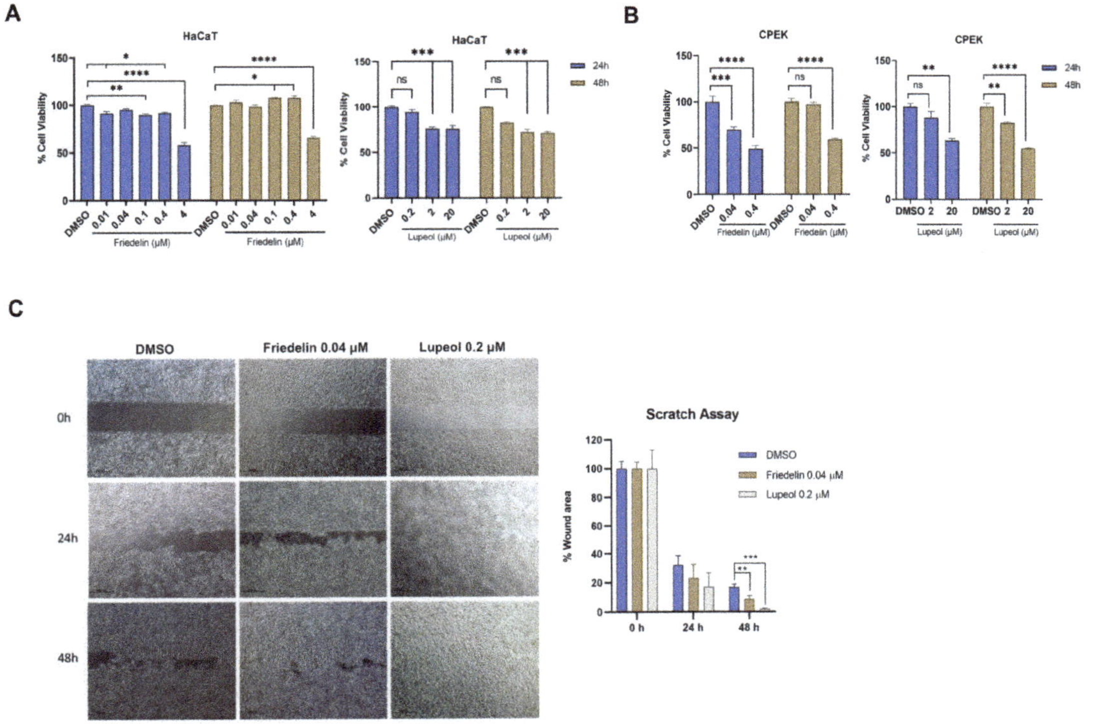

Figure 3. The cell viability of friedelin and lupeol on human and canine keratinocyte cells. Cell viability of human keratinocyte cells (**A**) and canine keratinocyte cells (**B**) treated with various concentrations of friedelin and lupeol. The percentage of cell viability was calculated as the percentage of DMSO treatment. The mean of cell viability is plotted with mean ± SEM. Statistical analysis was performed using one-way ANOVA, * $p < 0.05$, ** $p < 0.01$, *** $p < 0.001$, **** $p < 0.0001$. ns = not significant. (**C**) The effect of friedelin (0.04 μM) and lupeol (0.2 μM), on wound healing using scratch assay at 0, 24, and 48 h. The adjacent graph represents the percentage of the wound area. Statistical significance is indicated as ** $p < 0.01$, *** $p < 0.001$, ns = not significant compared to the control.

2.5. Effect of Friedelin/Lupeol on Gene Expression in Human Keratinocyte Re-Epithelialization

Desmosomes, adhesions, and structural components, cadherin 1 (CDH-1) and desmoglobin 1 (DSG-1), contribute to keratinocyte migration [13]. In addition, keratin-17 (KRT-17) has been associated with wound-healing activity in keratinocyte proliferation [14,15]. HaCaT cells were treated with 0.04 μM friedelin or 0.2 μM lupeol for 24 and 48 h, respectively, and the RNA expression of CDH-1, DSG-1, and KRT-17 was evaluated to determine the compound-altered gene expression. As shown in Figure 5, friedelin-treated cells significantly decreased CDH-1 gene expression by 0.27-fold ($p < 0.0001$) and 0.56-fold ($p < 0.001$) compared with the DMSO control at 24 and 48 h, respectively, whereas lupeol-treated cells at 48 h showed a significant decrease in CDH-1 expression to 0.29-fold ($p < 0.0001$). DSG-1 gene expression significantly decreased after 48 h of treatment with friedelin, however, not in lupeol-treated cells. The expression of KRT-17 was significantly increased by 2.75-fold ($p < 0.01$) at 48 h after lupeol treatment, compared with the DMSO control.

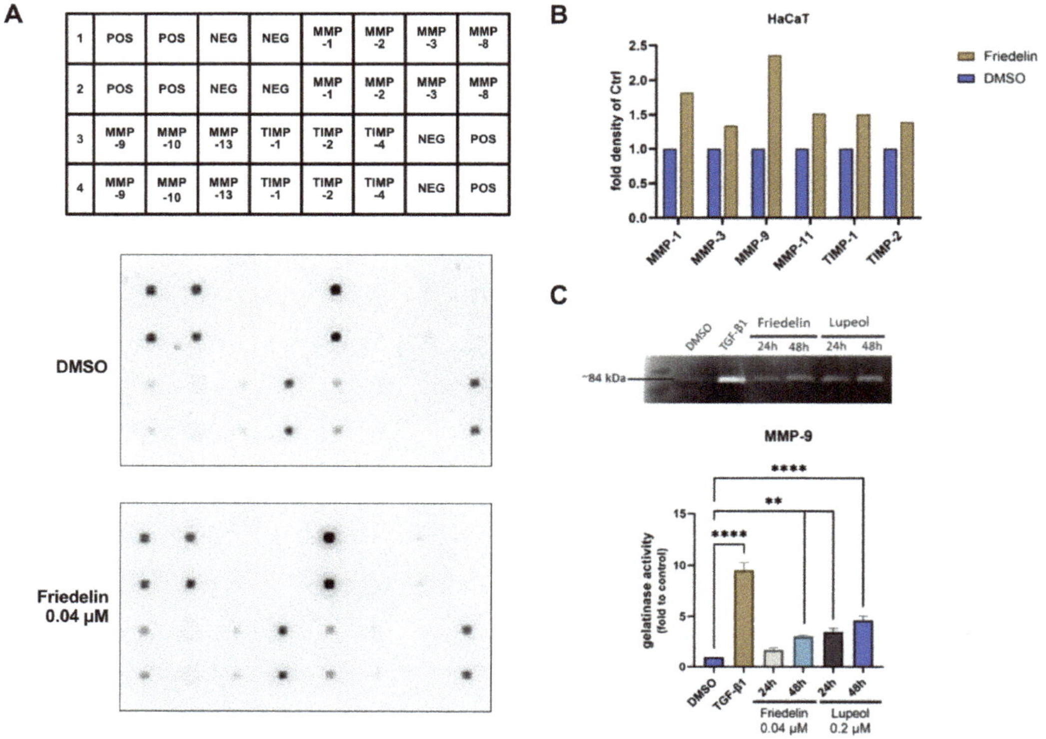

Figure 4. Antibody array analysis using friedelin-treated HaCaT cells. HaCaT cells were treated with 0.04 µM friedelin for 24 h; the conditioned media were harvested; and the antibody array was conducted using the human matrix-metalloproteinase (MMPs) family array. (**A**) The membrane indicates the original membrane after hybridization with conditioned media from either control-treated or friedelin-treated cell. (**B**) The graph represents the fold increase in MMPs during the MMP array, which were calculated compared to the control-treated sample. (**C**) The effect of friedelin and lupeol on MMP-9 activity was determined using gelatin zymography. The graph presents the MMP-9 activity compared to the control treatment. Statistical analysis was performed using one-way ANOVA, ** $p < 0.01$, **** $p < 0.0001$.

2.6. *Effect of H. integrifolia n-Hexane Extract and Friedelin/Lupeol on Anti-Inflammatory Activity in RAW 264.7 Cells*

Wound-healing activity is associated with the production of inflammatory cytokines and their clearance by macrophages [16]. To determine whether the extract and its bioactive compounds exhibited anti-inflammatory activity, LPS-induced RAW macrophage cells were used as a positive control. The n-hexane extracts (2.5 and 25 µg/mL) were used to treat the LPS-induced RAW cells. qRT-PCR results indicated that expression of genes involved in two inflammatory enzymes, COX-2 and iNOS, and two pro-inflammatory cytokines, TNF-α and IL-6, were downregulated in extract-treated cells compared to the positive control LPS-treated cells (Figure 6A). Similarly, 1 and 20 µM friedelin/lupeol treatment resulted in reduced COX-2, iNOS, TNF-α, and IL-6 gene expression (Figure 6B). To confirm the qRT-PCR data, the protein expression of the inflammatory enzymes, COX-2 and iNOS, was evaluated using Western blotting. As shown in Figure 6C, friedelin and lupeol treatment decreased LPS-induced COX-2 and iNOS expression in a dose-dependent manner.

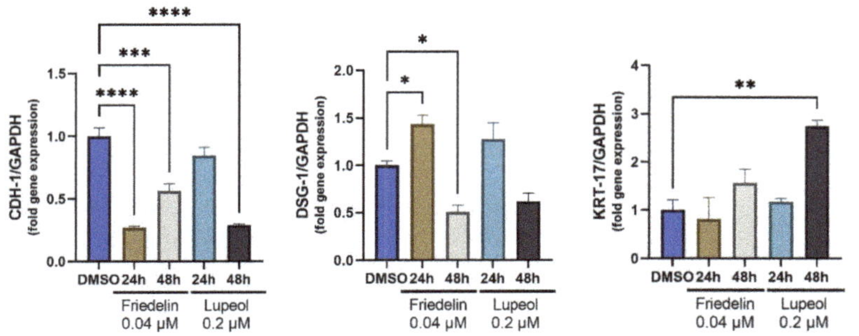

Figure 5. The effect of friedelin and lupeol on the expression of cell–cell adhesion genes (cadherin 1 [*CDH-1*] and desmoglein 1 [*DSG-1*]) and keratinocyte proliferative gene (keratin 17 [*KRT17*]). HaCaT cells were treated with indicated compounds for 24 and 48 h, respectively. The relative gene expression was calculated with housekeeping genes (*Gapdh)* as internalized control and the treated sample was compared with the DMSO control. Mean of fold expression to control are presented with Standard Error of the Mean (SEM). Statistical significance is indicated as * $p < 0.05$, ** $p < 0.01$, *** $p < 0.001$, **** $p < 0.0001$.

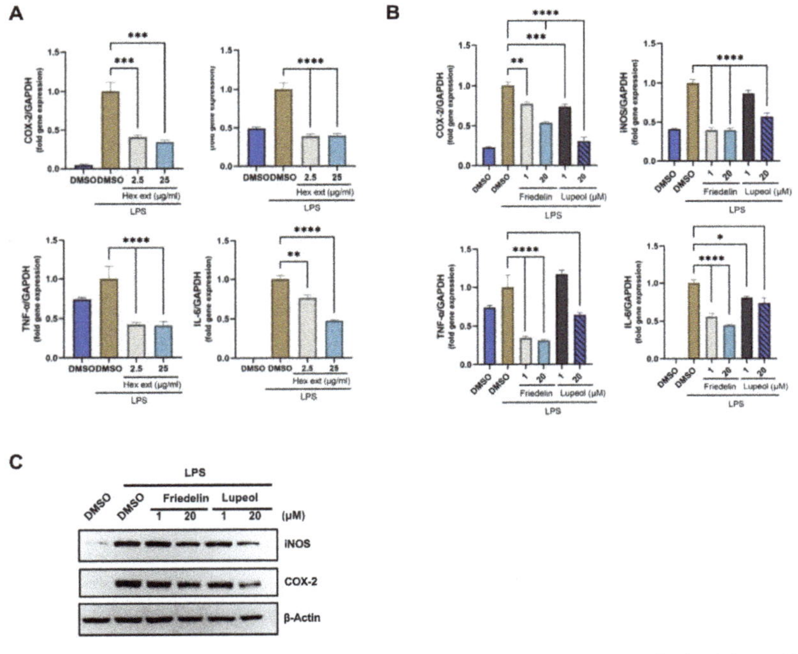

Figure 6. Effect of *H. integrifolia n*-hexane extract and its bioactive compounds, friedelin and lupeol, on anti-inflammatory activity using RAW 264.7 macrophage cells stimulated by LPS. The *H. integrifolia n*-hexane crude extracts (**A**) and its bioactive compounds friedelin and lupeol (**B**) were treated for 24 h. Anti-inflammatory gene expression, including COX-2, iNOS, TNF-α, and IL-6, was measured using qRT-PCR. The expression was plotted as mean ± SEM. (**C**) Western blot analysis of the anti-inflammatory effect on compound treated LPS induced RAW cells. After treatment with compounds for 24 h, immunoblotting was conducted using COX-2, iNOS antibodies. The results demonstrate the decrease in inflammatory marker proteins in dose-dependent manners. This blot is representative of three independent experiments. Statistical significance is indicated as * $p < 0.05$, ** $p < 0.01$, *** $p < 0.001$, **** $p < 0.0001$.

3. Discussion

The bark of *H. integrifolia* (Roxb.) Planch is used for the treatment of inflammatory diseases and applied dermally on wound lesions by Indian tribes [1]. Triterpenes/triterpenoids and sterols are the major bioactive compounds identified in this plant; however, many other compounds may be biologically active. Although *H. integrifolia* (Roxb.) Planch has potential for wound healing [7] and in vivo anti-inflammatory activity [17–19], the molecular mechanisms by which it affects the wound healing process remain unclear. Therefore, in this study, we demonstrated that increased MMP-9 activity, modulated gene expression involved in re-epithelization, and anti-inflammatory activity are key players in the wound-healing activity of *H. integrifolia*.

We identified friedelin and lupeol as the phytochemical components present in the *n*-hexane extract. Friedelin (or friedelan-3-one) has also been identified in the non-polar soluble fraction of *H. integrifolia* stem bark extract [1,9] and is a potent anti-inflammatory compound [20,21]. Lupeol is commonly found in many plants, including *H. integrifolia* stem bark extracts [3,22]. The biological activity of lupeol has anti-oxidizing, anti-cancerous, anti-inflammatory, and wound healing properties [23,24]. Thus, friedelin and lupeol may be promising compounds for wound-healing activity and elucidating their molecular mechanism may contribute to the development of wound healing applications.

A critical process in wound healing is re-epithelialization, which shows evident activity in the proliferative phase and is continuously active until the extracellular matrix remodeling phase. Chronic non-healing wounds can develop because of unsuccessful re-epithelialization. The scratch assay represents cell proliferation and migration [25], and our results suggest that *n*-hexane extract (1 μg/mL) does not significantly reduce wound scratches. However, friedelin and lupeol significantly reduced the in vitro wound scratches (Figure 3). This could be explained by the fact that 1 μg/mL of the extracts used in our study showed insufficient dose to exhibit wound-healing activity, whereas friedelin and lupeol are required at nanomolar concentrations to exhibit wound-healing activity.

During the early progression of wound healing, MMPs production, secreted by keratinocytes, is increased. MMPs initiate the degradation of the extracellular matrix to allow keratinocyte migration. We evaluated the effects of the extract and its active compounds on proteins in the MMP family using a human MMP array. Results revealed that friedelin treatment caused the highest increase in MMP-9 compared to the control treatment (Figure 4A,B). Several studies have indicated that MMP-1 and MMP-9 secreted from migrating keratinocytes promote cell migration and re-epithelialization by degrading cell–cell adhesion proteins [26]. Keratinocytes are the main constituents of the epidermis that re-form the epidermal layer at the wound site and secrete MMP-9 during wound healing progression [5]. Gelatin zymography results demonstrated that MMP-9 activity was increased by friedelin and lupeol treatment (Figure 4C), thereby enhancing re-epithelialization.

In addition, at the middle and later stages of wound repairing, keratinocytes detach from each other to allow cell migration and re-form the epidermal layer at the wound site. The expression of *KRT-17* increases cell growth, resulting in cell hyperproliferation during wound healing [27]. A previous study has suggested that *KRT-17* contributes to the restoration of epidermal permeability barrier in keratinocyte homeostasis [28]. Our results demonstrated altered expression of *KRT-17* with other genes (*CDH1* and *DSG-1*) in friedelin- and lupeol-induced keratinocytes, confirming the role of friedelin/lupeol in re-epithelialization during the healing process.

Inflammation is a critical complication of tissue injury and wound lesions, which can lead to serious conditions. Macrophages play crucial roles in inflammatory stimuli by producing proinflammatory cytokines. Downregulated transcription of proinflammatory genes results in a decrease in the inflammatory response [29]. Therefore, to investigate the effects of *H. integrifolia* *n*-hexane extract and its bioactive compounds on inflammation, we evaluated the inflammatory enzymes and cytokines. Our results revealed the beneficial effects of *n*-hexane extract and a single compound in LPS-induced RAW cells by downregulating the expression of associated inflammatory genes. Our results are consistent with those of previous studies, indicating that lupeol downregulates the gene expression of TNF-α and IL-6 [30,31]. In addition, 20 μM lupeol reduced the protein expression of COX-2 and iNOS, which induced an inflammatory state to increase prostaglandin production.

Similar to the observations of *n*-hexane extracts, the single bioactive compounds, friedelin and lupeol, illustrated notable benefits for wound healing and anti-inflammatory activity. A detailed study of the pharmacology and action mechanism of *H. integrifolia* may help in understanding the relationship between its pharmacological effects and traditional uses in the future.

In conclusion, our results demonstrated that the fresh stem bark extract of *H. integrifolia* and its single bioactive compounds, friedelin and lupeol, exhibited wound-healing activity via alteration of gene expression related to re-epithelialization, increased MMP-9 activity, and enhanced anti-inflammatory activity.

4. Materials and Methods

4.1. Reagents and Cell Lines

Lupeol (#L5632) and friedelin (#92187) were purchased from Sigma-Aldrich (St. Louis, MA, USA). Human immortalized keratinocyte (HaCaT) cells were purchased from Cell Line Service (Eppelhein, Germany), and canine immortalized keratinocyte (CPEK) cells were kindly provided by Dr. Cheol-Yong Hwang (College of Veterinary Medicine, Seoul National University). The murine macrophage cell line, RAW 264.7, was purchased from American Type Culture Collection (ATCC, Manassas, VA, USA). All cell lines were maintained in Dulbecco's modified Eagle medium (DMEM, Gibco, Grand Island, NY, USA) containing 10% fetal bovine serum (FBS), supplemented with 100 U/mL penicillin and 100 mg/mL streptomycin (Gibco), at 37 °C in an incubator with a humidified atmosphere of 5% CO_2.

4.2. Plant Preparation and Extraction

Fresh bark of *H. integrifolia* was collected from Soem Ngam, Lampang, in the northern part of Thailand from March 2019 to August 2019. All specimens were identified by Dr. Narin Printarakul, a taxonomist at the Department of Biology, Faculty of Science, Chiang Mai University. Voucher specimen numbers HI001-003 were maintained at the Chiang Mai University Herbarium. The plant samples were chopped and grounded into small pieces and then separately macerated with various solvents, including *n*-hexane, dichloromethane, ethanol, methanol, and deionized water in a ratio of 1:10 *w*/*v* on a shaker at room temperature overnight. The macerated solution was filtered using Whatman filter paper (Thermo Fisher Scientific, Waltham, MA, USA) before evaporation on a rotary evaporator or water bath for water extraction. The crude extracts were then stored at 4 °C.

4.3. Thin-Layer Chromatography (TLC)

TLC was performed with fresh stem bark extracts of *H. integrifolia* and compared to phytochemicals, such as terpenoids, flavonoids, and phenolic compounds; 1 mg/mL of each extract was spotted on a TLC sheet (TLC silica gel 60F245, Merck KGaA, Darmstadt, Germany) with toluene/ethyl acetate/ethanol (8:6:1, *v*/*v*). The TLC sheet was sprayed with anisaldehyde-sulfuric acid reagent (0.5% *p*-anisaldehyde, 10% glacial acetic acid, 85% methanol, and 5% sulfuric acid). After air-drying, the sheet was heated at 100 °C for 5–10 min.

4.4. Gas Chromatography Analysis (GC-FID)

The *n*-hexane extract and standard compounds were analyzed using a gas chromatograph (Agilent 7890 B, Agilent Technologies, Santa Clara, CA, USA). The sample (1 μL) was injected into the column and split at a ratio of 1:10 (*v*/*v*) at 250 °C. The injected volume was separated in HP-5 column (length 30 m, diameter 0.320 mm, and film thickness 0.25 μm) at 100 °C, then increased by 4 °C/min until 270 °C and held at 270 °C for 125 min. The flame ionization detector was programmed to operate at 270 °C. The chromatogram retention-time result of the plant extract and a single compound was compared.

4.5. Cell Viability Assay

All cell lines were seeded at 5000 cells/well in a 96 well-plate. The cells were then treated with various concentrations of plant extracts and single compounds for 24 h. The cell viability assay was performed using the CellTiter 96® Aqueous One Solution Cell Proliferation Assay kit (Promega, Madison, WI, USA). The optical density was measured using a Multiskan™ FC microplate spectrophotometer (Thermo Fisher Scientific) at a wavelength of 492 nm. Cell viability was then calculated and compared with that of the control group.

4.6. Immunoblotting Analysis

The treated cells were washed twice with cold phosphate-buffered saline (PBS), and proteins were extracted using RIPA cell lysis buffer (#R4100, GenDEPOT, TX, USA) with proteinase inhibitor (#BPI001, Biomax, Seoul, Republic of Korea). The protein concentration was measured using the Pierce™ BCA protein assay kit (#23225, Thermo Fisher Scientific), and 40 μg of each protein sample was heated at 98 °C for 5 min and then run through 8% SDS-PAGE gel. The protein samples were then transferred to a 0.22 μm PVDF membrane (#GE10600021, GE Healthcare Life Sciences, Solingen, Germany) and blocked with 5% non-fat milk in 0.1% TBS-T for 1 h. The blotted membrane was probed using specific antibodies against COX-2 (#sc-166475, Santa Cruz Biotechnology, Dallas, TX, USA) and iNOS (#13120s, Cell Signaling Technology, Danvers, MA, USA), 1:1000 dilution with 5% non-fat milk in 0.1% TBS-T at 4 °C overnight. This was followed by washing thrice with 0.1% TBS-T solution for 10 min and probing with secondary antibodies (goat anti-rabbit IgG [H + L] secondary antibody, HRP, #31460, 1:5000 dilution, Thermo Fisher Scientific) for 2 h at RT. The blots were developed using Pierce™ ECL Western blotting Substrate (#32106, Thermo Fisher Scientific), visualized using Alliance Q9 mini (Cambridge, UK), and quantified using ImageJ software 1.52a (National Institutes of Health, Bethesda, MD, USA).

4.7. Scratch Assay

HaCaT cells were seeded in six-well plates and incubated until they reached confluence. Scratches were developed using a sterile pipette tip, and the cells were washed twice with PBS to remove loose cells and debris. The cells were treated with *n*-hexane extract (1 μg/mL), friedelin (0.04 μM), and lupeol (0.2 μM). Wound area progression was observed using a light microscope (Nikon Ti-U; Nikon Instruments, Tokyo, Japan). The wound area was determined using the ImageJ software with the wound healing size tool plugin for ImageJ/Fiji® [32].

4.8. Antibody Array

The RayBio ® C-Series Human MMP Antibody Array C1 was purchased from RayBiotech (Peachtree Corners, GA, USA). HaCaT cells were seeded in six-well plates until they reached 80% confluency. Cells were treated with 0.04 μM friedelin and 1% DMSO in serum-free medium for 24 h. The conditioned medium was collected and centrifuged at 4500 rpm for 5 min to remove cell debris. The antibody array was prepared according to the manufacturer's instructions. The membranes were developed for chemiluminescence detection using the Alliance Q9 advanced imaging system, and the intensity was measured using ImageJ software. Positive control spots were used to normalize the signal intensity between treatments.

4.9. Gel zymography Assay

HaCaT cells were seeded in six-well plates until they reached 80% confluence. The cells were treated with friedelin (0.04 and 0.4 μM) and lupeol (0.2 and 2 μM) in DMEM with 5% FBS for 24 h and 48 h. Gelatinase activity was determined according to a previous study [33]. Conditioned media were collected and spun down to remove cell debris. The protein from the conditioned medium was separated using gel electrophoresis in a 0.5% gelatin–8% acrylamide gel and then washed twice with 2.5% Triton X-100 solution

(#T8787, Sigma-Aldrich) for 60 min. The gel was incubated with enzyme activating buffer (0.05 M Tris, 0.02 M NaCl, 0.005 M $CaCl_2$, and 0.02% sodium azide) at 37 °C overnight. It was then stained using 0.2% Coomassie blue (Coomassie blue R 250 solution, #CBC006, LPS solution, Daejeon, Republic of Korea) and washed with de-staining buffer (10% acetic acid in 40% methanol) until the clear band was visible. The image was captured using the Alliance Q9 advanced imaging system, and the gelatinase activity was measured using ImageJ software.

4.10. Quantitative Reverse Transcription Polymerase Chain Reaction (qRT-PCR)

Total RNA was extracted from HaCaT cells according to the manufacturer's protocol using TRIzol reagent (#15596018; Invitrogen, MA, USA). RNA (1 μg) was reverse transcribed using the Verso cDNA Synthesis Kit (Thermo Fisher Scientific). The cDNAs were amplified using a MiniAmp Plus Thermal Cycler (A37835; Applied Biosystems, Waltham, MA, USA). Target gene expression was examined through qRT-PCR using SYBR Green reagents (PowerUp SYBR Green Master Mix, A25741, Applied Biosystems) with a QuantStudio 1 real-time PCR system (Applied Biosystems), according to the manufacturer's instructions. The list of primer sequences for each specific gene is shown in Table 1.

Table 1. List of the human primers used in qRT-PCR.

human KRT-17	sense strand	5′-GAGATTGCCACCTACCGCC-3′
	anti-sense strand	5′-ACCTCTTCCACAATGGTACGC-3′
human DSG-1	sense strand	5′-GAGATTGCCACCTACCGCC-3′
	anti-sense strand	5′-ACCTCTTCCACAATGGTACGC-3′
human CDH-1	sense strand	5′-GCTGGACCGAGAGAGTTTCC-3′
	anti-sense strand	5′-CAAAATCCAAGCCCGTGGTG-3′
human GAPDH	sense strand	5′-GTCTCCTCTGACTTCAACAGCG-3′
	anti-sense strand	5′-ACCACCCTGTTGCTGTAGCCAA-3′

RAW 264.7 macrophage cells were seeded in a 35 mm cell culture dish until the confluency reached 80%. The cells were then treated with 1 μg/mL lipopolysaccharide (LPS; #L4391, Sigma-Aldrich) in serum-free medium for 1 h to induce inflammation. Subsequently, the cells were treated with 2.5 μg/mL and 20 μg/mL of *n*-hexane extract, friedelin, and lupeol for 18 h. The total RNA was extracted using a PureLink™ RNA Mini Kit (#12183020, Invitrogen). Two microgram of the total RNA was converted to cDNA using a Tetro cDNA Synthesis Kit (#BIO-65042, Bioline, Toronto, Canada) and a Thermal Cycler Labcycler (Senso Quest, Göttingen, Germany). To determine mRNA expression, cDNA was mixed with the SensiFAST™ SYBR No-ROX Kit (#BIO-98005, Bioline, Toronto, Canada) and amplification was performed using iCycler IQ5 (Bio-rad, Hercules, CA, USA). The mRNA expression levels of target genes were normalized to that of *Gapdh*. The relative gene expression was calculated using the comparative Ct ($2^{-\Delta\Delta Ct}$) method. The primer sequences for each specific gene are shown in Table 2.

Table 2. List of the mouse primers used in qRT-PCR.

mouse COX-2	sense strand	5′–CCCCCACAGTCAAAGACACT-3′
	anti-sense strand	5′–GAGTCCATGTTCCAGGAGGA-3′
mouse iNOS	sense strand	5′–GTCTTGCAAGCTGATGGTC-3′
	anti-sense strand	5′–CATGATGGTCACATTCTGC-3′
mouse TNF-α	sense strand	5′–GCCTCTTCTCATTCCTGCTTG-3′
	anti-sense strand	5′–CTGATGAGAGGGAGGCCATT-3′
mouse IL-6	sense strand	5′–TACCACTTCACAAGTCGGAGGC-3′
	anti-sense strand	5′–CTGCAAGTGCATCATCGTTGTTC-3′
mouse GAPDH	sense strand	5′–CAGGAGCGAGACCCCACTAACAT-3′
	anti-sense strand	5′–GTCAGATCCACGACGGACACATT-3′

4.11. Statistical Analysis

Data were expressed as mean ± SD or SEM from at least three independent experiments. Statistical analyses were performed using GraphPad Prism 9.0.0 software. Significant differences were indicated by * $p < 0.05$, ** $p < 0.01$, *** $p < 0.001$, and **** $p < 0.0001$.

Supplementary Materials: The following supporting information can be downloaded at: https://www.mdpi.com/article/10.3390/molecules27238540/s1, Figure S1: The TLC of phytochemical components of *H. integrifolia* extract with various solvents.

Author Contributions: Conceptualization, K.S., H.P., W.G. and S.J.B.; methodology, K.S., P.L., W.N., A.P. and Y.T.; data analysis, H.P., R.Y. and S.J.B.; writing—original draft preparation, K.S. and S.J.B.; All authors have read and agreed to the published version of the manuscript.

Funding: This research was funded by National Research Foundation of Korea (NRF) grant funded by the Korean government (Grant number NRF-2020R1A2B2002923), the Development and Promotion of Science and Technology Talents Project (DPST) Scholar in Thailand, and Technology to Industry Convergence, Northern Science Park, Science Park Promotion Agency, Thailand.

Institutional Review Board Statement: Not applicable.

Informed Consent Statement: Not applicable.

Data Availability Statement: Not applicable.

Acknowledgments: The authors would like to thank the members of the Signal Transduction Laboratory at the College of Veterinary Medicine, Seoul National University (Seoul, Republic of Korea) for their kind assistance and technical support.

Conflicts of Interest: The authors declare no conflict of interest.

References

1. Ganie, S.A.; Yadav, S.S. *Holoptelea integrifolia* (Roxb.) Planch: A review of its ethnobotany, pharmacology, and phytochemistry. *BioMed Res. Int.* **2014**, *2014*, 401213. [CrossRef] [PubMed]
2. Sutar, R.; Kasture, S.; Kalaichelvan, V. Isolation and identification of a new phytosterol from *Holoptelea integrifolia* (ROXB) planch leaves. *Int. J. Pharm. Sci. Res.* **2014**, *6*, 354–357.
3. Nadella, D.; Paarakh, P.; Vedamurthy, A. Isolation of phytoconstituents from the stem bark of *Holoptelea integrifolia* (Roxb.) Planch. *J. Pharm. Res.* **2012**, *5*, 532–533.
4. Rousselle, P.; Braye, F.; Dayan, G. Re-epithelialization of adult skin wounds: Cellular mechanisms and therapeutic strategies. *Adv. Drug Deliv. Rev.* **2019**, *146*, 344–365. [CrossRef] [PubMed]
5. Singer, A.J.; Clark, R.A. Cutaneous wound healing. *N. Engl. J. Med.* **1999**, *341*, 738–746. [CrossRef]
6. Wilkinson, H.N.; Hardman, M.J. Wound healing: Cellular mechanisms and pathological outcomes. *Open Biol.* **2020**, *10*, 200223. [CrossRef] [PubMed]
7. Reddy, B.S.; Reddy, R.K.K.; Naidu, V.; Madhusudhana, K.; Agwane, S.B.; Ramakrishna, S.; Diwan, P.V. Evaluation of antimicrobial, antioxidant and wound-healing potentials of *Holoptelea integrifolia*. *J. Ethnopharmacol.* **2008**, *115*, 249–256. [CrossRef]
8. Solanki, P.; Pandya, D.; Maitreya, B. Antimicrobial activity of methanolic & aqueous extracts of leaves & young stem bark of *Holoptelea integrifolia* L. *Int. J. Bot. Stud.* **2020**, *5*, 123–127.
9. Misra, G.; Bhatnagar, S.; Nigam, S. Constituents of *Holoptelea integrifolia* leaves and bark. *Planta Med.* **1974**, *26*, 394–396. [CrossRef]
10. Hassan, A.; Rasheed, M.; Ali, M.; Ishrat, G.; Ahmed, M. Identification of Five New Triterpenoids from Ethylacetate Bark Extract of *Holoptelea integrifolia* (Roxb.) Planch by GC-MS. *Nat. Prod. Chem. Res.* **2018**, *6*, 338. [CrossRef]
11. Sabino, F.; Keller, U. Matrix metalloproteinases in impaired wound healing. *Met. Med.* **2015**, *2*, 1–8. [CrossRef]
12. Caley, M.P.; Martins, V.L.; O'Toole, E.A. Metalloproteinases and wound healing. *Adv. Wound Care (New Rochelle)* **2015**, *4*, 225–234. [CrossRef] [PubMed]
13. Stone, R.C.; Pastar, I.; Ojeh, N.; Chen, V.; Liu, S.; Garzon, K.I.; Tomic-Canic, M. Epithelial-mesenchymal transition in tissue repair and fibrosis. *Cell Tissue Res.* **2016**, *365*, 495–506. [CrossRef] [PubMed]
14. Konop, M.; Rybka, M.; Drapała, A. Keratin biomaterials in skin wound healing, an old player in modern medicine: A mini review. *Pharmaceutics* **2021**, *13*, 2029. [CrossRef]
15. McGowan, K.M.; Coulombe, P.A. Onset of keratin 17 expression coincides with the definition of major epithelial lineages during skin development. *J. Cell Biol.* **1998**, *143*, 469–486. [CrossRef]
16. Raziyeva, K.; Kim, Y.; Zharkinbekov, Z.; Kassymbek, K.; Jimi, S.; Saparov, A. Immunology of acute and chronic wound healing. *Biomolecules* **2021**, *11*, 700. [CrossRef]

17. Sharma, P.; Nanjaian, M.; Kalra, S.; Chidambaram, K. In vivo anti-inflammatory and in vitro antioxidant potential of leaf and bark fractions of *Holoptelea integrifolia* (Roxb.) Planch. *J. Pharm. Pharmacogn. Res.* **2019**, *7*, 421–432.
18. Pratap, H.; Shakya, M.K.; Singh, T.; Agrawal, M.; Katiyar, N.S. A Study on Anti-inflammatory Activity of Stem Bark Extract of *Holoptelea integrifolia*. *Res. J. Pharm. Technol.* **2022**, *15*, 77–81. [CrossRef]
19. Munawar, T.M.; Rao, D.M.; Subramanyam, P. Antibacterial, antioxidant and anti-inflammatory potential of the different extracts of Holoptelia integrifolia. *Int. J. Eng. Adv. Technol.* **2019**, *9*, 5234–5240. [CrossRef]
20. Antonisamy, P.; Duraipandiyan, V.; Ignacimuthu, S. Anti-inflammatory, analgesic and antipyretic effects of friedelin isolated from *Azima tetracantha* Lam. in mouse and rat models. *J. Pharm. Pharmacol.* **2011**, *63*, 1070–1077. [CrossRef]
21. Nunes, R.; Broering, M.F.; De Faveri, R.; Goldoni, F.C.; Mariano, L.N.B.; Mafessoli, P.C.M.; Delle Monache, F.; Cechinel Filho, V.; Niero, R.; Santin, J.R. Effect of the metanolic extract from the leaves of *Garcinia humilis* Vahl (Clusiaceae) on acute inflammation. *Inflammopharmacology* **2021**, *29*, 423–438. [CrossRef] [PubMed]
22. Dixit, P.; Pal, M.; Upreti, D. Comparative studies on the analytical and antioxidant activities of the medicinally important stem bark of *Holoptelea integrifolia*. *JPC—J. Planar. Chromat.* **2014**, *27*, 162–165. [CrossRef]
23. Pereira Beserra, F.; Xue, M.; Maia, G.L.d.A.; Leite Rozza, A.; Helena Pellizzon, C.; Jackson, C.J. Lupeol, a pentacyclic triterpene, promotes migration, wound closure, and contractile effect in vitro: Possible involvement of PI3K/Akt and p38/ERK/MAPK pathways. *Molecules* **2018**, *23*, 2819. [CrossRef]
24. Pereira Beserra, F.; Sérgio Gushiken, L.F.; Vieira, A.J.; Augusto Bérgamo, D.; Luísa Bérgamo, P.; Oliveira de Souza, M.; Alberto Hussni, C.; Kiomi Takahira, R.; Henrique Nóbrega, R.; Monteiro Martinez, E.R. From inflammation to cutaneous repair: Topical application of lupeol improves skin wound healing in rats by modulating the cytokine levels, NF-κB, Ki-67, growth factor expression, and distribution of collagen fibers. *Int. J. Mol. Sci.* **2020**, *21*, 4952. [CrossRef] [PubMed]
25. Liang, C.-C.; Park, A.Y.; Guan, J.-L. In vitro scratch assay: A convenient and inexpensive method for analysis of cell migration in vitro. *Nat. Protoc.* **2007**, *2*, 329–333. [CrossRef]
26. Martins, V.L.; Caley, M.; O'Toole, E.A. Matrix metalloproteinases and epidermal wound repair. *Cell Tissue Res.* **2013**, *351*, 255–268. [CrossRef]
27. Zhang, X.; Yin, M.; Zhang, L.-j. Keratin 6, 16 and 17—Critical barrier alarmin molecules in skin wounds and psoriasis. *Cells* **2019**, *8*, 807. [CrossRef]
28. Pang, B.; Zhu, Z.; Xiao, C.; Luo, Y.; Fang, H.; Bai, Y.; Sun, Z.; Ma, J.; Dang, E.; Wang, G. Keratin 17 Is Required for Lipid Metabolism in Keratinocytes and Benefits Epidermal Permeability Barrier Homeostasis. *Front. Cell Dev. Biol.* **2022**, *9*, 779257. [CrossRef]
29. Lee, C.-H.; Choi, E.Y. Macrophages and inflammation. *J. Rheum. Dis.* **2018**, *25*, 11–18. [CrossRef]
30. Fernández, M.A.; de las Heras, B.; Garcia, M.D.; Sáenz, M.T.; Villar, A. New insights into the mechanism of action of the anti-inflammatory triterpene lupeol. *J. Pharm. Pharmacol.* **2001**, *53*, 1533–1539. [CrossRef]
31. Saha, S.; Profumo, E.; Togna, A.R.; Riganò, R.; Saso, L.; Buttari, B. Lupeol Counteracts the Proinflammatory Signalling Triggered in Macrophages by 7-Keto-Cholesterol: New Perspectives in the Therapy of Atherosclerosis. *Oxid. Med. Cell. Longev.* **2020**, *2020*, 1232816. [CrossRef] [PubMed]
32. Suarez-Arnedo, A.; Figueroa, F.T.; Clavijo, C.; Arbeláez, P.; Cruz, J.C.; Muñoz-Camargo, C. An image J plugin for the high throughput image analysis of in vitro scratch wound healing assays. *PLoS ONE* **2020**, *15*, e0232565. [CrossRef] [PubMed]
33. Toth, M.; Sohail, A.; Fridman, R. Assessment of gelatinases (MMP-2 and MMP-9) by gelatin zymography. *Methods Mol. Biol.* **2012**, *878*, 121–135. [CrossRef] [PubMed]

molecules

MDPI

Article

Angelica sinensis Polysaccharide and *Astragalus membranaceus* Polysaccharide Accelerate Liver Regeneration by Enhanced Glycolysis via Activation of JAK2/STAT3/HK2 Pathway

Xu-Dong Wen [1,†], Yao-Lei Zhang [2,†], Ling Yang [3], Zhen Ye [4], Guo-Chuan Fu [1], Yong-He Hu [2], Tao Pan [1,*] and Qiao-Bo Ye [4,*]

1 Department of Gastroenterology, Chengdu Integrated TCM&Western Medicine Hospital, Chengdu University of Traditional Chinese Medicine, Chengdu 610059, China
2 School of Materials Science and Engineering, Southwest Jiaotong University, Chengdu 610031, China
3 School of Clinical Medicine, Chengdu University of Traditional Chinese Medicine, Chengdu 610032, China
4 College of Basic Medical Sciences, Chengdu University of Traditional Chinese Medicine, Chengdu 610032, China
* Correspondence: pant414@163.com (T.P.); yeqiaobo@cdutcm.edu.cn (Q.-B.Y.)
† These authors contributed equally to this work.

Citation: Wen, X.-D.; Zhang, Y.-L.; Yang, L.; Ye, Z.; Fu, G.-C.; Hu, Y.-H.; Pan, T.; Ye, Q.-B. *Angelica sinensis* Polysaccharide and *Astragalus membranaceus* Polysaccharide Accelerate Liver Regeneration by Enhanced Glycolysis via Activation of JAK2/STAT3/HK2 Pathway. *Molecules* **2022**, *27*, 7890. https://doi.org/10.3390/molecules27227890

Academic Editors: Sokcheon Pak and Soo Liang Ooi

Received: 21 October 2022
Accepted: 9 November 2022
Published: 15 November 2022

Abstract: The promotion of liver regeneration is crucial to avoid liver failure after hepatectomy. *Angelica sinensis* polysaccharide (ASP) and *Astragalus membranaceus* polysaccharide (AMP) have been identified as being associated with hepatoprotective effects. However, their roles and specific mechanisms in liver regeneration remain to be elucidated. In the present study, it suggested that the respective use of ASP or AMP strikingly promoted hepatocyte proliferation in vitro with a wide range of concentrations (from 12.5 μg/mL to 3200 μg/mL), and a stronger promoting effect was observed in combined interventions. A significantly enhanced liver/body weight ratio (4.20%) on day 7 and reduced serum transaminase (ALT 243.53 IU/L and AST 423.74 IU/L) and total bilirubin (52.61 IU/L) levels on day 3 were achieved by means of ASP-AMP administration after partial hepatectomy in mice. Metabonomics showed that differential metabolites were enriched in glycolysis with high expression of beta-D-fructose 6-phosphate and lactate, followed by significantly strengthened lactate secretion in the supernatant (0.54) and serum (0.43) normalized to control. Upon ASP-AMP treatment, the knockdown of hexokinase 2 (HK2) or inhibited glycolysis caused by 2-deoxy-d-glucose decreased hepatocyte proliferation in vitro and in vivo. Furthermore, pathway analysis predicted the role of JAK2/STAT3 pathway in ASP-AMP-regulated liver regeneration, and phosphorylation of JAK2 and STAT3 was proven to be elevated in this promoting process. Finally, downregulated expression of HK2, an attenuated level of lactate secretion, and reduced hepatocyte proliferation were displayed when STAT3 was knocked out in vitro. Therefore, it can be concluded that ASP-AMP accelerated liver regeneration and exerted a hepatoprotective effect after hepatectomy, in which the JAK2/STAT3/HK2 pathway was actively involved in activating glycolysis.

Keywords: *Angelica sinensis* polysaccharide; *Astragalus membranaceus* polysaccharide; hexokinase 2; glycolysis; JAK2/STAT3 pathway; liver regeneration

1. Introduction

Liver failure is a potentially fatal complication after liver resection, with an incidence of about 7% in patients with healthy parenchyma and as high as 30% in those with cirrhosis [1,2]. Although the liver possesses an intrinsic regenerative capacity, it is challenging for the host to survive after major liver resection, with there being a high mortality rate [3]. Improving the liver's capacity to regenerate is crucial, as demonstrated by accumulating evidence surrounding this highly complex process [4]. Currently, there are no clinically available therapeutic agents that enhance liver regeneration, especially in a post-liver-resection setting [5].

To promote liver regeneration, we attempt to search for natural products from traditional Chinese medicines as an abundant resource to be utilized in drug development [6]. The Danggui Buxue Decoction (DBD) is composed of *Angelica sinensis* and *Astragalus membranaceus* at the ratio of 1:5 and has been widely used as a Chinese herbal medicine and functional food for thousands of years [7,8]. The DBD has also been proven to exert a protective effect on hepatic fibrosis [9] and have tissue regeneration capability [10]. As major active ingredients from *Angelica sinensis* and *Astragalus membranaceus*, *Angelica sinensis* polysaccharide (ASP) and *Astragalus membranaceus* polysaccharide (AMP) possess a variety of beneficial pharmacological activities towards the liver [11,12]. Evidence indicates that ASP and/or AMP exert anti-fibrosis [12] and hepatoprotection [13,14] activities. These results strongly support the potential for liver regeneration promotion by means of the combined use of ASP and AMP (ASP-AMP).

The liver maintains glucose homeostasis, in which gluconic biomolecules are dynamically regulated to provide energy as well as the anabolic molecules required for liver regeneration [15]. Metabonomics indicates that liver metabolites change remarkably before and after hepatectomy [16]. Interestingly, polysaccharides have been proven to activate liver glycometabolism-related signaling pathways in diabetes mellitus rats [17]. AMP showed beneficial effects in lowering blood glucose and triglyceride levels [18]. Meanwhile, during the regeneration process, hexokinase 2 (HK2) is actively involved in glycolysis, the inhibition of which leads to reduced proliferation of hepatic cells [19]. These observations suggest that HK2-mediated glycometabolism might be directly related to ASP-AMP-assisted liver regeneration. In addition, ASP and AMP have recently been proposed to be involved in various pathway regulations via triggering a variety of molecular responses in the liver [11,20]. ASP induces IL-22 secretion in liver tissue via the activation of the signal transducers and activators of transcription (STAT3) pathway to attenuate liver fibrosis [21]. AMP diminished the formation of free radicals and cytokines by the downregulation of STAT3 expression [22]. To date, the regulating effect of ASP and AMP on liver regeneration has not been fully elucidated, and the potential mechanism in metabolic alternations remains to be explored.

The present study aimed to reveal the potential mechanism by which ASP-AMP promoted liver regeneration (Scheme 1). First, the effect of ASP and/or AMP on cell proliferation was evaluated in vitro and in vivo. Second, metabonomics was performed on hepatocytes and liver tissue treated with ASP-AMP to investigate the differential metabolites. Finally, the predicted pathways from network pharmacology were verified and the specific mechanism by which ASP-AMP promoted liver regeneration was analyzed.

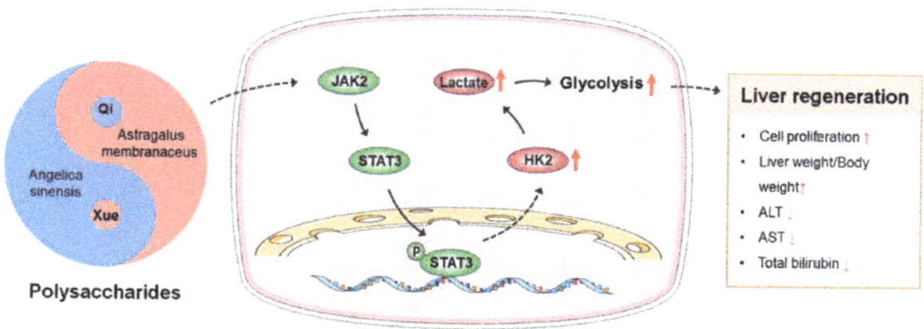

Scheme 1. Schematic illustration of combined use of *Angelica sinensis* polysaccharide and *Astragalus membranaceus* polysaccharide promoting liver regeneration via the activation of the JAK2/STAT3/HK2 pathway and enhanced glycolysis.

2. Results

2.1. ASP-AMP Promoted Hepatocytes Proliferation In Vitro

To confirm the combined effect of ASP and AMP on hepatocytes, two normal liver cell lines were first treated with ASP or AMP with different concentrations. NCTC clone 1469 was cultured with ASP at concentrations ranging from 12.5 μg/mL to 3200 μg/mL for 7 days. The ASP (12.5 μg/mL) significantly enhanced the proliferation of NCTC clone 1469 ($p < 0.01$), but higher concentrations did not promote or even reduce the proliferation from day 1 to day 7 (Figure 1a). Meanwhile, significant differences were observed in AMP-treated groups on day 1, day 3, day 5, and day 7 ($p < 0.001$, Figure 1b). The ability of AMP to promote proliferation improved with the increase in concentration, and this promoting ability reached its peak when the concentration was 100 μg/mL. Furthermore, the promoting effect at the optimal concentration (ASP 12.5 μg/mL and AMP 100 μg/mL) was verified by the EdU assay, and it demonstrated that the ratios of red to blue points were significantly higher in ASP (7.01%) and AMP (7.58%) groups than that in the control group (5.03%) on day 3 ($p < 0.05$, Figure 1c,d). Simultaneously, the proliferation effect of ASP or AMP was verified in BRL-3A. Similar to the results in NCTC clone 1469 in Figure 1a, a relatively low dose of ASP (100 μg/mL) had the most promoting effect on BRL-3A on day 3, day 5, and day 7 ($p < 0.01$, Figure S1A, see Supplementary Materials). With the increase in ASP concentration (from 100 μg/mL to 3200 μg/mL), inhibited cell viability was observed on day 5 and day 7 in BRL-3A. AMP with a concentration of 200 μg/mL began to work, and it exerted the greatest promoting function when the concentration reached 800 μg/mL ($p < 0.05$, Figure S1b). The EdU assay indicated that the number of hepatocytes increased more rapidly in the ASP (6.41%) and AMP (6.78%) groups than that in the control group on day 3 (4.53%) ($p < 0.001$, Figure S1c,d). These results reveal that respective use of ASP and AMP at specific concentrations could promote hepatocyte proliferation.

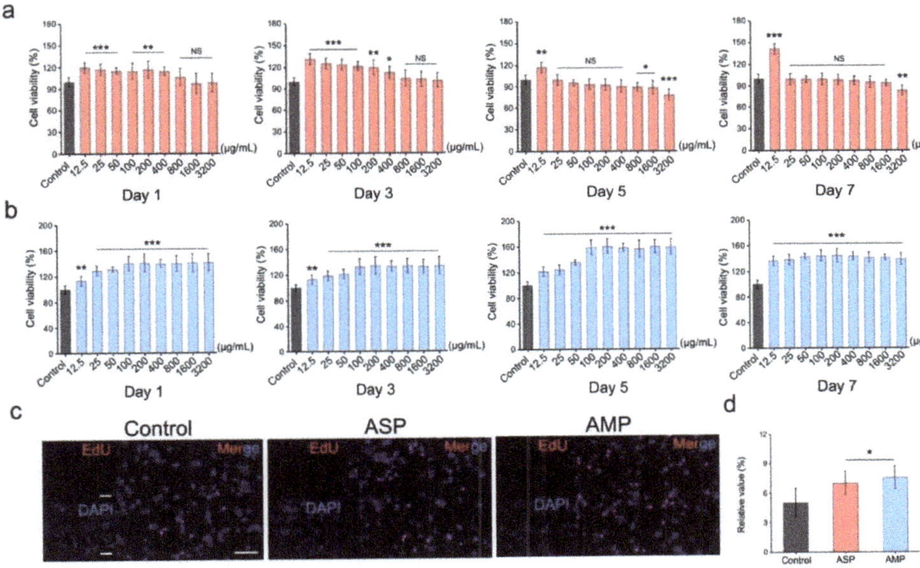

Figure 1. *Angelica sinensis* polysaccharide (ASP) and *Astragalus membranaceus* polysaccharide (AMP) promoted proliferation of NCTC clone 1469. The cell viability of hepatocytes treated by various concentrations of ASP (**a**) and AMP (**b**) for 7 days, which was determined by Alamar Blue Cell Viability Reagent ($n = 3$). (**c**) Click-iT EdU proliferation staining of hepatocytes treated with ASP and AMP for 3 days (scale bar = 100 μm). (**d**) Analysis of relative expression rates of positive EdU staining ($n = 3$). * $p < 0.05$, ** $p < 0.01$, and *** $p < 0.001$.

Furthermore, representative concentrations of ASP and AMP were selected in terms of the promoting effect presented in Figure 1 and Figure S1. NCTC clone 1469 cells that were cultured with both ASP (20 μg/mL) and AMP (100 μg/mL) proliferated significantly faster than cells solely cultivated with ASP at a concentration of 20 μg/mL ($p < 0.01$, Figure 2a) or AMP at a concentration of 100 μg/mL ($p < 0.05$, Figure 2b). Consistent with the results of cell viability in Figure 2a,b, EdU-positive hepatocytes were observed at a rate of 8.91% in the ASP-AMP group (20 μg/mL of ASP, 100 μg/mL of AMP, Figure 2c,d). The relative value in the ASP-AMP group was significantly higher than that in ASP or AMP group ($p < 0.05$). Meanwhile, ASP with a concentration of 160 μg/mL and AMP with a concentration of 800 μg/mL were chosen to verify the promoting effect of ASP-AMP on proliferation in BRL-3A cells (Figure S2). Significantly promoted cell proliferation by ASP-AMP was observed via cell viability analysis ($p < 0.05$, Figure S2a,b), as well as in the EdU assay results ($p < 0.05$, Figure S2c,d). It was demonstrated the upon the concentration above, combined use of ASP and AMP at a ratio of 1:5 enhanced the proliferation-promoting effect compared to their separate use.

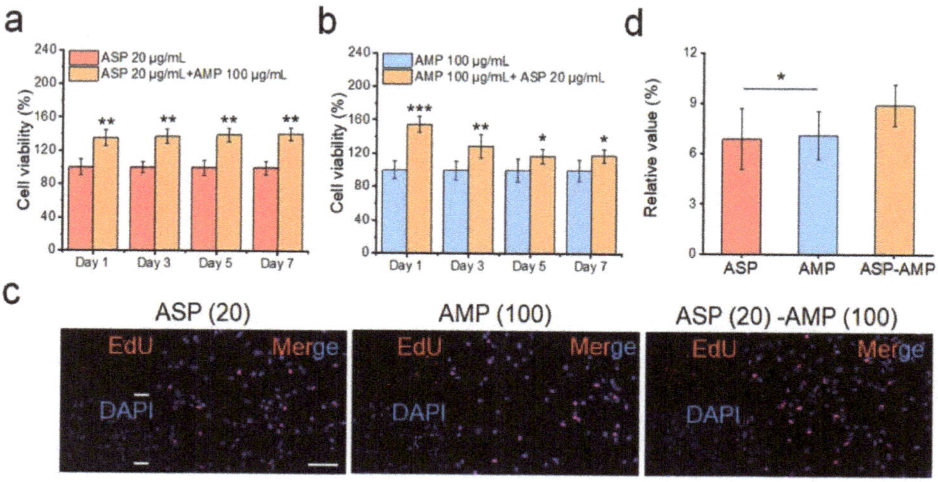

Figure 2. Combined treatment with both ASP and AMP enhanced the proliferation of NCTC clone 1469 cells. (**a**) The cell viability of hepatocytes treated with a combination of ASP and AMP compared to the separate use of ASP (**a**) or AMP (**b**) for 7 days ($n = 3$). (**c**) Click-iT EdU proliferation staining of hepatocytes treated with ASP, AMP, and ASP-AMP for 3 days (scale bar = 100 μm). (**d**) Relative expression rates of positive EdU staining ($n = 3$). * $p < 0.05$, ** $p < 0.01$, and *** $p < 0.001$.

2.2. *ASP-AMP Accelerated Liver Regeneration and Reduced Liver Injury after Hepatectomy In Vivo*

As shown in the schematic diagram of 70% hepatectomy in mice, after median incision (①) the left (②) and middle (③) lobes of livers were resected in sequence, and the right and caudate (④) lobes of livers were left in residue (Figure 3a). After hepatectomy, Balb/c mice treated with ASP-AMP for 7 days were sacrificed, and after the harvest of the isolated liver tissue, a larger volume was displayed in the ASP-AMP group (Figure 3b). A higher liver to body weight ratio was observed in the ASP-AMP group (4.20%), being 1.34 times higher than that of control group ($p < 0.05$, Figure 3c). Meanwhile, this ratio in the Sham group was 5.44%, indicating that the residual liver mass in the ASP-AMP group increased to 77.21% of the unresected liver. H&E analysis of the liver tissues showed no nuclear condensation or fragmentation in either group (Figure 3d). Strikingly, the ASP-AMP group exhibited increased hepatocyte proliferation after hepatectomy, as evidenced by increased Ki67 staining when compared with the control and sham groups ($p < 0.001$, Figure 3d,e). ASP-AMP-treated mice displayed a potent reduction in plasma levels of alanine aminotransferase (ALT, 243.53 IU/L), aspartate aminotransferase (AST, 423.74 IU/L), and total bilirubin (52.61 IU/L) three days after hepatectomy, demonstrating the hepatoprotection of liver injury ($p < 0.05$, Figure 3f–h).

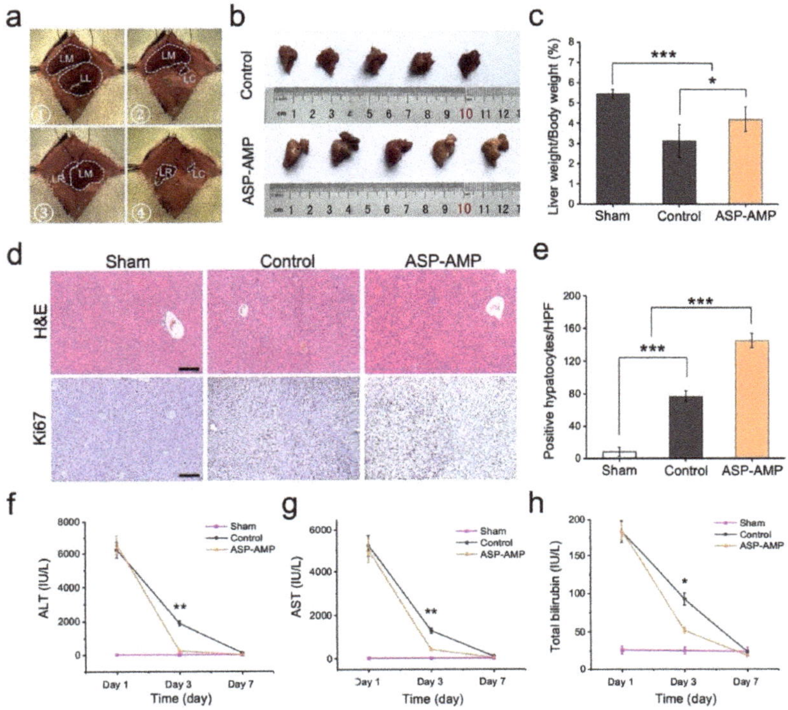

Figure 3. ASP-AMP enhanced liver regeneration after partial hepatectomy. (**a**) Schematic representation of 70% hepatectomy in mice. (**b**) The view of isolated liver tissues at day 7 (n = 5). (**c**) Ratio of liver weight to body weight 7 days after hepatectomy (n = 5). (**d**) Representative H&E and Ki67 staining of liver sections 3 days after hepatectomy (scale bar = 200 μm). (**e**) The quantification of Ki67 staining of liver sections 3 days after hepatectomy. Plasma ALT (**f**), AST (**g**), and total bilirubin (**h**) levels in mice 1 day, 3 days, and 7 days after hepatectomy with ASP-AMP intervention (n = 3). LM, liver middle lobe; LL, liver left lobe; LR, liver right lobe; LC, liver caudate lobe. ALT, alanine aminotransferase; AST, aspartate aminotransferase. * $p < 0.05$, ** $p < 0.01$, and *** $p < 0.001$.

2.3. ASP-AMP Enhanced HK2 Involved Glycolysis to Promote Liver Regeneration

To detect the variations of metabolites in ASP-AMP-promoted liver regeneration, metabonomics between the ASP-AMP treated group (test) and control group (Ctrl) was performed. A hierarchical clustering analysis was executed on all of the different metabolites, and the relative quantitative values of the difference metabolites were normalized, converted, and clustered. It can be intuitively observed that most of the metabolites were highly expressed in the ASP-AMP-treated groups (Figure 4a,b). These results are consistent with those in the volcano maps (Figure S3a,b). The enriched and differentially expressed metabolites in the KEGG pathway analysis are presented in bubble charts (top 20) in Figure 4c for hepatocytes and Figure 4d for liver tissue. The differential metabolites of the ASP-AMP intervention group and control group were enriched in glycolysis/gluconeogenesis, the citrate cycle (TCA cycle), and the pentose phosphate pathway (red lines). Two metabolites (beta-D-fructose 6-phosphate and lactate) with a p value < 0.05, and four metabolites (fructose 1, 6-bisphosphate, phosphoenolpyruvic acid, dihydroxyacetone phosphate, and succinic acid semialdehyde) with moderate elevation related to glycometabolism were displayed (Figure 4e,f). These results lay the foundation for understanding the ASP-AMP-intervened metabolic pathways and differential metabolites. Furthermore, it was verified in the hepatocyte supernatant (Figure 4g) and serum (Figure 4h) that ASP-AMP improved the lactate secretion ($p < 0.01$).

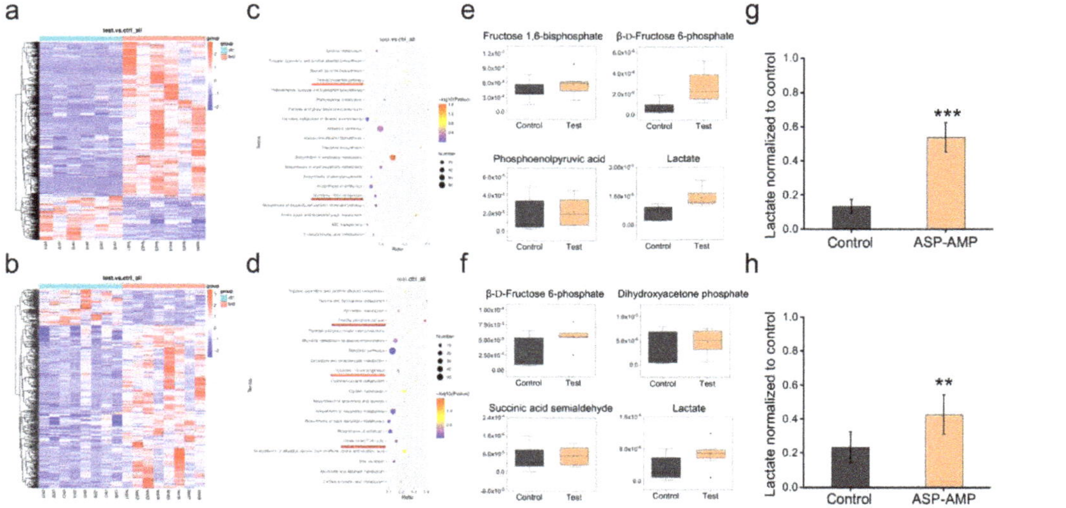

Figure 4. Differential metabolites found by metabolomics analysis and metabolic pathway analysis in NCTC clone 1469 cells ($n = 6$) and liver tissue ($n = 8$) after hepatectomy treated with ASP-AMP. Clustering heat map of total differential metabolites in hepatocytes (**a**) and liver tissue (**b**). Red refers to upregulated and blue refers to downregulated metabolites. KEGG pathway enrichment bubble diagram in hepatocytes (**c**) and liver tissue (**d**). The top 20 in terms of metabolite numbers are displayed, including glycolysis/gluconeogenesis, the citrate cycle (TCA cycle), and the pentose phosphate pathway. Six predictive metabolites related to glycometabolism are shown for hepatocytes (**e**) and liver tissue (**f**). Relative abundances were log10-transformed. Data were generated by LC-UHPLC. Lactate measurement in the hepatocyte supernatant (**g**) and mouse serum (**h**). Ctrl refers to the control group, and test refers to the experimental group treated with ASP-AMP for 3 days. ** $p < 0.01$, and *** $p < 0.001$.

The results above reveal that lactate-assisted glycolysis was actively involved in the process of ASP-AMP promoting liver regeneration. However, the exact enzyme dominating the process remains unclear. To elucidate the mechanism by which ASP-AMP regulates liver regeneration, several genes of rate-limiting enzymes demonstrated to be modulated in glycolysis were analyzed, including HK2, lactate dehydrogenase (LDHA), glyceraldehyde-3-phosphate dehydrogenase (GAPDH), 6-phosphofructo-2-kinase/Fructose-2,6-biphosphatase 3 (PFKB3), pyruvate kinase (PKM2), and GLUT1. ASP-AMP-administered NCTC clone 1469 cells demonstrated activated HK2 and GLUT1 at transcription level ($p < 0.01$, Figure 5a). Subsequently, we verified the protein expression of HK2 and GLUT1, and HK2 was upregulated, while the regulation of GLUT1 expression was insignificant (Figure 5b,c). To assess the impact of HK2 on lactate secretion and growth of hepatocytes, knockdown of HK2 in NCTC clone 1469 cells was performed by using an shRNA specific to HK2 (HK2 shRNA#2) (Figure 5d). Knockdown of HK2 decreased the lactate secretion in the supernatant ($p < 0.05$, Figure 5e), and inhibited the proliferation of NCTC clone 1469 cells for 7 days ($p < 0.01$, Figure 5f,g). The IHC staining of HK2 was lowly expressed in the ASP-AMP+2DG group compared with the ASP-AMP group ($p < 0.001$, Figure 5h), as well as the Ki67 staining ($p < 0.001$, Figure 5i). These results confirm that ASP-AMP accelerated liver regeneration via HK2-assisted glycolysis in vitro and in vivo.

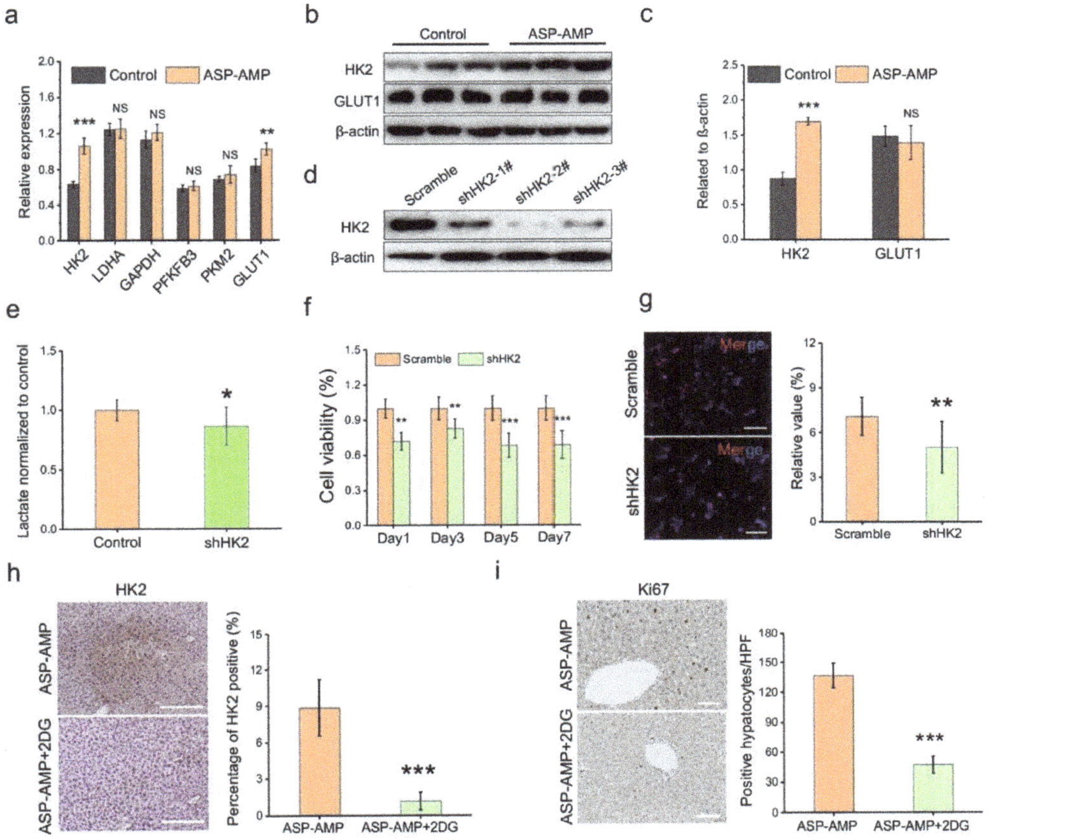

Figure 5. ASP-AMP accelerated the hexokinase 2 (HK2) involved glycolysis to promote liver regeneration. (**a**) RT-PCR analysis of rate-limiting enzymes related to glycolysis. (**b**) Western blot analysis for HK2 and GLUT1 after ASP-AMP treatment in NCTC clone 1469 cells ($n = 3$). (**c**) Relative protein was quantitatively expressed by densitometric analysis and is shown as the fold change relative to β-actin. (**d**) Efficiency of knockdown of HK2 in NCTC clone 1469 cells. (**e**) Lactate levels in supernatant of HK2 knockdown NCTC clone 1469 cells ($n = 3$). (**f**) The cell viability of HK2 knockdown NCTC clone 1469 cells treated with a combination of ASP and AMP for 7 days ($n = 3$). (**g**) Click-iT EdU proliferation staining of HK2 knockdown NCTC clone 1469 cells treated with ASP-AMP for 3 days and the relative expression rates of positive EdU staining (Scale bar = 100 μm). (**h**) Representative images of IHC staining for HK2 in the liver tissue of mice after hepatectomy when treated with ASP-AMP or ASP-AMP+2DG for 3 days and the analysis of percentage of HK2 positive area (Scale bar = 100 μm). (**i**) Ki67 staining of liver sections with 3 days of ASP-AMP+2DG after hepatectomy and the quantification of positive cells (scale bar = 100 μm). * $p < 0.05$, ** $p < 0.01$, and *** $p < 0.001$.

2.4. ASP-AMP Activated JAK2/STAT3 Signaling Pathway to Increase the Liver Regeneration

To elucidate the potential targets and molecular mechanisms of ASP-AMP-promoted liver regeneration, network pharmacology and bioinformatics were performed. The results show that ASP and AMP obtained 321 and 156 potential targets, respectively, in which a total of 417 targets were involved. A total of 1131 proteins were selected as the potential targets related to liver regeneration (Figure S4a). According to the above results, 417 putative targets of drugs (ASP and AMP) and the 1131 liver regeneration-related targets were imported to TBtools 1.038 to obtain 145 common targets (Figure S4b). Then, a PPI network reflected the function of the targets in liver regeneration and discovered the interactive effects (Figure 6a). Enrichment analysis involved GO and KEGG pathway analysis. We selected the top four GO items according to *p* values and counted 145 genes using bioinformatics (Figure 6b). For GO biological processes, the targets were highly enriched in the positive regulation of cell proliferation (GO: 0008284) in Biological Process (red line). As for pathway analysis, the JAK/STAT signaling pathway played a significant role in promoting liver regeneration (Figure 6c).

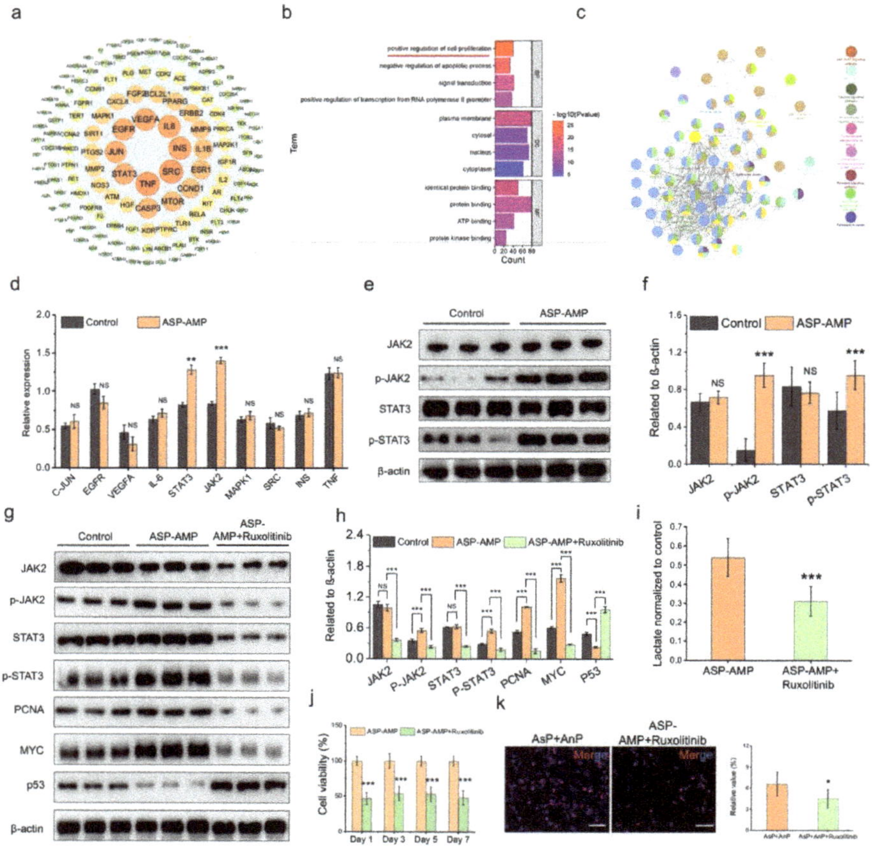

Figure 6. Activation of the JAK2/STAT3 signal pathway dominated the ASP-AMP-promoted liver regeneration. (**a**) PPI network of the common targets of drugs and liver regeneration. According to the degree value, the nodes were arranged from large to small and the color of the node changes from orange to yellow to green. (**b**) GO enrichment analysis. The top four terms in Biological Process, Cellular Component and Molecular Function are displayed. (**c**) The KEGG signaling pathway of ASP and AMP promoting liver regeneration. (**d**) RT-PCR analysis of predictive target factors in ASP-AMP promoted liver regeneration ($n = 3$). (**e**) The expression of JAK2, p-JAK2, STAT3, and p-STAT3 in hepatocytes were analyzed by Western blot ($n = 3$). (**f**) Densitometric analysis of grey value related to β-actin expression. (**g**) Pharmacologically inhibiting JAK2 downregulated the expression of p-STAT3, MYC, and PCNA, and upregulated p53 ($n = 3$). (**h**) Quantitative analysis of protein expression related to β-actin. (**i**) Lactate expression in the supernatant of hepatocytes with ASP-AMP and/or ruxolitinib treatment for 3 days ($n = 3$). (**j**) Cell viability of NCTC clone 1469 cells treated with ASP-AMP and/or ruxolitinib for 7 days ($n = 3$). (**k**) Click-iT EdU proliferation staining of hepatocytes treated with ASP-AMP and/or ruxolitinib for 3 days and relative expression rates of positive EdU staining (scale bar = 100 μm). * $p < 0.05$, ** $p < 0.01$, and *** $p < 0.001$.

According to the above predicted targets in Figure 6a, transcription levels of genes including C-JUN, epidermal growth factor receptor (EGFR), vascular endothelial growth factor A (VEGFA), interleukin 6 (IL-6), STAT3, JAK2, mitogen-activated protein kinase 1 (MAPK1), sarcoma gene (SRC), insulin, and tumor necrosis factor (TNF) were detected after the ASP-AMP treatment. STAT3 and JAK2 were upregulated in NCTC clone 1469 cells (Figure 6d). Western blot displayed that ASP-AMP increased phosphorylation of JAK2 and STAT3 (Figure 6e,f). Accordingly, associated downstream signaling factors, such as PCNA and MYC, were upregulated, and p53 was downregulated (Figure 6g,h). To verify that the JAK2/STAT3 pathway was the main pathway participating in the regulation of the liver by ASP-AMP, ruxolitinib was used to inhibit JAK2 activity pharmacologically (Figure 6g,h). The protein level of STAT3 and the phosphorylation of STAT3, PCNA, MYC, and p53 were reversed. Lactate in the supernatant decreased accordingly in the ASP-AMP+ruxolitinib group ($p < 0.001$, Figure 6i). The inhibition of JAK2 reduced cell proliferation for 7 days ($p < 0.05$, Figure 6j–l). These results demonstrate that JAK2/STAT3 dominated the process of the ASP-AMP-accelerating liver regeneration.

2.5. STAT3 Upregulated HK2 in the Process of the ASP-AMP Promoting Liver Regeneration

To evaluate the relationship between STAT3- and HK2-involved glycolysis in liver regeneration promoted by ASP-AMP, STAT3 was knocked down in NCTC clone 1469 cells (shSTAT3#3) (Figure 7a). The knockdown of STAT3 decreased the protein expression of HK2 (Figure 7b,c) and diminished the lactate secretion in the supernatant after ASP-AMP intervention ($p < 0.001$, Figure 7d). Subsequently, the effects on cell proliferation were analyzed. Cell viability and the EdU assay displayed that cell proliferation was significantly inhibited in STAT3 knockdown NCTC clone 1469 cells ($p < 0.001$, Figure 7e,f).

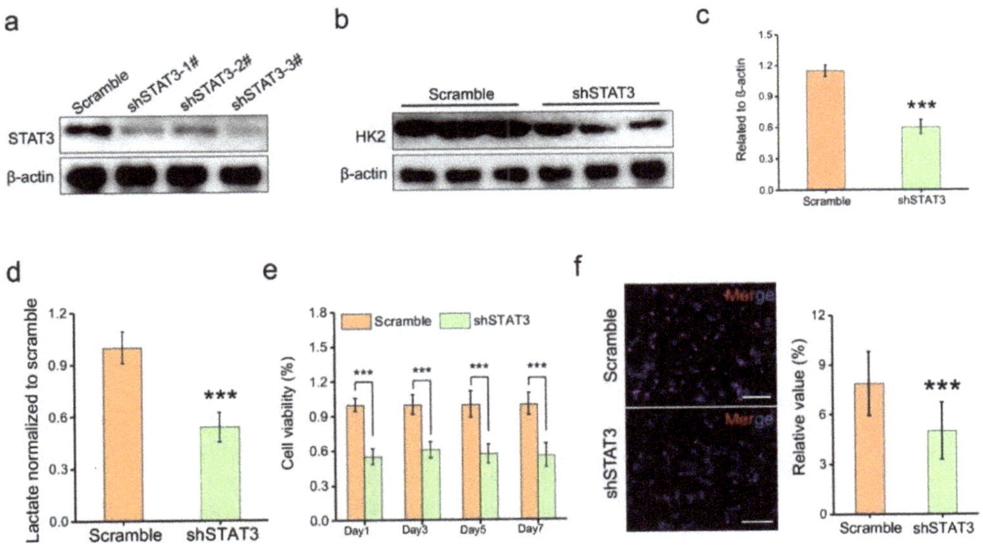

Figure 7. ASP-AMP accelerated HK2-assisted glycolysis via JAK2/STAT3 activation to enhance liver regeneration. (**a**) Efficiency of knockdown of STAT3 in NCTC clone 1469 cells. (**b**) HK2 was analyzed by Western blot in STAT3 knockdown hepatocytes ($n = 3$). (**c**) Densitometric analysis of protein expression related to β-actin. (**d**) Lactate in supernatant of STAT3 knockdown hepatocytes ($n = 3$). (**e**) The cell viability of STAT3 knockdown NCTC clone 1469 cells treated with ASP-AMP for 7 days ($n = 3$). (**f**) Click-iT EdU proliferation staining and relative positive EdU staining analysis of STAT3 knockdown NCTC clone 1469 cells treated with ASP-AMP for 3 days (Scale bar = 100 μm). *** $p < 0.001$.

3. Discussion

Improving liver regeneration after hepatectomy has clinically proven difficult. Although surgical strategies including portal vein embolization/ligation and associated liver partition and portal vein ligation for staged hepatectomy have been proven to be effective in liver regeneration, applications are limited due to strict inclusion criteria, severe complications, and high medical costs [23]. Agents including amiodarone [24], thalidomide [25], etanercept [26], and bicyclol [27] have been reported to enhance liver regeneration in animal models; however, they were rarely applied in clinical practice due to their side effects (kidney damage, gut reaction, perception abnormalities, or perception barriers, etc.) and toxicity. Physiologically, the cell proliferative response to hepatectomy lasted for 72 h, reaching a peak at 24–36 h, and then increasingly shifted to a termination phase [28]. Rapid administrations to regulate the proliferative process during the initial 72 h were of significance in enhancing liver regeneration [29].

Simultaneously, ASP and AMP—the main active polysaccharides in DBD—obtained essential implications in promoting regeneration [30]. By means of different extraction or structural synthesis processes, bioactive polysaccharides from plants have been tested for the biological engineering and regeneration of practically all tissues [6,31,32]. For instance, hyaluronic acid was proven to participate in articular pathologies, skin remodeling, vascular prosthesis, adipose tissue engineering, and nerve reconstruction [33]. Decorin accelerated liver regeneration after partial hepatectomy in fibrotic mice [34]. As for ASP, it has been proven to exert a hepatoprotective effect in CCl_4-induced hepatic dysfunction and tissue damage in mice [35]. Hepatic toxicity caused by 5-fluorouracil could be antagonized by ASP via apoptosis reduction through the Nrf2 pathway [36]. AMP reduced endoplasmic reticulum stress and restored glucose homeostasis in the liver [37]. Furthermore, the hepatoprotective effect of ASP-AMP was observed in CCl_4-induced intoxication in mice, and the effect was stronger than that of the respective use of ASP or AMP [14]. Based on the evidence, this study verified that ASP-AMP exerted a significant promoting effect on liver regeneration in vitro and in vivo, as well as hepatoprotection after liver resection, especially during the first 72 h. This provides evidence that supplementation of Qi and blood via ASP and AMP facilitated the regenerative capacity of the liver. Meanwhile, the promoting effect of ASP and AMP on cell proliferation was observed, which was within an effective range of concentrations from 12.5 to 3000 μg/mL. Importantly, these results support the safety of ASP and AMP application for further clinical use.

Major hepatectomy in mice metabolically led to a rapid fall in blood glucose levels in terms of a dramatic reduction in glycogen stores and gluconeogenic capacity [38]. However, the restoration of liver mass depended on increased energy production, which was indispensable for the recovery of liver function [39]. Increased hepatic glycometabolism was observed in an improved liver regeneration after fatty liver resection [40]. Thus, the metabolic process might be regulated for enhanced liver regeneration within the 72 h proliferative phase. With the promoting effect of ASP-AMP, metabonomics suggest that differential metabolites are enriched in the glycolysis process with related metabolites including beta-D-fructose, 6-phosphate, and lactate. As glucose was metabolized into lactate via glycolysis in the cytoplasm, the increased lactate levels in supernatant and serum verified the enhanced glucose mobilization in the initial period of liver regeneration. This was also consistent with a previous report which indicated that glycolysis actively participated in regulating the termination of liver regeneration [41]. Furthermore, we observed overexpression of HK2 in the transcription and protein level caused by ASP-AMP treatment, and conversely the knockdown of HK2 elicited decreased lactate secretion and suppressed cell proliferation in vitro and in vivo. As an essential rate-limiting enzyme in glycolysis, HK2 has been reported to be actively involved in glycolysis-related liver diseases, especially in hepatocellular carcinoma [42,43]. As for liver regeneration, the prolonged regaining of liver weight after hepatectomy could be observed to be parallel with decreased activities of the HK2-related glycolytic pathways by microarray analysis [44]. Therefore, this indicates that ASP-AMP mediates glycolysis via the regulation of HK2.

Regarding ASP-AMP's promoting role in liver regeneration, the specific mechanism remains to be explored. Through predictions of network pharmacology and further verification, we focused on the JAK2/STAT3 pathway activated by ASP-AMP. The JAK2/STAT3 pathway and the following cascade was considered to be a critical pathway in priming the hepatocytes for proliferation [45]. JAK2 is one of the JAKs which have been identified as being widely expressed in human cells and have molecular masses ranging from 120 to 140 kDa [46]. JAK2 mediates the activation of STAT3 in cells exposed to formulations. In detail, the proliferation process involved IL-6R activation and the engagement of JAK protein-bound gp130, the promotion of which resulted in phosphorylating STAT3, causing an expression of multiple target genes important for hepatocyte proliferation such as MYC, MCL1, and PCNA [3,47]. The JAK2/STAT3 activation played a pivotal role in the acceleration of liver hypertrophy in a rat hepatectomy model [48]. The suppressed mitogenic JAK2/STAT3 pathway inhibited cell cycle progression and constrained the ability of partial hepatectomy to stimulate hepatocyte replication [49]. These observations were consistent with our observation that the activation of the STAT3 pathway contributed to cell proliferation in a liver graft [50]. Simultaneously, the JAK2/STAT3 pathway has also been proven to be positively associated with glycolysis which was regulated in the malignant progression of tumors [51,52]. These above results support the assertion that JAK2/STAT3 plays a role in liver regeneration via glycolysis. Although evidence has indicated that ASP or AMP indirectly regulate the JAK2/STAT3 pathway in various diseases [53,54], the combined effect of ASP-AMP on liver regeneration is so far unreported, especially in metabolic pathways.

Based on the role of JAK2/STAT3 in glycolysis and liver regeneration, the regulatory effect between STAT3 and HK2 should be further clarified. Previous reports indicated that the upregulation of STAT3/HK2 contributed to the activation of dendritic cells by leptin-induced glycolytic metabolism [55], and circular RNA circCUL3 eliciting the Warburg effect of gastric cancer [56]. Meanwhile, suppressed proliferation in hepatocellular carcinoma cells or the inhibition of glycolysis was achieved by the blocking of the JAK2/STAT3 pathway [57,58]. Furthermore, STAT3 was reported to bind to the promoter of the HK2 gene to regulate proliferation in gastric epithelial cells [59]. In the therapy of porcine delta coronavirus by selenomethionine, the STAT3 could repress miR-125b-5p-1 expression, which targets and inhibits the expression of HK2 [60]. In line with these reports, we observed that by the knockdown of STAT3 in hepatocytes, HK2 expression was downregulated accordingly, followed by inhibited lactate secretion and cell proliferation. These results imply that the activation of STAT3 could regulate glycolysis and hepatocytes proliferation via HK2.

4. Materials and Methods

4.1. Cell Culture and Reagent

Normal liver cell line NCTC clone 1469 and BRL-3A cells were purchased from the Shanghai Institute of Cell Biology at the Chinese Academy of Sciences (Shanghai, China) and cultured in DMEM supplemented with 10% fetal bovine serum (BSA, 1099-141, Thermo Fisher Scientific, Inc., Waltham, MA, USA), 1×10^5 U/L penicillin and 100 mg/L streptomycin (15140122, Thermo Fisher Scientific, Inc., Waltham, MA, USA) in a humidified atmosphere with a 5% CO_2 incubator at 37 °C. Trypsin-EDTA (0.05%) (Thermo Fisher Scientific, Inc., Waltham, MA, USA) was used for cell passages. Cells were authenticated by STR profiling, and experiments were performed from 5 passages. shRNAs targeting HK2 or STAT3 were designed by OBiO Biotechnology Co. LTD (Shanghai, China). Three shRNAs targeting the coding sequence of mRNA were inserted into the pLKO.1 vector. The efficacy of each shRNA was assessed by Western blot of the NCTC clone 1469 cells, which were infected with the virus for 3 days. The shRNA with strongest knockdown efficacy was used for further experiments. ASP (DSTDDO28301) and AMP (DSTDH010901) were purchased from Desite Biotechnology Co., LTD (Chengdu, China). The information regarding ASP and AMP, including identification, ingredients, and physicochemical properties, etc., can be seen in http://www.028desite.com (accessed on 6 December 2021). Ruxolitinib (SD4740) and 2-deoxy-d-glucose (2DG, ST1024) were obtained from Beyotime Biotechnology (Shanghai, China).

4.2. Cell Viability

NCTC clone 1469 and BRL-3A cell lines were inoculated into 96-well plates with a concentration of 1×10^4 cells per well and incubated with a variety of concentrations (12.5 μg/mL, 25 μg/mL, 50 μg/mL, 100 μg/mL, 200 μg/mL, 400 μg/mL, 800 μg/mL, 1600 μg/mL, and 3200 μg/mL) of ASP and AMP for 7 days. Cell viability was determined by a AlamarBlue® Cell Viability Assay Kit (AB, DAL1100, Invitrogen Corporation, Waltham, MA, USA) on day 1, day 3, day 5, and day 7. The cells were supplemented with AB solution (medium199: fetal bovine serum: AB = 8:1:1 v/v) after the removal of the medium and PBS washing. For 4 h of incubation, the absorbance (optical density, OD) at 570 nm was measured and calculated according to the formula: Cell viability (%) = (OD experimental − OD blank)/(OD control − OD blank) × 100%. Experiments were performed in triplicate.

4.3. Click-iT Plus EdU Proliferation Staining

ASP (20 μg/mL) and/or AMP (100 μg/mL) were added to NCTC clone 1469 cells, and ASP (160 μg/mL) and/or AMP (800 μg/mL) were added to BRL-3A cells for 3 days' culture in 12-well plates. The EdU working solution (Beyotime Biotechnology, Shanghai, China) was added to the culture medium to obtain a final concentration of 10 μM. The cells were incubated with the solution for 2 h in an incubator at 37 °C. After removal of the solution, cells were treated with paraformaldehyde (1.5 M) for 15 min. Cells were then washed in PBS for 15 min, followed by 0.5% Triton-X 100 in PBS for 10 min. Click Additive Solution was prepared according to the manufacturer's instructions and added to each well, which was incubated for 30 min in the dark. Then, 5 μg/mL Hoechst 333,425 was added to the wells, and they were incubated for additional 30 min. During every step, cells were washed twice with 3% BSA in PBS. Finally, cells were examined using the Zeiss fluorescence microscope. Four views of each well were randomly recorded with triplicate samples (n = 3). The quantification of blue and red points was performed by ImageJ (v 1.53, National Institutes of Health, Stapleton, NY, USA).

4.4. Experimental Animals

The study was approved by the Animal Research Ethics Committee of Chengdu University of Traditional Chinese Medicine (Chengdu, China) and complied with the Guidelines for Animal Experiments. Thirty-six Balb/c mice aged 5 weeks old and weighing 18–22 g were purchased from Chengdu Dashuo Biotechnology Co., Ltd. (Chengdu, China; license no. SCXK 2008-24). All mice were maintained on a 12/12 h light/dark cycle, allowed free access to water and food, and acclimated for at least 5 days prior to surgery. Mice in the Sham group underwent the procedures of opening and closing the abdomen. Mice in the control group underwent 70% liver resection without any other treatment. Mice in the ASP-AMP group were orally administrated ASP (250 mg/kg/day) and AMP (1 g/kg/day) for 7 days after 70% hepatectomy. Mice in the ASP-AMP+2DG group were injected intraperitoneally with 2-DG (a hexokinase inhibitor, 1 g/kg, Sigma, Arklow, Ireland) on days 1, 2, 3, and 6 using sterile saline as a vehicle, based on the treatment in the ASP-AMP group. Inhalation of CO_2 (20% of the displacement volume per minute) was used for euthanasia. Harvested liver tissues were fixed in 4% paraformaldehyde for 24 h at room temperature and stored in 75% ethanol at 4 °C for further histology and immunohistochemistry (IHC) analysis.

4.5. Seventy Percent Partial Hepatectomy in a Murine Model

For the mouse model of partial hepatectomy, 2/3 of the liver was resected according to the previous description [61]. Briefly, mice were anesthetized with 2% isoflurane and 2 L/min oxygen flow to maintain anesthesia by isoflurane. A midline abdominal skin and muscle incision was created to expose the xiphoid process. The 4-0 silk thread was placed on the base of the left lateral lobe and the lobe was ligated over the top of it. The tied lobe was curved above the suture using microsurgery. Similarly, the median hepatic lobes were then ligated and resected. Finally, after closing the abdomen, the skin surrounding the suture was wiped with betadine, and mice were placed on a warming pad for recovery.

4.6. Enzyme-Linked Immunosorbent Assay (ELISA)

The whole blood of sacrificed mice was kept at room temperature for 2 h and centrifuged at 4000 rpm for 10 min to collect serum. Serum ALT (JL20911, Jonln Co., Ltd., Shanghai, China), AST (C010-2-1, Nanjing Jiancheng Bio Co., Ltd., Nanjing, China), and total bilirubin (C019-1-1, Nanjing Jiancheng Bio Co., Ltd., Nanjing, China). According to the instructions, the concentration gradient dilution standard was added to the 96-well plate (100 µL per well), 100 µL of mouse serum was then added to the 96-well plate in turn, and each sample was placed into 3 duplicate wells, which were incubated at 37 °C for 1 h. After discarding the liquid, 100 µL per well of biotinylated antibody was directly added without washing and incubated at 37 °C for 1 h. Then, the samples were washed 3 times with 1 × wash solution. After this, 100 µL per well of enzyme conjugate working solution was added and incubated at 37 °C for 30 min. Subsequently, 90 µL per well of substrate was added, and incubated at 37 °C for 15 min, followed by 10 µL per well of stop solution. We immediately measured the absorbance of the solution at 450 nm on a microplate (Multiskan GO, Thermo Fisher Scientific, Inc., Waltham, MA, USA).

4.7. Untargeted Metabolomics by Liquid Chromatography Coupled Mass Spectrometry (MS/MS)

See details in Methods S1.

4.8. Lactate Measurement

For lactate measurement in serum, the whole blood of treated mice was firstly collected and centrifuged at 4000 rpm for 10 min to obtain the serum. For the analysis of lactate in cell cultures, cells were seeded at approximately 3×10^5 in 6-well plates and treated with different concentrations of ASP and/or AMP for 3 days. After centrifugation, the culture medium and cells were collected. The lactate content was estimated using the L-lactate assay kit (Jiancheng Bioengineering, Nanjing, China) according to the manufacturer's protocol. Lactate dehydrogenase catalyzed the conversion of lactate to pyruvic acid, and the absorbance at 530 nm was determined spectrophotometrically after the addition of a color developing agent, which showed a linear relationship with lactate content. Three replicates were performed, and the experiments were repeated three times for accuracy.

4.9. Western Blot

Total cell protein was collected (KGP2100, KeyGEN BioTECH, Beijing, China), and its quantification was performed using the BCA (cat. no. KGP902, KeyGEN BioTECH, Beijing, China) method. Proteins were separated by SDS-PAGE and transferred through PVDF membranes. After blocking the membrane with 5% BSA for 2 h at room temperature, membranes were incubated overnight at 4 °C with one of the following primary antibodies: rabbit monoclonal anti-STAT3 (ab68153, Abcam, Cambridge, UK), rabbit monoclonal anti-phospho STAT3 (ab32143, Abcam, Cambridge, UK), rabbit monoclonal anti-Janus kinase 2 (JAK2, ab108596, Abcam, Cambridge, UK), rabbit monoclonal anti-phospho JAK2 (ab32101, Abcam, Cambridge, UK), rabbit monoclonal anti-proliferating cell nuclear antigen (PCNA, ab92552, Abcam, Cambridge, UK), rabbit monoclonal anti-MYC (ab32072, Abcam), mouse monoclonal anti-P53 (ab90363, Abcam, Cambridge, UK), rabbit monoclonal anti-β-actin (Bsm-33036M, Biosynthesis, Beijing, China), rabbit monoclonal anti-HK2 (ab209847, Abcam, Cambridge, UK), and rabbit monoclonal anti-glucose transporter 1 (GLUT1, ab115730, Abcam, Cambridge, UK). All antibodies were diluted at 1:1000 in TBST (T1086, Solarbio, Beijing, China). After washing with PBS, membranes were incubated with secondary antibodies, including goat anti-mouse IgG antibody (bs-0296G, Bioss, Beijing, China) and goat anti-rabbit IgG antibody (bs-0295G; Bioss, Beijing, China), which were diluted to 1:4000 by TBST, for 1 h at room temperature. After washing off the excess secondary antibody, protein intensity was determined with Clarity Western ECL Substrate (Bio-Rad Laboratories Co. Ltd., Hercules, CA, USA) and measured by Image Lab software (5.2.1 Version, Bio-Rad Laboratories Co. Ltd., Hercules, CA, USA). Proteins were quantified by densitometric analysis of ImageJ (version 1.8.0.172, Bethesda, MA, USA).

4.10. Histology and Immunohistochemistry (IHC) Analysis

Murine liver tissues were collected, fixed in 4% paraformaldehyde for 48 h, dehydrated, and embedded in paraffin. The paraffin tissue blocks were cut into 4 μm-thick sections, deparaffinized with xylene and passed through graded alcohol (10009218, ChengFeng Co., Ltd., Hangzhou, China). To detect the expression of HK2 and Ki-67 in liver tissue, sections were incubated with anti-HK2 antibody (1:200, ab209847, Abcam, Cambridge, UK) or anti-Ki67 antibody (1:200, bs-2130R, Bioss, Beijing, China). Sections were incubated overnight at 4 °C and rinsed 3 times with 1 × PBS for 5 min each time. Reagents were added according to the instructions of the secondary antibody kit (sp-9000; ZSGB-Bio, Beijing, China), and finally DAB (ZLI-9018; ZSGB-Bio, Beijing, China) were added for a 2 min reaction. Four views were randomly collected from each section, and images were acquired using an optical microscope (DM3000, Leica Co., Ltd., Weztlar, Germany).

4.11. Real-Time Quantitative PCR Analysis

Total cellular RNA was extracted by using Trizol Reagent (Invitrogen, Thermo Fisher Scientific, Inc., Waltham, MA, USA). Reverse transcription was performed using PrimeScript (Takara Bio, Inc., Nojihigashi Kusatsu, Japan), first removing gDNA by using the following conditions: 42 °C for 2 min. The conditions used for reverse transcription were as follows: 37 °C for 15 min and 85 °C for 5 s. The cDNA was subjected to PCR by adding SYBR Green (Takara Bio, Inc., Nojihigashi Kusatsu, Japan), and the sequences of the primers used are shown in Table 1. Real-time quantitative PCR detection was performed by using a CFX96 Deep well (Bio-Rad Laboratories, Inc., Hercules, CA, USA) under the following cycling conditions: 95 °C for 30 s, 95 °C for 5 s, and 60 °C for 30 s, with 40 cycles. Experiments normalized the mRNA expression of each gene with β-actin as an internal control. All samples were repeated three times. Finally, the relative gene expression results were calculated using the $2^{-\Delta\Delta Ct}$ method.

Table 1. Primers which have been used for real time PCR.

Primer Name	Sequence 5′-3′	Product Length
HK2	Forward: ATGATCGCCTGCTTATTCACG Reverse: CGCCTAGAAATCTCCAGAAGGG	110 bp
LDHA	Forward: ACATTGTCAAGTACAGTCCACAC Reverse: TTCCAATTACTCGGTTTTTGGGA	114 bp
GAPDH	Forward: TGACCTCAACTACATGGTCTACA Reverse: CTTCCCATTCTCGGCCTTG	85 bp
PFKFB3	Forward: CAACTCCCCAACCGTGATTGT Reverse: TGAGGTAGCGAGTCAGCTTCT	83 bp
PKM2	Forward: CGCCTGGACATTGACTCTG Reverse: GAAATTCAGCCGAGCCACATT	135 bp
GLUT1	Forward: TCAAACATGGAACCACCGCTA Reverse: AAGAGGCCGACAGAGAAGGAA	123 bp
C-JUN	Forward: ACTCGGACCTTCTCACGTC Reverse: GGTCGGTGTAGTGGTGATGT	110 bp
EGFR	Forward: ATGAAAACACCTATGCCTTAGCC Reverse: TAAGTTCCGCATGGGCAGTTC	83 bp

Table 1. *Cont.*

Primer Name	Sequence 5′-3′	Product Length
VEGFA	Forward: GCACATAGAGAGAATGAGCTTCC Reverse: CTCCGCTCTGAACAAGGCT	105 bp
IL-6	Forward: CTGCAAGAGACTTCCATCCAG Reverse: AGTGGTATAGACAGGTCTGTTGG	131 bp
STAT3	Forward: CACCTTGGATTGAGAGTCAAGAC Reverse: AGGAATCGGCTATATTGCTGGT	112 bp
JAK2	Forward: GGAATGGCCTGCCTTACAATG Reverse: TGGCTCTATCTGCTTCACAGAAT	108 bp
MAPK1	Forward: GGTTGTTCCCAAATGCTGACT Reverse: CAACTTCAATCCTCTTGTGAGGG	84 bp
SRC	Forward: TTTGGCAAGATCACTAGACGGG Reverse: GAGGCAGTAGGCACCTTTTGT	111 bp
INS	Forward: CACTTCCTACCCCTGCTGG Reverse: ACCACAAAGATGCTGTTTGACA	177 bp
TNF	Forward: CAGGCGGTGCCTATGTCTC Reverse: CGATCACCCCGAAGTTCAGTAG	89 bp
β-actin	Forward: GGCTGTATTCCCCTCCATCG Reverse: CCAGTTGGTAACAATGCCATGT	154 bp

4.12. Network Pharmacology and Bioinformatics

See details in Methods S2.

4.13. Statistical Analysis

Statistical analysis was performed using SPSS version 22.0 (IBM, Corp., Armonk, NY, USA). All the data are expressed as mean ± standard deviation. One-way ANOVA or unpaired two-tail Student's t-test was used to determine differences among groups. A statistically significant difference was considered significant at $p = 0.05$, * $p < 0.05$, ** $p < 0.01$, and *** $p < 0.001$.

5. Conclusions

The present investigations demonstrated that ASP-AMP acted as a promoting factor for liver regeneration. ASP-AMP intervention in hepatocytes enhanced proliferation by accelerating HK2-associated glycolysis. Moreover, ASP-AMP-induced glycolysis involved liver regeneration by JAK2/STAT3/HK2 signaling. These findings demonstrate that ASP and AMP, as natural agents, are feasible and experimentally recommended in the treatment of liver regeneration in patients who undergo hepatectomy. To impel the use of ASP-AMP in clinic therapy, the pharmacokinetics and toxicity should be further investigated. In addition, several missing links remain to be elucidated in future studies, including the molecular mechanism responsible for the ASP-AMP-mediated upregulation of JAK2 and the STAT3-induced HK2 activation.

Supplementary Materials: The following supporting information can be downloaded at: https://www.mdpi.com/article/10.3390/molecules27227890/s1, Method S1: Untargeted metabolomics by liquid chromatography coupled mass spectrometry (MS/MS); Method S2: Network pharmacology and bioinformatics; Figure S1: *Angelica sinensis* polysaccharide (ASP) and *Astragalus membranaceus* polysaccharide (AMP) promoted the proliferation of BRL-3A; Figure S2: Combined use of both ASP and AMP enhanced the proliferation of BRL-3A; Figure S3: Volcanic map of differential metabolites; Figure S4: Potential targets analysis of drugs (ASP and AMP) and liver regeneration.

Author Contributions: Conceptualization, Y.-H.H., T.P. and Q.-B.Y.; data curation, X.-D.W.; formal analysis, X.-D.W., Y.-L.Z. and L.Y.; funding acquisition, X.-D.W. and Q.-B.Y.; investigation, Z.Y. and Q.-B.Y.; methodology, X.-D.W., Y.-L.Z., L.Y. and Z.Y.; resources, Y.-L.Z.; software, G.-C.F.; validation, L.Y. and Z.Y.; writing—original draft, X.-D.W. and Y.-L.Z.; writing—review and editing, T.P. and Q.-B.Y. All authors have read and agreed to the published version of the manuscript.

Funding: This research was funded by the National Natural Science Foundation of China, grant number 81973742; Experimental Formulary Sichuan Youth Science and technology Innovation research team, grant number 2020JDTD0022; XingLin Scholars Program of Chengdu University of TCM, grant number QJJJ2022002, YYZX2020036 and Sichuan Provincial Science and Technology Department, grant number 2021YJ0198.

Institutional Review Board Statement: The study was conducted in accordance with the Declaration of Helsinki, and approved by the Animal Research Ethics Committee of Chengdu University of Traditional Chinese Medicine (protocol code 2022-0342, 14 March 2022) for studies involving animals.

Informed Consent Statement: Not applicable.

Data Availability Statement: The data presented in this study are available upon request from the corresponding author.

Conflicts of Interest: The authors declare no conflict of interest.

Sample Availability: Samples of ASP and AMP are available from the corresponding author.

References

1. Golse, N.; Bucur, P.O.; Adam, R.; Castaing, D.; Sa Cunha, A.; Vibert, E. New paradigms in post-hepatectomy liver failure. *J. Gastrointest. Surg.* **2013**, *17*, 593–605. [CrossRef]
2. Rassam, F.; Olthof, P.B.; Takkenberg, B.; Besselink, M.G.; Busch, O.R.; Erdmann, J.I.; Swijnenburg, R.J.; van Lienden, K.P.; Beuers, U.H.; Bennink, R.J.; et al. Functional assessment of liver regeneration after major hepatectomy. *Hepatobiliary Surg. Nutr.* **2022**, *11*, 530–538.
3. van Mierlo, K.M.; Schaap, F.G.; Dejong, C.H.; Olde Damink, S.W. Liver resection for cancer: New developments in prediction, prevention and management of postresectional liver failure. *J. Hepatol.* **2016**, *65*, 1217–1231.
4. Ober, E.A.; Lemaigre, F.P. Development of the liver: Insights into organ and tissue morphogenesis. *J. Hepatol.* **2018**, *68*, 1049–1062.
5. Ray, S.; Mehta, N.N.; Golhar, A.; Nundy, S. Post hepatectomy liver failure—A comprehensive review of current concepts and controversies. *Ann. Med. Surg.* **2018**, *34*, 4–10.
6. Herburger, K.; Glazowska, S.; Mravec, J. Bricks out of the wall: Polysaccharide extramural functions. *Trends Plant Sci.* **2022**, *1385*, 00188-1. [CrossRef]
7. Lin, H.Q.; Gong, A.G.; Wang, H.Y.; Duan, R.; Dong, T.T.; Zhao, K.J.; Tsim, K.W. Danggui Buxue tang (Astragali Radix and Angelicae Sinensis Radix) for menopausal symptoms: A review. *J. Ethnopharmacol.* **2017**, *199*, 205–210. [CrossRef] [PubMed]
8. Ma, C.-C.; Jiang, Y.-H.; Wang, Y.; Xu, R.-R. The latest research advances of Danggui Buxue Tang as an effective prescription for various diseases: A comprehensive review. *Curr. Med. Sci.* **2022**, *42*, 913–924. [PubMed]
9. Wang, P.; Liang, Y.Z. Chemical composition and inhibitory effect on hepatic fibrosis of Danggui Buxue Decoction. *Fitoterapia* **2010**, *81*, 793–798. [PubMed]
10. Wang, W.L.; Sheu, S.Y.; Chen, Y.S.; Kao, S.T.; Fu, Y.T.; Kuo, T.F.; Chen, K.Y.; Yao, C.H. Evaluating the bone tissue regeneration capability of the Chinese herbal decoction Danggui Buxue Tang from a molecular biology perspective. *Biomed. Res. Int.* **2014**, *2014*, 853234. [CrossRef] [PubMed]
11. Nai, J.; Zhang, C.; Shao, H.; Li, B.; Li, H.; Gao, L.; Dai, M.; Zhu, L.; Sheng, H. Extraction, structure, pharmacological activities and drug carrier applications of Angelica sinensis polysaccharide. *Int. J. Biol. Macromol.* **2021**, *183*, 2337–2353. [PubMed]
12. Zheng, Y.; Ren, W.; Zhang, L.; Zhang, Y.; Liu, D.; Liu, Y. A Review of the Pharmacological action of Astragalus Polysaccharide. *Front. Pharmacol.* **2020**, *11*, 349. [CrossRef]
13. Hua, Y.; Xue, W.; Zhang, M.; Wei, Y.; Ji, P. Metabonomics study on the hepatoprotective effect of polysaccharides from different preparations of Angelica sinensis. *J. Ethnopharmacol.* **2014**, *151*, 1090–1099. [PubMed]
14. Pu, X.; Fan, W.; Yu, S.; Li, Y.; Ma, X.; Liu, L.; Ren, J.; Zhang, W. Polysaccharides from Angelica and Astragalus exert hepatoprotective effects against carbon-tetrachloride-induced intoxication in mice. *Can. J. Physiol. Pharmacol.* **2015**, *93*, 39–43.
15. Rui, L. Energy metabolism in the liver. *Compr. Physiol.* **2014**, *4*, 177–197.
16. Saito, Y.; Morine, Y.; Iwahashi, S.; Ikemoto, T.; Imura, S.; Yamanaka-Okumura, H.; Hirayama, A.; Soga, T.; Tomita, M.; Shimada, M. Changes of liver metabolites following hepatectomy with ischemia reperfusion towards liver regeneration. *Ann. Gastroenterol. Surg.* **2018**, *2*, 204–211. [PubMed]
17. Yang, B.; Luo, Y.; Wei, X.; Kan, J. Polysaccharide from Hovenia dulcis (Guaizao) improves pancreatic injury and regulates liver glycometabolism to alleviate STZ-induced type 1 diabetes mellitus in rats. *Int. J. Biol. Macromol.* **2022**, *214*, 655–663. [PubMed]
18. Ye, Y.; Deng, T.; Wan, X.Y.; Ouyang, J.P.; Liu, M.; Mao, X.Q. The role of quantitative changes in the epxression of insulin receptor substrate-1 and nuclear ubiquitin in abnormal glycometabolism in the livers of KKay mice and the relative therapeutic mechanisms of Astragalus polysaccharide. *Int. J. Mol. Med.* **2014**, *33*, 341–350. [PubMed]
19. Wang, P.; Cong, M.; Liu, T.; Li, Y.; Liu, L.; Sun, S.; Sun, L.; Zhu, Z.; Ma, H.; You, H.; et al. FoxA2 inhibits the proliferation of hepatic progenitor cells by reducing PI3K/Akt/HK2-mediated glycolysis. *J. Cell Physiol.* **2020**, *235*, 9524–9537. [PubMed]

20. Zhang, J.; Feng, Q. Pharmacological effects and molecular protective mechanisms of Astragalus polysaccharides on nonalcoholic fatty liver disease. *Front. Pharmacol.* **2022**, *13*, 854674. [PubMed]
21. Wang, K.; Wang, J.; Song, M.; Wang, H.; Xia, N.; Zhang, Y. Angelica sinensis polysaccharide attenuates CCl_4-induced liver fibrosis via the IL-22/STAT3 pathway. *Int. J. Biol. Macromol.* **2020**, *162*, 273–283. [PubMed]
22. Wu, T.H.; Yeh, K.Y.; Wang, C.H.; Wang, H.; Li, T.L.; Chan, Y.L.; Wu, C.J. The Combination of Astragalus membranaceus and Angelica sinensis inhibits lung cancer and cachexia through its immunomodulatory function. *J. Oncol.* **2019**, *2019*, 9206951.
23. Wen, X.D.; Xiao, L. Associating liver partition and portal vein ligation for staged hepatectomy in the treatment of colorectal cancer liver metastases. *World J. Gastrointest. Surg.* **2021**, *13*, 814–821. [PubMed]
24. Lin, C.W.; Chen, Y.S.; Lin, C.C.; Chen, Y.J.; Lo, G.H.; Lee, P.H.; Kuo, F.L.; Dai, C.Y.; Huang, J.F.; Chung, W.L.; et al. Amiodarone as an autophagy promoter reduces liver injury and enhances liver regeneration and survival in mice after partial hepatectomy. *Sci. Rep.* **2015**, *5*, 15807. [CrossRef] [PubMed]
25. Hung, K.C.; Hsieh, P.M.; Yang, K.L.; Lin, K.J.; Chen, Y.S.; Hung, C.H. Effect of thalidomide on the expression of vascular endothelial growth factor in a rat model of liver regeneration. *Oncol. Lett.* **2013**, *5*, 852–856. [PubMed]
26. Viswanathan, P.; Kapoor, S.; Kumaran, V.; Joseph, B.; Gupta, S. Etanercept blocks inflammatory responses orchestrated by TNF-alpha to promote transplanted cell engraftment and proliferation in rat liver. *Hepatology* **2014**, *60*, 1378–1388. [CrossRef]
27. Yao, X.M.; Zhao, J.; Li, Y.; Li, Y. Effects of bicyclol on liver regeneration after partial hepatectomy in rats. *Dig. Dis. Sci.* **2009**, *54*, 774–781.
28. Sadri, A.R.; Jeschke, M.G.; Amini-Nik, S. Advances in liver regeneration: Revisiting hepatic stem/progenitor cells and their origin. *Stem Cells Int.* **2016**, *2016*, 7920897. [CrossRef] [PubMed]
29. Margagliotti, S.; Clotman, F.; Pierreux, C.E.; Beaudry, J.B.; Jacquemin, P.; Rousseau, G.G.; Lemaigre, F.P. The Onecut transcription factors HNF-6/OC-1 and OC-2 regulate early liver expansion by controlling hepatoblast migration. *Dev. Biol.* **2007**, *311*, 579–589.
30. Xie, J.-H.; Jin, M.-L.; Morris, G.A.; Zha, X.-Q.; Chen, H.-Q.; Yi, Y.; Li, J.-E.; Wang, Z.-J.; Gao, J.; Nie, S.-P. Advances on bioactive polysaccharides from medicinal plants. *Crit. Rev. Food Sci. Nutr.* **2016**, *56*, S60–S84.
31. Kazachenko, A.S.; Malyar, Y.N.; Vasilyeva, N.Y.; Borovkova, V.S.; Issaoui, N. Optimization of guar gum galactomannan sulfation process with sulfamic acid. *Biomass Convers. Biorefin.* **2021**, *9*, 1–10.
32. Zeng, P.; Li, J.; Chen, Y.; Zhang, L. The structures and biological functions of polysaccharides from traditional Chinese herbs. *Prog. Mol. Biol. Transl. Sci.* **2019**, *163*, 423–444. [PubMed]
33. Abatangelo, G.; Vindigni, V.; Avruscio, G.; Pandis, L.; Brun, P. Hyaluronic acid: Redefining its role. *Cells* **2020**, *9*, 1743. [CrossRef] [PubMed]
34. Ma, R.; Chen, J.; Li, Z.; Tang, J.; Wang, Y.; Cai, X. Decorin accelerates the liver regeneration after partial hepatectomy in fibrotic mice. *Chin. Med. J.* **2014**, *127*, 2679–2685.
35. Gao, Z.; Zhang, C.; Tian, W.; Liu, K.; Hou, R.; Yue, C.; Wu, Y.; Wang, D.; Liu, J.; Hu, Y.; et al. The antioxidative and hepatoprotective effects comparison of Chinese angelica polysaccharide (CAP) and selenizing CAP (sCAP) in CCl4 induced hepatic injury mice. *Int. J. Biol. Macromol.* **2017**, *97*, 46–54.
36. Zeng, D.; Wang, Y.; Chen, Y.; Li, D.; Li, G.; Xiao, H.; Hou, J.; Wang, Z.; Hu, L.; Wang, L.; et al. Angelica polysaccharide antagonizes 5-FU-induced oxidative stress injury to reduce apoptosis in the liver through Nrf2 pathway. *Front. Oncol.* **2021**, *11*, 720620. [CrossRef]
37. Mao, X.Q.; Wu, Y.; Wu, K.; Liu, M.; Zhang, J.F.; Zou, F.; Ou-Yang, J.P. Astragalus polysaccharide reduces hepatic endoplasmic reticulum stress and restores glucose homeostasis in a diabetic KKAy mouse model. *Acta Pharmacol. Sin.* **2007**, *28*, 1947–1956.
38. Alvarez-Guaita, A.; Blanco-Munoz, P.; Meneses-Salas, E.; Wahba, M.; Pollock, A.H.; Jose, J.; Casado, M.; Bosch, M.; Artuch, R.; Gaus, K.; et al. Annexin A6 is critical to maintain glucose homeostasis and survival during liver regeneration in mice. *Hepatology* **2020**, *72*, 2149–2164.
39. Abu Rmilah, A.; Zhou, W.; Nelson, E.; Lin, L.; Amiot, B.; Nyberg, S.L. Understanding the marvels behind liver regeneration. *Wiley Interdiscip. Rev. Dev. Biol.* **2019**, *8*, e340.
40. Valdecantos, M.P.; Pardo, V.; Ruiz, L.; Castro-Sanchez, L.; Lanzon, B.; Fernandez-Millan, E.; Garcia-Monzon, C.; Arroba, A.I.; Gonzalez-Rodriguez, A.; Escriva, F.; et al. A novel glucagon-like peptide 1/glucagon receptor dual agonist improves steatohepatitis and liver regeneration in mice. *Hepatology* **2017**, *65*, 950–968. [CrossRef]
41. Tang, N.; Zhang, J.; Fu, X.; Xie, W.; Qiu, Y. PP2Aalpha inhibits PFKFB2-induced glycolysis to promote termination of liver regeneration. *Biochem. Biophys. Res. Commun.* **2020**, *526*, 1–7. [CrossRef]
42. Feng, J.; Li, J.; Wu, L.; Yu, Q.; Ji, J.; Wu, J.; Dai, W.; Guo, C. Emerging roles and the regulation of aerobic glycolysis in hepatocellular carcinoma. *J. Exp. Clin. Cancer Res.* **2020**, *39*, 126. [CrossRef]
43. Du, D.; Liu, C.; Qin, M.; Zhang, X.; Xi, T.; Yuan, S.; Hao, H.; Xiong, J. Metabolic dysregulation and emerging therapeutical targets for hepatocellular carcinoma. *Acta Pharm. Sin. B* **2022**, *12*, 558–580. [CrossRef]
44. Shimizu, T.; Togo, S.; Kumamoto, T.; Makino, H.; Morita, T.; Tanaka, K.; Kubota, T.; Ichikawa, Y.; Nagasima, Y.; Okazaki, Y.; et al. Gene expression during liver regeneration after partial hepatectomy in mice lacking type 1 tumor necrosis factor receptor. *J. Surg. Res.* **2009**, *152*, 178–188.
45. Hata, S.; Namae, M.; Nishina, H. Liver development and regeneration: From laboratory study to clinical therapy. *Dev. Growth Differ.* **2007**, *49*, 163–170. [CrossRef]

46. Johnson, D.E.; O'Keefe, R.A.; Grandis, J.R. Targeting the IL-6/JAK/STAT3 signalling axis in cancer. *Nat. Rev. Clin. Oncol.* **2018**, *15*, 234–248. [CrossRef]
47. Ibrahim, S.; Weiss, T.S. Augmenter of liver regeneration: Essential for growth and beyond. *Cytokine Growth Factor Rev.* **2019**, *45*, 65–80.
48. Otsuka, N.; Yoshioka, M.; Abe, Y.; Nakagawa, Y.; Uchinami, H.; Yamamoto, Y. Reg3alpha and Reg3beta expressions followed by JAK2/STAT3 activation play a pivotal role in the acceleration of liver hypertrophy in a rat ALPPS model. *Int J. Mol. Sci* **2020**, *21*, 4077. [CrossRef]
49. Xiong, Y.; Torsoni, A.S.; Wu, F.; Shen, H.; Liu, Y.; Zhong, X.; Canet, M.J.; Shah, Y.M.; Omary, M.B.; Liu, Y.; et al. Hepatic NF-kB-inducing kinase (NIK) suppresses mouse liver regeneration in acute and chronic liver diseases. *eLife* **2018**, *7*, e34152.
50. Wen, X.; Huan, H.; Wang, X.; Chen, X.; Wu, L.; Zhang, Y.; Liu, W.; Bie, P.; Xia, F. Sympathetic neurotransmitters promote the process of recellularization in decellularized liver matrix via activating the IL-6/Stat3 pathway. *Biomed. Mater.* **2016**, *11*, 065007.
51. Chen, J.; Gao, P.; Peng, L.; Liu, T.; Wu, F.; Xu, K.; Chen, L.; Tan, F.; Xing, P.; Wang, Z.; et al. Downregulation of STK25 promotes autophagy via the Janus kinase 2/signal transducer and activator of transcription 3 pathway in colorectal cancer. *Mol. Carcinog.* **2022**, *61*, 572–586. [PubMed]
52. Zheng, X.; Gou, Y.; Jiang, Z.; Yang, A.; Yang, Z.; Qin, S. Icaritin-induced FAM99A affects GLUT1-mediated glycolysis via regulating the JAK2/STAT3 pathway in hepatocellular carcinoma. *Front. Oncol.* **2021**, *11*, 740557.
53. Zhao, Q.; Bai, J.; Chen, Y.; Liu, X.; Zhao, S.; Ling, G.; Jia, S.; Zhai, F.; Xiang, R. An optimized herbal combination for the treatment of liver fibrosis: Hub genes, bioactive ingredients, and molecular mechanisms. *J. Ethnopharmacol.* **2022**, *297*, 115567. [PubMed]
54. Ren, F.; Li, J.; Wang, Y.; Wang, Y.; Feng, S.; Yuan, Z.; Qian, X. The Effects of Angelica sinensis polysaccharide on tumor growth and iron metabolism by regulating hepcidin in tumor-bearing mice. *Cell. Physiol. Biochem.* **2018**, *47*, 1084–1094. [CrossRef] [PubMed]
55. Bai, Z.; Ye, Y.; Ye, X.; Yuan, B.; Tang, Y.; Wei, J.; Jin, M.; Wang, G.; Li, X. Leptin promotes glycolytic metabolism to induce dendritic cells activation via STAT3-HK2 pathway. *Immunol. Lett.* **2021**, *239*, 88–95. [PubMed]
56. Pu, Z.; Xu, M.; Yuan, X.; Xie, H.; Zhao, J. Circular RNA circCUL3 accelerates the Warburg effect progression of gastric cancer through regulating the STAT3/HK2 axis. *Mol. Ther. Nucleic Acids* **2020**, *22*, 310–318. [CrossRef] [PubMed]
57. Li, Y.; Wang, Y.; Liu, Z.; Guo, X.; Miao, Z.; Ma, S. Atractylenolide I induces apoptosis and suppresses glycolysis by blocking the JAK2/STAT3 signaling pathway in colorectal cancer cells. *Front. Pharmacol.* **2020**, *11*, 273. [CrossRef] [PubMed]
58. Li, M.; Jin, R.; Wang, W.; Zhang, T.; Sang, J.; Li, N.; Han, Q.; Zhao, W.; Li, C.; Liu, Z. STAT3 regulates glycolysis via targeting hexokinase 2 in hepatocellular carcinoma cells. *Oncotarget* **2017**, *8*, 24777–24784.
59. Zhou, Y.; Chen, S.; Yang, F.; Zhang, Y.; Xiong, L.; Zhao, J.; Huang, L.; Chen, P.; Ren, L.; Li, H.; et al. Rabeprazole suppresses cell proliferation in gastric epithelial cells by targeting STAT3-mediated glycolysis. *Biochem. Pharmacol.* **2021**, *188*, 114525.
60. Ren, Z.; Ding, T.; He, H.; Wei, Z.; Shi, R.; Deng, J. Mechanism of selenomethionine inhibiting of PDCoV replication in LLC-PK1 cells based on STAT3/miR-125b-5p-1/HK2 signaling. *Front. Immunol.* **2022**, *13*, 952852.
61. Mitchell, C.; Willenbring, H. A reproducible and well-tolerated method for 2/3 partial hepatectomy in mice. *Nat. Protoc.* **2008**, *3*, 1167–1170. [PubMed]

molecules MDPI

Review

Mechanism of Citri Reticulatae Pericarpium as an Anticancer Agent from the Perspective of Flavonoids: A Review

Li Song [1], Peiyu Xiong [1], Wei Zhang [1], Hengchang Hu [1], Songqi Tang [2], Bo Jia [1] and Wei Huang [1,*]

[1] College of Basic Medicine, Chengdu University of Traditional Chinese Medicine, Chengdu 610000, China
[2] College of Traditional Chinese Medicine, Hainan Medical University, Haikou 571199, China
* Correspondence: gracehw@126.com

Abstract: Citri Reticulatae Pericarpium (CRP), also known as "chenpi", is the most common qi-regulating drug in traditional Chinese medicine. It is often used to treat cough and indigestion, but in recent years, it has been found to have multi-faceted anti-cancer effects. This article reviews the pharmacology of CRP and the mechanism of the action of flavonoids, the key components of CRP, against cancers including breast cancer, lung cancer, prostate cancer, hepatic carcinoma, gastric cancer, colorectal cancer, esophageal cancer, cervical cancer, bladder cancer and other cancers with a high diagnosis rate. Finally, the specific roles of CRP in important phenotypes such as cell proliferation, apoptosis, autophagy and migration–invasion in cancer were analyzed, and the possible prospects and deficiencies of CRP as an anticancer agent were evaluated.

Keywords: Citri Reticulatae Pericarpium; flavonoids; anticancer; mechanism; phenotype

Citation: Song, L.; Xiong, P.; Zhang, W.; Hu, H.; Tang, S.; Jia, B.; Huang, W. Mechanism of Citri Reticulatae Pericarpium as an Anticancer Agent from the Perspective of Flavonoids: A Review. *Molecules* **2022**, *27*, 5622. https://doi.org/10.3390/molecules27175622

Academic Editors: Sokcheon Pak and Soo Liang Ooi

Received: 29 July 2022
Accepted: 25 August 2022
Published: 31 August 2022

1. Introduction

Citri Reticulatae Pericarpium is a commonly used traditional Chinese medicine derived from the ripe peel of the Rutaceae plant *Citrus reticulate Blanco* and its cultivars [1], which was first recorded in *Shen Nong Ben Cao Jing* and has a history of thousands of years in China. As a botanical medicine with the same origin of medicine and food, CRP has various pharmacological effects, which can be used alone or combined with other traditional Chinese medicines to form many well-known classical prescriptions. It is widely used in the clinical treatment of diseases of various systems and has outstanding advantages for diseases of the respiratory and digestive system, especially diseases with cough, expectoration, nausea and vomiting as the main symptoms.

The clinical efficacy of CRP has been affirmed, especially its contribution in cancer treatment. Researchers began to explore its mechanism, trying to find out its countermeasures against cancer in cancer cell lines and animal models, and concluded that the natural compounds flavonoids contained in CRP are the implementers of its anti-cancer effect. According to reports, flavonoids of CRP have outstanding performance in terms of regulating key signaling pathways and related effectors, blocking the cancer cell cycle to resist proliferation, inducing cell apoptosis, enhancing autophagy and inhibiting cell migration and invasion. Thanks to these research results and data, the core code of CRP as an anticancer agent has gradually been revealed.

As a novel anticancer agent, CRP has received extensive attention, and many studies have confirmed that CRP and its active ingredients have inhibitory effects on cancer. This review aims to collect and introduce the mechanism of CRP and its components in inhibiting cancer, focusing on breast cancer, lung cancer, prostate cancer, hepatic carcinoma, gastric cancer, colorectal cancer, esophageal cancer, cervical cancer, bladder cancer and other cancers with a high diagnosis rate.

2. Pharmacological Effects and Chemical Composition of CRP

Pharmacological studies have found that when applied to the digestive system, the efficacy of CRP is to ameliorate gastrointestinal smooth muscle activity, accelerate gastric emptying and intestinal push, alter gut microbiota, protect the esophagus and gastrointestinal mucosa and resist peptic ulcer [2–7]. To date, CRP has proven effects in the respiratory system, including but not limited to inhibiting airway inflammation, fighting acute lung injury and pulmonary fibrosis [8–13]. In terms of the cardiovascular system, CRP can improve cardiac insufficiency, alleviate cardiac hypertrophy and myocardial fibrosis, prevent hyperlipidemia and keep cardiomyocytes away from the damage of hypercholesterolemia [14–18]. Experiments have found that CRP's neuroprotective properties confer the ability to prevent neurodegeneration caused by Parkinson's, Alzheimer's, Huntington's disease and multiple sclerosis [19–22]. Unlike the above systems, CRP has relatively few studies in the related direction of the urinary system and has been reported to be anti-inflammatory and enhance renal function [23–25]. It is worth mentioning that CRP exhibits anti-inflammatory activity not only in the urinary system, but also in various systems of the body. Nobiletin is considered to be a marker of anti-inflammatory effect of CRP, while Xue N. et al. suggested that naringenin inhibits reactive oxygen species (ROS) and inflammatory cytokines involved in the process of CRP against inflammatory damage [26,27]. The antioxidant capacity of CRP is also an important part of its pharmacological action, and it is positively correlated with the aging of CRP [28]. Data proves that long-term storage benefits the quality of CRP, in which phenolic acids play a key role [29]. Moreover, the antioxidant activity of CRP is closely related to the fungi represented by Aspergillus niger growing on its surface, which can promote the transformation of flavonoids in CRP to increase its content [30].

The composition of natural compounds of CRP is complex. Through supercritical CO_2 extraction [31,32], Soxhlet extraction [1,33,34], pressurized liquid extraction (Wei L) and high-speed countercurrent separation [35], 161 constituents from CRP have been extracted and identified, including 65 flavonoids, 51 phenolic acids, 27 fatty acids and 18 amino acids [29,36]. Certainly, flavonoids are the most relevant class of all compounds with the pharmacological effects of CRP, and flavanone, flavone, and polymethoxyflavone aglycones, flavanone-and flavone-O-glycosides and flavone-C-glycosides are all the compounds it contains [37–39]. Due to its abundant content, hesperidin has been used as a chemical reference for CRP quality control by the *Chinese Pharmacopoeia*. Among them, in terms of exerting medical effects, the representative flavonoids are tangeretin and nobiletin, belonging to the polymethoxyflavones, and naringin and hesperidin, belonging to the flavanone O-glacosides (Table 1, Figures 1 and 2).

Figure 1. (**A**) Citrus reticulata Blanco tree (**B**) fresh ripe citrus (**C**) fresh mature pericarps (**D**) CRP.

Table 1. The Flavonoids in CRP.

Classification	Chemical Component	Molecular Formula	Molecular Mass	References
Polymethoxyflavones	Nobiletin	$C_{21}H_{22}O_8$	402	[40]
	3,5,6,7,8,3′,4′-Heptamethoxyflavone	$C_{22}H_{24}O_9$	433	[40]
	5-Hydroxy-6,7,8,3′,4′-pentamethoxyflavone	$C_{20}H_{20}O_8$	388	[40]
	Tangeretin	$C_{20}H_{20}O_7$	373	[40]
	Monohydroxy-trimethoxyflavone	$C_{18}H_{16}O_6$	329	[40]
	Monohydroxy-tetramethoxyflavone	$C_{19}H_{18}O_7$	359	[40]
	Monohydroxy-pentamethoxyflavone	$C_{20}H_{20}O_8$	389	[40]
	Trihydroxy-dimethoxyflavone	$C_{17}H_{14}O_7$	331	[40]
	Trihydroxy-trimethoxyflavone	$C_{18}H_{16}O_8$	361	[40]
	Isosinensetin	$C_{20}H_{20}O_7$	373	[40]
	Monohydroxy-hexamethoxyflavone	$C_{21}H_{22}O_9$	419	[40]
	Tetramethoxyflavone	$C_{19}H_{18}O_6$	343	[40]
	Hexamethoxyflavone	$C_{21}H_{22}O_8$	403	[40]
	Sinensetin	$C_{20}H_{20}O_7$	373	[40]
	Tetramethyl-O-isoscutellarein	$C_{19}H_{18}O_6$	343	[40]
	Dihydroxy-trimethoxyflavone	$C_{18}H_{16}O_7$	345	[40]
	Trimethoxyflavone	$C_{18}H_{16}O_5$	313	[40]
	Pentamethoxyflavone	$C_{20}H_{20}O_7$	373	[40]
	Tetramethyl-O-scutellarein	$C_{19}H_{18}O_6$	343	[40]
	Dihydroxy-tetramethoxyflavone	$C_{19}H_{18}O_8$	375	[40]
	Dihydroxy-pentmethpxyflavone	$C_{20}H_{20}O_9$	405	[39]
	Natsudaidai	$C_{19}H_{18}O_7$	359	[40]
Flavanone O-glycosides	Hesperidin	$C_{28}H_{34}O_{15}$	611	[40]
	Naringin	$C_{27}H_{32}O_{14}$	581	[40]
	Eriocitrin	$C_{27}H_{32}O_{15}$	597	[40]
	Neohesperidin	$C_{27}H_{32}O_{15}$	597	[40]
	Narirutin	$C_{27}H_{32}O_{14}$	581	[40]
	Prunin	$C_{21}H_{22}O_{10}$	435	[40]
	Neohesperidin	$C_{28}H_{34}O_{15}$	611	[40]
	Poncirin	$C_{28}H_{32}O4_{14}$	595	[40]
	Didymin	$C_{28}H_{34}O_{14}$	595	[40]
	Melitidin	$C_{33}H_{40}O_{18}$	725	[40]
Flavone O-glycosides	Rhoifolin	$C_{27}H_{30}O_{14}$	579	[40]
	Hexamethoxyflavone-o-glucoside	$C_{27}H_{33}O_{13}$	565	[40]
	Luteolin-7-O-rutinoside	$C_{27}H_{30}O_{15}$	595	[40]
	Diosmin	$C_{28}H_{32}O_{15}$	609	[40]
	Neodiosmin	$C_{28}H_{32}O_{15}$	609	[40]
	Sudachiin C or B	$C_{30}H_{34}O_{17}$	667	[40]
Flavone C-glycosides	Vicenin-2	$C_{27}H_{30}O_{15}$	595	[40]
	Diosmetin-6,8-di-C-glucoside	$C_{28}H_{32}O_{16}$	625	[40]
	Lucenin-2	$C_{27}H_{30}O_{16}$	611	[40]
	Apigenin-8-C-glucoside	$C_{21}H_{20}O_{10}$	433	[40]
	Diosmetin-6-C-glucoside	$C_{22}H_{22}O_{11}$	463	[40]
	Apigenin-6,8-di-C-glucoside	$C_{27}H_{30}O_2$	595	[39]
	Chysoeriol-6,8-di-C-glucoside	$C_{28}H_{32}O_{16}$	625	[39]
Flavanone aglycones	Naringenin	$C_{15}H_{13}O_5$	273	[40]
	Hesperetin	$C_{16}H_{14}O_6$	303	[40]

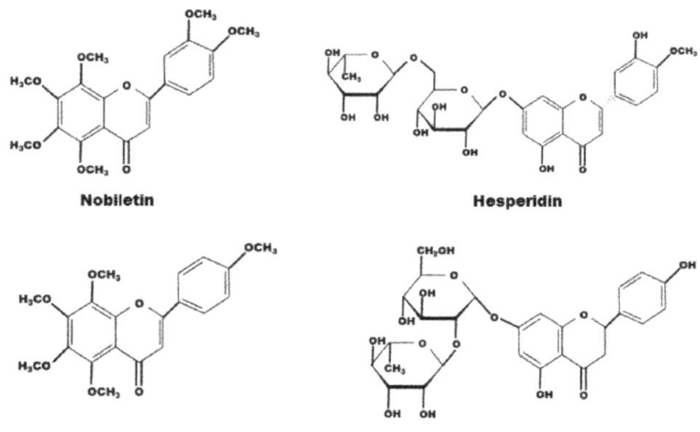

Figure 2. The chemical structure of nobiletin, hesperidin, tangeretin and naringin contained in CRP.

3. Epidemiological Investigation of Cancer

Cancer develops from the clonal expansion of abnormal cells inside the body [41]. As changes in the prevalence and distribution of the main risk factors, cancer incidence and mortality are rapidly growing worldwide. Cancer was the leading cause of premature death in 57 countries until 2020 [42]. Based on cancer incidence and mortality from the GLOBOCAN 2020 and The World Health Organization database, the most commonly diagnosed cancers worldwide were female breast cancer, lung and prostate cancer; the most common cause of cancer death were lung, liver and stomach cancers [43,44]. Female breast cancer has now surpassed lung cancer as the leading cause of global cancer incidence in 2020 with more than 6 million deaths [45]. The causes of cancer are multifaceted, with multiple external factors combined with internal genetic changes leading to cancer, but it mainly originates from both environment and genetics [46]. Unhealthy lifestyles, such as cigarette smoking, alcohol abuse, excess fat and red meat intake and lack of fiber, can also affect the development of cancer [47,48]. Conventional treatment modalities for cancer include surgery, radiation therapy, chemotherapy, targeted therapy, hormonal therapy and immunotherapy [49]. However, current treatments are often accompanied by various physical and psychological side effects, which severely impact the prognosis and life expectancy of patients. Throughout the past few years, both clinical and laboratory studies of the treatment of cancer through traditional Chinese medicine have gained great attention. The efficacy and safety of herbal medicine make it unique in the treatment of cancer. At the same time, Chinese medicine can also be used as an adjuvant to reduce the side effects of conventional cancer treatment.

4. The Performance of CRP in the Typical Phenotype of Cancer

Broadly speaking, cancer is a disease caused by dysregulation of cell proliferation [50]. The manifestation of cell proliferation is cell division, and mitosis is the main means of cell division in eukaryotes, which is periodic. The cell cycle goes through 4 phases: G1, S, G2 and M; and 3 checkpoints: G1/S, G2/M and SAC [51]. Apparently, this program is regulated by proliferation signals, cyclins and corresponding cyclin-dependent kinases (CDKs) [52]. Inactivation or mutation of growth suppressor genes and overexpression of oncogenes lead to uncontrolled proliferation of cancer cells and upregulation of cyclin-CDKs' expression, which in turn affects cell cycle progression and mitosis [53]. However, mTOR, which coordinates growth metabolism, suppresses the PI3K/AKT pathway to attenuate its anti-proliferative effect [54]. From the perspective of cancer metabolism, one of the biological markers of cancer, the Warburg effect, a form of energy metabolism manifested in aerobic glycolysis, enhances the proliferation and division, invasion and anti-apoptosis of cancer cells [55]. It is worth noting that in the G1 phase, on the one hand, cyclin D1 inactivates the inhibition of a mitochondria through the action of a series of factors, such as NRF-1, PPARγ and PGC-1α [56–58]. On the other hand, during G1 phase in cancer cells, PKM2 affected by PFK1, Ras, HIF-1 and PI3K/AKT/mTOR upregulates the gene CCND1 encoding cyclin D1 by increasing c-Myc expression and promoting β-catenin transactivation [59–61]. Conversely, PKM2 activated under the action of proliferative signals, such as EGFR, NF-κB and AKT, can also upregulate the expression of HIF-1, STAT3 and c-Myc through the STAT3 signaling pathway to maintain cell cycle progression [62–64]. Naringin and tangeretin were determined to affect cyclin D and beta-catenin, resulting in cell cycle arrest, which in turn counteracts the disordered proliferation of cancer cells. In addition, in the anti-proliferation process of CRP, effectors on JAK/STAT, PI3K/AKT/mTOR pathways are also involved (Figure 3).

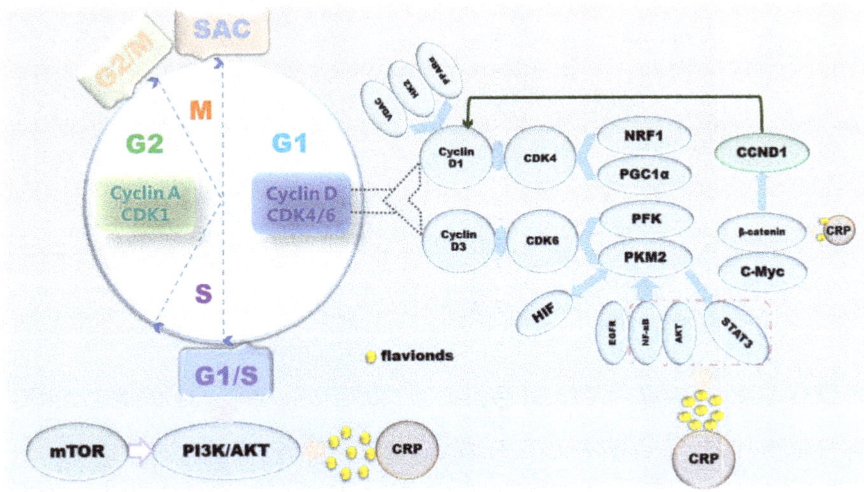

Figure 3. The mechanism of CRP causing cancer cell cycle arrest.

Interestingly, some proteins are involved in both cell proliferation and apoptosis. For example, p53 not only acts on its key effectors PUMA and p21 to induce cell cycle arrest to inhibit proliferation, but also acts as an upstream activator of pro-apoptotic proteins to initiate apoptosis [65,66]. It is reasonable to think that the cycle arrest caused by the regulation of P53 by hesperidin and tangeretin is the result of this.

The morphological changes of apoptosis are cytoplasmic shrinkage, chromatin pyknosis, nuclear fragmentation and plasma membrane blebbing, which finally form an apoptotic body [67]. Biochemical changes that can be observed during this process include caspase activation, DNA and protein breakdown, membrane changes and recognition by phagocytes [68]. The intrinsic apoptotic pathway is due to growth factor loss, DNA damage, ER stress, ROS overload, replication stress, microtubule alterations and mitotic defects, resulting in irreversible enhancement of mitochondrial outer membrane permeability and the release of pro-apoptotic factors [67,69]. The BCL-2 family is an indispensable regulatory protein in the internal apoptosis pathway, including the pro-apoptotic BAX, BAK, BAD, BCL-XS, BID, BIK, BIM and Hrk, and the anti-apoptotic BCL-2, BCL -Xl, BCL-W, BFL-1 and MCL-1 [70]. Under the action of cytotoxic signals, BH3-like proteins, such as BIM, BAD, PUMA and Noxa, transmit the signals to the downstream BAX and BAK. Under normal conditions, BAX located in the cytoplasm is enriched in mitochondria and becomes an activated state [71–73]. During this process, BAK is also activated. The apoptosis-inhibiting proteins BCL-2, BCL-XL, MCL-1, etc., can combine with BAX and BAK to suppress their activity [74]. Cytochrome *c* is released under the action of these two diametrically opposed proteins and then binds to Apaf-1 under the induction of apoptotic factors, such as Smac,

DIABLO and Omi/HtrA2, to form apoptosome to further activate caspase-3 [75]. The apoptosis inhibitor IAP family binds to and prevents the activation of caspase, forming a negative feedback mechanism [76]. Another apoptosis-inducing factor induces apoptosis in a caspase-independent manner. Excessive DNA damage leads to the onset of the PPAR-1-dependent cell death program; AIF then translocates from the mitochondria to the nucleus, followed by nuclear condensation, phosphatidylserine exposure at the plasma membrane and mitochondrial transmembrane collapse, a process unaffected by caspase inhibitors [77]. The extrinsic apoptosis pathway is completed by death receptors that are members of the TNFR superfamily, and the death receptors that receive cytotoxic signals bind to their ligands. For instance, Fas/FasL, the most representative one, activates pro-caspase-8 into caspase-8 through the combination of FADD and DED and then activates other executioners of the caspase family in turn [78–80]. Activated caspase-8 activates BAX and BAK by cleaving BID to release cytochrome c, a link that links the extrinsic and intrinsic apoptotic pathways. Activated caspase-3 and caspase-7 cleave ICAD, which is an inhibitor of caspase-activated DNase, and then release CAD, resulting in DNA fragmentation and ultimately, complete apoptosis [78]. CRP is involved in the release of related proteins on the internal/external apoptotic pathway, as well as upstream and downstream signaling pathways, including but not limited to ERK and PPARγ-dependent or independent (Figure 4).

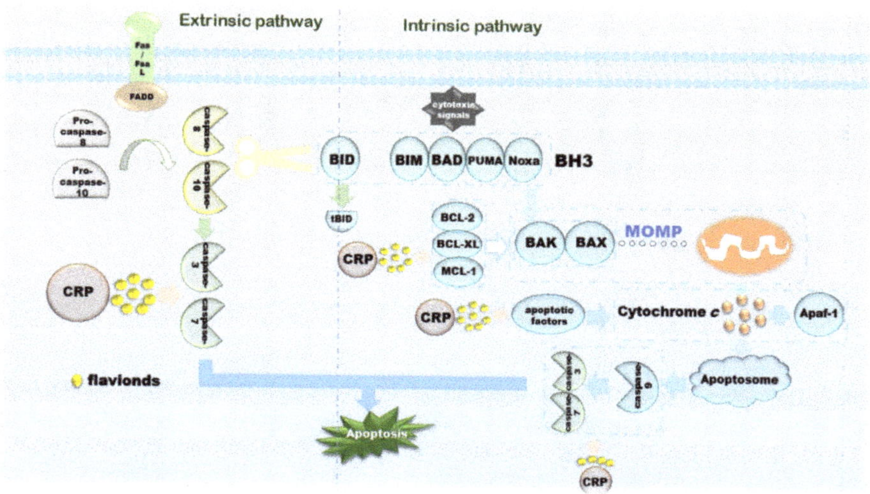

Figure 4. The mechanism of CRP promoting apoptosis of cancer cells.

Autophagy suppresses tumors by eliminating oncogenic substrate proteins, toxic unfolded proteins and damaged organelles, preventing chronic tissue damage and cancer development; conversely, autophagy-mediated intracellular recycling promotes tumor growth metabolism in cancer [81]. Autophagy was originally thought to be tumor-suppressive, as the absence of the autophagy-related protein ATG6 has been observed in human breast, ovarian and other cancers [82]. The mTOR and ULK complexes mediated by AMPK, p53, PI3K/AKT and MAPK/ERK cooperate with ATG on the autophagic membrane to form autophagosomes, which in turn control the entire autophagy process [83–86]. The formation of autophagosomes involves a series of proteins and multimolecular complexes, involving BECLIN 1, LC3, Rab complex, etc. [86]. The damaged organelles are captured by the autophagosome membrane, recognized by the autophagy substrate p62, and then degraded by lysosomes [87]. Autophagy inhibition is beneficial to tumor growth, as its deficiency leads to the accumulation of p62, and the binding of p62 to mTORC1 inhibits autophagy and activates NF-κB and NRF-2, which further promotes tumor cell proliferation [88–90].

From a metabolic point of view, the existence of autophagy allows cells to still metabolize in a starved state, which means that the activation of autophagy is sufficient to maintain tumor cell survival [87,91]. CRP has not been researched in the field of autophagy for a long time, but it is currently certain that its effects on autophagy include but are not limited to regulating the PI3K/AKT/GSK-3c/mTOR pathway (Figure 5).

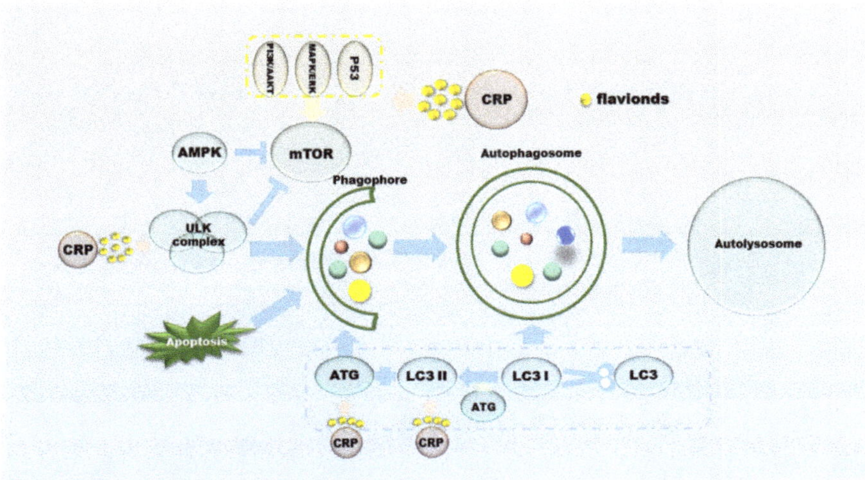

Figure 5. The role of CRP in the process of autophagy in cancer cells.

Cancer metastasis is the spread of cancer cells to distant organs, and the invasion–migration of cancer cells during this process is controlled by multiple factors. The process of cell invasion–migration consists of the following aspects: (a) local invasion and cell migration of the basement membrane; (b) intravascular deep vasculature and/or lymphatic system; (c) survival in the circulation; (d) arrest and extravasation at distant organ sites; and (e) colonization at metastatic sites [92]. Protocols for single cell invasion involve protease, actin cytoskeleton, integrin-dependent, integrin-independent and Rho- and Rock/MLCK-dependent modes of migration into mesenchymal migration and glide-like "amoematoid-migration" [93,94]. Epithelial–mesenchymal transitions (EMT) during cell migration and invasion is regulated by transcription factors such as slug, Twist, ZEB1 and ZEB2, which inhibit E-cadherin, the cornerstone of the epithelial state [95,96]. In addition, the double negative feedback loop formed by miR-200 and ZEB1/ZEB2 is another important mechanism for regulating EMT [96]. The invasion–metastasis cascade of cancer cells begins with the destruction of BM, and its damage is the result of MMPs as enzymatically hydrolyzed active proteins [97,98]. Indeed, the effects of MMPs on cancer are multifaceted, including inflammation, cell proliferation, extracellular matrix (ECM) degradation, cell migration, resistance to cell death, replicative immortality and metastatic niches; the most critical role is the degradation and remodeling of the ECM during cancer cell invasion and metastasis [99]. One of the sources of TGF-β, which has the function of inhibiting tumor cell differentiation, is the inactive precursor after proteolysis of MMP-9. At the same time, the regulation of VEGF by MMP-9 can promote tumor angiogenesis, which is beneficial to tumor colonization [100–102]. In addition, TGF-β1 can also be activated by MMP-14 and MMP-2 [103]. All three of the above proteins cleave LTBP-1 of the ECM to indirectly regulate the activity of TGF-β [104,105]. NF-κB enhances the production of MMPs, and its ability is restricted by TIMPs, with which it can form complexes that inhibit proteolysis [106,107]. In particular, MMP-7 inhibits apoptosis by cleaving Fas ligands, making it a place in the process of apoptosis [108]. The ability of hesperidin, naringin to downregulate MMPs has been demonstrated, which allows it to successfully reverse EMT (Figure 6).

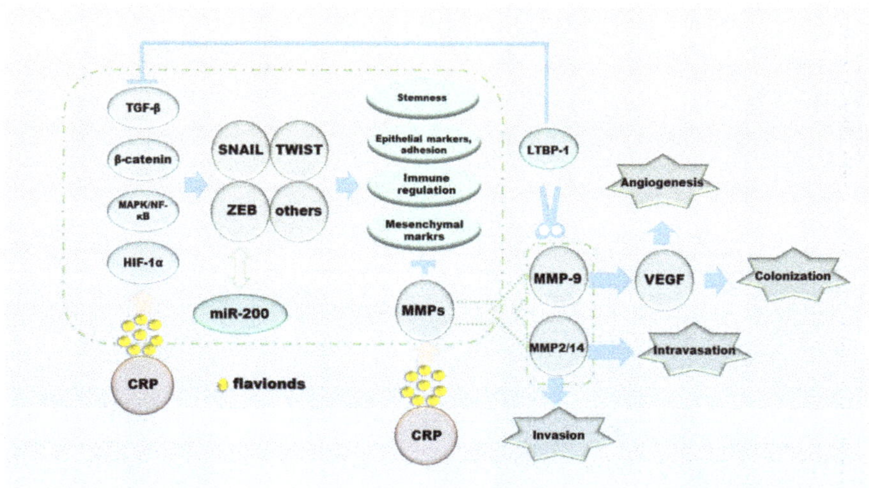

Figure 6. The effect of CRP during the migration of cancer cells.

5. Inhibitory Effect of CRP and Its Active Components on Cancer

5.1. Breast Cancer

With the highest diagnosis rate in women, breast cancer severely threatens the survival and quality of life of women around the world. The mechanism of CRP in the treatment of breast cancer is still under study. According to the existing results, flavonoids such as nobiletin, tangeretin, hesperidin and naringin play a crucial role. The study found that nobiletin can simultaneously inhibit the ERK1/2 and PI3K/AKT pathways to suppress the growth of TNBC MDA-MB-468 cells and perform anti-tumor effects through anti-proliferation and induction of apoptosis [109]. Tangeretin inhibits breast cancer cell metastasis by targeting TP53, PTGS2, MMP9 and PIK3CA and regulating the PI3K/AKT pathway [110]. Tangeretin inhibits the formation of BCSCs and targets BCSCs by inhibiting the Stat3/Sox2 signaling pathway, thereby treating breast cancer and BCSCs [111]. Tangeretin also attenuates DMBA-induced oxidative stress, reduces kidney DNA damage and has chemo preventive activity against DMBA-induced breast cancer with cellular engraftment [112,113]. Hesperidin exhibits a concentration-dependent cytotoxicity effect on human breast cancer cell line MCF-7, which induces apoptosis and causes DNA damage [114,115]. The combination of hesperidin and chlorogenic acid modulates mitochondrial and ATP production via the estrogen receptor pathway and synergistically inhibits the growth of MCF-7 [116]. Naringin inhibits cell proliferation and promotes apoptosis and G1 cycle arrest through regulating the β-catenin pathway, thereby suppressing the growth potential of TNBC cells [117]. What's more, hesperidin performed the inhibitory activity of the proliferation of MCF-7-GFP-Tubulin cells, fought against drug-resistant cancer cells and amelioratde the cell migration of MDA-MB 231 cells [118–120].

5.2. Lung Cancer

As the second most common cancer worldwide, lung cancer is the leading cause of cancer death in 2020, with incidence and mortality rates approximately twice as high in men as in women [45]. Under the current situation of extremely low survival rate, how to improve the rate and the quality of life of patients diagnosed with lung cancer has become the main subject of medical research. CRP is good at treating respiratory system diseases, and pharmacological research has carried out in-depth exploration on the treatment of lung cancer. Smoking greatly increases the risk of lung cancer. Hesperidin is able to down-regulate the expression of MMPs and enhance antioxidant status to combat nicotine toxicity and suppression of smoking-induced lung cancer [121]. The antioxidant capacity of hesperidin also inhibits tumor cell proliferation in benzo(a)pyrene-induced lung cancer mouse models [122]. Another study found that inhibition of NSCLC cells proliferation and promotion of apoptosis through the miR-132/ZEB2 signaling pathway may be one of the mechanisms by which hesperidin alleviates NSCLC [123]. Hesperidin also induces apoptosis through the mitochondrial pathway, up-regulates the expression of P21 and P53 to triggers G0/G1 phase arrest in A549 cells and down-regulates cyclin D1 for anti-proliferation [124,125]. Another set of experiments by Xia R found that blocking the SDF-1/CXCR-4 pathway to inhibit the migration of A549 cells and the suppression of EMT phenotype transformation are also the approach for hesperidin to prevent tumors and its metastasis [126]. Taking A549 cells as the research object, another important natural compound of CRP, naringin, attenuates the EGF-induced MUC5AC mucin and mRNA overexpression by inhibiting the synergistic activity of MAPKs/AP-1 and IKKs/IκB/NF-κB signaling pathways [127]. Chen M found that naringin exhibited the capacity to inhibit PI3K/AKT/mTOR and NF-κB pathways and activate the expression of miR-126 in H69 cells, thereby preventing cell growth and inducing apoptosis in SCLC cells [128].

5.3. Prostate Cancer

Prostate cancer is the most common malignancy among men worldwide, with 1.4 million cases diagnosed in 2016 and more than 380,000 deaths [129]. A series of experimental studies have demonstrated that flavonoids in CRP have positive effects on weakening cell viability and inducing cytotoxicity in prostate cancer. The expressions of NF-κB and HIF-1α were down-regulated in nobiletin-treated prostate cell lines DU145 and PC-3 cells, accompanied by decreased phosphorylation of AKT, which in turn impairs cell viability [130]. Coincidentally, tangeretin also down-regulated the expression of AKT and AR in C4-2 cells and synergistically antagonized the resistance of CRPC cells to sorafenib or cisplatin [131]. In PC-3 cells, the pathways of tangeretin-induced cytotoxicity include not only caspase-30mediated apoptosis, but also inhibition of PI3K/AKT/mTOR pathway to reverse the EMT process [132].

5.4. Liver Cancer

Primary liver cancer, including hepatocellular carcinoma (HCC) and intrahepatic cholangiocarcinoma, was the sixth most commonly diagnosed cancer and the third leading cause of cancer death globally in 2020, with HCC accounting for 75–85% of these [45,133]. Hence, the exploration of CRP in the treatment of liver cancer mainly focuses on the direction of HCC. Zheng J. et al. verified that the regulation of JNK/Bcl-2/BECLIN1 pathway-mediated autophagy is the mechanism by which tangeretin antagonizes the proliferation and migration of HepG2 cells [134]. Hesperidin' fight against HCC cells' invasiveness is achieved by suppressing the activities of NF-κB and AP-1 to down-regulate the expression and secretion of MMP-9 in acetaldehyde- and TPA-induced HCC [135,136]. The pro-apoptotic protein BAX is the key to cell apoptosis, and up-regulation of BAX to induce apoptosis in HepG2 cells is an effective way for hesperidin and naringin to inhibit liver cancer [137,138]. The fact that naringin reduces cell proliferation in DEN-induced hepatocarcinoma rats is manifested by a marked decrease in AgNOR/nuclear and PCNA levels, as well as altered DNA fragmentation in liver tissue [139].

5.5. Gastric Cancer

According to statistics, in 2020, there were more than 1 million new cases of gastric cancer, causing 769,000 deaths, ranking fourth and fifth in the world in morbidity and mortality, respectively [45]. Relying on the long-term clinical experience of CRP in the treatment of digestive system diseases, related experiments have thus explored multiple mechanisms of anti-gastric cancer. Endoplasmic reticulum stress mediates apoptosis and autophagy. Nobiletin down-regulates AKT/mTOR signaling pathway to promote endoplasmic reticulum stress response in SNU-16 cells, which may be an integral part of its anticancer activity [140]. Mitochondrial apoptosis mediated by activated caspase-9 and Fas/Fas L synergistically enables tangeretin to fulfill its mission of inhibiting AGS cells [141]. Mitochondrial-mediated apoptosis is also applicable to the killing effect of hesperidin on AGS cells. The level of reactive oxygen species in AGS cells after hesperidin intervention increases, and the MAPK signaling pathway is regulated to induce cell apoptosis [142]. Radiation therapy is one of the main methods for the treatment of tumors at present, through downregulating the expression of Notch-, Jagged1/2, Hey-1 and Hes-1, downregulating the expression of miR-410, causing attenuated invasion and migration in GC cells; tangeretin greatly enhanced the radiosensitivity of GC cells [143].

5.6. Colorectal Cancer

As the third most common cancer in terms of incidence and the second in mortality, colorectal cancer accounts for one tenth of all diagnosed cancers and deaths [45]. Similar to gastric cancer, the advantages of CRP in digestive diseases are also reflected in colorectal cancer. Nobiletin and its major metabolites M1, M2 and M3 in the colon have established roles in cell cycle arrest and apoptosis, thus effectively inhibiting AOM/DSS induced colitis-associated colon cancer in CD-1 male mice [144]. In rectal cancer, the mechanism of nobiletin is downregulating MMP-7 gene expression to inhibit the invasion and metastasis of cancer cells [145]. Tangeretin and 5-FU synergistically up-regulate P21 in HCT-116 cells, which in turn activates the P53-mediated DNA damage response and triggers apoptosis via the JNK pathway, suggesting that the combination of tangeretin and 5-FU suppresses the autophagy pathway, enabling cancer cells susceptible to oxidative stress-induced programmed cell death [146]. Inhibition of colon cancer by hesperidin involves multiple alterations, including caspase-3-mediated apoptosis, the autophagy program initiated by PI3K/Akt/GSK-3c and mTOR pathway and, down-regulation of NF-κB and its target molecules iNOS and COX-2 to attenuate oxidative stress and enhance antioxidants to fight tumor-induced inflammation [147–150]. The above phenomenon also occurred in naringin-treated CRC cells, where the suppression of the PI3K/AKT/mTOR pathway resulted in ameliorated abnormal proliferation and apoptosis [151].

5.7. Esophageal Cancer

Esophageal cancer, another intractable disease of the digestive system, is also accompanied by a particularly high fatality rate, with about 1 in every 18 cancer patients dying from it [45]. The way that naringin combats esophageal cancer is to suppress the proliferation and colony formation and the invasion of Eca109 cells by regulating related proteins to block the JAK/STAT signaling pathway, so as to promote cell apoptosis [152]. In nude mice, synephrine has a significant inhibitory effect on ESCC xenografts, and in vitro experiments have observed that it down-regulates Galectin-3 to inactivate the AKT/ERK pathway in ESCC cells [153].

5.8. Cervical Cancer

Although cervical cancer is the fourth leading cause of cancer death in women, the most common cancer and the leading cause of cancer death in many countries worldwide, it is considered a preventable cancer [45]. The inhibition of NEU3 activity in HeLa cells and A549 cells mediated by naringin resulted in the accumulation of GM3 ganglioside, which further led to the weakening of EGFR signaling, and finally resulted in cell growth restriction. At the same time, the down-regulation of phosphorylation of EGFR and ERK was also involved in the process [154]. Lin R's data demonstrated that naringin abolishes Wnt/β-catenin signaling and ultimately triggers cell cycle arrest at G0/G1 phase in Cervical cancer cells, while ER stress-induced cell killing is also a pathway for naringin to act on [155]. Naringin also induces cell cycle arrest in the G2/M phase, inhibits cell growth and induces apoptosis via the NF-κB/COX-2-caspase-1 pathway, thereby exerting its anticancer activity on SiHa cells and HeLa cells [156,157].

5.9. Bladder Cancer

The incidence of bladder cancer in men is much higher than women, making it the sixth most common cancer and the ninth cause of cancer death [45]. Apoptosis is an important mechanism of CRP against bladder cancer discovered in current research. Nobiletin-induced apoptosis is accomplished through the regulation of endoplasmic reticulum stress via PERK/elF2α/ATF4/CHOP pathway and PI3K/AKT/mTOR pathway, and its inhibitory effect on BFTC cell growth is positively correlated with concentration range [158]. Anti-tumor formation through apoptosis is also suitable for tangeretin by inducing the release of pro-apoptotic factors such as cytochrome c to form an apoptotic complex with activated caspase-9, thereby initiating the apoptotic response and disrupting mitochondrial function, resulting in BFTC-905 cells being cytotoxic [159].

5.10. Other Cancers with High Diagnosis Rate

In addition to the above-mentioned cancers, CRP is quite successful in the treatment of other common cancers. The mechanism of nobiletin against osteosarcoma metastasis is to down-regulate the expression of MMP-2 and MMP-9 via ERK/JNK pathways and inhibit the movement, migration and invasion of U2OS and HOS cells through activation of NF-κB, CREB and SP-1 proteins [160]. Moreover, naringin suppresses the migration and invasion of human chondrosarcoma by up-regulating the expression of miR-126 and downregulating VCAM-1 expression [161]. By inhibiting MAPK and AKT/protein kinase B signaling pathway and downregulating cell cycle-related factors, nobiletin achieves the purpose of combating glioma cell proliferation and migration [162]. Suppression of the cyclin-D/cdc-2 complex formation leads to cell cycle arrest at G2/M arrest, decreased glioblastoma cell growth after tangeretin treatment, increased G2/M phase cells and induces apoptosis [163]. Blocking the MAPKs signaling pathway, downregulating the activity and expression of MMPs, thereby inhibiting the invasion, migration and adhesion of U87 cells, is a non-negligible characteristic of naringin's anti-metastatic properties [164]. Negative effects on cell proliferation by inhibiting the FAK/cyclin D1 pathway, promoting apoptosis via affecting the FAK/BADs pathway and attenuating cell invasion and metastasis through the FAK/MMPs pathway are another example of naringin in the treatment of glioblastoma cells [165]. Naringin-treated Walker 256 carcinosarcoma rats inhibited tumor growth, down-regulated the expression of IL-6 and TNF-α and significantly prolonged the survival rate without the occurrence of cachexia [166]. Induction of apoptosis by activating caspase-3 and up-regulation of intracellular ROS and blocking cell cycle progression in G2 phase are the main ways that hesperidin reduces the viability of gallbladder cancer cells [167]. Inhibition of MAPK pathway, STAT3 activation and down-regulation of ER signaling pathway in the genome seems to be an important mechanism by which hesperidin induces apoptosis or autophagy in ECC-1 cancer cells [168]. Hesperidin inhibits mesothelioma cell growth by inducing apoptosis by downregulating the mRNA and protein expression levels of Sp1 and its regulatory proteins [169]. Hesperidin triggers apoptosis in lymphocyte lineages in a PPARγ-dependent or PPARγ-independent manner and inactivates NF-κB, which in turn sensitizes Ramos cells to chemotherapeutic agent-induced apoptosis [170]. The mechanism of action of hesperidin in NALM-6 cells is manifested in multiple aspects. It can not only play pro-apoptotic and anti-proliferative effects via PPARγ-dependent and PPARγ-independent pathways, but also affect apoptosis and cytotoxicity through PI3K/AKT/IKK signaling pathway [171,172]. As early as 1998, some scholars have found that hesperidin has an inhibitory effect on 12-O-tetradecanoyl-13-phorbo lactate-induced skin tumor [173]. In vitro, hesperidin affected PD-L1 expression in HN6 cells and HN15 cells by reducing the phosphorylation of STAT1 and STAT3, thereby inhibiting cancer cell survival and avoiding evasion of antitumor immunity. Correspondingly, in vivo experiments found that hesperidin had a negative effect on 4-NQO-induced proliferation of rat oral cancer [174–176]. Nobiletin inhibited the proliferation of TCA-8113 cells and CAL-27 cells through cell cycle arrest in G1 phase, accompanied by changes in intracellular levels of acidified PKA and phosphorylated CREB, impaired mitochondrial function, glucose consumption and pyruvate and lactate production [177]. The antagonism of nobiletin on LPS- and INF-γ-induced PGE2, COX-2, NO endows it with the property of preventing inflammation-related tumors [178]. Nobiletin has positive and negative regulatory effects on MMPs and TMP-1, which is a specific manifestation of its interference with PI3K signaling pathway to suppress tumors [179].

6. Conclusions

To sum up, the performance of CRP, especially its flavonoids, in the fight against cancer, is worthy of recognition because, while it does not prevent the process of a certain aspect of cancer alone, it reverses or suppresses the development of cancer through various pathways, which is a characteristic that traditional anti-cancer agents lack compared to traditional Chinese medicine. Because of the multiple functions of CRP, its ability has not been thoroughly studied. In addition, CRP is also commendable as an advantage in that it is a natural product that is homologous to medicine and food, which makes it inexpensive and easy to obtain. However, the most prominent defect of CRP as an anti-cancer agent is the lack of clinical research, which will be an important content of CRP to be explored next.

Author Contributions: Conceptualization, L.S.; methodology, L.S.; resources, L.S., P.X., W.Z. and H.H.; writing—review and editing, L.S.; Supervision, W.H., S.T. and B.J. All authors have read and agreed to the published version of the manuscript.

Funding: This research was funded by National Natural Science Foundation Of China [Grant No.81860816].

Institutional Review Board Statement: Not applicable.

Informed Consent Statement: Not applicable.

Data Availability Statement: Not applicable.

Conflicts of Interest: The authors declare no conflict of interest.

References

1. Committee of National Pharmacopoeia. *Pharmacopoeia of RP China*; Chemical Industry Press: Beijing, China, 2015; p. 1329.
2. Li, W.; Zheng, T.Z.; Qu, S.Y.; Tian, Z.F.; Qiu, X.Q.; Ding, G.H.; Wei, Y.L. Effects of tangerine peel on gastric emptying and intestinal propulsion in mice. *Chin. Mater. Med. Clin.* **2002**, *18*, 22–23.
3. Chen, B.; Luo, J.K.; Han, Y.H.; Du, H.J.; Liu, J.; Wei, H.; Zhu, J.H.; Xiao, J.; Wang, J.; Cao, Y.; et al. Dietary Tangeretin Alleviated Dextran Sulfate Sodium-Induced Colitis in Mice via Inhibiting Inflammatory Response, Restoring Intestinal Barrier Function, and Modulating Gut Microbiota. *J. Agric. Food Chem.* **2021**, *69*, 7663–7674. [CrossRef] [PubMed]
4. Zhang, M.; Zhang, X.; Zhu, J.Y.; Zhao, D.G.; Ma, Y.Y.; Li, L.D.; He, Z.T.; Huang, Q.R. Bidirectional interaction of nobiletin and gut microbiota in mice fed with a high-fat diet. *Food Funct.* **2021**, *12*, 3516–3526. [CrossRef] [PubMed]
5. Wu, M.; Li, Y.; Gu, Y. Hesperidin Improves Colonic Motility in Loeramide-Induced Constipation Rat Model via 5-Hydroxytryptamine 4R/cAMP Signaling Pathway. *Digestion* **2020**, *101*, 692–705. [CrossRef]
6. Lee, J.A.; Shin, M.R.; Park, H.J.; Roh, S.S. Scutellariae Radix and Citri Reticulatae Pericarpium Mixture Regulate PPARγ/RXR Signaling in Reflux Esophagitis. *Evid. Based Complement Altern. Med.* **2022**, *2022*, 6969241. [CrossRef]
7. Bae, E.A.; Han, M.J.; Kim, D.H. In vitro anti-Helicobacter pylori activity of some flavonoids and their metabolites. *Planta Med.* **1999**, *65*, 442–443. [CrossRef]
8. Hao, Y.; Cheung, C.S.; Yip, W.C.; Ko, W.H. Nobiletin Stimulates Chloride Secretion in Human Bronchial Epithelia via a cAMP/PKA-Dependent Pathway. *Cell Physiol. Biochem.* **2015**, *37*, 306–320. [CrossRef]
9. Wei, D.J.; Ci, X.X.; Chu, X.; Wei, M.M.; Hua, S.C.; Deng, X.M. Hesperidin suppresses ovalbumin-induced airway inflammation in a mouse allergic asthma model. *Inflammation* **2012**, *35*, 114–121. [CrossRef]
10. Xu, J.J.; Liu, Z.; Tang, W.; Wang, G.C.; Hou, Z.X.; Liu, Q.Y.; Ling, Z.; Li, Y.L. Tangeretin from Citrus reticulate Inhibits Respiratory Syncytial Virus Replication and Associated Inflammation in vivo. *J. Agric. Food Chem.* **2015**, *63*, 9520–9527. [CrossRef]
11. Li, M.Q.; Zhao, Y.; Qi, D.; Jing, H.; Wang, D.X. Tangeretin attenuates lipopolysaccharide-induced acute lung injury through Notch signaling pathway via suppressing Th17 cell response in mice. *Microb. Pathog.* **2020**, *138*, 103826. [CrossRef]
12. Ding, Z.; Sun, G.; Zhu, Z. Hesperidin attenuates influenza A virus (H1N1) induced lung injury in rats through its anti-inflammatory effect. *Antivir. Ther.* **2018**, *23*, 611–615. [CrossRef]

13. Zhou, Z.; Kandhare, A.D.; Kandhare, A.A.; Bodhankar, S.L. Hesperidin ameliorates bleomycin-induced experimental pulmonary fibrosis via inhibition of TGF-beta1/Smad3/AMPK and IkappaBalpha/NF-kappaB pathways. *EXCLI J.* **2019**, *18*, 723–745. [PubMed]
14. Cheng, H.L.; Wu, X.D.; Ni, G.H.; Wang, S.Q.; Peng, W.J.; Zhang, H.F.; Gao, J.; Li, X.L. *Citri Reticulatae* Pericarpium protects against isoproterenol-induced chronic heart failure via activation of PPARγ. *Ann. Transl. Med.* **2020**, *8*, 1396. [CrossRef]
15. Chen, M.; Zhu, H.; Zhu, Q.; Wu, X.; Zhou, Y.; Gao, R.; Shi, M.; Zhang, T.; Yin, T.; Zhang, H. Citri Reticulatae Pericarpium alleviates postmyocardial infarction heart failure by upregulating PPARγ expression. *Clin. Exp. Pharmacol. Physiol.* **2022**, *49*, 661–673. [CrossRef] [PubMed]
16. Green, C.O.; Wheatley, A.O.; McGrowder, D.A.; Dilworth, L.L.; Asemota, H.N. Citrus peel polymethoxylated flavones extract modulates liver and heart function parameters in diet induced hypercholesterolemic rats. *Food Chem. Toxicol.* **2013**, *51*, 306–309. [CrossRef] [PubMed]
17. Zeng, W.; Huang, K.E.; Luo, Y.; Dong, X.L.; Chen, W.; Yu, X.Q.; Ke, X.H. Nontargeted urine metabolomics analysis of the protective and therapeutic effects of Citri Reticulatae Chachiensis Pericarpium on high-fat feed-induced hyperlipidemia in rats. *Biomed. Chromatogr.* **2020**, *34*, e4795. [CrossRef]
18. DU, Y.Z.; Su, J.; Yan, M.Q.; Chen, S.H.; Lyu, G.Y.; Yu, J.J. Improvement effect and mechanism of ethanol extract from Citri Reticulatae Pericarpium on triglyceride in hyperlipidemia model rat. *Zhongguo Zhong Yao Za Zhi* **2021**, *46*, 190–195.
19. Hwang, S.L.; Shih, P.H.; Yen, G.C. Neuroprotective effects of citrus flavonoids. *J. Agric. Food Chem.* **2012**, *60*, 877–885. [CrossRef]
20. Hajialyani, M.; Hosein, F.M.; Echeverría, J.; Nabavi, S.M.; Uriarte, E.; Sobarzo-Sánchez, E. Hesperidin as a Neuroprotective Agent: A Review of Animal and Clinical Evidence. *Molecules* **2019**, *24*, 648. [CrossRef]
21. Cirmi, S.; Ferlazzo, N.; Lombardo, G.E.; Ventura-Spagnolo, E.; Gangemi, S.; Calapai, G.; Navarra, M. Neurodegenerative Diseases: Might Citrus Flavonoids Play a Protective Role? *Molecules* **2016**, *21*, 1312. [CrossRef]
22. Kim, J.; Wie, M.B.; Ahn, M.; Tanaka, A.; Matsuda, H.; Shin, T. Benefits of hesperidin in central nervous system disorders: A review. *Anat. Cell Biol.* **2019**, *52*, 369–377. [CrossRef] [PubMed]
23. Sun, P.; Huang, R.; Qin, Z.; Liu, F. Influence of Tangeretin on the Exponential Regression of Inflammation and Oxidative Stress in Streptozotocin-Induced Diabetic Nephropathy. *Appl. Biochem. Biotechnol.* **2022**, *194*, 3914–3929. [CrossRef]
24. Wu, J.; Zhao, Y.M.; Deng, Z.K. Tangeretin ameliorates renal failure via regulating oxidative stress, NF-κB-TNF-α/iNOS signalling and improves memory and cognitive deficits in 5/6 nephrectomized rats. *Inflammopharmacology* **2018**, *26*, 119–132. [CrossRef] [PubMed]
25. Ali, A.M.; Gabbar, M.A.; Abdel-Twab, S.M.; Fahmy, E.M.; Ebaid, H.; Alhazza, I.M.; Ahmed, O.M. Antidiabetic Potency, Antioxidant Effects, and Mode of Actions of *Citrus reticulata* Fruit Peel Hydroethanolic Extract, Hesperidin, and Quercetin in Nicotinamide/Streptozotocin-Induced Wistar Diabetic Rats. *Oxid. Med. Cell Longev.* **2020**, *2020*, 1730492. [CrossRef]
26. Chen, X.M.; Tait, A.R.; Kitts, D.D. Flavonoid composition of orange peel and its association with antioxidant and anti-inflammatory activities. *Food Chem.* **2017**, *218*, 15–21. [CrossRef] [PubMed]
27. Xue, N.; Wu, X.; Wu, L.; Li, L.; Wang, F. Antinociceptive and anti-inflammatory effect of Naringenin in different nociceptive and inflammatory mice models. *Life Sci.* **2019**, *217*, 148–154. [CrossRef]
28. Yu, Q.; Tao, Y.; Huang, Y.; Zogona, D.; Wu, T.; Liu, R.; Pan, S.; Xu, X. Aged *Pericarpium Citri Reticulatae 'Chachi'* Attenuates Oxidative Damage Induced by *tert*-Butyl Hydroperoxide (*t*-BHP) in HepG2 Cells. *Foods* **2022**, *11*, 273. [CrossRef]
29. Bian, X.; Xie, X.; Cai, J.; Zhao, Y.; Miao, W.; Chen, X.; Xiao, Y.; Li, N.; Wu, J.L. Dynamic changes of phenolic acids and antioxidant activity of Citri Reticulatae Pericarpium during aging processes. *Food Chem.* **2022**, *373*, 131399. [CrossRef]
30. Wang, F.; Chen, L.; Li, F.Q.; Liu, S.J.; Chen, H.P.; Liu, Y.P. The Increase of Flavonoids in Pericarpium Citri Reticulatae (PCR) Induced by Fungi Promotes the Increase of Antioxidant Activity. *Evid. Based Complement Altern. Med.* **2018**, *2018*, 2506037. [CrossRef]
31. Lv, X.J.; XU, Y.; Dong, P.F.; Xu, L.Y.; Li, S.M.; Long, T.; Wang, Y.L. Optimization of Supercritical CO_2 Extraction of Polymethoxyflavonoes from Citri Reticulate Pericarpium by Response Surface Methodology. *Food Res. Develop.* **2019**, *40*, 16–20.
32. Long, T.; Lv, X.; Xu, Y.; Yang, G.; Xu, L.Y.; Li, S. Supercritical fluid CO_2 extraction of three polymethoxyflavones from Citri reticulatae pericarpium and subsequent preparative separation by continuous high-speed counter-current chromatography. *J. Chromatogr. B Analyt. Technol. Biomed. Life Sci.* **2019**, *1124*, 284–289. [CrossRef] [PubMed]
33. Luo, H.J.; Yang, Y.T.; Huang, S.G.; Pan, H.J.; Luo, M.X.; Zheng, G.D. Analysis and comparison of flavonoids extracted from tangerine peel by ultrasonic extraction and Soxhlet extraction. *Chinese Mater. Med.* **2016**, *39*, 371–374.
34. Anagnostopoulou, M.A.; Kefalas, P.; Kokkalou, E.; Assimopoulou, A.N.; Papageorgiou, V.P. Analysis of antioxidant compounds in sweet orange peel by HPLC-diode array detection-electrospray ionization mass spectrometry. *Biomed. Chromatogr.* **2005**, *19*, 138–148. [CrossRef] [PubMed]
35. Zheng, G.D.; Zhou, F.; Jiang, L.; Yang, D.P.; Yang, X.; Lin, L.W. Isolation and purification of polymethoxylated flavonoids from Pericarpium Citri Reticulatae by high-speed counter-current chromatography. *Chin. Tradit. Herbal Drugs* **2010**, *41*, 52–55.
36. Yu, X.; Sun, S.; Guo, Y.; Liu, Y.; Yang, D.; Li, G.; Lü, S. Citri Reticulatae Pericarpium (Chenpi): Botany, ethnopharmacology, phytochemistry, and pharmacology of a frequently used traditional Chinese medicine. *J. Ethnopharmacol.* **2018**, *220*, 265–282. [CrossRef] [PubMed]
37. Roowi, S.; Crozier, A. Flavonoids in tropical citrus species. *J. Agric. Food Chem.* **2011**, *59*, 12217–12225. [CrossRef]

38. Gattuso, G.; Barreca, D.; Gargiulli, C.; Leuzzi, U.; Caristi, C. Flavonoid composition of Citrus juices. *Molecules* **2007**, *12*, 1641–1673. [CrossRef]
39. Zheng, G.D.; Zhou, P.; Yang, H.; Li, Y.S.; Li, P.; Liu, E.H. Rapid resolution liquid chromatography-electrospray ionisation tandem mass spectrometry method for identification of chemical constituents in Citri Reticulatae Pericarpium. *Food Chem.* **2013**, *136*, 604–611. [CrossRef]
40. Duan, L.; Guo, L.; Liu, K.; Liu, E.H.; Li, P. Characterization and classification of seven citrus herbs by liquid chromatography-quadrupole time-of-flight mass spectrometry and genetic algorithm optimized support vector machines. *J. Chromatogr. A* **2014**, *1339*, 118–127. [CrossRef]
41. Martincorena, I.; Campbell, P.J. Somatic mutation in cancer and normal cells. *Science* **2015**, *349*, 1483–1489. [CrossRef]
42. Bray, F.; Laversanne, M.; Weiderpass, E.; Soerjomataram, I. The ever-increasing importance of cancer as a leading cause of premature death worldwide. *Cancer* **2021**, *127*, 3029–3030. [CrossRef] [PubMed]
43. Bray, F.; Ferlay, J.; Soerjomataram, I.; Siegel, R.; Torre, L.A.; Jemal, A. Global cancer statistics 2018: GLOBOCAN estimates of incidence and mortality worldwide for 36 cancers in 185 countries. *CA Cancer J. Clin.* **2018**, *68*, 394–424. [CrossRef] [PubMed]
44. Ferlay, J.; Colombet, M.; Soerjomataram, I.; Parkin, D.M.; Piñeros, M.; Znaor, A.; Bray, F. Cancer statistics for the year 2020: An overview. *Int. J. Cancer* **2021**, *149*, 778–789. [CrossRef]
45. Sung, H.; Ferlay, J.; Siegel, R.L.; Laversanne, M.; Soeriomataram, I.; Jemal, A.; Bray, F. Global Cancer Statistics 2020: GLOBOCAN Estimates of Incidence and Mortality Worldwide for 36 Cancers in 185 Countries. *CA Cancer J. Clin.* **2021**, *71*, 209–249. [CrossRef]
46. Parsa, N. Environmental factors inducing human cancers. *Iran. J. Public Health* **2012**, *41*, 1–9.
47. Mbemi, A.; Khanna, S.; Njiki, S.; Yedjou, C.G.; Tchounwou, P.B. Impact of Gene-Environment Interactions on Cancer Development. *Int. J. Environ. Res. Public Health* **2020**, *17*, 8089. [CrossRef] [PubMed]
48. Bagot, R.C.; Meaney, M.J. Epigenetics and the biological basis of gene x environment interactions. *J. Am. Acad. Child. Adolesc. Psychiatry* **2010**, *49*, 752–771.
49. Miller, K.D.; Nogueira, L.; Mariotto, A.B.; Rowland, J.H.; Yabroff, K.R.; Alfano, C.M.; Jemal, A.; Kramer, J.L.; Siegel, R.L. Cancer treatment and survivorship statistics, 2019. *CA Cancer J. Clin.* **2019**, *69*, 363–385. [CrossRef]
50. Evan, G.I.; Vousden, K.H. Proliferation, cell cycle and apoptosis in cancer. *Nature* **2001**, *411*, 342–348. [CrossRef]
51. Liu, J.; Peng, Y.; Wei, W. Cell cycle on the crossroad of tumorigenesis and cancer therapy. *Trends Cell Biol.* **2022**, *32*, 30–44. [CrossRef]
52. Malumbres, M. Cyclin-dependent kinases. *Genome Biol.* **2014**, *15*, 122. [CrossRef] [PubMed]
53. Icard, P.; Fournel, L.; Wu, Z.; Alifano, M.; Lincet, H. Interconnection between Metabolism and Cell Cycle in Cancer. *Trends Biochem. Sci.* **2019**, *44*, 490–501. [CrossRef]
54. O'Reilly, K.E.; Rojo, F.; She, Q.B.; Solit, D.; Mills, G.B.; Smith, D.; Lane, H.; Hofmann, F.; Hicklin, D.J.; Ludwig, D.L.; et al. mTOR inhibition induces upstream receptor tyrosine kinase signaling and activates Akt. *Cancer Res.* **2006**, *66*, 1500–1508. [CrossRef] [PubMed]
55. Hanahan, D.; Weinberg, R.A. Hallmarks of cancer: The next generation. *Cell* **2011**, *144*, 646–674. [CrossRef] [PubMed]
56. Dominy, J.E., Jr.; Lee, Y.; Gerhart-Hines, Z.; Puigserver, P. Nutrient-dependent regulation of PGC-1alpha's acetylation state and metabolic function through the enzymatic activities of Sirt1/GCN5. *Biochim. Biophys. Acta* **2010**, *1804*, 1676–1683. [CrossRef] [PubMed]
57. Wang, C.; Li, Z.; Lu, Y.; Du, R.; Katiyar, S.; Yang, J.; Fu, M.; Leader, J.E.; Quong, A.; Novikoff, P.M.; et al. Cyclin D1 repression of nuclear respiratory factor 1 integrates nuclear DNA synthesis and mitochondrial function. *Proc. Natl. Acad. Sci. USA* **2006**, *103*, 11567–11572. [CrossRef]
58. Lee, Y.; Dominy, J.E.; Choi, Y.J.; Jurczak, M.; Tolliday, N.; Camporez, J.P.; Chim, H.; Lim, J.H.; Ruan, H.B.; Yang, X.; et al. Cyclin D1-Cdk4 controls glucose metabolism independently of cell cycle progression. *Nature* **2014**, *510*, 547–551. [CrossRef]
59. Courtnay, R.; Ngo, D.C.; Malik, N.; Ververis, K.; Tortorella, S.M.; Karagiannis, T.C. Cancer metabolism and the Warburg effect: The role of HIF-1 and PI3K. *Mol. Biol. Rep.* **2015**, *42*, 841–851. [CrossRef]
60. Yang, W.; Xia, Y.; Ji, H.; Zheng, Y.; Liang, J.; Huang, W.; Gao, X.; Aldape, K.; Lu, Z. Nuclear PKM2 regulates β-catenin transactivation upon EGFR activation. *Nature* **2011**, *480*, 118–122. [CrossRef]
61. Icard, P.; Shulman, S.; Farhat, D.; Steyaert, J.M.; Alifano, M.; Lincet, H. How the Warburg effect supports aggressiveness and drug resistance of cancer cells? *Drug Resist. Updat.* **2018**, *38*, 1–11. [CrossRef]
62. Lu, Z.; Hunter, T. Metabolic Kinases Moonlighting as Protein Kinases. *Trends Biochem. Sci.* **2018**, *43*, 301–310. [CrossRef] [PubMed]
63. Prakasam, G.; Iqbal, M.A.; Bamezai, R.N.K.; Mazurek, S. Posttranslational Modifications of Pyruvate Kinase M2: Tweaks that Benefit Cancer. *Front. Oncol.* **2018**, *8*, 22. [CrossRef] [PubMed]
64. Liang, J.; Cao, R.; Zhang, Y.; Xia, Y.; Zheng, Y.; Li, X.; Wang, L.; Yang, W.; Lu, Z. PKM2 dephosphorylation by Cdc25A promotes the Warburg effect and tumorigenesis. *Nat. Commun.* **2016**, *7*, 12431. [CrossRef] [PubMed]
65. Chan, T.A.; Hwang, P.M.; Hermeking, H.; Kinzler, K.W.; Vogelstein, B. Cooperative effects of genes controlling the G(2)/M checkpoint. *Genes Dev.* **2000**, *14*, 1584–1588. [CrossRef]
66. Yu, J.; Zhang, L. The transcriptional targets of p53 in apoptosis control. *Biochem. Biophys. Res. Commun.* **2005**, *331*, 851–858. [CrossRef]

67. Galluzzi, L.; Vitale, I.; Aaronson, S.A.; Abrams, J.M.; Adam, D.; Agostinis, P.; Alnemri, E.S.; Altucci, L.; Amelio, I.; Andrews, D.W.; et al. Molecular mechanisms of cell death: Recommendations of the Nomenclature Committee on Cell Death 2018. *Cell Death Differ.* **2018**, *25*, 486–541. [CrossRef]
68. Kumar, V.; Abbas, A.K.; Fausto, N.; Aster, J.C. *Robins and Cotran: Pathologic Basis of Disease*; Saunders Elsevier: Philadelphia, PA, USA, 2010; pp. 25–32.
69. Danial, N.N.; Korsmeyer, S.J. Cell death: Critical control points. *Cell* **2004**, *116*, 205–219. [CrossRef]
70. Wong, R.S. Apoptosis in cancer: From pathogenesis to treatment. *J. Exp. Clin. Cancer Res.* **2011**, *30*, 87. [CrossRef]
71. Czabotar, P.E.; Lessene, G.; Strasser, A.; Adams, J.M. Control of apoptosis by the BCL-2 protein family: Implications for physiology and therapy. *Nat. Rev. Mol. Cell Biol.* **2014**, *15*, 49–63. [CrossRef]
72. Willis, S.N.; Fletcher, J.I.; Kaufmann, T.; Delft, M.F.; Chen, L.; Czabotar, P.E.; Lerino, H.; Lee, E.F.; Fairlie, W.D.; Bouillet, P.; et al. Apoptosis initiated when BH3 ligands engage multiple Bcl-2 homologs, not Bax or Bak. *Science* **2007**, *315*, 856–859. [CrossRef]
73. Kim, H.; Rafiuddin-Shah, M.; Tu, H.C.; Jeffers, J.R.; Zambetti, G.P.; Hsieh, J.J.D.; Cheng, E.H.Y. Hierarchical regulation of mitochondrion-dependent apoptosis by BCL-2 subfamilies. *Nat. Cell Biol.* **2006**, *8*, 1348–1358. [CrossRef] [PubMed]
74. Ku, B.; Liang, C.; Jung, J.U.; Oh, B.H. Evidence that inhibition of BAX activation by BCL-2 involves its tight and preferential interaction with the BH3 domain of BAX. *Cell Res.* **2011**, *21*, 627–641 [CrossRef]
75. Kroemer, G.; Galluzzi, L.; Brenner, C. Mitochondrial membrane permeabilization in cell death. *Physiol. Rev.* **2007**, *87*, 99–163. [CrossRef] [PubMed]
76. Varfolomeev, E.; Vucic, D. Inhibitor of apoptosis proteins: Fascinating biology leads to attractive tumor therapeutic targets. *Future Oncol.* **2011**, *7*, 633–648. [CrossRef] [PubMed]
77. Joza, N.; Pospisilik, J.A.; Hangen, E.; Hanada, T.; Modjtahedi, N.; Penninger, J.M.; Kroemer, G. AIF: Not just an apoptosis-inducing factor. *Ann. N. Y. Acad. Sci.* **2009**, *1171*, 2–11. [CrossRef]
78. Toda, S.; Nishi, C.; Yanagihashi, Y.; Segawa, K.; Nagata, S. Clearance of Apoptotic Cells and Pyrenocytes. *Curr. Top Dev. Biol.* **2015**, *114*, 267–295.
79. Strasser, A.; Jost, P.J.; Nagata, S. The many roles of FAS receptor signaling in the immune system. *Immunity* **2009**, *30*, 180–192. [CrossRef]
80. Schneider, P.; Tschopp, J. Apoptosis induced by death receptors. *Pharm. Acta Helv.* **2000**, *74*, 281–286. [CrossRef]
81. White, E. Deconvoluting the context-dependent role for autophagy in cancer. *Nat. Rev. Cancer* **2012**, *12*, 401–410. [CrossRef]
82. Liang, X.H.; Jackson, S.; Seaman, M.; Brown, K.; Kempkes, B.; Hibshoosh, H.; Levine, B. Induction of autophagy and inhibition of tumorigenesis by beclin 1. *Nature* **1999**, *402*, 672–676. [CrossRef]
83. White, E.; Mehnert, J.M.; Chan, C.S. Autophagy, Metabolism, and Cancer. *Clin. Cancer Res.* **2015**, *21*, 5037–5046. [CrossRef] [PubMed]
84. Lamb, C.A.; Yoshimori, T.; Tooze, S.A. The autophagosome: Origins unknown, biogenesis complex. *Nat. Rev. Mol. Cell Biol.* **2013**, *14*, 759–774. [CrossRef] [PubMed]
85. Rubinsztein, D.C.; Shpilka, T.; Elazar, Z. Mechanisms of autophagosome biogenesis. *Curr. Biol.* **2012**, *22*, R29–R34. [CrossRef] [PubMed]
86. Tanida, I. Autophagosome formation and molecular mechanism of autophagy. *Antioxid. Redox Signal.* **2011**, *14*, 2201–2214. [CrossRef]
87. Rabinowitz, J.D.; White, E. Autophagy and metabolism. *Science* **2010**, *330*, 1344–1348. [CrossRef]
88. Duran, A.; Amanchy, R.; Linares, J.F.; Joshi, J.; Abu-Baker, S.; Porollo, A.; Hansen, M.; Moscat, J.; Diaz-Meco, M.T. P62 is a key regulator of nutrient sensing in the mTORC1 pathway. *Mol. Cell* **2011**, *44*, 134–146. [CrossRef]
89. Duran, A.; Linares, J.F.; Galvez, A.S.; Wikenheiser, K.; Flores, J.M.; Diaz-Meco, M.T.; Moscat, J. The signaling adaptor p62 is an important NF-kappaB mediator in tumorigenesis. *Cancer Cell* **2008**, *13*, 343–354. [CrossRef]
90. Komatsu, M.; Kurokawa, H.; Waguri, S.; Taguchi, K.; Kobayashi, A.; Ichimura, Y.; Sou, Y.S.; Ueno, I.; Sakamoto, A.; Tong, K.I.; et al. The selective autophagy substrate p62 activates the stress responsive transcription factor Nrf2 through inactivation of Keap1. *Nat. Cell Biol.* **2010**, *12*, 213–223. [CrossRef]
91. Degenhardt, K.; Mathew, R.; Beaudoin, B.; Bray, K.; Anderson, D.; Chen, G.; Mukherjee, C.; Shi, Y.; Gélinas, C.; Fan, Y.; et al. Autophagy promotes tumor cell survival and restricts necrosis, inflammation, and tumorigenesis. *Cancer Cell* **2006**, *10*, 51–64. [CrossRef]
92. Zanotelli, M.R.; Zhang, J.; Reinhart-King, C.A. Mechanoresponsive metabolism in cancer cell migration and metastasis. *Cell Metab.* **2021**, *33*, 1307–1321. [CrossRef]
93. Friedl, P.; Wolf, K. Tumour-cell invasion and migration: Diversity and escape mechanisms. *Nat. Rev. Cancer* **2003**, *3*, 362–374. [CrossRef] [PubMed]
94. Wolf, K.; Friedl, P. Molecular mechanisms of cancer cell invasion and plasticity. *Br. J. Dermatol.* **2006**, *154* (Suppl. S1), 11–15. [CrossRef] [PubMed]
95. Thiery, J.P.; Acloque, H.; Huang, R.Y.; Nieto, M.A. Epithelial-mesenchymal transitions in development and disease. *Cell* **2009**, *139*, 871–890. [CrossRef] [PubMed]
96. Lu, W.; Kang, Y. Epithelial-Mesenchymal Plasticity in Cancer Progression and Metastasis. *Dev. Cell* **2019**, *49*, 361–374. [CrossRef]
97. Valastyan, S.; Weinberg, R.A. Tumor metastasis: Molecular insights and evolving paradigms. *Cell* **2011**, *147*, 275–292. [CrossRef]

98. Kessenbrock, K.; Plaks, V.; Werb, Z. Matrix metalloproteinases: Regulators of the tumor microenvironment. *Cell* **2010**, *141*, 52–67. [CrossRef]
99. Siddhartha, R.; Garg, M. Molecular and clinical insights of matrix metalloproteinases into cancer spread and potential therapeutic interventions. *Toxicol. Appl. Pharmacol.* **2021**, *426*, 115593. [CrossRef]
100. Yu, Q.; Stamenkovic, I. Cell surface-localized matrix metalloproteinase-9 proteolytically activates TGF-beta and promotes tumor invasion and angiogenesis. *Genes Dev.* **2000**, *14*, 163–176. [CrossRef]
101. Bergers, G.; Brekken, R.; McMahon, G.; Vu, T.H.; Itoh, T.; Tamaki, K.; Tanzawa, K.; Thorpe, P.; Itohara, S.; Werb, Z.; et al. Matrix metalloproteinase-9 triggers the angiogenic switch during carcinogenesis. *Nat. Cell Biol.* **2000**, *2*, 737–744. [CrossRef]
102. Kaplan, R.N.; Riba, R.D.; Zacharoulis, S.; Bramley, A.H.; Vincent, L.; Costa, C.; Macdonald, D.D.; Jin, D.K.; Shido, K.; Kerns, S.A.; et al. VEGFR1-positive haematopoietic bone marrow progenitors initiate the pre-metastatic niche. *Nature* **2005**, *438*, 820–827. [CrossRef]
103. Mu, D.; Cambier, S.; Fjellbirkeland, L.; Baron, J.L.; Munger, J.S.; Kawakatsu, H.; Sheppard, D.; Broaddus, V.C.; Nishimura, S.L. The integrin alpha(v)beta8 mediates epithelial homeostasis through MT1-MMP-dependent activation of TGF-beta1. *J. Cell Biol.* **2002**, *157*, 493–507. [CrossRef] [PubMed]
104. Dallas, S.L.; Rosser, J.L.; Mundy, G.R.; Bonewald, L.F. Proteolysis of latent transforming growth factor-beta (TGF-beta)-binding protein-1 by osteoclasts. A cellular mechanism for release of TGF-beta from bone matrix. *J. Biol. Chem.* **2002**, *277*, 21352–21360. [CrossRef]
105. Tatti, O.; Vehviläinen, P.; Lehti, K.; Keski-Oja, J. MT1-MMP releases latent TGF-beta1 from endothelial cell extracellular matrix via proteolytic processing of LTBP-1. *Exp. Cell Res.* **2008**, *314*, 2501–2514. [CrossRef] [PubMed]
106. Deryugina, E.I.; Quigley, J.P. Matrix metalloproteinases and tumor metastasis. *Cancer Metastasis Rev.* **2006**, *25*, 9–34. [CrossRef] [PubMed]
107. Bond, M.; Chase, A.J.; Baker, A.H.; Newby, A.C. Inhibition of transcription factor NF-kappaB reduces matrix metalloproteinase-1, -3 and -9 production by vascular smooth muscle cells. *Cardiovasc. Res.* **2001**, *50*, 556–565. [CrossRef]
108. Wang, W.S.; Chen, P.M.; Wang, H.S.; Liang, W.Y.; Su, Y. Matrix metalloproteinase-7 increases resistance to Fas-mediated apoptosis and is a poor prognostic factor of patients with colorectal carcinoma. *Carcinogenesis* **2006**, *27*, 1113–1120. [CrossRef] [PubMed]
109. Chen, C.; Ono, M.; Takeshima, M.; Nakano, S. Antiproliferative and apoptosis-inducing activity of nobiletin against three subtypes of human breast cancer cell lines. *Anticancer Res.* **2014**, *34*, 1785–1792. [PubMed]
110. Hermawan, A.; Putri, H.; Hanif, N.; Ikawati, M. Integrative Bioinformatics Study of Tangeretin Potential Targets for Preventing Metastatic Breast Cancer. *Evid. Based Complement Alternat. Med.* **2021**, *2021*, 2234554. [CrossRef]
111. Ko, Y.C.; Choi, H.S.; Liu, R.; Kim, J.H.; Kim, S.L.; Yun, B.S.; Lee, D.S. Inhibitory Effects of Tangeretin, A Citrus Peel-Derived Flavonoid, on Breast Cancer Stem Cell Formation through Suppression of Stat3 Signaling. *Molecules* **2020**, *25*, 2599. [CrossRef]
112. Lakshmi, A.; Subramanian, S.P. Tangeretin ameliorates oxidative stress in the renal tissues of rats with experimental breast cancer induced by 7, 12-dimethylbenz[a]anthracene. *Toxicol. Lett.* **2014**, *229*, 333–348. [CrossRef]
113. Gul, H.F.; Ilhan, N.; Ilhan, N.; Ozercan, I.H.; Kuloglu, T. The combined effect of pomegranate extract and tangeretin on the DMBA-induced breast cancer model. *J. Nutr. Biochem.* **2021**, *89*, 108566. [CrossRef] [PubMed]
114. Natarajan, N.; Thamaraiselvan, R.; Lingaiah, H.; Srinivasan, P.; Periyasamy, B.M. Effect of flavonone hesperidin on the apoptosis of human mammary carcinoma cell line MCF-7. *Biomed. Prev. Nutr.* **2011**, *1*, 207–215. [CrossRef]
115. Sulaiman, G.M.; Waheeb, H.M.; Jabir, M.S.; Khazaal, S.H.; Dewir, Y.H.; Naidoo, Y. Hesperidin Loaded on Gold Nanoparticles as a Drug Delivery System for a Successful Biocompatible, Anti-Cancer, Anti-Inflammatory and Phagocytosis Inducer Model. *Sci. Rep.* **2020**, *10*, 9362. [CrossRef] [PubMed]
116. Hsu, P.H.; Chen, W.H.; Juan-Lu, C.; Hsieh, S.C.; Lin, S.C.; Mai, R.T.; Chen, S.Y. Hesperidin and Chlorogenic Acid Synergistically Inhibit the Growth of Breast Cancer Cells via Estrogen Receptor/Mitochondrial Pathway. *Life* **2021**, *11*, 950. [CrossRef] [PubMed]
117. Li, H.; Yang, B.; Huang, J.; XIang, T.; Yin, X.; Wan, J.; Luo, F.; Zhang, L.; Li, H.; Ren, G. Naringin inhibits growth potential of human triple-negative breast cancer cells by targeting β-catenin signaling pathway. *Toxicol. Lett.* **2013**, *220*, 219–228. [CrossRef] [PubMed]
118. Lee, C.J.; Wilson, L.; Jordan, M.A.; Nguyen, V.; Tang, J.; Smiyun, G. Hesperidin suppressed proliferations of both human breast cancer and androgen-dependent prostate cancer cells. *Phytother. Res.* **2010**, *24*, S15–S19. [CrossRef]
119. El-Sisi, A.E.; Sokkar, S.S.; Ibrahim, H.A.; Hamed, M.F.; Abu-Risha, S.E. Targeting MDR-1 gene expression, BAX/BCL2, caspase-3, and Ki-67 by nanoencapsulated imatinib and hesperidin to enhance anticancer activity and ameliorate cardiotoxicity. *Fundam. Clin. Pharmacol.* **2020**, *34*, 458–475. [CrossRef]
120. Kongtawelert, P.; Wudtiwai, B.; Shwe, T.H.; Pothacharoen, P.; Phitak, T. Inhibitory Effect of Hesperidin on the Expression of Programmed Death Ligand (PD-L1) in Breast Cancer. *Molecules* **2020**, *25*, 252. [CrossRef]
121. Balakrishnan, A.; Menon, V.P. Effect of hesperidin on matrix metalloproteinases and antioxidant status during nicotine-induced toxicity. *Toxicology* **2007**, *238*, 90–98. [CrossRef]
122. Kamaraj, S.; Ramakrishnan, G.; Anandakumar, P.; Jagan, S.; Devaki, T. Antioxidant and anticancer efficacy of hesperidin in benzo(a)pyrene induced lung carcinogenesis in mice. *Investig. New Drugs* **2009**, *27*, 214–222. [CrossRef]
123. Tan, S.; Dai, L.; Tan, P.; Liu, W.; Mu, Y.; Wang, J.; Huang, X.; Hou, A. Hesperidin administration suppresses the proliferation of lung cancer cells by promoting apoptosis via targeting the miR-132/ZEB2 signalling pathway. *Int. J. Mol. Med.* **2020**, *46*, 2069–2077. [CrossRef]

124. Xia, R.; Sheng, X.; Xu, X.; Yu, C.; Lu, H. Hesperidin induces apoptosis and G0/G1 arrest in human non-small cell lung cancer A549 cells. *Int. J. Mol. Med.* **2018**, *41*, 464–472. [CrossRef] [PubMed]
125. Kamaraj, S.; Anandakumar, P.; Jagan, S.; Ramakrishnan, G.; Periyasamy, P.; Asokkumar, S.; Subramanian, R.; Devaki, T. Hesperidin inhibits cell proliferation and induces mitochondrial-mediated apoptosis in human lung cancer cells through down regulation of β-catenin/c-myc. *Biocatal. Agric. Biotechnol.* **2019**, *18*, 101065. [CrossRef]
126. Xia, R.; Xu, G.; Huang, Y.; Sheng, X.; Xu, X.; Lu, H. Hesperidin suppresses the migration and invasion of non-small cell lung cancer cells by inhibiting the SDF-1/CXCR-4 pathway. *Life Sci.* **2018**, *201*, 111–120. [CrossRef] [PubMed]
127. Nie, Y.C.; Wu, H.; Li, P.B.; Xie, L.M.; Luo, Y.L.; Shen, J.G.; Su, W.W. Naringin attenuates EGF-induced MUC5AC secretion in A549 cells by suppressing the cooperative activities of MAPKs-AP-1 and IKKs-IκB-NF-κB signaling pathways. *Eur. J. Pharmacol.* **2012**, *690*, 207–213. [CrossRef]
128. Chen, M.; Peng, W.; Hu, S.; Deng, J. miR-126/VCAM-1 regulation by naringin suppresses cell growth of human non-small cell lung cancer. *Oncol. Lett.* **2018**, *16*, 4754–4760. [CrossRef]
129. Global Burden of Disease Cancer Collaboration; Fitzmaurice, C.; Akinyemiju, T.F.; Lami, F.H.A.; Alam, T.; Alizadeh-Navaei, R.; Allen, C.; Alsharif, U.; Alvis-Guzman, N.; Amini, E.; et al. Global, Regional, and National Cancer Incidence, Mortality, Years of Life Lost, Years Lived With Disability, and Disability-Adjusted Life-Years for 29 Cancer Groups, 1990 to 2016: A Systematic Analysis for the Global Burden of Disease Study. *JAMA Oncol.* **2018**, *4*, 1553–1568. [CrossRef]
130. Chen, J.; Creed, A.; Chen, A.Y.; Huang, H.; Li, Z.; Rankin, G.O.; Ye, X.; Xu, G.; Chen, Y.C. Nobiletin suppresses cell viability through AKT pathways in PC-3 and DU-145 prostate cancer cells. *BMC Pharmacol. Toxicol.* **2014**, *15*, 59. [CrossRef]
131. Zhang, N.; Wu, W.; Huang, Y.; An, L.; He, Z.; Chang, Z.; He, Z.; Lai, Y. Citrus Flavone Tangeretin Inhibits CRPC Cell Proliferation by Regulating Cx26, AKT, and AR Signaling. *Evid. Based Complement. Alternat. Med.* **2022**, *2022*, 6422500. [CrossRef]
132. Zhu, W.B.; Xiao, N.; Liu, X.J. Dietary flavonoid tangeretin induces reprogramming of epithelial to mesenchymal transition in prostate cancer cells by targeting the PI3K/Akt/mTOR signaling pathway. *Oncol. Lett.* **2018**, *15*, 433–440. [CrossRef]
133. Petrick, J.L.; Florio, A.A.; Znaor, A.; Ruggieri, D.; Laversanne, M.; Alvarez, C.S.; Ferlay, J.; Valery, P.C.; Bray, F.; McGlynn, K.A. International trends in hepatocellular carcinoma incidence, 1978–2012. *Int. J. Cancer* **2020**, *147*, 317–330. [CrossRef] [PubMed]
134. Zheng, J.; Shao, Y.; Jiang, Y.; Chen, F.; Liu, S.; Yu, N.; Zhang, D.; Liu, X.; Zou, L. Tangeretin inhibits hepatocellular carcinoma proliferation and migration by promoting autophagy-related BECLIN1. *Cancer Manag. Res.* **2019**, *11*, 5231–5242. [CrossRef] [PubMed]
135. Yeh, M.H.; Kao, S.T.; Hung, C.M.; Liu, C.J.; Lee, K.H.; Yeh, C.C. Hesperidin inhibited acetaldehyde-induced matrix metalloproteinase-9 gene expression in human hepatocellular carcinoma cells. *Toxicol. Lett.* **2009**, *184*, 204–210. [CrossRef] [PubMed]
136. Lee, K.H.; Yeh, M.H.; Kao, S.T.; Hung, C.M.; Liu, C.J.; Huang, Y.Y.; Yeh, C.C. The inhibitory effect of hesperidin on tumor cell invasiveness occurs via suppression of activator protein 1 and nuclear factor-kappaB in human hepatocellular carcinoma cells. *Toxicol. Lett.* **2010**, *194*, 42–49. [CrossRef] [PubMed]
137. Naz, H.; Tarique, M.; Ahamad, S.; Alajmi, M.F.; Hussain, A.; Rehman, M.T.; Luqman, S.; Hassan, M.I. Hesperidin-CAMKIV interaction and its impact on cell proliferation and apoptosis in the human hepatic carcinoma and neuroblastoma cells. *J. Cell Biochem.* **2019**, *120*, 15119–15130. [CrossRef]
138. Banjerdpongchai, R.; Wudtiwai, B.; Khawon, P. Induction of Human Hepatocellular Carcinoma HepG2 Cell Apoptosis by Naringin. *Asian Pac. J. Cancer Prev.* **2016**, *17*, 3289–3294.
139. Prabu, T.; Manju, V. Antiproliferative and apoptotic effects of naringin on diethylnitrosamine induced hepatocellular carcinoma in rats. *Biomed. Aging Pathol.* **2013**, *3*, 59–64.
140. Moon, J.Y.; Cho, S.K. Nobiletin Induces Protective Autophagy Accompanied by ER-Stress Mediated Apoptosis in Human Gastric Cancer SNU-16 Cells. *Molecules* **2016**, *21*, 914. [CrossRef]
141. Dong, Y.; Cao, A.; Shi, J.; Yin, P.; Wang, L.; Ji, G.; Xie, J.; Wu, D. Tangeretin, a citrus polymethoxyflavonoid, induces apoptosis of human gastric cancer AGS cells through extrinsic and intrinsic signaling pathways. *Oncol. Rep.* **2014**, *31*, 1788–1794. [CrossRef]
142. Yu, W.; Xie, X.; Yu, Z.; Jin, Q.; Wu, H. Mechanism of hesperidin-induced apoptosis in human gastric cancer AGS cells. *Trop. J. Pharm. Res.* **2019**, *50*, 2363–2369.
143. Zhang, X.; Zheng, L.; Sun, Y.; Wang, T.; Wang, B. Tangeretin enhances radiosensitivity and inhibits the radiation-induced epithelial-mesenchymal transition of gastric cancer cells. *Oncol. Rep.* **2015**, *34*, 302–310. [CrossRef] [PubMed]
144. Wu, X.; Song, M.; Wang, M.; Zheng, J.; Gao, Z.; Xu, F.; Zhang, G.; Xiao, H. Chemopreventive effects of nobiletin and its colonic metabolites on colon carcinogenesis. *Mol Nutr Food Res.* **2015**, *59*, 2383–2394. [CrossRef]
145. Kawabata, K.; Murakami, A.; Ohigashi, H. Nobiletin, a citrus flavonoid, down-regulates matrix metalloproteinase-7 (matrilysin) expression in HT-29 human colorectal cancer cells. *Biosci. Biotechnol. Biochem.* **2005**, *69*, 307–314. [CrossRef] [PubMed]
146. Dey, D.K.; Chang, S.N.; Vadlamudi, Y.; Park, J.G.; Kang, S.C. Synergistic therapy with tangeretin and 5-fluorouracil accelerates the ROS/JNK mediated apoptotic pathway in human colorectal cancer cell. *Food Chem. Toxicol.* **2020**, *143*, 111529. [CrossRef] [PubMed]
147. Park, H.J.; Kim, M.J.; Ha, E.; Chung, J.H. Apoptotic effect of hesperidin through caspase3 activation in human colon cancer cells, SNU-C4. *Phytomedicine* **2008**, *15*, 147–151. [CrossRef] [PubMed]

148. Yang, Z.; Yang, H.; Dong, X.; Pu, M.; Ji, F. Hesperidin loaded Zn^{2+}@ SA/PCT nanocomposites inhibit the proliferation and induces the apoptosis in colon cancer cells (HCT116) through the enhancement of pro-apoptotic protein expressions. *J. Photochem. Photobiol. B* **2020**, *204*, 111767. [CrossRef] [PubMed]
149. Saiprasad, G.; Chitra, P.; Manikandan, R.; Sudhandiran, G. Hesperidin induces apoptosis and triggers autophagic markers through inhibition of Aurora-A mediated phosphoinositide-3-kinase/Akt/mammalian target of rapamycin and glycogen synthase kinase-3 beta signalling cascades in experimental colon carcinogenesis. *Eur. J. Cancer* **2014**, *50*, 2489–2507. [CrossRef]
150. Saiprasad, G.; Chitra, P.; Manikandan, R.; Sudhandiran, G. Hesperidin alleviates oxidative stress and downregulates the expressions of proliferative and inflammatory markers in azoxymethane-induced experimental colon carcinogenesis in mice. *Inflamm. Res.* **2013**, *62*, 425–440. [CrossRef]
151. Cheng, H.; Jiang, X.; Zhang, Q.; Alma, J.; Cheng, R.; Yong, H.; Shi, H.; Zhou, X.; Ge, L.; Gao, G. Naringin inhibits colorectal cancer cell growth by repressing the PI3K/AKT/mTOR signaling pathway. *Exp. Ther. Med.* **2020**, *19*, 3798–3804. [CrossRef]
152. Yang, F.; Jiang, T.; Zhang, L.; Qi, Y.; Yang, Z. Naringin inhibits the proliferation and invasion of Eca109 esophageal cancer cells and promotes its apoptosis by blocking JAK/STAT signal pathway. *Xi Bao Yu Fen Zi Mian Yi Xue Za Zhi* **2021**, *37*, 1085–1091.
153. Xu, W.W.; Zheng, C.C.; Huang, Y.N.; Chen, W.Y.; Yang, Q.S.; Ren, J.Y.; Wang, Y.M.; He, Q.Y.; Liao, H.X.; Li, B. Synephrine Hydrochloride Suppresses Esophageal Cancer Tumor Growth and Metastatic Potential through Inhibition of Galectin-3-AKT/ERK Signaling. *J. Agric. Food Chem.* **2018**, *66*, 9248–9258. [CrossRef] [PubMed]
154. Yoshinaga, A.; Kajiya, N.; Oishi, K.; Kamada, Y.; Ikeda, A.; Chiwechokha, P.K.; Kibe, T.; Kishida, M.; Kishida, S.; Komatsu, M.; et al. NEU3 inhibitory effect of naringin suppresses cancer cell growth by attenuation of EGFR signaling through GM3 ganglioside accumulation. *Eur. J. Pharmacol.* **2016**, *782*, 21–29. [CrossRef] [PubMed]
155. Lin, R.; Hu, X.; Chen, S.; Shi, Q.; Chen, H. Naringin induces endoplasmic reticulum stress-mediated apoptosis, inhibits β-catenin pathway and arrests cell cycle in cervical cancer cells. *Acta Biochim. Pol.* **2020**, *67*, 181–188. [PubMed]
156. Ramesh, E.; Alshatwi, A.A. Naringin induces death receptor and mitochondria-mediated apoptosis in human cervical cancer (SiHa) cells. *Food Chem. Toxicol.* **2013**, *51*, 97–105. [CrossRef]
157. Zeng, L.; Zhen, Y.; Chen, Y.; Zou, L.; Zhang, Y.; Hu, F.; Feng, J.; Shen, J.; Wei, B. Naringin inhibits growth and induces apoptosis by a mechanism dependent on reduced activation of NF-κB/COX-2-caspase-1 pathway in HeLa cervical cancer cells. *Int. J. Oncol.* **2014**, *45*, 1929–1936. [CrossRef]
158. Goan, Y.G.; Wu, W.T.; Liu, C.I.; Neoh, C.A.; Wu, Y.J. Involvement of Mitochondrial Dysfunction, Endoplasmic Reticulum Stress, and the PI3K/AKT/mTOR Pathway in Nobiletin-Induced Apoptosis of Human Bladder Cancer Cells. *Molecules* **2019**, *24*, 2881. [CrossRef]
159. Lin, J.J.; Huang, C.C.; Su, Y.L.; Luo, H.L.; Lee, N.L.; Sung, M.T.; Wu, Y.J. Proteomics Analysis of Tangeretin-Induced Apoptosis through Mitochondrial Dysfunction in Bladder Cancer Cells. *Int. J. Mol. Sci.* **2019**, *20*, 1017. [CrossRef]
160. Cheng, H.L.; Hsieh, M.J.; Yang, J.S.; Lin, C.W.; Lue, K.H.; Lu, K.H.; Yang, S.F. Nobiletin inhibits human osteosarcoma cells metastasis by blocking ERK and JNK-mediated MMPs expression. *Oncotarget* **2016**, *7*, 35208–35223. [CrossRef]
161. Tan, T.W.; Chou, Y.E.; Yang, W.H.; Hsu, C.J.; Fong, Y.C.; Tang, C.H. Naringin suppress chondrosarcoma migration through inhibition vascular adhesion molecule-1 expression by modulating miR-126. *Int. Immunopharmacol.* **2014**, *22*, 107–114. [CrossRef]
162. Lien, L.M.; Wang, M.J.; Chen, R.J.; Chiu, H.C.; Wu, J.L.; Shen, M.Y.; Chou, D.S.; Sheu, J.R.; Lin, K.H.; Lu, W.J. Nobiletin, a Polymethoxylated Flavone, Inhibits Glioma Cell Growth and Migration via Arresting Cell Cycle and Suppressing MAPK and Akt Pathways. *Phytother. Res.* **2016**, *30*, 214–221. [CrossRef]
163. Ma, L.L.; Wang, D.W.; Yu, X.D. Tangeretin induces cell cycle arrest and apoptosis through upregulation of PTEN expression in glioma cells. *Biomed. Pharmacother.* **2016**, *81*, 491–496. [CrossRef] [PubMed]
164. Sonia, A.; Bakhta, A.; Yassine, C. Naringin suppresses cell metastasis and the expression of matrix metalloproteinases (MMP-2 and MMP-9) via the inhibition of ERK-P38-JNK signaling pathway in human glioblastoma. *Chem.-Biolog. Interact.* **2016**, *244*, 195–203.
165. Li, J.; Dong, Y.; Hao, G.; Wang, B.; Wang, J.; Liang, Y.; Liu, Y.; Zhen, E.; Feng, D.; Liang, G. Naringin suppresses the development of glioblastoma by inhibiting FAK activity. *J. Drug Target.* **2017**, *25*, 41–48. [CrossRef] [PubMed]
166. Camargo, C.A.; Gomes-Marcondes, M.C.; Wutzki, N.C.; Aoyama, H. Naringin inhibits tumor growth and reduces interleukin-6 and tumor necrosis factor α levels in rats with Walker 256 carcinosarcoma. *Anticancer Res.* **2012**, *32*, 129–133. [PubMed]
167. Pandey, P.; Sayyed, U.; Tiwari, R.K.; Siddiqui, M.H.; Pathak, N.; Bajpai, P. Hesperidin Induces ROS-Mediated Apoptosis along with Cell Cycle Arrest at G2/M Phase in Human Gall Bladder Carcinoma. *Nutr. Cancer.* **2019**, *71*, 676–687. [CrossRef]
168. Cincin, Z.B.; Kiran, B.; Baran, Y.; Cakmakoglu, B. Hesperidin promotes programmed cell death by downregulation of nongenomic estrogen receptor signalling pathway in endometrial cancer cells. *Biomed. Pharmacother.* **2018**, *103*, 336–345. [CrossRef]
169. Lee, K.A.; Lee, S.H.; Lee, Y.J.; Baeg, S.M.; Shim, J.H. Hesperidin induces apoptosis by inhibiting Sp1 and its regulatory protein in MSTO-211H cells. *Biomol. Ther.* **2012**, *20*, 273–279. [CrossRef]
170. Nazari, M.; Ghorbani, A.; Hekmat-Doost, A.; Jeddi-Tehrani, M.; Zand, H. Inactivation of nuclear factor-κB by citrus flavanone hesperidin contributes to apoptosis and chemo-sensitizing effect in Ramos cells. *Eur. J. Pharmacol.* **2011**, *650*, 526–533. [CrossRef]
171. Ghorbani, A.; Nazari, M.; Jeddi-Tehrani, M.; Zand, H. The citrus flavonoid hesperidin induces p53 and inhibits NF-κB activation in order to trigger apoptosis in NALM-6 cells: Involvement of PPARγ-dependent mechanism. *Eur. J. Nutr.* **2012**, *51*, 39–46. [CrossRef]

172. Shahbazi, R.; Cheraghpour, M.; Homayounfar, R.; Nazari, M.; Nasrollahzadeh, J.; Davoodi, S.H. Hesperidin inhibits insulin-induced phosphoinositide 3-kinase/Akt activation in human pre-B cell line NALM-6. *J. Cancer Res. Ther.* **2018**, *14*, 503–508.
173. Berkarda, B.; Koyuncu, H.; Soybir, G.; Baykut, F. Inhibitory effect of Hesperidin on tumor initiation and promotion in mouse skin. *Res. Exp. Med.* **1998**, *198*, 93–99. [CrossRef] [PubMed]
174. Wudtiwai, B.; Makeudom, A.; Krisanaprakornkit, S.; Pothacharoen, P.; Kongtawelert, P. Anticancer Activities of Hesperidin via Suppression of Up-Regulated Programmed Death-Ligand 1 Expression in Oral Cancer Cells. *Molecules* **2021**, *26*, 5345. [CrossRef] [PubMed]
175. Tanaka, T.; Makita, H.; Ohnishi, M.; Hirose, Y.; Wang, A.; Mori, H.; Satoh, K.; Hara, A.; Ogawa, H. Chemoprevention of 4-nitroquinoline 1-oxide-induced oral carcinogenesis by dietary curcumin and hesperidin: Comparison with the protective effect of beta-carotene. *Cancer Res.* **1994**, *54*, 4653–4659. [PubMed]
176. Tanaka, T.; Makita, H.; Ohnishi, M.; Mori, H.; Satoh, K.; Hara, A.; Sumida, T.; Fukutani, K.; Tanaka, T.; Ogawa, H. Chemoprevention of 4-nitroquinoline 1-oxide-induced oral carcinogenesis in rats by flavonoids diosmin and hesperidin, each alone and in combination. *Cancer Res.* **1997**, *57*, 246–252.
177. Lin, C.X.; Tu, C.W.; Ma, Y.K.; Ye, P.C.; Shao, X.; Yang, Z.A.; Fang, Y.M. Nobiletin inhibits cell growth through restraining aerobic glycolysis via PKA-CREB pathway in oral squamous cell carcinoma. *Food Sci. Nutr.* **2020**, *8*, 3515–3524. [CrossRef]
178. Murakami, A.; Nakamura, Y.; Torikai, K.; Tanaka, T.; Koshiba, T.; Koshimizu, K.; Kuwahara, S.; Takahashi, Y.; Ogawa, K.; Yano, M.; et al. Inhibitory effect of citrus nobiletin on phorbol ester-induced skin inflammation, oxidative stress, and tumor promotion in mice. *Cancer Res.* **2000**, *60*, 5059–5066.
179. Sato, T.; Koike, L.; Miyata, Y.; Hirata, M.; Mimaki, Y.; Sashida, Y.; Yano, M.; Ito, A. Inhibition of activator protein-1 binding activity and phosphatidylinositol 3-kinase pathway by nobiletin, a polymethoxy flavonoid, results in augmentation of tissue inhibitor of metalloproteinases-1 production and suppression of production of matrix metalloproteinases-1 and -9 in human fibrosarcoma HT-1080 cells. *Cancer Res.* **2002**, *62*, 1025–1029.

molecules

MDPI

Article

Walnut Prevents Cognitive Impairment by Regulating the Synaptic and Mitochondrial Dysfunction via JNK Signaling and Apoptosis Pathway in High-Fat Diet-Induced C57BL/6 Mice

Jong Hyun Moon [1,†], Jong Min Kim [1,†], Uk Lee [2], Jin Yong Kang [1,3], Min Ji Kim [1], Hyo Lim Lee [1], Hye Rin Jeong [1], Min Ji Go [1], Hyun-Jin Kim [1], Hye Won Park [2], Chul-Woo Kim [2], Sung Jin Park [4] and Ho Jin Heo [1,*]

1 Division of Applied Life Science (BK21), Institute of Agriculture and Life Science, Gyeongsang National University, Jinju 52828, Korea
2 Division of Special Forest Resources, Department of Forest Bio-resources, National Institute of Forest Science (NIFoS), Suwon 16631, Korea
3 World Institute of Kimchi an Annex of Korea Food Research Institute, Gwangju 61755, Korea
4 Research Institute for Advanced Industrial Technology, Korea University, Sejong 30019, Korea
* Correspondence: hjher@gnu.ac.kr
† These authors contributed equally to this work.

Abstract: This study was conducted to evaluate the protective effect of *Juglans regia* (walnut, Gimcheon 1ho cultivar, GC) on high-fat diet (HFD)-induced cognitive dysfunction in C57BL/6 mice. The main physiological compounds of GC were identified as pedunculagin/casuariin isomer, strictinin, tellimagrandin I, ellagic acid-O-pentoside, and ellagic acid were identified using UPLC Q-TOF/MS analysis. To evaluate the neuro-protective effect of GC, 3-(4,5-dimethylthiazol-2-yl)-2,5-diphenyltetrazolium bromide (MTT), 2′,7′-dichlorodihydrofluorecein diacetate (DCF-DA) analysis were conducted in H_2O_2 and high glucose-induced neuronal PC12 cells and hippocampal HT22 cells. GC presented significant cell viability and inhibition of reactive oxygen species (ROS) production. GC ameliorated behavioral and memory dysfunction through Y-maze, passive avoidance, and Morris water maze tests. In addition, GC reduced white adipose tissue (WAT), liver fat mass, and serum dyslipidemia. To assess the inhibitory effect of antioxidant system deficit, lipid peroxidation, ferric reducing antioxidant power (FRAP), and advanced glycation end products (AGEs) were conducted. Administration of GC protected the antioxidant damage against HFD-induced diabetic oxidative stress. To estimate the ameliorating effect of GC, acetylcholine (ACh) level, acetylcholinesterase (AChE) activity, and expression of AChE and choline acetyltransferase (ChAT) were conducted, and the supplements of GC suppressed the cholinergic system impairment. Furthermore, GC restored mitochondrial dysfunction by regulating the mitochondrial ROS production and mitochondrial membrane potential (MMP) levels in cerebral tissues. Finally, GC ameliorated cerebral damage by synergically regulating the protein expression of the JNK signaling and apoptosis pathway. These findings suggest that GC could provide a potential functional food source to improve diabetic cognitive deficits and neuronal impairments.

Keywords: walnut; *Juglans regia*; high-fat diet; insulin resistance; inflammation; cognitive function; JNK/NFκB pathway

Citation: Moon, J.H.; Kim, J.M.; Lee, U.; Kang, J.Y.; Kim, M.J.; Lee, H.L.; Jeong, H.R.; Go, M.J.; Kim, H.-J.; Park, H.W.; et al. Walnut Prevents Cognitive Impairment by Regulating the Synaptic and Mitochondrial Dysfunction via JNK Signaling and Apoptosis Pathway in High-Fat Diet-Induced C57BL/6 Mice. *Molecules* **2022**, *27*, 5316. https://doi.org/10.3390/molecules27165316

Academic Editor: Arjun H. Banskota

Received: 18 July 2022
Accepted: 19 August 2022
Published: 20 August 2022

1. Introduction

Diabetes, one of the major metabolic diseases, has increased in prevalence over the past few decades, and more than 150 million people currently suffer from diabetes worldwide [1]. The causes of diabetes are reported to obesity, hyperglycemia, insulin resistance, and hormone changes derived from changes in diet such as high-fat and high-carbohydrate diets [2]. A high-fat diet (HFD) impairs energy homeostasis in the whole body due to comprehensive factors and lifestyle changes and contributes to obesity [3]. Obesity is related to type 2 diabetes by causing insulin resistance, increasing fasting plasma insulin levels, and impairing glucose tolerance [4]. Hyperglycemia by glucose intolerance leads to an increase in the formation of advanced glycation end products (AGEs) and activation of nuclear factor-kappa B (NFκB), leading to the activation of inflammation [5]. Furthermore, brain tissue is easily damaged by the overproduction of reactive oxygen species (ROS) due to the metabolic abnormalities accompanying insulin resistance. Oxidative stress caused by HFD leads to the activation of c-Jun N-terminal kinase (JNK), which precedes cell death by apoptosis and inflammation through the downregulation of protein kinase B (Akt) phosphorylation [6,7]. Inhibition of Akt activity ultimately promotes the hyperphosphorylation of tau protein and the production of amyloid β (Aβ) peptides, leading to synaptic dysfunction [8,9]. Moreover, the expression of pro-inflammatory cytokines due to insulin resistance increases the permeability of the blood–brain barrier (BBB) [10–12]. It causes neuronal inflammation, neurodegenerative diseases, and cognitive dysfunction in the hippocampal areas [13,14]. Therefore, HFD is associated with type 2 diabetes and leads to Alzheimer's disease (AD) by increasing diabetic cognitive dysfunction due to insulin resistance [15]. Thus, it is important to evaluate the protective or ameliorating effect of natural resources against HFD-induced diabetic disease. In previous studies, research was conducted to assess the diabetic pathology in an alloxan (ALX) and/or streptozotocin (STZ)-induced animal model without nutritional supplements [16]. However, in contrast to the HFD-induced model, ALX and STZ can cause toxic reactions in other organs and lead to a different aspect of the expression of hyperglycemic status [17]. Therefore, an HFD-induced diabetic model was used for evaluation in this study.

Walnut (*Juglans regia*) is a crop grown worldwide that has various nutrients such as high unsaturated fatty acid, amino acid, mineral contents, fat-soluble vitamins containing α- and γ-tocopherol, and unsaturated fatty acids, such as oleic acid, linoleic acid and linolenic acid [18,19]. In addition, walnuts have various physiologically active compounds such as ellagitannin-based tannins with physiologically bioactive substances [20]. Ellagitannins are polyphenols in which hexahydroxydiphenic acid (HHDP) esters, or their metabolites, are mostly combined with glucose [21]. When ingested into the human body, ellagitannin is hydrolyzed to ellagic acid and/or gallic acid, and those metabolites have physiological activities such as anti-diabetic activity, regulation of hepatic steatosis and serum lipid composition, anti-amnesic effect, and inhibition of inflammasome activation [22–25]. Based on these bioactive substances, walnut showed antioxidant activity, hyperlipidemia, anti-diabetic activity, and anti-inflammatory activity [18,26–28]. Furthermore, walnut showed a protective effect on BBB damage and improved cognitive dysfunction in Aβ-induced mice [29]. However, there are few studies related to the protective effect of walnut against HFD-induced cognitive deficit. Therefore, this study was conducted to evaluate the protective effect of Gimcheon 1ho cultivar walnut (GC) on cerebral disorder by insulin resistance, oxidative stress, and inflammation in HFD-induced diabetic disorder mice.

2. Materials and Methods

2.1. Chemicals

3-(4,5-dimethylthiazol-2-yl)-2,5-diphenyltetrazolium bromide (MTT), 2′,7′-dichlorodihydrofluorecein diacetate (DCF-DA), fetal bovine serum (FBS), calf serum (CS), dimethyl sulfoxide (DMSO), penicillin, streptomycin, Roswell Park Memorial Institute medium 1640 (RPMI 1640), Dulbecco's modified Eagle's medium (DMEM), bovine serum albumin (BSA), hydrogen peroxide (H_2O_2) D-glucose, phosphoric acid, thiobarbituric acid (TBA), trichloroacetic acid (TCA), sodium hydroxide, hydroxylamine, hydrogen chloride (HCl), 2,4,6-tris(2-pyridyl)-s-triazine, ferric chloride hexahydrate, sodium acetate, sodium phosphate, sodium azide, fructose, mannitol, sucrose, HEPES sodium salt, digitonin, egtazic acid (EGTA), hydroxylamine, 5,5′-dithiobis-(2-nitrobenzoic acid) (DTNB), tris base, acetic acid, pyruvic acid, malate, 5,5,6,6-tetrachloro-1,1,3,3-tetraethylbenzimidazolylcarbocyanine iodide (JC-1), potassium chloride (KCl), potassium dihydrogen phosphate, magnesium chloride, pyruvate, hydroxylamine, and sodium hydroxide and solvents were obtained from Sigma-Aldrich Chemical Corporation (St. Louis. Mo, USA).

2.2. Sample Preparation

The walnut (*Juglans regia*) used in this experiment was obtained from Gimcheon 1ho cultivar (GC) walnut in an experimental forest (production in 2018) of Gimcheon City (Department of Forestry and Greenery) (Gimcheon, Korea) and verified by the National Institute of Forest Science (NIFoS) (Suwon, Korea). According to the previous study, GC presented significant contents of unsaturated fatty acids, total tocopherols, and essential amino acids than other Korean cultivars [30]. The sample was prepared using vacuum lyophilization (Operon, Gimpo, Korea) and stored at −20 °C. The lyophilized sample was extracted with 50-fold 80% ethanol concentrations at 40 °C for 2 h. The extraction sample was filtered through No.2 filter (Whatman Inc, Kent, UK) and concentrated using a vacuum rotary evaporator (N-N series, Eyela Co., Tokyo, Japan). The concentrated extract of GC was lyophilized and stored at −20 °C until use.

2.3. UPLC Q-TOF/MS

The main physiological compounds in the GC were analyzed using ultra-performance liquid chromatography–ion mobility separation–quadrupole time of flight/tandem mass spectrometry (UPLC IMS Q-TOF/MS, Vion, Waters Corp., Milford, MA, USA). UPLC separation was investigated with an ACQUITY UPLC BEH C_{18} column (2.1 × 100 mm, 1.7 µm particle size; Waters Corp.). The sample was analyzed using distilled water and acetonitrile (ACN) containing 0.1% formic acid at a flow rate of 0.35 mL/min for 10 min. The mass spectrometer used electrospray ionization (ESI) in the negative ion mode. The capillary and cone voltages were set at 2.5 kV and 40 V, respectively, and the source and desolvation temperatures were performed at 100 °C and 400 °C, respectively. Mass spectral data were collected from m/z 50 to 1500 and processed using MarkerLynx software (Waters Corp.).

2.4. In Vitro Cells Study

2.4.1. Cell Culture

PC12 cells with the characteristics of adrenal gland blastoma cell lines were purchased by the Koran Cell Line Bank (Seoul, Korea) and cultured in RPMI 1640 medium with 10% FBS, 50 units/mL penicillin and 100 µg/mL streptomycin. HT22 hippocampal cells were supplied in October 2017 from the Department of Anatomy of the College of Veterinary Medicine, Gyeongsang National University. HT22 cells were cultured in a DMEM medium containing 10% CS, 50 units/mL penicillin and 100 µg/mL streptomycin. Cells were cultured at 37 °C in 5% CO_2.

2.4.2. Neuronal Cell Viability and Intracellular ROS

The cell viability was performed using the MTT assay [31]. PC12 and HT22 cells were seeded in 96 well plates (1 × 10^4 cells/well). After 24 h, the samples and Vitamin C as the positive control were treated. After 24 h, H_2O_2 was treated for 3 h. MTT stock solution (10 mg/mL) was added to the pretreated cells for 3 h. The production of violet formazan crystals was determined using a microplate reader (Epoch 2, BioTek, Winooski, VT, USA). The absorbance was measured at 570 nm (determination wavelength) and 655 nm (reference wavelength).

The intracellular ROS contents were performed using the DCF-DA method [31]. After sampling and H_2O_2 treatment, a DCF-DA reagent was added to cells and incubated for 50 min. After that, fluorescence levels were performed using a fluorometer (Infinite F200, TECAN, Männedorf, Switzerland) at 485 nm (excitation wavelengths) and 525 nm (emission wavelengths).

2.5. Animal Experimental Design

The male C57BL/6 mice (4 weeks old) were obtained from Samtako (Osan, Korea). The experimental animals were randomly assigned to 4 groups: NC group (Normal chow), HFD group, and HFD with sample groups (GC20 and GC50; 20 and 50 mg/kg of body weight, respectively). The mice were selected by body weight and oral glucose tolerance test (OGTT) to confirm the diabetic condition and randomly divided into four groups (n = 8; total n = 32). After HFD feed was supplied for 12 weeks, samples were orally ingested for 4 weeks to GC20 and GC50 groups. The NC and HFD groups were orally administered with the same amount of water. All animal experiments in the study were in compliance with the Institutional Animal Care and Use Committee (IACUC) of Gyeongsang National University (certificate: GNU-190530-M0028; approved on 30 May 2019).

2.6. Glucose Tolerance Test

Fasting blood glucose was measured once a week during sample intake at 15 weeks old. To measure the fasting blood glucose concentration, mice were fasted for 8 h. After 4 weeks, D-glucose (2 g/kg of body weight) was orally administered to all mice at 19 weeks old to evaluate the OGTT. Blood glucose level collected from the tail vein was measured using an Accu-Chek glucose meter at 0, 15, 30, 60, 90, and 120 min (Roche Diagnostics, Basel, Switzerland).

2.7. Behavioral Tests

2.7.1. Y-Maze Test

The Y-maze was designed with an internal dimension of 470 mm × 160 mm and 460 mm in height for testing mice. Each mouse was placed at the end of the arm of the maze and allowed to explore freely in the maze for 8 min [32]. The movement and path tracing were recorded using a video system (Smart 3.0, Panlab, Barcelona, Spain).

2.7.2. Passive Avoidance Test

Passive avoidance equipment consists of two compartments as the bright and dark parts, and a door that can pass between them was located between compartments [33]. The compartments were composed of a dark chamber with electrical shock capability. In the first experiment, the mice were located in a bright chamber. When the mice's four feet entered the dark compartment, a mild foot electrical shock was applied at 0.5 mA for 3 s, and the latency time of the bright compartment was recorded. On the following day, the step-through latency to reenter the dark compartment was measured.

2.7.3. Morris Water Maze (MWM) Test

MWM circular pool designed in diameter of 900 mm × 300 mm height was split into 4 zones as N, S, E, and W by marks on the outside of the pool. In the center of the W quadrant, a platform was submerged below the surface. The pool water was diluted using non-toxic tempera paint. Experimental animals swam to find the platform for a maximum of 1 min and were trained repeatedly for 4 days. After 4 days of training, the platform was removed and retention time in the W zone was measured using a video system (Smart 3.0, Panlab) [34].

2.8. Blood Serum Biochemical

After the behavioral tests, mice were fasted for 8 h and sacrificed using exposure to CO_2. The blood sample was collected at the postcaval vein to evaluate blood serum biochemical analysis. The collected blood samples were centrifuged at 10,000× g for 10 min at 4 °C to obtain supernatants of blood. Lactate dehydrogenase (LDH), TG, total cholesterol (TCHO) and high-density lipoprotein cholesterol (HDLC) contents were measured using a clinical chemistry analyzer (Fuji dri-chem 4000i, Fujifilm Co., Tokyo, Japan). Low-density lipoprotein cholesterol (LDLC) level and the ratio of HDLC to TCHO (HTR) were calculated as follows [35].

$$LDLC\ (\text{mg/dL}) = TCHO - \left(HDLC + \frac{TG}{5}\right)$$

$$HTR(\%) = \frac{HDLC}{TCHO} \times 100 \quad (1)$$

2.9. Preparation of Tissue

Before evaluating the ex vivo test, brain, liver, perirenal white adipose tissue (WAT) fat, retroperitoneal WAT fat, epididymal WAT fat and mesenteric WAT fat tissues were collected, and organ weight was measured. The collected brain tissues were homogenized in a bullet blender (Next Advance Inc., AverillPark, NY, USA). The collected sample was quantitated to calculated using Bradford protein assay [36].

2.10. Antioxidant Activity

2.10.1. Malondialdehyde (MDA) Level

To determine MDA level, the supernatant was obtained from brain and liver tissues by centrifugation (5000 rpm, 10 min, 4 °C). The supernatants reacted with 1% phosphoric acid and 0.67% TBA at 95 °C using a water bath for 60 min. The mixtures were spun down at 600× *g* for 10 min and measured at 532 nm (Epoch 2, BioTek) [37].

2.10.2. Ferric Reducing Antioxidant Power (FRAP) in Serum

The serum supernatants were obtained as previously described to determine serum antioxidant levels. The serum level of FRAP was measured according to the modified method of Benzie [38]. FRAP solution mixed in 300 mM sodium acetate buffer (pH 3.6), 10 mM TPTZ in 40 mM HCl and 20 mM iron (III) chloride. After 30 min in the dark, the absorbance was measured at 593 nm.

2.10.3. AGEs Formation in Serum

The serum AGEs level was measured according to the method of Sampath [39]. To confirm the serum of AGEs formation, serum was diluted to 1:20 with PBS (pH 7.4). The fluorescence was measured using a fluorometer (Infinite F200, TECAN) at 360 nm (excitation wavelengths) and 460 nm (emission wavelengths) and calculated as a relative unit of controls.

2.11. Cerebral Cholinergic System

2.11.1. Acetylcholine (ACh) Level

The ACh level experiment was measured according to the method of Vincent [40]. The supernatant was obtained from brain tissues by centrifugation (12,000× *g*, 30 min, 4 °C). The supernatant was mixed with an alkaline hydroxylamine reagent (2 M hydroxylamine in 0.1 M HCl and 3.5 N NaOH). After mixing, 0.5 N HCl and 0.3 M $FeCl_3$ were added and immediately performed using a microplate reader (Epoch 2, BioTek).

2.11.2. Acetylcholinesterase (AChE) Activity

To assess the AChE activity, the supernatant from brain tissues was mixed with 50 mM sodium phosphate buffer and incubated at 37 °C for 15 min [41]. After, the reactants were added to Ellman's reaction mixture and detected at 405 nm (Epoch 2, BioTek).

2.12. Mitochondrial Activity

2.12.1. Extraction of Mitochondria from Brain Tissues

To extract the cerebral mitochondria, brain tissues were homogenized with 10-times as much mitochondrial isolation (MI) buffer (215 mM mannitol, 75 mM sucrose, 0.1% BSA, 20 mM HEPES sodium salt, pH 7.2) with 1 mM EGTA using a bullet blender (Next Advance Inc.). The homogenates were centrifuged at 1300× *g* for 5 min. After, the supernatants of homogenates were centrifuged at 13,000× *g* for 10 min to obtain the pellet. The pellets were mixed with MI buffer containing 0.1% digitonin in DMSO and incubated on ice for 5 min. After, the supernatants were centrifuged at 13,000× *g* for 10 min. Then, the pellets were re-suspended in MI buffer and centrifuged again at 13,000× *g* for 15 min. Finally, the pellets were suspended in MI buffer.

2.12.2. Mitochondrial ROS Contents

Mitochondrial ROS contents were estimated by the DCF-DA assay. The mitochondrial extraction solution was reacted with DCF-DA with respiration buffer (125 mM potassium chloride, 2 mM potassium phosphate, 20 mM HEPES, 1 mM magnesium chloride, and 500 μM EGTA). After incubation for 20 min, the reactant was detected at 535 nm (excitation wavelength) and 458 nm (emission wavelength) using a fluorescence microplate reader (Infinite 200, TECAN) [42].

2.12.3. Mitochondrial Membrane Potential

The level of mitochondrial membrane potential was measured by the JC-1 assay. The mitochondria isolation extract was mixed with JC-1 dye in MI buffer containing 5 mM pyruvate and 5 mM malate and incubated at room temperature for 20 min in the dark. The reactant was detected at 535 nm (excitation wavelength) and 590 nm (emission wavelength) using a fluorescence microplate reader (Infinite 200, TECAN) [42].

2.13. Western Blot

The brain tissues were homogenized in lysis buffer (GeneAll Biotechnology, Seoul, Korea) containing 1% protease inhibitor. The homogenates were centrifuged at 13,000× *g* for 10 min at 4 °C to obtain supernatants of homogenates and reacted by loading buffer at 95 °C for 5 min. The protein is separated in electrophoresis through the SDS-PAGE on 8 or 12% SDS gels and transferred to PVDF membranes. The membranes were blocked with 5% skim milk solution in with Tris-buffered saline with 0.5% Tween-20 (TBST), and reacted with primary antibodies at 4 °C. After reaction overnight, membranes were reacted with horseradish peroxidase (HRP)-conjugated secondary antibodies (1:5000) for 1 h. Finally, the membrane reacted with chemiluminescence reagent (TLP-112, TransLab., Daejeon, Korea) was analyzed using the iBright™ CL1000 Imaging System (Thermo Fischer Scientific, Rockford, IL, USA). Strip procedure was conducted using strip buffer (TLP-116.1, TransLab.) at 25 °C for 30 min and re-blotted for evaluation of other protein expressions. β-actin was used as a loading control marker. The density of expression was calculated using ImageJ software (National Institutes of Health, Bethesda, MD, USA). Antibody information is presented in Table 1.

Table 1. List of primary antibody information used in this study.

Antibody	Catalog	Conc.	Manufacturer
β-actin	sc-69879	1:1000	Santa Cruz Biotech (Dallas, TX, USA)
AChE	sc-373901	1:1000	Santa Cruz Biotech (Dallas, TX, USA)
p-JNK	sc-6254	1:1000	Santa Cruz Biotech (Dallas, TX, USA)
p-Akt	sc-514032	1:1000	Santa Cruz Biotech (Dallas, TX, USA)
p-tau	sc-12952	1:1000	Santa Cruz Biotech (Dallas, TX, USA)
IDE	sc-393887	1:1000	Santa Cruz Biotech (Dallas, TX, USA)
Aβ	sc-28365	1:1000	Santa Cruz Biotech (Dallas, TX, USA)
BAX	sc-7480	1:1000	Santa Cruz Biotech (Dallas, TX, USA)
Caspase-1	sc-392736	1:1000	Santa Cruz Biotech (Dallas, TX, USA)
TNF-α	sc-393887	1:1000	Santa Cruz Biotech (Dallas, TX, USA)
IL-1β	sc-4592	1:1000	Santa Cruz Biotech (Dallas, TX, USA)
HO-1	sc-136960	1:1000	Santa Cruz Biotech (Dallas, TX, USA)
ChAT	20747-1AP	1:1000	Bioneer (Daejeon, Korea)
p-NF-κB	#3033	1:1000	Cell Signaling Tech (Danvers, MA, USA)
Caspase-3	CSB-PA05689A0Rb	1:1000	Cusabio (Hubei, China)

2.14. Statistical Analysis

The results for the whole data were proposed as mean ± SD and analyzed by using a one-way analysis of variance (ANOVA) with Duncan's multiple range test ($p < 0.05$) of SAS 9.4 (SAS Institute Inc., Cary, NC, USA). All experimental data were evaluated for normality and variance homogeneity with the Shapiro–Wilk and Levene's variance homogeneity tests. Data were statistically represented as significantly different from the NC group (*) and significantly different from the PM group (#), respectively (* and # $p < 0.05$, ** and ## $p < 0.01$).

3. Results

3.1. UPLC Q-TOF/MS

To identify physiological compounds, GC was qualitatively confirmed using UPLC Q-TOF/MS analysis (Figure 1, Table 2). The LC/MS spectrum was compared to that of candidate compounds found in previous reports, especially when compounds in walnut were reported [20,24,27,41]. The ESI-MS^E spectra were continuously collected in negative ion mode $(M-H)^-$, and the seven main fragments were identified as pedunculagin/casuariin isomer I (bis-HHDP–glucose) (783.06 m/z), strictinin (633.07 m/z), pedunculagin/casuariin isomer II (bis-HHDP–glucose) (783.06 m/z), (-)-epicatechin (289.07 m/z), tellimagrandin I isomer (digalloyl-HHDP–glucose) (735.07 m/z), ellagic acid-*O*-pentoside (433.03 m/z), and ellagic acid (300.99 m/z) [43].

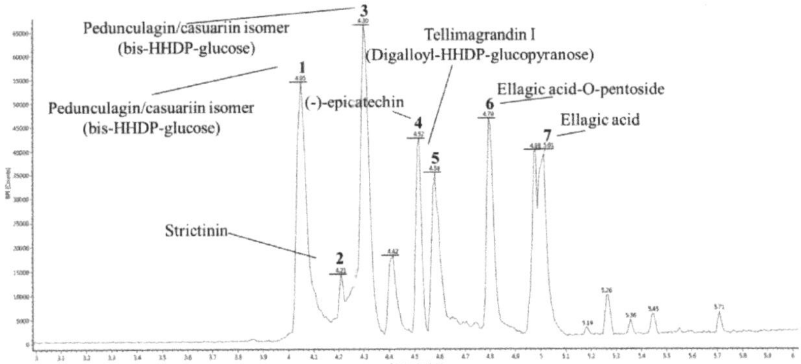

Figure 1. UPLC Q-TOF/MS^E chromatography in negative ion mode of ethyl acetate fraction from Gimcheon 1ho (GC) cultivar walnut (*Juglans regia*).

Table 2. Identification of main compounds of Gimcheon 1ho (GC) cultivar walnut (*Juglans regia*).

Peak No.	RT [a] (min)	Negative Ion Mode (m/z)	MS^E Fragments (m/z)	Proposed Compound	Compounds Formula
1	4.05	783.06	481.06, 300.99, 275.02	Pedunculagin/casuariin isomer (bis-HHDP-glucose) I	$C_{34}H_{24}O_{22}$
2	4.21	633.07	481.06, 300.99	Strictinin	$C_{27}H_{22}O_{18}$
3	4.30	783.06	481.06, 300.99, 275.02	Pedunculagin/casuariin isomer (bis-HHDP-glucose) II	$C_{34}H_{24}O_{22}$
4	4.52	289.07	245.08	(−)-Epicatechin	$C_{15}H_{14}O_6$
5	4.58	785.07	483.07, 300.99, 275.02, 169.01	Tellimagrandin I (Digalloyl-HHDP-glucopyranose)	$C_{34}H_{26}O_{22}$
6	4.79	433.03	300.99, 299.98	Ellagic acid-O-pentoside	$C_{19}H_{14}O_{12}$
7	5.01	300.99	302.00	Ellagic acid	$C_{14}H_6O_8$

[a] RT means retention time. All results were detected in negative ion mode using UPLC Q-TOF/MS^E.

3.2. Protective Effect against H_2O_2 and High Glucose-induced Neurotoxicity in PC12 and HT22 Cells

The neuroprotective effects of GC against H_2O_2 and high glucose in PC12 and HT22 cells are shown in Figures 2 and 3. In the results of cell viability in PC12 cells, the H_2O_2-treated group (55.08%) and high glucose-treated group (85.23%) showed cytotoxicity that decreased by 44.92% and 14.77%, respectively, compared to the control group (100%) (Figure 2a,b). On the other hand, treatment of the GC groups with 20 μg/mL (69.32% and 108.89%) and 50 μg/mL (72.50% and 111.33%) showed significantly increased cell viability compared to the H_2O_2 and high glucose-treated groups. In addition, the hippocampal HT22 cells were evaluated (Figure 2c,d). Both the 20 μg/mL (84.73% and 91.34%) and 50 μg/mL (94.91% and 94.96%) GC groups showed an increase in cell viability from neurotoxicity caused by H_2O_2 (69.64%) compared to the high glucose (89.26%)-treated group.

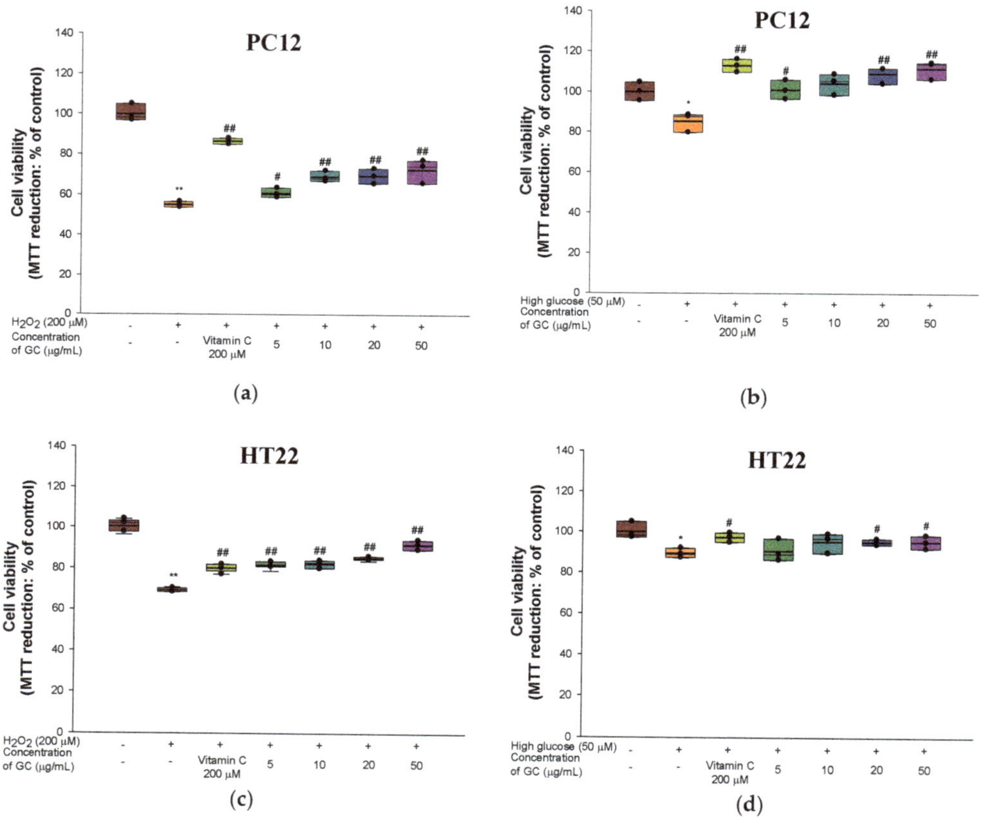

Figure 2. Protective effect of 80% ethanolic extract from Gimcheon 1ho (GC) cultivar walnut (*Juglans regia*): (**a**) cell viability from H_2O_2-induced cytotoxicity in PC12 cells; (**b**) cell viability from high-glucose-induced cytotoxicity in PC12 cells; (**c**) cell viability from H_2O_2-induced cytotoxicity in HT22 cells; (**d**) cell viability from high glucose-induced cytotoxicity in HT22 cells. Results shown are mean ± SD (n = 3). Data are statistically represented with * = significantly different from the NC group, and # = significantly different from PM group; * and # $p < 0.05$, ** and ## $p < 0.01$. Bold line indicates mean.

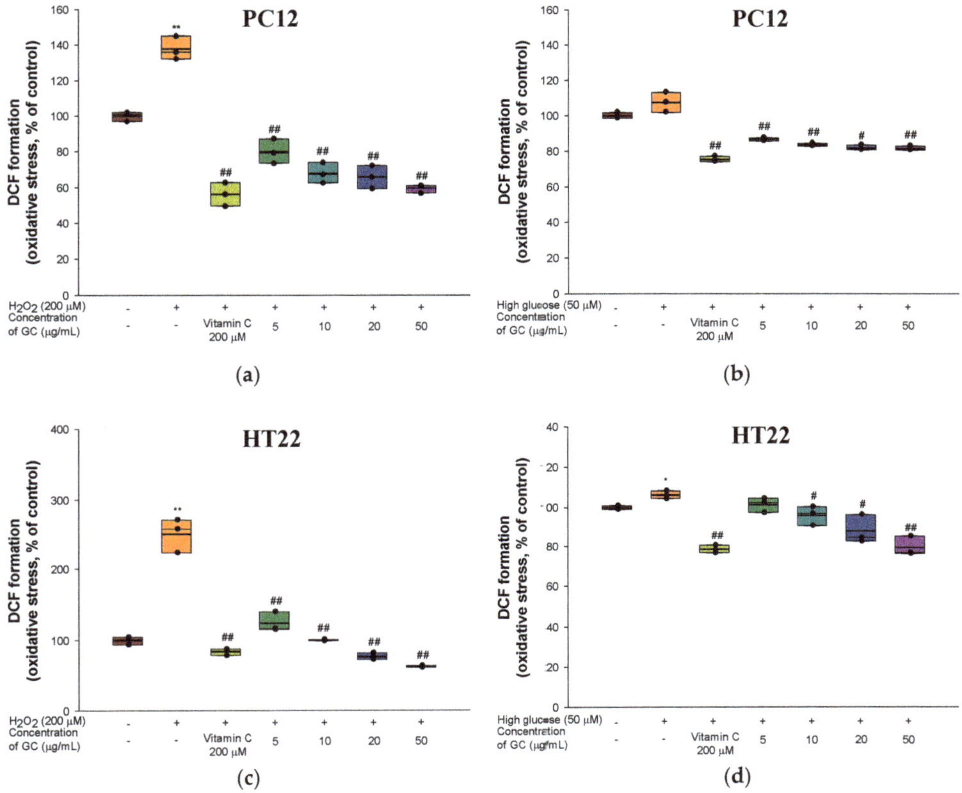

Figure 3. Protective effect of 80% ethanolic extract from Gimcheon 1ho (GC) cultivar walnut (*Juglans regia*): (**a**) ROS production from H_2O_2-induced cytotoxicity in PC12 cells; (**b**) ROS production from high glucose-induced cytotoxicity in PC12 cells; (**c**) ROS production from H_2O_2-induced cytotoxicity in HT22 cells; (**d**) ROS production from high glucose-induced cytotoxicity in HT22 cells. Results shown are mean ± SD (n = 3). Data are statistically represented with * = significantly different from the NC group, and # = significantly different from PM group; * and # $p < 0.05$, ** and ## $p < 0.01$. Bold line indicates mean.

Intracellular ROS in the H_2O_2 (137.83%) and high-glucose (107.52%)-treated groups increased by 37.83% and 7.52%, respectively, compared to the control group (100%) in PC12 cells (Figure 3a,b). However, the 20 μg/mL (65.88% and 81.78%) and 50 μg/mL (59.67% and 81.60%) GC groups had a decreased level of ROS production. In addition, intracellular ROS production in HT22 increased with H_2O_2 (251.46%) and high glucose (106.27%) compared to the control group (100.00%). In contrast, GC treatment showed a remarkable reduction in oxidative stress induced by high glucose at concentrations of 20 μg/mL (75.34% and 87.74%) and 50 μg/mL (61.02% and 79.27%) (Figure 3c,d).

3.3. Glucose Tolerance Test

To confirm the induction of type 2 diabetes through the HFD, fasting blood glucose was measured at 15 weeks old. The HFD group (210.50 mg/dL) was confirmed to have HFD-induced glucose intolerance, and the fasting glucose level was 1.51 times higher compared to the NC group (139.50 mg/dL) (Figure 4a). The fasting blood glucose level of the HFD group (208.44 mg/dL) significantly increased compared to the NC group (147.57 mg/dL) at 19 weeks old. On the other hand, the GC20 (180.45 mg/dL) and GC50 (186.52 mg/dL) groups showed a significant decrease compared to the HFD group. OGTT was conducted at 0, 15, 30, 60, 90, and 120 min, and the above experimental results were expressed as area under the curve (AUC) (Figure 4b,c). Compared to the normal control group (7090.31 dL/mL∗min), the AUC of the HFD group (17,134.41 dL/mL∗min) showed a 2.42-fold increase, confirming that glucose tolerance increased. However, the GC20 and GC50 groups showed AUC levels of 5413.12 and 4963.79 dL/mL∗min, respectively, and a significantly lower AUC level compared to the HFD group.

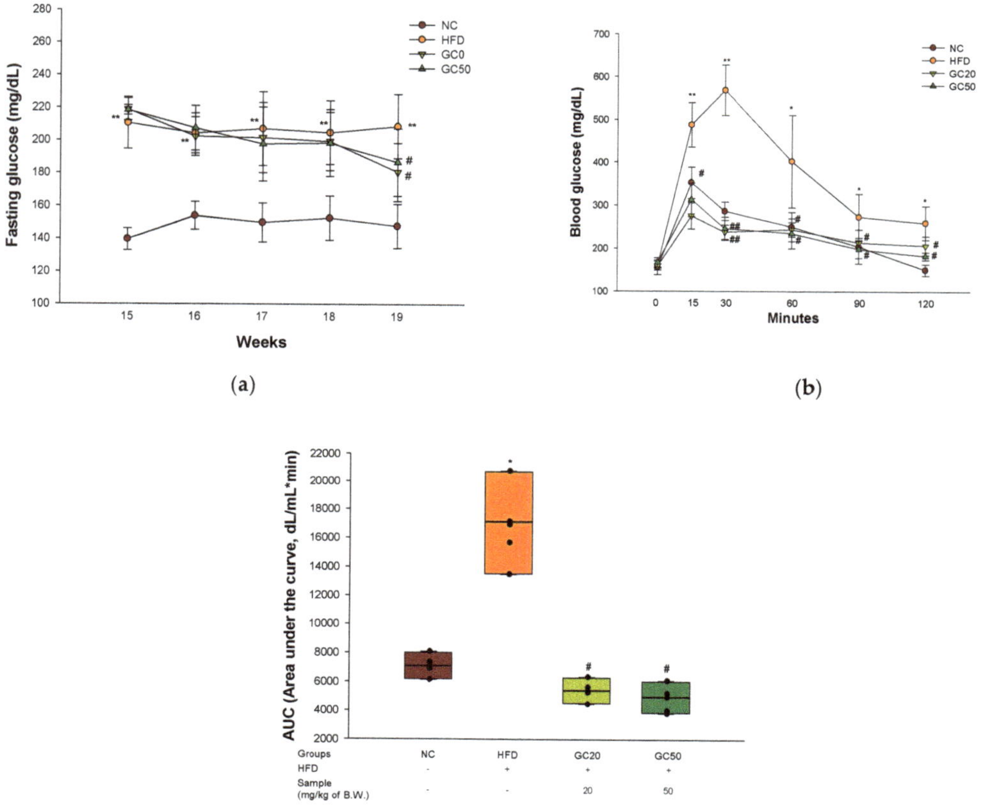

(a) (b) (c)

Figure 4. Protective effect of 80% ethanolic extract from Gimcheon 1ho (GC) cultivar walnut (*Juglans regia*) in HFD-induced mice: (**a**) fasting glucose; (**b**) oral glucose tolerance test (OGTT) at 19 weeks old; (**c**) area under the curve (AUC) of OGTT. Results shown are mean ± SD (n = 5). Data are statistically represented with * = significantly different from the NC group, and # = significantly different from PM group; * and # $p < 0.05$, ** and ## $p < 0.01$. Bold line indicates mean.

3.4. Behavioral Tests

To investigate spontaneous alternation behavior, a Y-maze test was conducted for 8 min with HFD-induced diabetics [32]. The total number of arm entries (n) in the maze did not show a significant difference between all animals in exercise ability affecting spatial behavior (Figure 5a). The HFD group (45.99%) showed decreased spontaneous alternation behavior compared to that of the NC group (57.63%) (Figure 5b). However, the behavior of the GC groups (GC20, 58.67% and GC50, 62.69%) improved compared to that of the HFD group. In particular, it was confirmed that the above two groups significantly restored ability, similar to the NC group. In an image showing the path tracing of each group of mice as a 3D schematic diagram, the HFD group showed reduced spontaneous alternation behavior. In contrast, the NC and GC groups were similarly indicated in each arm.

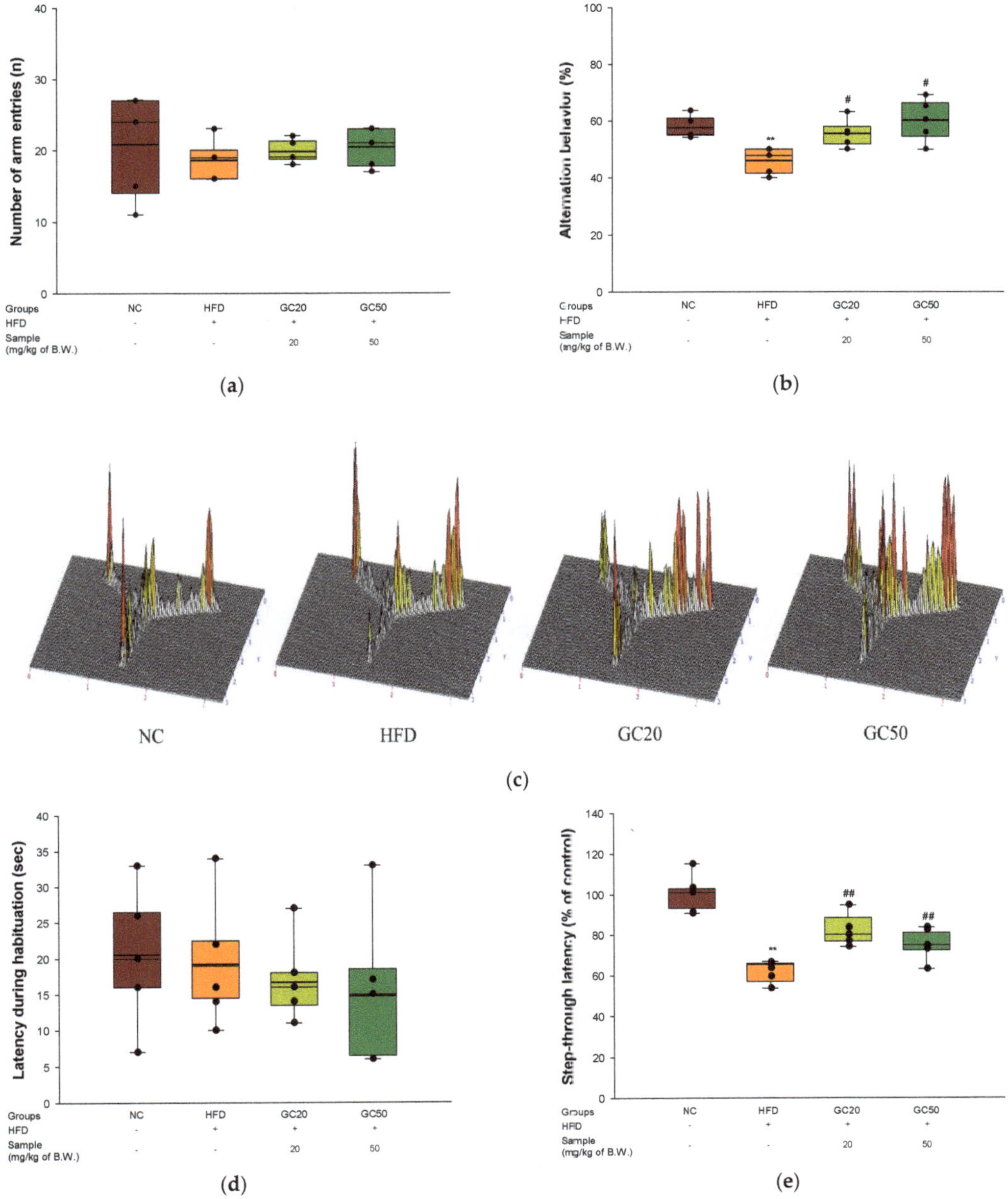

Figure 5. *Cont.*

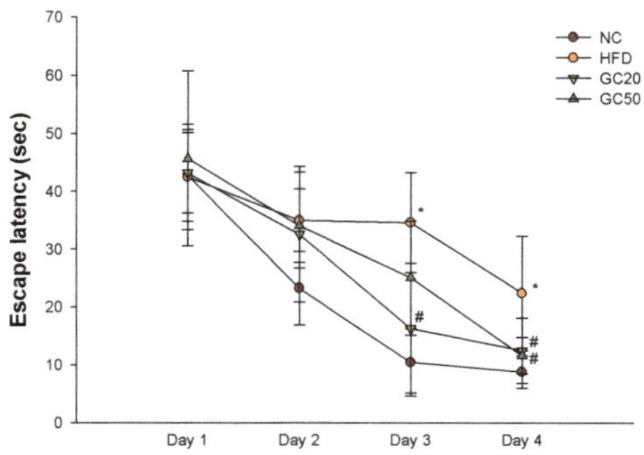

(f)

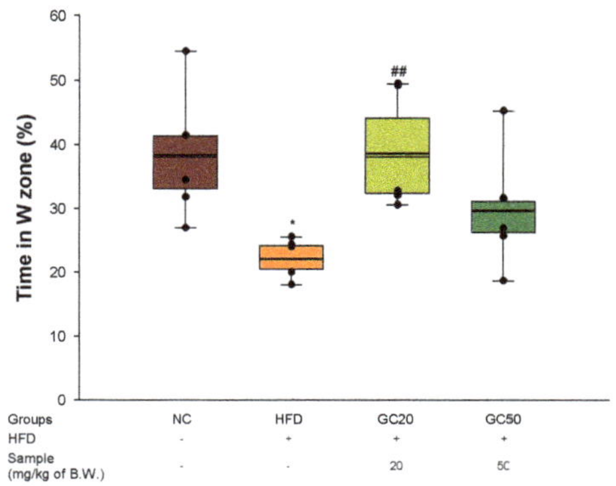

(g)

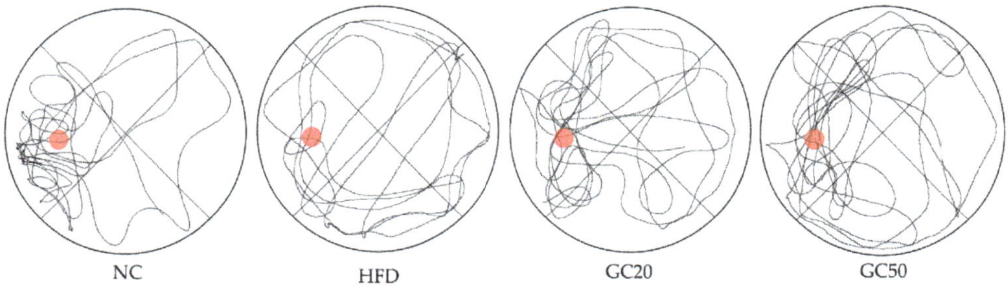

(h)

Figure 5. Protective of 80% ethanolic extract from Gimcheon 1ho (GC) cultivar walnut (*Juglans regia*) in HFD-induced mice: (**a**) a number of arm entries; (**b**) spontaneous alternation behavior; (**c**) 3D moving routes in Y-maze test; (**d**) latency during habituation; (**e**) step-through latency in passive avoidance test; (**f**) escape latency in the hidden test; (**g**) retention time in W zone; (**h**) path tracing of each group in Morris water maze (MWM) test. Results shown are mean ± SD (n = 5). Data are statistically represented with * = significantly different from the NC group, and # = significantly different from PM group; * and # $p < 0.05$, ** and ## $p < 0.01$. Bold line indicates mean.

A passive avoidance test was performed to measure short-term working memory ability associated with the amygdala [33]. The first step-through latency showed no significant differences between all groups (Figure 5d). There is no prior memory of the average time entering the trial test. In the trial test, the step-through latency of the HFD group (61.80%) was reduced in short-term memory by 38.2% compared with the NC group (100%) (Figure 5e). On the other hand, the GC20 and GC50 groups (82.10% and 75.47%, respectively) were considerably ameliorated compared to the HFD group.

To measure spatial learning acquisition and long-term memory, an MWM test was conducted [34]. In the hidden trial on the fourth day, the escape latency time of the HFD group (22.33 s) was delayed compared to the NC group (8.78 s) (Figure 5f). On the other hand, the GC groups had improved escape latency time following GC administration (GC20, 12.47 s, and GC50, 11.61 s). In the probe test, the retention time in the W zone of the HFD group (22.19%) decreased compared with that of the NC group (38.37%) (Figure 5g). However, the GC groups showed significantly increased retention times (GC20, 38.59% and GC50, 29.71%) compared to the HFD group. In a schematic diagram of the swimming path (Figure 5h), it was seen that the movement in the W zone of the HFD group decreased compared with the NC group. However, the GC groups showed improved movement in the W zone.

3.5. Blood Serum Biochemicals and Changes in Weight of Organs

3.5.1. Blood Serum Biochemicals

Serum biomarkers are shown in Table 3. The levels of LDH (782.14 U/L), TG (122.00 mg/dL), TCHO (215.57 mg/dL), and LDLC (59.74 mg/dL) of the HFD group increased higher than those of the NC group (LDH, 280.29 U/L; TG, 95.00 mg/dL; TCHO, 124.14 mg/dL; LDLC, 22.00 mg/dL). The GC groups showed no significantly decreased TG (GC20, 118.14 mg/dL) and TCHO (GC20, 215.00 mg/dL, GC50 201.29 mg/dL) compared to the HFD group. However, the levels of LDH (660.43 U/L and 537.57 U/L, respectively) and LDLC (39.53 mg/dL and 33.33 mg/dL, respectively) in the GC groups seemed to significantly decrease compared to the HFD group. In particular, the GC50 group had reduced LDH (537.57 U/L) and TG (109.29 mg/dL) levels compared to the HFD group. In addition, the HFD groups increased HDLC (131.43 mg/dL) and decreased HTR (60.77%) compared to the NC group (83.86 mg/dL, 67.63%, respectively). On the other hand, the GC groups had improved HDLC (157.43 mg/dL, 150.86 mg/dL, respectively) and HTR (70.80% and 69.61%, respectively) compared to the HFD group.

Table 3. Effect of 80% ethanolic extract from Gimcheon 1ho (GC) cultivar walnut (*Juglans regia*) on serum biomarkers.

Groups	NC	HFD	GC20	GC50
LDH (U/L)	280.29 ± 79.82	782.14 ± 231.88 **	660.43 ± 244.87	537.57 ± 127.19 $^{\#}$
TG (mg/dL)	95.00 ± 19.17	122.00 ± 11.49 *	118.14 ± 12.93	109.29 ± 12.72 $^{\#}$
TCHO (mg/dL)	124.14 ± 14.3	215.57 ± 21.16 **	215.00 ± 35.72	201.29 ± 37.75
LDLC (mg/dL)	22.00 ± 4.26	59.74 ± 5.24 **	39.53 ± 10.24 $^{\#\#}$	33.33 ± 20.71 $^{\#\#}$
HDLC (mg/dL)	83.86 ± 9.19	131.43 ± 18.18 **	157.43 ± 25.83	150.86 ± 25.39
HTR (%)	67.63 ± 2.14	60.77 ± 3.35 **	70.80 ± 9.67	69.61 ± 8.23

Results shown are mean ± SD (n = 5). Data are statistically represented with * = significantly different from the NC group, and # = significantly different from PM group; * and # $p < 0.05$, ** and ## $p < 0.01$.

3.5.2. Changes in Weight of Organs

To estimate fat accumulation, organs mass is presented in Table 4. The weight of the brain showed no significant differences between all groups. However, the weight of the liver (2.67 g), hepatic lipid (19.58 mg/g), perirenal WAT fat (0.35 g), retroperitoneal WAT fat (1.23 g), epididymal WAT fat (2.22 g), mesenteric WAT fat (0.90 g) and total WAT fat (4.39 g) in the HFD group increased compared to the NC group (liver, 1.28 g; hepatic lipid, 6.83 mg/g; perirenal WAT fat, 0.12 g; retroperitoneal WAT fat, 0.29 g; epididymal WAT fat, 1.25 g; mesenteric WAT fat, 0.25 g and total WAT fat, 2.00 g). The liver weight and hepatic lipid in the GC20 (1.70 g, 15.28 mg/g, respectively) and GC50 (1.54 g, 10.28 mg/g, respectively) groups were reduced compared to the HFD group. In addition, the perirenal (GC20, 0.40 g and GC50, 0.40 g), retroperitoneal (GC20, 0.71 g and GC50, 0.75 g), and epididymal (GC20, 1.65 g and GC50, 0.94 g) WAT fat and total WAT fat (GC20, 3.50 g and GC50, 3.26 g) of the GC groups were suppressed compared to the HFD group. However, mesenteric WAT fat showed no significant difference between the GC groups (GC20, 0.91 g and GC50, 0.95 g) compared to the HFD group.

Table 4. Effect of 80% ethanolic extract from Gimcheon 1ho (GC) cultivar walnut (*Juglans regia*) on organ weights.

Group	NC	HFD	GC20	GC50
Brain (g)	0.36 ± 0.04	0.36 ± 0.03	0.36 ± 0.04	0.35 ± 0.02
Liver (g)	1.28 ± 0.27	2.67 ± 0.31 **	1.70 ± 0.16 ##	1.54 ± 0.34 ##
Hepatic lipid (mg/g)	6.83 ± 0.13	19.58 ± 2.93 **	15.28 ± 0.86 #	10.28 ± 2.75 ##
Perirenal WAT fat (g)	0.12 ± 0.10	0.35 ± 0.11 **	0.40 ± 0.06	0.40 ± 0.07
Retroperitoneal WAT fat (g)	0.29 ± 0.19	1.23 ± 0.33 **	0.71 ± 0.28 ##	0.75 ± 0.17 ##
Epididymal WAT fat (g)	1.25 ± 0.39	2.22 ± 0.54 **	1.65 ± 0.49 #	0.94 ± 0.73 ##
Mesenteric WAT fat (g)	0.25 ± 0.13	0.90 ± 0.22 **	0.91 ± 0.27	0.95 ± 0.09
Total WAT fat (g)	2.00 ± 0.45	4.39 ± 0.29 **	3.50 ± 0.05 #	3.26 ± 0.40 ##

Results shown are mean ± SD (n = 5). Data are statistically represented with * = significantly different from the NC group, and # = significantly different from PM group; # $p < 0.05$, ** and ## $p < 0.01$.

3.6. Inhibition of Lipid Peroxidation

To estimate the inhibition of lipid peroxidation in hepatic and cerebral tissues, the hepatic and cerebral levels of MDA were assessed. The MDA level in the hepatic and cerebral tissues of the HFD group (1.73 and 3.57 mmole/mg of protein, respectively) increased compared to the NC group (1.09 and 2.69 mmole/mg of protein, respectively) (Figure 6a,b). Conversely, the levels in the GC20 (1.09 and 3.20 mmole/mg of protein) and GC50 (1.20 and 3.04 mmole/mg of protein) groups decreased more than the HFD group in hepatic and cerebral tissues, respectively.

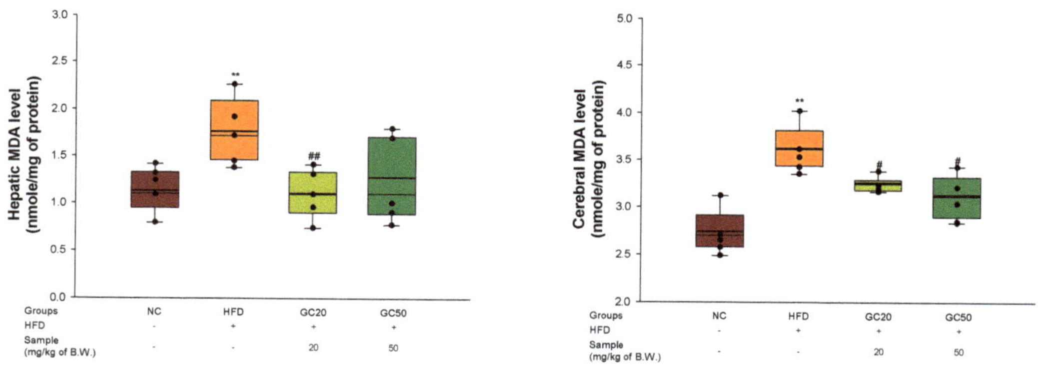

(**a**) (**b**)

Figure 6. Protective effect of 80% ethanolic extract from Gimcheon 1ho (GC) cultivar walnut (*Juglans regia*) on malondialdehyde (MDA) contents in HFD-induced mice biochemical changes related with antioxidant system: (**a**) hepatic MDA level; (**b**) cerebral MDA level Results shown are mean ± SD (n = 5). Data are statistically represented with * = significantly different from the NC group, and # = significantly different from PM group; # $p < 0.05$, ** and ## $p < 0.01$. Bold line indicates mean.

3.7. Serum Level of FRAP and AGEs Formation

To examine the serum level of FRAP and AGEs formation in HFD-induced mice, a FRAP and AGEs formation assay was performed. There was no difference in FRAP levels in serum between the NC group (0.52) and the HFD group (0.53) (Figure 7a). However, the GC groups (GC20, 0.61 and GC50, 0.63) were increased. AGEs formation in serum was evaluated. The HFD group (121.02%) had accumulated AGEs compared to the NC group (100%) (Figure 7b). Conversely, the GC groups (GC20, 110% and GC50, 105%) inhibited AGEs formation in serum.

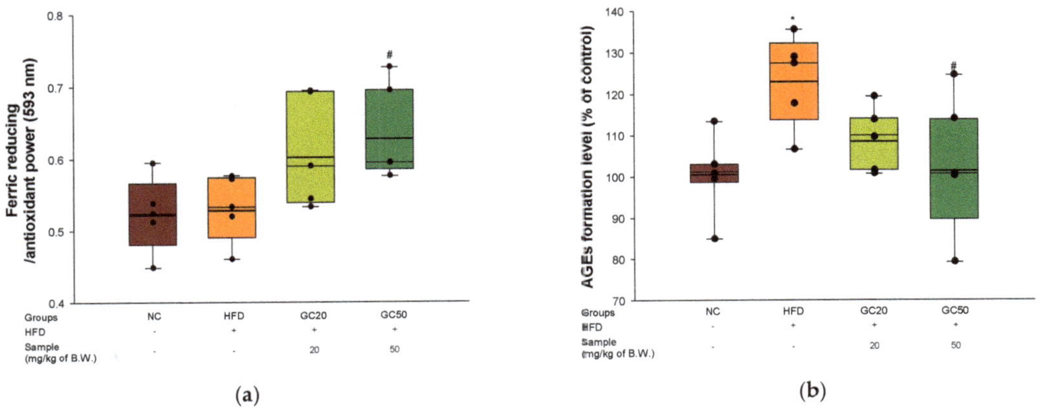

Figure 7. Protective of 80% ethanolic extract from Gimcheon 1ho (GC) cultivar walnut (*Juglans regia*) in HFD-induced mice: (**a**) serum level of FRAP; (**b**) serum level of AGEs formation. Results shown are mean ± SD (n = 5). Data are statistically represented with * = significantly different from the NC group, and # = significantly different from PM group; * and # $p < 0.05$. Bold line indicates mean.

3.8. Cholinergic System

To assess the cognitive functions of the cholinergic system in cerebral tissue, ACh levels, AChE activity, and expression levels of AChE and choline acetyltransferase (ChAT) were evaluated. The ACh levels of the HFD group (4.30 mmol/mg of protein) decreased compared to the NC group (6.47 mmol/mg of protein) (Figure 8a). However, the ingestion of GC showed considerably increased ACh levels (6.26 mmol/mg of protein and 5.84 mmol/mg of protein, respectively). The AChE activity of the HFD group (130.27%) increased compared to the NC group (100%) (Figure 8b). However, the GC groups (GC20, 112.07% and GC50, 115.24%) reduced AChE activity compared to the HFD group.

The expression of AChE and ChAT is shown in Figure 8c–e. The AChE expression level in the HFD group (1.16) was up-regulated compared to the NC group (1.00). On the other hand, the GC groups (GC20, 0.93 and GC50, 0.93) suppressed AChE expression compared to the HFD group. The ChAT expression level in the HFD group (0.81) was down-regulated compared to the NC group (1.00) (Figure 8e). However, the GC groups (GC20, 1.00 and GC50, 0.92) showed up-regulated ChAT expression compared to the HFD group.

3.9. Mitochondrial Activity

To substantiate mitochondrial function in cerebral tissues, ROS production and MMP levels were evaluated. The ROS production of the HFD group in cerebral tissues (114.93%) increased compared to the NC group (100%) (Figure 9a). However, the GC20 (69.53%) and GC50 (78.40%) groups showed considerably decreased ROS production compared with the HFD group. The MMP of the HFD group in cerebral tissues (68.61%) decreased compared to the NC group (100%) (Figure 9b). In contrast, the GC20 (94.88%) and GC50 (97.50%) groups showed restored MMP in cerebral tissues compared to the HFD group.

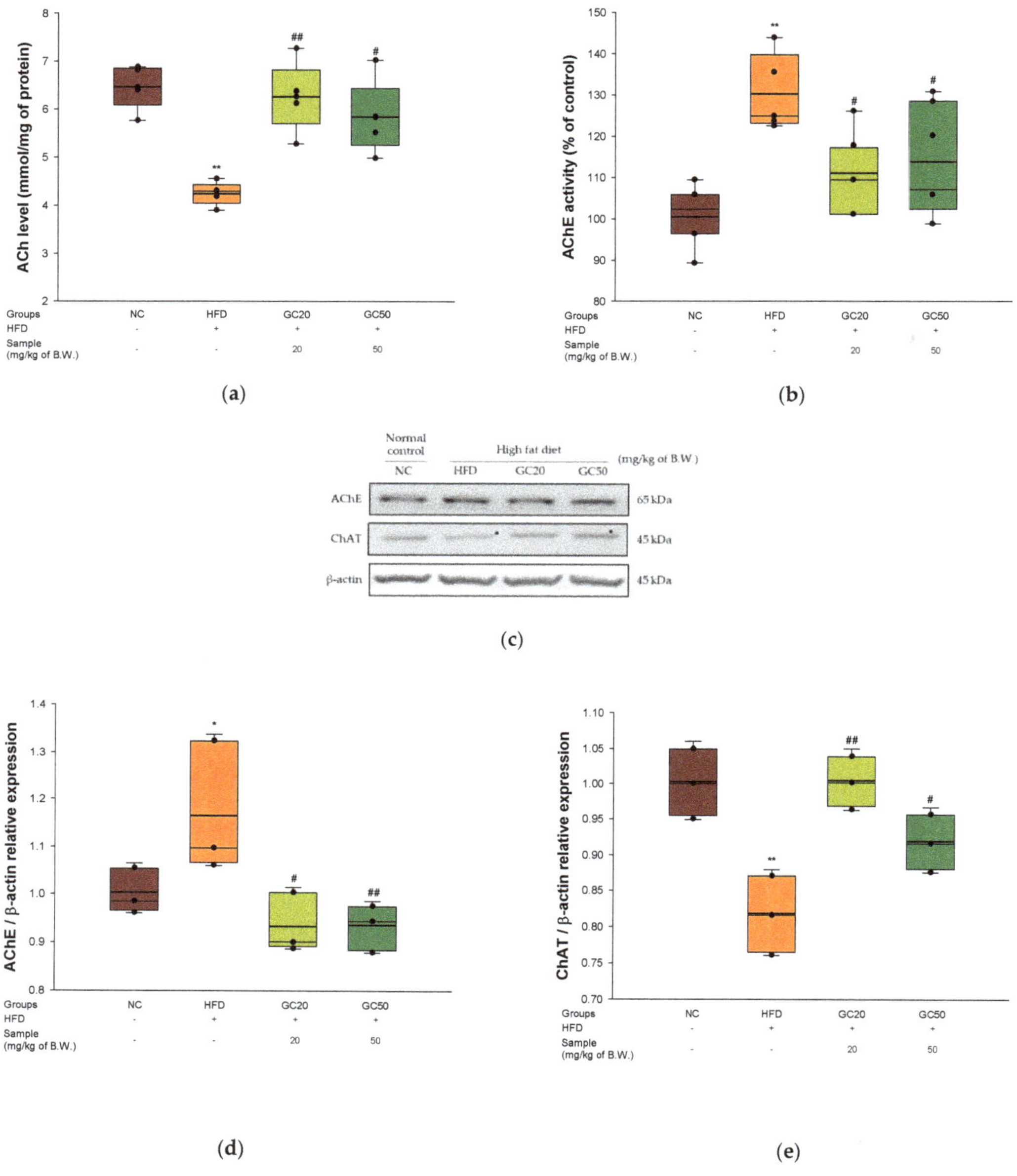

Figure 8. Protective effect of 80% ethanolic extract from Gimcheon 1ho (GC) cultivar walnut (*Juglans regia*) on HFD-induced mice: (**a**) ACh level; (**b**) AChE activity; (**c**) representative western blots for total protein and expression of AChE, ChAT and β-actin; (**d**) protein expression levels of AChE; (**e**) protein expression levels of ChAT. Results shown are mean ± SD (**a**,**b**: n = 5, **c**–**e**: n = 3). Data are statistically represented with * = significantly different from the NC group, and # = significantly different from PM group; * and # $p < 0.05$, ** and ## $p < 0.01$. Bold line indicates mean.

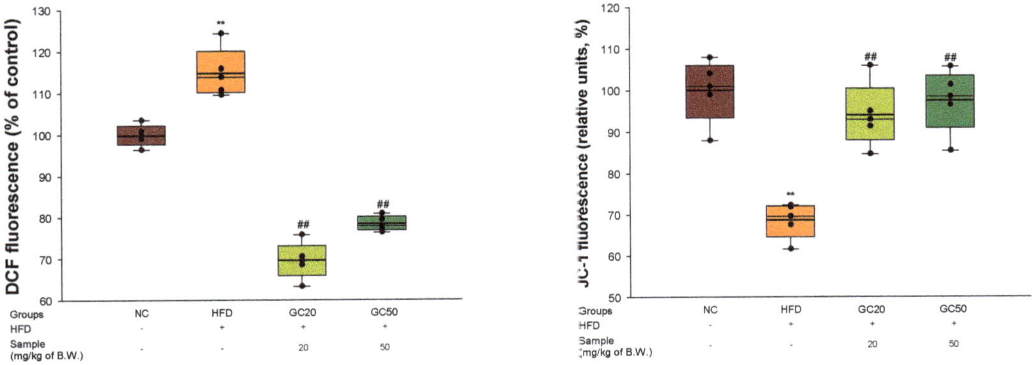

(a) (b)

Figure 9. Protective effect of 80% ethanolic extract from Gimcheon 1ho (GC) cultivar walnut (*Juglans regia*) on mitochondrial dysfunction in HFD-induced mice: (**a**) cerebral ROS contents; (**b**) cerebral MMP levels Results shown are mean ± SD (n = 5). Data are statistically represented with * = significantly different from the NC group, and # = significantly different from PM group; ** and ## $p < 0.01$. Bold line indicates mean.

3.10. Protein Expression in Cerebral Tissue

3.10.1. Synaptic Disorders and Neuronal Apoptosis

Cerebral protein expressions associated with synaptic disorders and neuronal apoptosis are shown in Figure 10a. *p*-JNK (1.20), *p*-tau (1.19), and Aβ (1.20) expression levels in the HFD group significantly increased compared to the NC group (Figure 10b). For the GC20 and GC50 groups *p*-JNK (0.98 and 1.00, respectively), *p*-tau (0.74 and 0.45, respectively), and Aβ (1.16 and 0.94, respectively) expression levels were significantly decreased. In addition, the HFD group showed a decreased expression level of *p*-Akt (Ser 473) (0.85) and insulin-degrading enzyme (IDE) (0.53) compared to the NC group (1.00) (Figure 10b). However, the GC20 and GC50 groups statistically up-regulated *p*-Akt (Ser 473) (0.82 and 1.11, respectively) and IDE (0.65 and 0.82, respectively) expression levels compared to the HFD group. Furthermore, the expression levels of B-cell lymphoma (BCl)-2 associated X protein (BAX) (1.43) and caspase-3 (1.25) were significantly up-regulated in the HFD group compared to the NC group (1.00) (Figure 10b). In contrast, the GC20 and GC50 groups had significantly down-regulated BAX (0.88 and 0.69, respectively) and caspase-3 (0.91 and 0.69, respectively) expression levels compared to the HFD group.

3.10.2. Neuroinflammation

Cerebral protein expressions related to neuronal inflammation are shown in Figure 11a. The expression levels of inflammation-related factors TNF-α (1.13), IL-1β (2.11), *p*-NFκB (1.29) and caspase-1 (1.29) were significantly increased in the HFD group compared to the NC group (1.00) (Figure 11b). On the other hand, the GC20 and GC50 groups statistically ameliorated TNF-α (0.90 and 0.85, respectively), IL-1β (1.30 and 1.14, respectively), *p*-NFκB (0.95 and 0.83, respectively) and caspase-1 (0.87 and 0.60, respectively) expression levels compared to the HFD group. Moreover, heme oxygenase-1 (HO-1) expression levels in the HFD group (0.53) were significantly down-regulated compared to the NC group (1.00) (Figure 11b). HO-1 (1.07 and 1.01) expression levels were significantly up-regulated in the GC20 and GC50 groups compared to the HFD group.

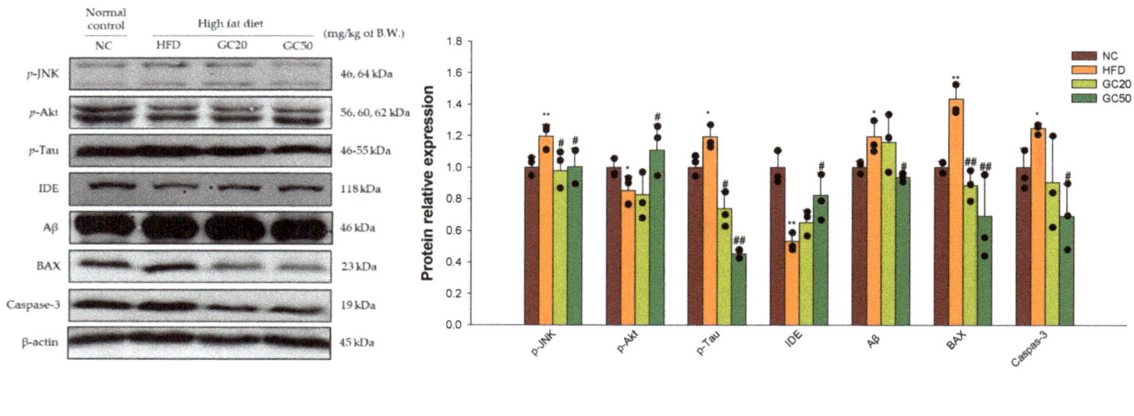

(a) (b)

Figure 10. Protective effect of 80% ethanolic extract from Gimcheon 1ho (GC) cultivar walnut (*Juglans regia*) on HFD-induced synaptic disorders and neuronal apoptosis in mice brain tissues: (**a**) representative Western blots for total protein and expression; (**b**) protein expression levels of *p*-JNK, *p*-Akt, *p*-tau, IDE, Aβ, BAX, caspase-3. Results shown are mean ± SD (n = 3). Data are statistically represented with * = significantly different from the NC group, and # = significantly different from PM group; * and # $p < 0.05$, ** and ## $p < 0.01$. Bold line indicates mean.

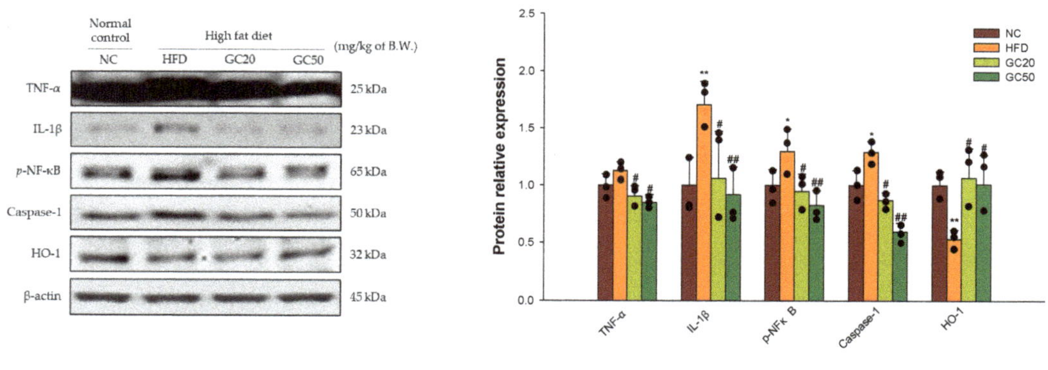

(a) (b)

Figure 11. Protective effect of 80% ethanolic extract from Gimcheon 1ho (GC) cultivar walnut (*Juglans regia*) on HFD-induced neuroinflammation in mice brain tissues: (**a**) representative Western blots for total protein and expression; (**b**) protein expression levels of TNF-α, IL-1β, *p*-NFκB, caspase-1, and HO-1. Results shown are mean ± SD (n = 3). Data are statistically represented with * = significantly different from the NC group, and # = significantly different from PM group; * and # $p < 0.05$, ** and ## $p < 0.01$. Bold line indicates mean.

4. Discussion

Diabetes is a metabolic disease caused by inflammation in a complex immunological process and is related to HFD intake, leading to obesity [3,44]. In general, HFD-induced systemic oxidative stress is associated with reduced insulin sensitivity and the promotion of inflammation following damage in a variety of organs [45]. Insulin resistance and inflammation are the major factors contributing to the degree of severity of the progression of cognitive impairment [10]. Therefore, this study was conducted to evaluate the protective effect of GC on increased oxidative stress, inflammation, and insulin resistance in HFD-induced diabetic disorder.

Type 2 diabetes is characterized by hyperglycemia related to impaired glucose tolerance [46]. High glucose causes reactive oxygen and nitrogen species, primarily mitochondrial dysfunction [47]. Therefore, the abnormal metabolism of glucose, such as hyperglycemia, leads to excess free radical and oxidative stress generation [48]. Chronic hyperglycemia is considered a cause of brain injury by increasing neuroinflammation and apoptotic neuronal death [49]. This study showed that GC effectively protected neuronal cells against H_2O_2 and high glucose-induced neurotoxicity and reduced ROS production in PC12 and HT22 cells. Walnut, which has an abundance of polyunsaturated fatty acids (PUFAs) and polyphenolic compounds, reduced cerebral oxidative stress and inflammatory reaction by enhancing neuronal signaling [30,50]. According to Muthaiyah et al., walnut extract protected neuronal cells against Aβ-mediated cytotoxicity by increasing the capacity of endogenous antioxidant defenses and modulating the cellular redox state [51]. Therefore, GC might be effective in ameliorating ROS production and neurodegeneration.

Increased levels of circulating FFAs are implicated in pancreatic β-cell dysfunction, leading to glucose intolerance [15]. High levels of serum FFA are an important cause of obesity-associated insulin resistance and complications of dyslipidemia [52]. HFD caused dyslipidemic changes by increasing serum levels such as triacylglycerol, TCHO, LDLC, and very-low-density lipoprotein (VLDLC) and decreasing HDLC levels [53]. Dyslipidemia increases oxidative stress through a lipid chain reaction related to endothelial dysfunction and inflammation [45]. Increased oxidative stress induced by insulin resistance, dyslipidemia, and impaired glucose tolerance ultimately lead to type 2 diabetes [54,55]. On the other hand, GC intake improved glucose tolerance and aberrant lipid profiles in serum and significantly increased HDLC content. According to previous studies, the administration of walnut extract for 6 weeks in streptozotocin (STZ)-induced diabetic CD rats reduced blood glucose levels and increased insulin sensitivity by ameliorating insulin resistance [56]. Walnut oil-derived PUFA administration decreased fasting blood glucose and increased hepatic glycogen levels in pregnant diabetic rats [57]. In addition, walnut significantly reduced serum cholesterol, LDLC, TG, and VLDLC levels and increased HDLC compared to STZ-induced diabetic rats [58]. Moreover, the linoleic acid and α-linolenic acid in walnut significantly lowered TCHO, LDL, and TG in obese females [58]. According to Shi et al., the administration of walnut polyphenol extracts inhibited intestinal lipid absorption in HFD-induced mice [26]. Tellimagrandin I, one of the walnut polyphenols, inhibited the TG mechanism in the hyperglycemia model [43]. Consequently, GC consumption plays an important role in improving glucose tolerance and serum lipid levels and preventing type 2 diabetes.

HFD contributes to chronic inflammation and hyperglycemia, resulting in increased oxidative stress [44]. Exposure to hyperglycemia conditions results in irreversible AGEs by rearrangement without degradation of reversible Amadori-type initial glycation products [59]. The accumulation of AGEs induces oxidative stress and contributes to insulin resistance and tissue damage [60]. High levels of oxidative stress destroy cellular membranes, resulting in the overproduction of cytotoxic aldehyde byproducts such as MDA [10]. The production of lipid peroxidation can react with cellular proteins or DNA to form adducts, causing biomolecular damage [37]. In addition, oxidative stress increases the permeability of the BBB, alters the morphology of the brain, damages the central nervous system (CNS), and ultimately promotes neurodegenerative disorders [61]. This study showed that GC significantly reduced AGEs and inhibited MDA in the brain. Furthermore, GC with significantly increased FRAP ability in serum shows antioxidant activity in HFD. According to previous studies, intake of walnut in mice increased plasma antioxidant capacity [62]. Walnut prevented oxidative damage in tissues by reducing lipid oxidation in an ethanol-induced rat model [63]. In addition, it was reported that the unsaturated fatty acids in walnuts effectively inhibit lipid peroxidation [21,62,64]. In conclusion, GC inhibits hepatic and cerebral lipid peroxidation and the formation of serum AGEs and increases serum antioxidant activity to improve HFD-induced oxidative stress.

High levels of FFA are oxidized by β-oxidation in liver mitochondria or accumulated to TG by esterification [65]. The production of electron donors such as reduced nicotinamide adenine dinucleotide (NADH) and dihydroflavine-adenine dinucleotide ($FADH_2$) by β-oxidation produces ATP according to the mitochondrial electron transport chain (ETC) [47]. However, increased β-oxidation of FFA results in significant loss of ETC due to overproduction of ROS in mitochondria [66]. Abnormally damaged ETC activity reduces mitochondrial membrane potential and ATP synthesis [67]. Brain mitochondrial dysfunction occurring in association with neuronal insulin resistance could lead to the development of neuronal damage [10]. However, the administration of GC improved mitochondrial membrane potential function and suppressed oxidative stress in the brain. Similar to this, ellagitannin of walnut polyphenol effectively improved hyperlipidemia and metabolic syndrome by enhancing peroxisomal β-oxidation [42]. In addition, walnut showed mitochondrial ROS scavenging activity and improvement of neuronal energy metabolism in Aβ-injected mice [29]. Therefore, GC may protect against neuronal loss by improving the function of cerebral mitochondria.

HFD-induced diabetes is a risk factor for cognitive impairment in AD [68]. Hippocampus, amygdala, and cerebellar granule cells are susceptible to oxidative stress, and oxidative stress-induced brain damage significantly affects behavioral and cognitive decline [60]. Increased oxidative stress derived from HFD causes neuronal inflammation in the hippocampal region and impairs hippocampal synaptic plasticity, learning, and memory [14]. When spatial memory in a rodent model is assessed, hippocampal-dependent memory/learning deficits are highly correlated with cognitive–behavioral impairments [69]. In this study, the HFD group showed significant memory impairment in behavioral tests [27,70]. On the other hand, the administration of GC showed improvement in spatial learning and memory function in HFD-induced mice. According to previous studies, walnut could improve the memory and cognition of d-galactose-induced aging mice in behavioral tests [70]. In addition, walnut extract has been shown to ameliorate behavioral disorders and memory deficits in an $A\beta_{1\text{-}42}$-induced mouse model [29]. Similar to previous studies, it is suggested that GC reduces memory deficits in HFD-induced behavioral disorders.

Cholinergic neural circuits play essential roles in memory dysfunction [41]. However, alterations of cholinergic function are associated with memory impairment in animals [71]. Ach, a cholinergic neurotransmitter, is released from a broad range of cortical and subcortical sites at the end of synapses and plays an essential role in cognitive functions such as learning and memory [72]. Normally, ACh is synthesized by ChAT by mediating the transfer of the acetyl group from acetyl CoA to choline at the synaptic endings of cholinergic neurons [73]. However, excessive intake of HFD results in cholinergic dysfunction such as abnormal changes in ChAT and AChE expression in many brain regions, including the hypothalamus, hippocampus, amygdala, and cortex [74]. In this study, GC intake restored the cholinergic system in brain tissue and significantly improved the protein expression levels of ChAT and AChE. According to previous studies, walnut inhibited cerebral AChE activity, indicating improvement in learning and memory in D-galactose-induced aging mice [71]. In addition, ellagitannins such as tellimagradin I purified from *Trapa taiwanensis* Nakai hulls showed AChE inhibitory activities in scopolamine-induced amnesia mice [75]. In conclusion, GC containing various ellagitannins has an effect on the cholinergic system by protecting against cognitive dysfunction in HFD-induced mice.

Excessive intake of HFD causes various damages to brain tissue, which is similar to the pathology of AD [9]. HFD impairs JNK/Akt signaling related to brain insulin resistance and leads to cognitive dysfunction [10]. Inhibited Akt activity increases the phosphorylation of GSK-3β and continuously induces hyperphosphorylation of tau protein and aggregation of neurofibrillary tangles (NFTs), leading to synaptic dysfunction and apoptosis [8]. Another pathological feature of HFD-induced insulin resistance is the presence of islet amyloid deposits by the $A\beta_{1\text{-}40}$ and $A\beta_{1\text{-}42}$ peptides due to increased γ-secretase activity in the brain [5]. IDE is one of the Aβ-degrading enzymes involved in the selective cleavage of Aβ peptides in combination with insulin degradation and prevents the formation of amyloid deposits [76]. However, when insulin is increased in the blood, IDE does not cleavage Aβ effectively and causes Aβ neurotoxicity, which eventually leads to AD amyloidosis [9,77]. Neurotoxicity via the accumulation of Aβ peptides and tau NFTs down-regulates BCl-2 at synapses and dendrites and up-regulates BAX to activate downstream signals of apoptosis [78]. Activation of BAX at the mitochondrial surface forms macropores on the mitochondrial outer membrane, leading to the release of cytochrome c into the cytosol [79]. In addition, the release of cytochrome c from mitochondria induces caspase-3 activation, which stimulates signaling pathways and leads to synaptic loss and neuronal apoptosis [80]. In the present study, GC caused a significantly improved HFD-induced JNK/Akt signaling pathway and suppressed the expression of tau and Aβ. In addition, GC inhibited BAX and caspase-3 activation and protected against HFD-induced cerebral apoptosis. According to Ma et al., walnut supplementation significantly reduced the activated cell death-associated expression of *p*-p38 mitogen-activated protein kinase (p38K) and *p*-JNK [62]. In addition, it was confirmed that ingestion of walnut extract inhibited Aβ and tau production in brain tissue by improving the Akt signaling pathway in Aβ-induced mice [29]. Additionally, ellagic acid dose-dependently decreased pathogenic Aβ oligomers and Aβ cytotoxicity in origin SH-SY5Y cells [81]. Moreover, walnut protected against apoptosis by reducing BAX protein levels and cytochrome c release in UVB-induced origin HaCaT cells [82]. Furthermore, walnut peptides inhibited apoptosis by inhibiting the expression of cytochrome c and caspase-3 [83]. These results suggest that GC regulates JNK/Akt signaling to protect against significantly suppressed insulin resistance-mediated synaptic disorders and apoptosis.

HFD-mediated excessive FFA activates inflammation by releasing inflammatory cytokines in brain tissue [10,84]. Increased expression of various pro-inflammatory cytokines such as TNF-α and IL-1β activates NFκB, a transcriptional activator [62]. The active NFκB promotes the transcription of NFκB-dependent genes, such as leucine-rich repeat (NLR) pyrin domain containing 3 (NLRP3), pro-IL-1β, and TNF- α in the nucleus [85]. The NLRP3 family member forms, triggering autocatalytic activation of caspase-1 and suppresses nuclear factor erythroid 2–related factor 2 (Nrf2) expression [86]. This reaction reduces the expression of HO-1 with a strong antioxidant effect and stimulates the production of IL-1β and TNF-α [87,88]. Thus, excessive inflammatory cytokine expression contributes to the pathogenesis of inflammatory responses and cognitive impairment [11,89]. GC restored neuroinflammation by suppressing protein expression levels of TNF-α, IL-1β, *p*-NFκB, caspase-1, and HO-1. In a previous study, walnut peptides inhibited NFκB pathway activation and attenuated the neurotoxic cascade by overexpression of IL-1β and TNF-α [83]. In addition, walnut reduced the production of TNF-α, IL-1β, and IL-6 by suppressing their mRNA expressions in LPS-induced mice [89]. Furthermore, walnut-derived peptides protected insulin resistance and decreased oxidative stress by activating HO-1 in high glucose-induced origin HepG2 cells [27]. Ellagic acid protected against inflammation such as nitric oxide (NO), MDA, IL-1β, TNF-α, cyclooxygenase 2 (COX-2), and NFκB expression in carrageenan-induced rats [90]. Therefore, GC may improve cognitive function by ameliorating neurodegenerative disorders by reducing neuroinflammation.

Overall, walnut showed a protective effect against HFD-induced diabetic symptoms such as glucose tolerance, diabetic cognitive dysfunction, hyperlipidemia, mitochondrial deficit, apoptosis, and neuronal inflammation. Walnut contains various physiological compounds such as PUFA, α-linolenic acid, ellagitannins, ellagic acid, and bioactive peptides. According to previous studies, walnut-derived PUFA decreased fasting blood glucose, TCHO, LDL, and TG levels and inhibited lipid peroxidation [21,57,58,62]. Walnut ellagitannin, such as tellimagradin I and ellagic acid, enhanced peroxisomal β-oxidation and reduced the TG mechanism and inflammatory responses [42,90]. In addition, ellagitannins protected against synaptic dysfunction by regulating AChE activity and pathogenic Aβ oligomers [75,81]. Various bioactive peptides of walnut inhibited apoptosis by regulating mitochondrial apoptosis and suppressed insulin resistance by increasing the HO-1 pathway [27,83]. In conclusion, based on these physiological activities, GC showed protective effects against HFD-induced diabetic dysfunctions through complex and diverse pathways (Figure 12).

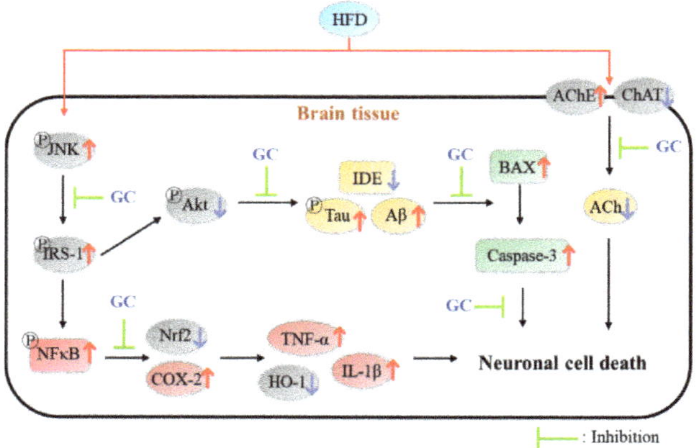

Figure 12. A schematic illustration shows the immunomodulatory effect of 80% ethanolic extract from Gimcheon 1ho (GC) cultivar walnut (*Juglans regia*) on HFD-induced neuroinflammation in mice brain tissues.

5. Conclusions

In summary, based on this study, GC showed a significant neuroprotective effect in neuronal cells and hippocampal cells induced by H_2O_2 and high glucose stresses. GC restored behavioral dysfunction in HFD-induced C57BL/6 mice. In addition, GC protected antioxidant and cholinergic systems and improved mitochondrial dysfunction. Furthermore, the administration of GC suppressed synaptic disorders by regulating AChE and ChAT expression. In addition, GC inhibited cerebral cytotoxicity via JNK cascade signaling. GC down-regulated inflammatory responses by regulating the protein expression of TNF-α, IL-1β, *p*-NFκB, caspase-1, and HO-1. In conclusion, it is suggested that GC extract can be used as a material for functional foods that improve memory loss and cognitive dysfunction by regulating synaptic function and inflammation.

Author Contributions: Conceptualization, J.H.M., H.J.H.; methodology, J.Y.K., M.J.K.; software, H.L.L., H.R.J.; validation, J.M.K., H.J.H.; formal analysis, J.H.M.; investigation, J.H.M., M.J.G.; resources, U.L., H.-J.K., H.W.P., C.-W.K., S.J.P.; data curation, J.H.M.; writing—original draft preparation, J.H.M.; writing—review and editing, J.M.K., H.J.H.; visualization, J.H.M.; supervision, H.J.H.; project administration, H.J.H. All authors have read and agreed to the published version of the manuscript.

Funding: This research was funded by Gimcheon-si through Research on An Efficacy Evaluation of Gimcheon Walnut [grant number: 20190300160-00] and Basic Science Research Program through the National Research Foundation (NRF) of Korea [NRF 2018R1D1A3B07043398] funded by the Ministry of Education, Republic of Korea.

Institutional Review Board Statement: All animal experiments in the study were in compliance with the Institutional Animal Care and Use Committee (IACUC) of Gyeongsang National University (certificate: GNU-190530-M0028; approved on 30 May 2019).

Informed Consent Statement: Not applicable.

Data Availability Statement: The data underlying this article are shared upon reasonable request to the corresponding author.

Conflicts of Interest: The authors declare no conflict of interest.

References

1. Shaw, J.E.; Sicree, R.A.; Zimmet, P.Z. Global estimates of the prevalence of diabetes for 2010 and 2030. *Diabetes Rec. Clin. Pract.* **2010**, *87*, 4–14. [CrossRef] [PubMed]
2. Asmat, U.; Abad, K.; Ismail, K. Diabetes mellitus and oxidative stress—A concise review. *Saudi Pharm. J.* **2016**, *24*, 547–553. [CrossRef] [PubMed]
3. Lizarbe, B.; Cherix, A.; Duarçte, J.M.N.; Cardinaux, J.; Gruetter, R. High-fat diet consumption alters energy metabolism in the mouse hypothalamus. *Int. J. Obes.* **2019**, *43*, 1295–1304. [CrossRef] [PubMed]
4. Kopelman, P.G. Obesity as a medical problem. *Nature* **2000**, *404*, 635–643. [CrossRef] [PubMed]
5. Monnier, L.; Mas, E.; Ginet, C.; Michel, F.; Villon, L.; Cristol, J.; Colette, C. Activation of oxidative stress by acute glucose fluctuations compared with sustained chronic hyperglycemia in patients with type 2 diabetes. *JAMA* **2006**, *295*, 1681–1687. [CrossRef]
6. Kaneto, H.; Matsuoka, T.A.; Nakatani, Y.; Kawamori, D.; Matsuhisa, M.; Yamasaki, Y. Oxidative stress and the JNK pathway in diabetes. *Curr. Diabetes Rev.* **2005**, *1*, 65–72. [CrossRef] [PubMed]
7. Malhi, H.; Gores, G.J. Molecular mechanisms of lipotoxicity in non-alcoholic fatty liver disease. *Semin. Liver Dis.* **2008**, *28*, 360–369. [CrossRef]
8. Vivanco, I.; Sawyers, C.L. The phosphatidylinositol 3-kinase–AKT pathway in human cancer. *Nat. Rev. Cancer* **2002**, *2*, 489–501. [CrossRef] [PubMed]
9. Ho, L.; Qin, W.; Pompl, P.N.; Xiang, Z.; Wang, J.; Zhao, Z.; Peng, Y.; Cambareri, G.; Rocher, A.; Mobbs, C.V.; et al. Diet-induced insulin resistance promotes amyloidosis in a transgenic mouse model of Alzheimer's disease. *FASEB* **2004**, *18*, 902–904. [CrossRef] [PubMed]
10. Sripetchwandee, J.; Chattipakorn, N.; Chattipakorn, S.C. Links between obesity-induced brain insulin resistance, brain mitochondrial dysfunction, and dementia. *Front. Endocrinol.* **2018**, *9*, 496. [CrossRef]
11. Wieser, V.; Moschen, A.R.; Tilg, H. Inflammation, cytokines and insulin resistance: A clinical perspective. *Arch. Immunol. Ther. Exp. (Warsz.)* **2013**, *61*, 119–125. [CrossRef] [PubMed]
12. Tu, Y.F.; Tsai, Y.S.; Wang, L.W.; Wu, H.C.; Huang, C.C.; Ho, C.J. Overweight worsens apoptosis, neuroinflammation and blood-brain barrier damage after hypoxic ischemia in neonatal brain through JNK hyperactivation. *J. Neuroinflamm.* **2011**, *8*, 40. [CrossRef] [PubMed]
13. Alkan, I.; Altunkaynak, B.Z.; Gültekin, G.İ.; Bayçu, C. Hippocampal neural cell loss in high-fat diet-induced obese rats–exploring the protein networks, ultrastructure, biochemical and bioinformatical markers. *J. Chem. Neuroanat.* **2021**, *114*, 101947. [CrossRef] [PubMed]
14. Maciejczyk, M.; Żebrowska, E.; Chabowski, A. Insulin resistance and oxidative stress in the brain: What's new? *Int. J. Mol. Sci.* **2019**, *20*, 874. [CrossRef] [PubMed]
15. Fraulob, J.C.; Ogg-Diamantino, R.; Fernandes-Santos, C.; Aguila, M.B ; Mandarim-de-Lacerda, C.A. A mouse model of metabolic syndrome: Insulin resistance, fatty liver and non-alcoholic fatty pancreas disease (NAFPD) in C57BL/6 mice fed a high fat diet. *J. Clin. Biochem. Nutr.* **2010**, *46*, 212–223. [CrossRef]
16. Verspohl, E.J. Recommended testing in diabetes research. *Planta Med.* **2002**, *68*, 581–590. [CrossRef] [PubMed]
17. Srinivasan, K.; Ramarao, P. Animal model in type 2 diabetes research: An overview. *Indian J. Med. Res.* **2007**, *125*, 451–472. [PubMed]
18. Li, L.; Tsao, R.; Yang, R.; Liu, C.; Zhu, H.; Young, J.C. Polyphenolic profiles and antioxidant activities of heartnut (*Juglans ailanthifolia* Var. *cordiformis) and persian walnut (Juglans regia L.). J. Agric. Food Chem.* **2006**, *54*, 8033–8040. [CrossRef] [PubMed]
19. Li, L.; Tsao, R.; Yang, R.; Kramer, J.K.; Hernandez, M. Fatty acid profiles, tocopherol contents, and antioxidant activities of heartnut (*Juglans ailanthifolia* Var. *cordiformis) and persian walnut (Juglans regia L.). J. Agric. Food Chem.* **2007**, *55*, 1164–1169. [CrossRef] [PubMed]
20. Samaranayaka, A.G.; John, J.A.; Shahidi, F. Antioxidant activity of english walnut (*Juglans Regia* L.). *J. Food Lipids* **2008**, *15*, 384–397. [CrossRef]
21. Okuda, T.; Yoshida, T.; Hatano, T. Ellagitannins as active constituents of medicinal plants. *Planta Med.* **1989**, *55*, 117–122. [CrossRef] [PubMed]
22. Abdel-Moneim, A.; Yousef, A.I.; El-Twab, A.; Sanaa, M.; Abdel Reheim, E.S.; Ashour, M.B. Gallic acid and p-coumaric acid attenuate type 2 diabetes-induced neurodegeneration in rats. *Metab. Brain Dis.* **2017**, *32*, 1279–1286. [CrossRef] [PubMed]
23. Yoshimura, Y.; Nishii, S.; Zaima, N.; Moriyama, T.; Kawamura, Y. Ellagic acid improves hepatic steatosis and serum lipid composition through reduction of serum resistin levels and transcriptional activation of hepatic ppara in obese, diabetic KK-Ay mice. *Biochem. Biophys. Res. Commun.* **2013**, *434*, 486–491. [CrossRef] [PubMed]
24. Yu, M.; Chen, X.; Liu, J.; Ma, Q.; Zhuo, Z.; Chen, H.; Zhou, L.; Yang, S.; Zheng, L.; Ning, C.; et al. Gallic acid disruption of $A\beta_{1-42}$ aggregation rescues cognitive decline of APP/PS1 double transgenic mouse. *Neurobiol. Dis.* **2019**, *124*, 67–80. [CrossRef] [PubMed]
25. He, X.M.; Zhou, Y.Z.; Sheng, S.; Li, J.J.; Wang, G.Q.; Zhang, F. Ellagic acid protects dopamine neurons via inhibition of NLRP3 inflammasome activation in microglia. *Oxid. Med. Cell Longev.* **2020**, *2020*, 2963540. [CrossRef]
26. Shi, D.; Chen, C.; Zhao, S.; Ge, F.; Liu, D.; Song, H. Walnut polyphenols inhibit pancreatic lipase activity in vitro and have hypolipidemic effect on high-fat diet-induced obese mice. *Food Nutr. Res.* **2014**, *2*, 757–763. [CrossRef]

27. Wang, J.; Wu, T.; Fang, L.; Liu, C.; Liu, X.; Li, H.; Shi, J.; Li, M.; Min, W. Peptides from walnut (*Juglans mandshurica* Maxim.) protect hepatic HepG2 cells from high glucose-induced insulin resistance and oxidative stress. *Food Funct.* **2020**, *11*, 8112–8121. [CrossRef] [PubMed]
28. Sun, B.; Yan, H.; Li, C.; Yin, L.; Li, F.; Zhou, L.; Han, X. Beneficial effects of walnut (*Juglans regia* L.) oil-derived polyunsaturated fatty acid prevents a prooxidant status and hyperlipidemia in pregnant rats with diabetes. *Nutr. Metab.* **2020**, *17*, 1–11. [CrossRef]
29. Kim, J.M.; Lee, U.; Kang, J.Y.; Park, S.K.; Shin, E.J.; Kim, H.; Kim, C.; Kim, M.; Heo, H.J. Anti-amnesic effect of walnut via the regulation of BBB function and neuro-inflammation in $A\beta_{1\text{-}42}$-induced mice. *Antioxidants* **2020**, *9*, 976. [CrossRef]
30. Kim, G.H.; Kim, J.M.; Park, S.K.; Kang, J.Y.; Han, H.J.; Shin, E.J.; Moon, J.H.; Kim, C.W.; Lee, U.; Shin, E.C.; et al. Nutritional composition of domestic and imported walnuts (*Juglans regia* L.). *J. Korean Soc. Food Sci. Nutr.* **2020**, *49*, 608–616. [CrossRef]
31. Kim, J.M.; Park, S.K.; Kang, J.Y.; Park, S.B.; Yoo, S.K.; Han, H.J.; Cho, K.H.; Kim, J.C.; Heo, H.J. Green tea seed oil suppressed $A\beta_{1\text{-}42}$-induced behavioral and cognitive deficit via the Aβ-related Akt pathway. *Int. J. Mol. Sci.* **2019**, *20*, 1865. [CrossRef] [PubMed]
32. Van der Borght, K.; Havekes, R.; Bos, T.; Eggen, B.J.; Van der Zee, E.A. Exercise improves memory acquisition and retrieval in the Y-maze task: Relationship with hippocampal neurogenesis. *Behav. Neurosci.* **2007**, *121*, 324–334. [CrossRef] [PubMed]
33. Newman, J.P.; Kosson, D.S. Passive avoidance learning in psychopathic and nonpsychopathic offenders. *J. Abnorm. Psychol.* **1986**, *95*, 252–256. [CrossRef] [PubMed]
34. Morris, R. Developments of a water-maze procedure for studying spatial learning in the rat. *J. Neurosci. Methods* **1984**, *11*, 47–60. [CrossRef]
35. Friedewald, W.T.; Levy, R.I.; Fredrickson, D.S. Estimation of the concentration of low-density lipoprotein cholesterol in plasma, without use of the preparative ultracentrifuge. *Clin. Chem.* **1972**, *18*, 499–502. [CrossRef] [PubMed]
36. Bradford, M.M. A rapid and sensitive method for the quantitation of microgram quantities of protein utilizing the principle of protein-dye binding. *Anal. Biochem.* **1976**, *72*, 248–254. [CrossRef]
37. Ayala, A.; Muñoz, M.F.; Argüelles, S. Lipid peroxidation: Production, metabolism, and signaling mechanisms of malondialdehyde and 4-hydroxy-2-nonenal. *Oxid. Med. Cell. Longev.* **2014**, *2014*, 360438. [CrossRef] [PubMed]
38. Benzie, I.F.; Strain, J.J. The ferric reducing ability of plasma (FRAP) as a measure of "antioxidant power": The FRAP assay. *Anal. Biochem.* **1996**, *239*, 70–76. [CrossRef]
39. Sampath, C.; Rashid, M.R.; Sang, S.; Ahmedna, M. Green tea epigallocatechin 3-gallate alleviates hyperglycemia and reduces advanced glycation end products via nrf2 pathway in mice with high fat diet-induced obesity. *Biomed. Pharmacother.* **2017**, *87*, 73–81. [CrossRef]
40. Vincent, D.; Segonzac, G.; Vincent, M.C. Colorimetric determination of acetylcholine by the Hestrin hydroxylamine reaction and its application in pharmacy. *Ann. Pharm. Fr.* **1958**, *16*, 179–185.
41. Ellman, G.L.; Courtney, K.D.; Andres jr, V.; Featherstone, R.M. A new and rapid colorimetric determination of acetylcholinesterase activity. *Biochem. Pharmacol.* **1961**, *7*, 88–95. [CrossRef]
42. Brown, M.R.; Geddes, J.W.; Sullivan, P.G. Brain region-specific, age-related, alterations in mitochondrial responses to elevated calcium. *J. Bioenerg. Biomembr.* **2004**, *36*, 401–406. [CrossRef] [PubMed]
43. Shimoda, H.; Tanaka, J.; Kikuchi, M.; Fukuda, T.; Ito, H.; Hatano, T.; Yoshida, T. Effect of polyphenol-rich extract from walnut on diet-induced hypertriglyceridemia in mice via enhancement of fatty acid oxidation in the liver. *J. Agric. Food Chem.* **2009**, *57*, 1786–1792. [CrossRef] [PubMed]
44. Berbudi, A.; Rahmadika, N.; Tjahjadi, A.I.; Ruslami, R. Type 2 diabetes and its impact on the immune system. *Curr. Diabetes Rev.* **2020**, *16*, 442–449. [CrossRef] [PubMed]
45. Kesh, S.; Sarkar, D.; Manna, K. High-fat diet-induced oxidative stress and its impact on metabolic syndrome: A review. *Asian J. Pharm. Clin. Res.* **2016**, *9*, 47–52.
46. World Health Organization. *Classification of Diabetes Mellitus*; World Health Organization: Geneva, Switzerland, 2019.
47. Brownlee, M. Biochemistry and molecular cell biology of diabetic complications. *Nature* **2001**, *414*, 813–820. [CrossRef]
48. Allen, D.A.; Yaqoob, M.M.; Harwood, S.M. Mechanisms of high glucose-induced apoptosis and its relationship to diabetic complications. *J. Nutr. Biochem.* **2005**, *16*, 705–713. [CrossRef] [PubMed]
49. Hamed, S.A. Brain injury with diabetes mellitus: Evidence, mechanisms and treatment implications. *Expert Rev. Clin. Pharmacol.* **2017**, *10*, 409–428. [CrossRef] [PubMed]
50. Poulose, S.M.; Miller, M.G.; Shukitt-Hale, B. Role of walnuts in maintaining brain health with age. *J. Nutr.* **2014**, *144*, 561S–566S. [CrossRef] [PubMed]
51. Muthaiyah, B.; Essa, M.; Chauhan, V.; Chauhan, A. Protective effects of walnut extract against amyloid beta peptide-induced cell death and oxidative stress in PC12 cells. *Neurochem. Res.* **2011**, *36*, 2096–2103. [CrossRef]
52. Boden, G. 45Obesity, insulin resistance and free fatty acids. *Curr. Opin. Endocrinol. Diabetes Obes.* **2011**, *18*, 139–143. [CrossRef] [PubMed]
53. Noeman, S.A.; Hamooda, H.E.; Baalash, A.A. Biochemical study of oxidative stress markers in the liver, kidney and heart of high fat diet induced obesity in rats. *Diabetol. Metab. Syndr.* **2011**, *3*, 1–8. [CrossRef] [PubMed]
54. Leopold, J.A.; Loscalzo, J. Oxidative mechanisms and atherothrombotic cardiovascular disease. *Drug Discov. Today Ther. Strateg.* **2008**, *5*, 5–13. [CrossRef] [PubMed]

55. Tangvarasittichai, S. Oxidative stress, insulin resistance, dyslipidemia and type 2 diabetes mellitus. *World J. Diabetes* **2015**, *6*, 456–480. [CrossRef] [PubMed]
56. Li, Y.; Chen, D.; Zhang, F.; Lin, Y.; Ma, Y.; Zhao, S.; Chen, C.; Wang, X.; Liu, J. Preventive effect of pressed degreased walnut meal extracts on T2DM rats by regulating glucolipid metabolism and modulating gut bacteria flora. *J. Funct. Foods* **2020**, *64*, 103694. [CrossRef]
57. Ebrahim Abbasi, O.; Arash Noori, S.; Ali, R. Effects of walnut on lipid profile as well as the expression of sterol-regulatory element binding protein-1c (SREBP-1c) and peroxisome proliferator activated receptors α (PPARα) in diabetic rat. *Food Nutr. Sci.* **2012**, *3*, 2. [CrossRef]
58. Mushtaq, R.; Mushtaq, R.; Khan, Z.T. Effect of walnut on lipid profile in obese female in different ethnic groups of Quetta, Pakistan. *J. Nutr.* **2009**, *8*, 1617–1622. [CrossRef]
59. Voziyan, P.A.; Khalifah, R.G.; Thibaudeau, C.; Yildiz, A.; Jacob, J.; Serianni, A.S.; Hudson, B.G. Modification of proteins in vitro by physiological levels of glucose: Pyridoxamine inhibits conversion of Amadori intermediate to advanced glycation end-products through binding of redox metal ions. *J. Biol. Chem.* **2003**, *278*, 46616–46624. [CrossRef] [PubMed]
60. Unoki, H.; Yamagishi, S. I Advanced glycation end products and insulin resistance. *Curr. Pharm. Des.* **2008**, *14*, 987–989. [CrossRef] [PubMed]
61. Salim, S. Oxidative stress and the central nervous system. *J. Pharmacol. Exp. Ther.* **2017**, *360*, 201–205. [CrossRef] [PubMed]
62. Ma, Z.F.; Ahmad, J.; Khan, I.; Wang, C.W.; Jiang, P.; Zhang, Y. Interaction of phytochemicals from walnut on health: An updated comprehensive review of reported bioactivities and medicinal properties of walnut. *J. Biol. Act. Prod. Nat.* **2019**, *9*, 410–425. [CrossRef]
63. Bati, B.; Celik, I.; Dogan, A. Determination of hepatoprotective and antioxidant role of walnuts against ethanol-induced oxidative stress in rats. *Cell Biochem. Biophys.* **2015**, *71*, 1191–1198. [CrossRef] [PubMed]
64. Fukuda, T.; Ito, H.; Yoshida, T. Antioxidative polyphenols from walnuts (*Juglans regia* L.). *Phytochemistry.* **2003**, *63*, 795–801. [CrossRef]
65. de Mello, A.H.; Costa, A.B.; Engel, J.D.G.; Rezin, G.T. Mitochondrial dysfunction in obesity. *Life Sci.* **2018**, *192*, 26–32. [CrossRef] [PubMed]
66. Gusdon, A.M.; Song, K.; Qu, S. Nonalcoholic fatty liver disease: Pathogenesis and therapeutics from a mitochondria-centric perspective. *Oxid. Med. Cell. Longev.* **2014**, *2014*, e637027. [CrossRef]
67. Brownlee, M. The pathobiology of diabetic complications: A unifying mechanism. *Diabetes* **2005**, *54*, 1615–1625. [CrossRef] [PubMed]
68. Bagkos, G.; Koufopoulos, K.; Piperi, C. A new model for mitochondrial membrane potential production and storage. *Med. Hypotheses* **2014**, *83*, 175–181. [CrossRef]
69. Park, H.R.; Park, M.; Choi, J.; Park, K.; Chung, H.Y.; Lee, J. A high-fat diet impairs neurogenesis: Involvement of lipid peroxidation and brain-derived neurotrophic factor. *Neurosci. Lett.* **2010**, *482*, 235–239. [CrossRef]
70. Wahl, D.; Coogan, S.C.; Solon-Biet, S.M.; de Cabo, R.; Haran, J.B.; Raubenheimer, D.; Cogger, V.C.; Mattson, M.P.; Simpson, S.J.; Le Couteur, D.G. Cognitive and behavioral evaluation of nutritional interventions in rodent models of brain aging and dementia. *Clin. Interv. Aging* **2017**, *12*, 1419–1428. [CrossRef]
71. Liu, J.; Chen, D.; Wang, Z.; Chen, C.; Ning, D.; Zhao, S. Protective effect of walnut on d-galactose-induced aging mouse model. *Food Sci. Nutr.* **2019**, *7*, 969–976. [CrossRef] [PubMed]
72. Heydemann, A. An overview of murine high fat diet as a model for type 2 diabetes mellitus. *J. Diabetes Res.* **2016**, *2016*, e2902351. [CrossRef] [PubMed]
73. Newman, E.L.; Gupta, K.; Climer, J.R.; Monaghan, C.K.; Hasselmo, M E. Cholinergic modulation of cognitive processing: Insights drawn from computational models. *Front. Behav. Neurosci.* **2012**, *6*, 24. [CrossRef] [PubMed]
74. Oda, Y. Choline acetyltransferase: The structure, distribution and pathologic changes in the central nervous system. *Pathol. Int.* **1999**, *49*, 921–937. [CrossRef] [PubMed]
75. Chang, E.H.; Chavan, S.S.; Pavlov, V.A. Cholinergic control of inflammation, metabolic dysfunction, and cognitive impairment in obesity-associated disorders: Mechanisms and novel therapeutic opportunities. *Front. Neurosci.* **2019**, *13*, 263. [CrossRef]
76. Chen, L.; Lin, S.; Lee, Y.; Wang, C.; Hou, W. Hydrolysable tannins exhibit acetylcholinesterase inhibitory and anti-glycation activities in vitro and learning and memory function improvements in scopolamine-induced amnesiac mice. *Biomedicines* **2021**, *9*, 1066. [CrossRef] [PubMed]
77. Malito, E.; Hulse, R.E.; Tang, W. Amyloid β-degrading cryptidases: Insulin degrading enzyme, neprilysin, and presequence peptidase. *Cell. Mol. Life Sci.* **2008**, *65*, 2574–2585. [CrossRef] [PubMed]
78. Qiu, W.Q.; Folstein, M.F. Insulin, insulin-degrading enzyme and amyloid-β peptide in Alzheimer's disease: Review and hypothesis. *Neurobiol. Aging* **2006**, *27*, 190–198. [CrossRef] [PubMed]
79. Paradis, E.; Douillard, H.; Koutroumanis, M.; Goodyer, C.; LeBlanc, A. Amyloid β peptide of Alzheimer's disease downregulates Bcl-2 and upregulates Bax expression in human neurons. *J. Neurosci.* **1996**, *16*, 7533–7539. [CrossRef]
80. Singh, R.; Letai, A.; Sarosiek, K. Regulation of apoptosis in health and disease: The balancing act of BCL-2 family proteins. *Nat. Rev. Mol. Cell Biol.* **2019**, *20*, 175–193. [CrossRef] [PubMed]

81. Wang, L.; Wei, Y.; Ning, C.; Zhang, M.; Fan, P.; Lei, D.; Du, J.; Gale, M.; Ma, Y.; Yang, Y. Ellagic acid promotes browning of white adipose tissues in high-fat diet-induced obesity in rats through suppressing white adipocyte maintaining genes. *Endocr. J.* **2019**, *66*, 923–936. [CrossRef] [PubMed]
82. Garrido, C.; Galluzzi, L.; Brunet, M.; Puig, P.E.; Didelot, C.; Kroemer, G. Mechanisms of cytochrome c release from mitochondria. *Cell Death Differ.* **2006**, *13*, 1423–1433. [CrossRef] [PubMed]
83. Park, G.; Kim, H.G.; Hong, S.-P.; Kim, S.Y.; Oh, M.S. Walnuts (Seeds of *Juglandis sinensis* L.) protect human epidermal keratinocytes against UVB-induced mitochondria-mediated apoptosis through upregulation of ROS elimination pathways. *Skin Pharmacol. Physiol.* **2014**, *27*, 132–140. [CrossRef] [PubMed]
84. Liu, C.; Guo, Y.; Zhao, F.; Qin, H.; Lu, H.; Fang, L.; Wang, J.; Min, W. Potential mechanisms mediating the protective effects of a peptide from walnut (*Juglans mandshurica* Maxim.) against hydrogen peroxide induced neurotoxicity in PC12 cells. *Food Funct.* **2019**, *10*, 3491–3501. [CrossRef] [PubMed]
85. Liu, T.; Zhang, L.; Joo, D.; Sun, S.C. NFκB signaling in inflammation. *Signal. Transduct. Target. Ther.* **2017**, *2*, 17023. [CrossRef] [PubMed]
86. Shimada, K.; Crother, T.; Karlin, J.; Dagvadorj, J.; Chiba, N.; Chen, S.; Ramanujan, V.K.; Wolf, A.; Vergnes, L.; Ojcius, D.; et al. Oxidized mitochondrial DNA activates the NLRP3 inflammasome during apoptosis. *Immunity* **2012**, *36*, 401–414. [CrossRef] [PubMed]
87. Liang, J.; Li, L.; Sun, Y.; He, W.; Wang, X.; Su, Q. The protective effect of activating Nrf2/HO-1 signaling pathway on cardiomyocyte apoptosis after coronary microembolization in rats. *BMC Cardiovasc. Disord.* **2017**, *17*, 272. [CrossRef] [PubMed]
88. Sah, S.K.; Lee, C.; Jang, J.; Park, G.H. Effect of high-fat diet on cognitive impairment in triple-transgenic mice model of Alzheimer's disease. *Biochem. Biophys. Res. Commun.* **2017**, *493*, 731–736. [CrossRef]
89. Wang, S.; Zheng, L.; Zhao, T.; Zhang, Q.; Liu, Y.; Sun, B.; Su, G.; Zhao, M. Inhibitory effects of walnut (*Juglans regia*) peptides on neuroinflammation and oxidative stress in lipopolysaccharide-induced cognitive impairment mice. *J. Agric. Food Chem.* **2020**, *68*, 2381–2392. [CrossRef]
90. Sánchez-González, C.; Ciudad, C.J.; Noé, V.; Izquierdo-Pulido, M. Health benefits of walnut polyphenols: An exploration beyond their lipid profile. *Crit. Rev. Food Sci. Nutr.* **2017**, *57*, 3373–3383. [CrossRef]

molecules

MDPI

Article

Molineria recurvata Ameliorates Streptozotocin-Induced Diabetic Nephropathy through Antioxidant and Anti-Inflammatory Pathways

Prasanta Dey [1,†], Amit Kundu [1], Ha Eun Lee [1], Babli Kar [2], Vineet Vishal [3], Suvakanta Dash [4], In Su Kim [1], Tejendra Bhakta [4,*] and Hyung Sik Kim [1,*]

1 School of Pharmacy, Sungkyunkwan University, 2066, Seobu-ro, Jangan-gu, Suwon 16419, Korea
2 Bengal Homoeopathic Medical College and Hospital, Asansol 713301, India
3 Department of Botany, Bangabasi Evening College, Kolkata 700009, India
4 Regional Institute of Pharmaceutical Science & Technology, Agartala 799006, India
* Correspondence: tbhakta2010@gmail.com (T.B.); hkims@skku.edu (H.S.K.)
† Current address: Rutgers Cancer Institute of New Jersey, Rutgers University, New Brunswick, NJ 08901, USA.

Abstract: *Molineria recurvata* (MR) has been traditionally used to manage diabetes mellitus in India. However, the molecular mechanism of MR on the diabetic-induced nephropathy has not been clearly investigated. Thus, this study investigates the protective effects of the MR extract on nephropathy in streptozotocin (STZ)-induced diabetic rats. Diabetes was instigated by a single intraperitoneal injection of STZ (45 mg/kg) in male Sprague-Dawley rats. Once the diabetes was successfully induced, the MR extract (200 mg/kg/day) or metformin (200 mg/kg/day) was orally administered for 14 days. Renal function, morphology changes and levels of inflammatory cytokines were measured. Blood glucose concentrations were considerably reduced in STZ-induced diabetic rats following treatment with the MR extract. The administration of the MR extract substantially restored the abnormal quantity of the oxidative DNA damage marker 8-hydroxy-2′-deoxy-guanosine (8-OHdG), malondialdehyde, glutathione, oxidized glutathione, superoxide dismutase, catalase, interleukin (IL)-1β, IL-6, IL-10, and transforming growth factor-β (TGF-β). The urinary excretion of kidney injury molecule-1 (KIM-1), neutrophil gelatinase-associated lipocalin (NGAL), selenium binding protein 1 (SBP1), and pyruvate kinase M2 (PKM2) was significantly reduced in diabetes rats after administration of the MR extracts. In the kidneys of STZ-induced diabetic rats, the MR extracts markedly downregulated the expression of fibronectin, collagen-1, and α-smooth muscle actin (α-SMA). In particular, the MR extracts markedly increased the level of SIRT1 and SIRT3 and reduced claudin-1 in the kidney. These results suggest that the MR extracts exhibits therapeutic activity in contrast to renal injury in STZ-induced diabetic rats through repressing inflammation and oxidative stress.

Keywords: *Molineria recurvata*; diabetic nephropathy; urinary biomarkers; inflammation; oxidative stress

Citation: Dey, P.; Kundu, A.; Lee, H.E.; Kar, B.; Vishal, V.; Dash, S.; Kim, I.S.; Bhakta, T.; Kim, H.S. *Molineria recurvata* Ameliorates Streptozotocin-Induced Diabetic Nephropathy through Antioxidant and Anti-Inflammatory Pathways. *Molecules* **2022**, 27, 4985. https://doi.org/10.3390/molecules27154985

Academic Editors: Sokcheon Pak and Soo Liang Ooi

Received: 30 May 2022
Accepted: 1 August 2022
Published: 5 August 2022

1. Introduction

Diabetic patients have developed different microvascular disorders including nephropathy, retinopathy, and neuropathy [1,2]. Around 40% of newly diagnosed patients with diabetes in both Asian and Western countries have diabetic nephropathy (DN), which is one of the major causes of end-stage renal disease (ESRD) [3]. Hypertension, obesity, and sustained hyperglycemia are the major risk factors for the initiation or progression of DN and induce glomerular malfunction and kidney damage. Since the progression of DN to ESRD is irreversible, it is important to take precise therapeutic approaches to ameliorate kidney injury. Progress has been made in our understanding of the pharmacology of DN, and effective drugs are currently being introduced for the treatment of DN in diabetic patients [4,5].

In patients with diabetes, the accumulation of advanced glycation end products (AGEs), which play a major role in the development of diabetic nephropathy, is one of the main causes of chronic kidney damage [6]. Oxidative stress and chronic inflammation are strongly connected with the development and progression of diabetic nephropathy [7]. The high excretion of urinary microalbumin, thickening of the basement membrane, and mesangial expansion are some of the primary characteristics of diabetic nephropathy [8]. The deposition of extracellular matrix (ECM) proteins also has an important role in the progression of diabetic nephropathy [9,10]. Diabetic nephropathy can also lead to glomerulosclerosis and renal fibrosis, reflecting the fact that that the epithelial-mesenchymal transition (EMT) occurs owing to the loss of E-cadherin and the deposition of α-smooth muscle actin (α-SMA) [11,12]. It has been also established that expression of transforming growth factor-β (TGF-β) is highly upregulated and glomerular function capacity is reduced in patients with diabetic nephropathy [10,13].

The plant *Molineria recurvata* (of the family Hypoxidaceae) is grown widely in Tripura and other hotter regions of India. The leaves, which have fewer or no side effects compared to marketed drugs for diabetes, have been utilized for the management of diabetic mellitus for many years in the Northeastern part of India. The leaf extract of *Molineria recurvata* (MR) possesses anticoagulant and anthelmintic activity [14,15], but the exact molecular mechanism underlying the protective effects against DN is still undetermined. Here, we aimed to evaluate the protective impacts of the MR extract on DN in streptozotocin (STZ)-induced diabetic rats. This study will also highlight the underlying molecular mechanism modulated by the MR extract on STZ-induced diabetic rats.

2. Results

2.1. Phytochemical Properties of the MR Extract

A yield of 12.01% MR methanolic extract was obtained from 100 g of leaves. As indicated in Table 1, initially phytochemical examination discovered the existence of alkaloids, carbohydrates, steroids, flavonoids, and hydroxy-anthraquinone glycosides in the methanolic leaf extract of MR [16].

Table 1. Photochemical properties of the MR extracts from the leaf of *Molineria recurvata*.

Phytochemicals	Methanolic Extract
Alkaloid	+++
Carbohydrate	+
Saponin	−
Steroid	++
Sulphate	−
Tannin	−
Flavonoid	++
Hydroxy-anthraquinone glycoside	+
Starch	−
Dextrin	+

+++: Copiously present; ++: moderately present; +: slightly present; −: absent.

2.2. Acute Toxicity Study of the MR Extract

We measured the toxicity of the MR extract in rats. No indications of any unusual emaciation, respiratory depression, behavior, posture, or mortality at a highest dosage (1000 mg/kg) of the MR extract were observed. No mortality was observed in the groups treated with the MR extract. Therefore, we decided 1000 mg/kg was safe to conduct our further study. Here, based on the solubility, volume, and safety of the MR extract that could be simply dispensed orally, we selected 200 mg/kg as an efficient dose.

2.3. Protective Impact of the MR Extract on Blood Glucose Concentrations in STZ-Administered Rats

The oral glucose tolerance test and insulin tolerance test was initially performed in different groups of rats (Supplementary Figure S1A,B). In STZ-administered rats, the blood glucose concentration was significantly enhanced compared with the control group. After 14 days of oral administration of the MR extract and standard compounds, the fasting blood glucose amount was noticeably reduced in STZ-treated rats treated by the MR extract (200 mg/kg), to a level that was comparable with that achieved in Mef (200 mg/kg)-treated rats. Thus, it was shown that the MR extract could successfully synchronize the blood glucose amounts in STZ-stimulated diabetic rats (Figure 1A).

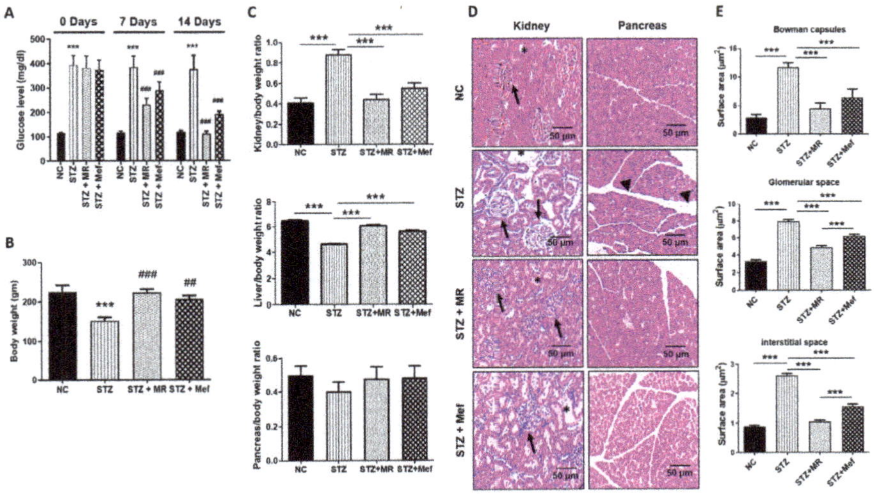

Figure 1. Effects of the MR extract on blood glucose level, body and organ weight, and histopathological modifications in STZ-treated rats. (**A**) Blood glucose level (fasting). (**B**) Changes in body weight. (**C**) Relative change of the kidney, liver, and pancreas weight. The abnormal alterations in these factors were restored by MR treatment. All values are the mean ± S.D. (n = 6). Statistical calculation was completed by one-way ANOVA subsequently by Tukey's honest significant difference (HSD) post hoc test for multiple comparisons. (*** $p < 0.001$). ## $p < 0.01$, ### $p < 0.001$ compared with the STZ group. (**D**) Alteration of the histopathology in the kidney and pancreas (H&E stained). The STZ-treated diabetic rats displayed enlarged cortex with expansion (asterisk), glomerular sclerosis (arrowheads), and dilatation. The medulla showed interstitial nodular sclerosis, tubular dilatation, and fibroplasia (asterisk). The treatment of STZ-treated rats with the MR extract induced minor occurrence of the medulla and tubular injury, normal-sized renal cortex, and normal histological structure of thin tubules (arrows). Original magnification: 200×, scale bar: 50 µm. In contrast, the pancreas is composed of exocrine components that are tightly packed by acinar cells and organized into small lobules divided by intact intralobular and interlobular connective tissue septa. The pathology of both exocrine and endocrine components has changed after STZ treatment. Most of the acinar cells were enlarged, and tiny vacuoles were noted. The interlobular ducts were covered with a flattened epithelium (black arrowheads). STZ treatment destroyed almost all the islet β-cells. However, the MR extract restored the general pathology of the pancreas of diabetic rats. Atrophic changes in the acinar cells were negligible. The border between the exocrine and endocrine portions became more distinctive. All images are symbolic of three rats per investigational group. Original magnification: 200×, scale bar: 100 µm. (**E**) Quantitative investigation of the Bowman capsule size and glomerular and interstitial spaces of H&E-stained kidney sections in the experimental rats. NC: Normal control, STZ: streptozotocin-treated group, STZ + MR: Streptozotocin-treated rats received MR extract, STZ + Mef: Streptozotocin-treated rats received Metformin.

2.4. Protective Effect of the MR Extract on Organ and Body Weight Alterations in STZ-Administered Rats

The total body weight of STZ-administered rats decreased drastically contrasted to that of the normal control group. After 14 days of oral administration with the MR extract, the body weight of diabetic rats treated with the MR extract (200 mg/kg) was normalized to levels comparable to those of rats treated with Mef (200 mg/kg; Figure 1B). Additionally, the average weight of the major organs (the liver and pancreas) was boosted distinctly in STZ-administered rats after administration of the MR extract, but the MR extract caused a marked reduction in kidney weight (Figure 1C).

2.5. Protective Effect of the MR Extract on Histopathological Injury in STZ-Administered Rats

The histopathology of the kidney tissues was examined by Hematoxylin & eosin staining. In STZ-administered rats, both the kidneys showed distinct hydropic variation in the proximal tubules and enlarged glomerular size. Moreover, morphological changes in the interstitial space and Bowman capsules were observed, and glomerular space was enhanced in STZ-administered diabetic rats (Figure 1D,E). Nonetheless, the deformity in the kidney morphology was restored markedly after the MR extract or Mef administration (Figure 1D,E).

Moreover, a histological inspection was also performed to estimate any changes in the pancreatic acini and islets of Langerhans. STZ-administered rats exhibited few islets with numerous vacuous areas; further, they had randomly shaped islets in the pancreas which are comparable with the transparent and round borders across the islets in the pancreas of normal rats. These malformations were reestablished successfully after administration with the MR extract (200 mg/kg) or Mef (200 mg/kg), proving that the MR extract successfully recovered STZ-induced pancreatic damage (Figure 1D). Therefore, the MR extract could repair diabetes-induced damage to the pancreatic structure.

2.6. Protective Effect of the MR Extract on Renal Injury Biomarkers in STZ-Administered Rats

In the case of diabetes-induced kidney injury, the microalbuminuria level is a representative parameter in the urine [17]. We found that STZ-treated rats increased microalbuminuria and serum creatinine concentration in the urine. The concentration of microalbumin excreted in urine and serum creatinine was markedly restored after 14 days of administration of the MR extract (200 mg/kg) or Mef (200 mg/kg) (Figure 2A,B). The urinary volume and urinary creatinine level was also restored after administration of the MR extract to the STZ-treated rats (Supplementary Figure S2A,B). To further confirm the protective effect of MR extract against STZ-induced kidney toxicity, western blotting was performed for urinary excretion of biomarkers (KIM-1, SBP-1, NGAL, and PKM2). The oral administration of the MR extract or Mef reduced the expression of nephrotoxicity biomarkers, which was upregulated markedly in STZ-treated rats (Figure 2C,D). The concentration of the kidney injury marker (3-IS) was also determined [18]. STZ increased 3-IS concentration in the urine, serum, and kidney, and the resulting high 3-IS level was normalized after administration of the MR extract or Mef (Figure 2E).

2.7. MR Extract Reduced AGE Levels in STZ-Treated Rats

In hyperglycemia, the extensive development of AGEs is responsible for the progress of diabetic nephropathy [19]. In patients with chronic renal disease, the AGE concentration in serum is not correlated appropriately with diabetic events, possibly because serum concentrations of AGEs are not associated with its deposition in the target tissues. Hence, we measured the AGE amounts in the kidney of STZ-treated rats. STZ treatment drastically increased AGE levels in the kidney but was normalized after administration of the MR extract (200 mg/kg) or Mef (200 mg/kg) (Figure 3A).

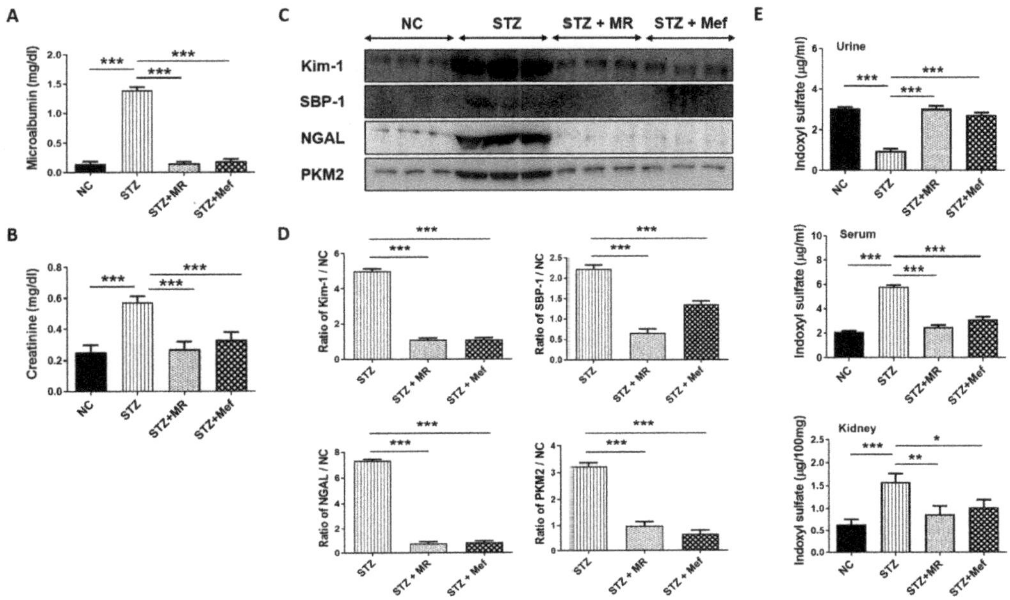

Figure 2. Effects of MR extract administration on biochemical and urinary parameters in STZ-induced diabetic rats. Alteration of (**A**) Urinary excretion of microalbumin and (**B**) serum creatinine level. Values are the mean ± S.D. ($n = 6$). Statistical evaluation was achieved by one-way ANOVA subsequently by Tukey's HSD post hoc test for multiple comparisons (*** $p < 0.001$). (**C**) The expression pattern of kidney damage biomarkers (KIM-1, SBP-1, NGAL, and PKM2) in the urine. (**D**) The band intensity was evaluated densitometrically using ImageJ software (*** $p < 0.001$). (**E**) Alteration of 3-indoxyl sulfate concentration in STZ-treated rats in the urine, serum, and kidney tissues was quantified by high-performance liquid chromatography (HPLC). Values are the mean ± S.D. ($n = 6$). Statistical evaluation was completed by one-way ANOVA subsequently by Tukey's HSD post hoc test for multiple comparisons (*** $p < 0.001$, ** $p < 0.01$, * $p < 0.05$). NC: Normal control, STZ: streptozotocin-treated group, STZ + MR: Streptozotocin-treated rats received MR extract, STZ + Mef: Streptozotocin-treated rats received Metformin.

2.8. Effect of the MR Extract on Oxidative Biomarkers in STZ-Administered Rats

We quantified the concentration of ROS and MDA in the renal tissues of STZ-treated rats, as augmentation of oxidative stress is directly related to elevated AGE concentration in the case of patients with diabetes. After the administration of the MR extract (Figure 3B,C), the levels of ROS and MDA were markedly reduced; these amounts were elevated in STZ-treated rats as compared to the control group. The 8-OHdG (important factor in oxidative DNA damage) concentration was drastically elevated in diabetic-induced rats, which was restored after administration of the MR extract and Mef (Figure 3D). GSH (a non-enzymatic antioxidant) level was reduced markedly in diabetic rats; however, the GSSG level was elevated in diabetic rats compared to the normal control group. GSH and GSSG amounts were restored in STZ-treated rats after administration of the MR extract or Mef (Figure 3E,F). The concentration of SOD (an essential oxidative protein for maintaining mitochondrial functions) was distinctly decreased in STZ-treated rats. However, administration of the MR extract or Mef significantly ameliorated SOD levels (Figure 3G). Furthermore, the concentration of CAT in STZ-administration rats was restored after administration of the MR extract or Mef (Figure 3H).

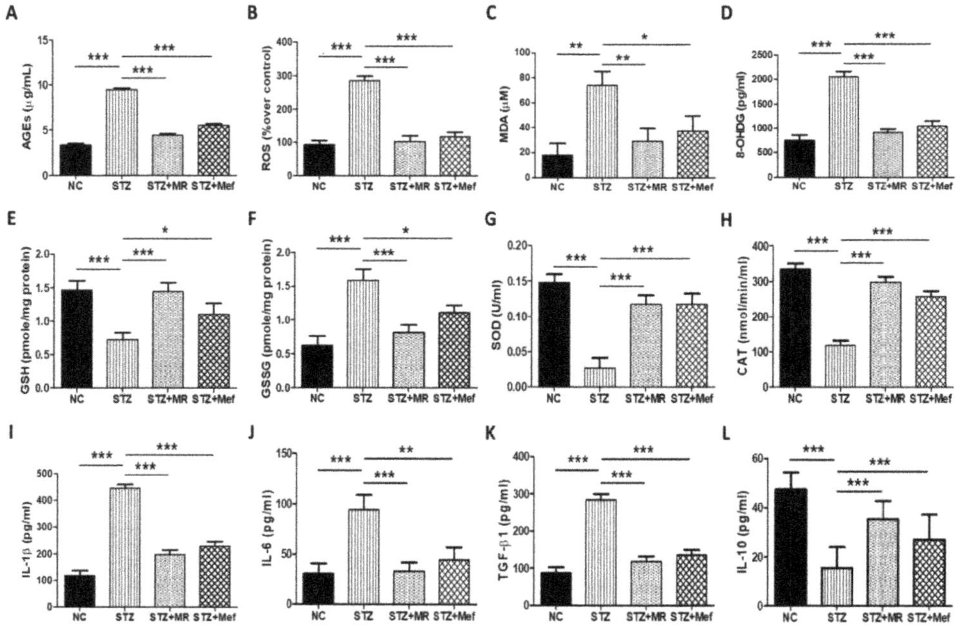

Figure 3. Effects of MR extract on oxidative and inflammatory stress in STZ-treated rats. Levels of (**A**) AGEs, (**B**) intracellular ROS, (**C**) MDA, and (**D**) the oxidative DNA damage marker (8-OHdG) were measured in the kidneys of diabetic animals. (**E–H**) Variations in oxidative biomarkers (GSH, GSSG, SOD, and CAT) were evaluated in the kidneys of diabetic animals. Data are expressed as the mean ± S.D. of duplicate experiments (n = 6). (**I–L**) Changes in the level of the pro-inflammatory cytokines in STZ-treated rats. Data are expressed as the mean ± S.D. of duplicate experiments (n = 6). Statistical analysis was completed by one-way ANOVA subsequently by Tukey's HSD post hoc test for multiple comparisons (*** $p < 0.001$, ** $p < 0.01$, * $p < 0.05$). NC: Normal control, STZ: streptozotocin-treated group, STZ + MR: Streptozotocin-treated rats received MR extract, STZ + Mef: Streptozotocin-treated rats received Metformin.

2.9. Protective Effect of the MR Extract on Inflammatory Cytokines in Diabetic Rats

The impact of the MR extract on the maintenance of inflammatory cytokine levels was examined by ELISA. Significant increases in the serum concentrations of inflammatory cytokines (IL-1β, IL-6, and TGF-β1) were observed in diabetic rats, whereas the amount of the anti-inflammatory cytokine IL-10 was reduced (Figure 3I–L). After administration of the MR extracts or Mef, these inflammatory cytokine levels were normalized in diabetes rats (Figure 3I–L).

2.10. Protective Effect of the MR Extract on Renal Fibrosis in Diabetic Rats

The deposition of ECM proteins, which leads to renal fibrosis, increased markedly in glomerular and tubular cells in diabetic-induced hypertrophy. Therefore, the expression of ECM-related proteins was measured in the kidneys of rats. The expression of TGF-β, α-tubulin, vimentin, fibronectin, collagen-1, and α-SMA was upregulated in STZ-induced rats (Figure 4A and Supplementary Figure S3). The administration of the MR extracts or Mef ameliorated the ECM-proteins deposition in the kidney (Figure 4A and Supplementary Figure S3). Furthermore, we confirmed the expression of collagen-1, fibronectin, and α-SMA in the kidney by IHC staining. As shown in Figure 4B, the expressions of these proteins (stained as brownish granules) were higher in STZ-treated rats and were markedly reduced after the administration of the MR extracts and Mef.

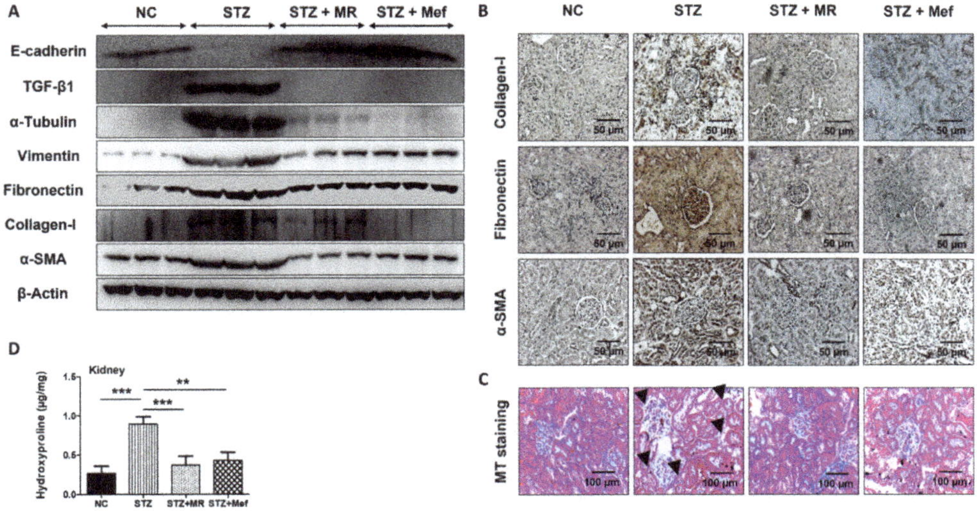

Figure 4. Effects of the MR extract on expression of renal fibrosis biomarkers in STZ-treated rats. (**A**) Expression pattern of E-cadherin, TGF-β1, α-tubulin, vimentin, fibronectin, collagen-1, and α-SMA in the kidney of rats, as determined by western blotting. β-Actin was used as the loading control. The results represent three independent experiments. (**B**) Immunohistochemical analysis of collagen-I, fibronectin, and α-SMA in the kidneys of animals from all groups. Original magnification: 200×, scale bar: 50 μm. (**C**) Images of kidney sections stained with Masson's trichrome, which indicated renal collagen deposition (blue). Original magnification: 200×, scale bar: 100 μm. (**D**) Level of 4-hydroxyproline content in the serum of STZ-treated rats. Data are expressed as the mean ± S.D. of duplicate experiments ($n = 6$). Statistical analysis was performed by one-way ANOVA followed by Tukey's HSD post hoc test for multiple comparisons (*** $p < 0.001$, ** $p < 0.01$). NC: Normal control, STZ: streptozotocin-treated group, STZ + MR: Streptozotocin-treated rats received MR extract, STZ + Mef: Streptozotocin-treated rats received Metformin.

In addition, renal fibrosis was investigated by MT staining to determine the degree of collagen deposition in diabetic kidneys. Diabetic rats resulted in the deposition of a high level of collagen in renal tissues followed by nodular formation (Figure 4C), which was reduced after administration of the MR extract or Mef. The hydroxyproline concentration was also drastically increased in the kidney tissues of STZ-treated diabetic rats. However, administration of the MR extracts or Mef reduced hydroxyproline concentration in the kidney to a similar level as in normal control rats (Figure 4D).

2.11. *Effects of the MR Extract on SIRTs and Claudin-1 Expression in the Renal Cortex of STZ-Treated Rats*

SIRTs and claudin-1 are involved in metabolic disorders resulting from diabetic nephropathy [20,21]. In this study, the SIRT1, SIRT3, and SIRT4 expression was markedly downregulated in the renal cortex of STZ-treated rats compared to the normal control group. However, administration of the MR extracts markedly increased the expression of SIRT1, SIRT3 and SIRT4 in the kidney of STZ-treated rats. Claudin-1 expression was upregulated in STZ-treated rats, but noticeably downregulated after administration of the MR extract or Mef (Figure 5A and Supplementary Figure S4). Furthermore, SIRT1 and claudin-1 expression was also assessed using IHC staining. A robust higher expression of SIRT1 was observed in the kidneys of glomeruli of the normal control group, whereas claudin-1 levels were increased in the renal interstitial tubules of the STZ-treated group. Administration of the MR extracts restored the expression of SIRT1 and claudin-1 in the glomeruli and interstitial tubules (Figure 5B).

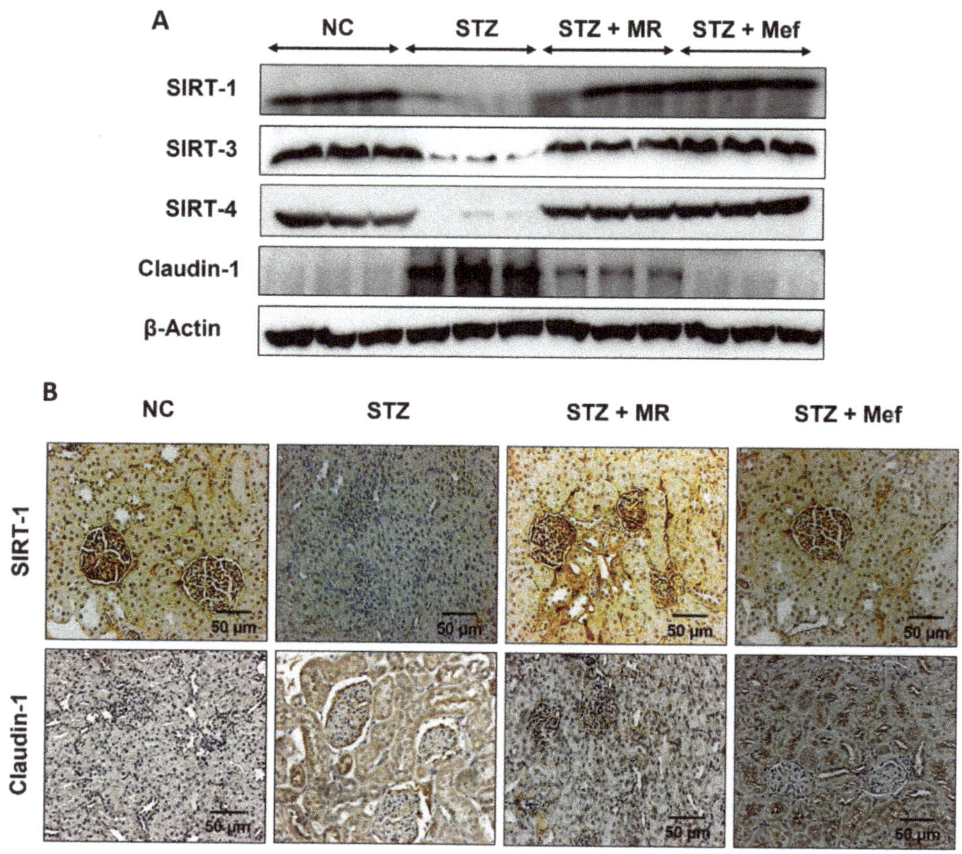

Figure 5. Effects of MR extract on SIRTs and claudin-1 expression in the kidneys of diabetic rats. (**A**) western blotting analysis of SIRT1, SIRT3, SIRT4, and claudin-1 expression. β-Actin expression was used as the loading control. The results represent three independent experiments. (**B**) Immunohistochemical analysis of SIRT-1 and claudin-1 in the kidneys of mice with STZ-induced diabetes. Original magnification: 200×, scale bar: 50 μm. NC: Normal control, STZ: streptozotocin-treated group, STZ + MR: Streptozotocin-treated rats received MR extract, STZ + Mef: Streptozotocin-treated rats received Metformin.

3. Discussion

Since primordial times, natural medicinal plants containing different phytochemicals such as alkaloids, glycosides, steroids, flavonoids, tannins, and polysaccharides, which help to cure various diseases, have been used [22–28]. Previous studies have shown that the methanol extracts from the leaves of MR contain several compounds including steroids, alkaloids, and flavonoids. Owing to their pharmacological activities such as antioxidant [29,30] and anti-inflammatory properties [31,32], MR extract has been used traditionally for the treatment of kidney damage. Here, we examined the defensive role of the MR extract against STZ-induced DN.

STZ produces a necrotic effect on pancreatic beta cells and decreases insulin secretion. Hence, it is used widely to establish the type I diabetic model [33,34]. Blood glucose level was markedly increased after 5 days of STZ treatment. Nonetheless, the blood glucose level was restored after 14 days of treatment with the MR extract (200 mg/kg) or Mef (200 mg/kg) administration, which proved that the MR extract restrains the adverse effects of hyperglycemia, thereby aiding in the management of renal abnormalities. Prolonged hyperglycemia caused by STZ treatment reduced the body weight of diabetic rats, which was restored after MR treatment; this showed the protective effect of MR on the muscle tissues damaged by hyperglycemia. Owing to hypertrophy, the weight of the kidney in STZ-treated rats was increased compared to normal rats, but the weight was normalized after the administration of the MR extract or Mef. The administration of the MR extract protected the renal structural and functional parameters, which reflected the protective role of the MR extract against tubular damage by STZ-induced diabetic nephrotoxicity. The MR extract successfully restored all the structural and functional abnormalities including interstitial fibrosis, glomerular collapse, hemorrhage within tubules and glomeruli, and tubular degeneration in STZ-induced diabetic rats [6,35].

The urinary level of microalbumin is a critical marker of diabetic nephropathy [36,37]. STZ treatment resulted in an abnormal increase in microalbumin and creatinine levels, which was suppressed by treatment with the MR extract or Mef. Tubular and glomerular damage was detected by the analysis of urinary protein-based biomarkers [38–42]. The urinary secretion of KIM-1, SBP-1, NGAL, and PKM2 was markedly increased in the STZ-treated group and was normalized after the administration of the MR extract. This reflects that diabetic renal injury develops by complex mechanisms of hyperglycemia, oxidative stress, and inflammation [43,44].

According to Won et al., 2016, 3-IS is a vital renal injury biomarker present in the serum, urine, and kidney tissues, and it was markedly increased in both the serum and renal tissue in diabetes model [18]. In this study, we also observed a significantly elevated level of 3-IS in diabetic rats, which was recovered after MR treatment. The accretion of 3-IS in the kidney was linked to secretion ability and signified tubular damage. Moreover, ROS production in the tubular region was increased and NF-κB was activated at a high concentration of 3-IS, which was directly correlated with plasminogen activator inhibitor-1 [45]. In the case of patients with chronic kidney disease, 3-IS concentration in serum was increased [46,47]. During kidney injury, a high concentration of 3-IS reduces superoxide scavenging activity [48]. Our results showed that the MR extract may have the ability to prevent ROS generation by hindering the accumulation of 3-IS in the kidney of STZ-induced diabetic rats.

Relatively high concentrations of AGEs play an important function in the progression of diabetic nephropathy [19]. In diabetic nephropathy, a high concentration of AGEs is not only associated with the structural abnormalities of renal cortex and blood vessels. Therefore, targeting AGEs can leads to a therapeutic regime against diabetic nephropathy [35,49]. The kidneys are highly prone to the deposition and development of AGEs [50], and elevated accumulation of AGEs can damage the kidneys and their surrounding blood vessels [49]. The concentration of AGEs is directly associated with the degree of abnormality in the kidney structure [35]. During hyperglycemia, AGEs are highly deposited in the kidney, enhancing ROS generation and ultimately injuring the kidney. These make AGEs an attractive therapeutic target for the prevention of diabetic nephropathy. Here, we showed that the MR extract markedly inhibited AGE formation in the kidneys of STZ-treated diabetic rats. This result aligned well with a previous study, where it was found that the overproduction of AGEs increased ROS generation in diabetic rats [51,52]. Thus, the modification of antioxidant enzymes is a possible therapeutic regimen for diabetic nephropathy.

Insulin resistance increases the oxidative stress that simultaneously increased glyco-oxidation and lipid peroxidation [53]. When the concentration of unstable ROS surpasses the cellular defense mechanism, it interacts with vital cellular macromolecules that leads to tissue damage and functional abnormalities [54]. Apart from the biological functional changes, increased level of oxidative stress also reduced serum- antioxidant enzyme activities, increased cellular leakage, and lead to the loss of functional integrity of renal membrane, which in turn increased the glomerular filtration rate and glomerular sclerosis [55,56]. Antioxidants neutralize the harmful free radicals and neutralize the oxidative cellular damage. Physiological changes in concentration can effect on the resistance of cellular DNA, proteins and lipids to oxidative damage [55,56]. In presence of high glucose concentration, MDA level is increased in vascular smooth muscle cells, proximal tubule cells, and mesangial cells. In diabetic patients, a high concentration of MDA is found in the renal cortex, plasma, and aorta [57]. Hence, antioxidant controlling opens up a new therapeutic strategy for the patients with diabetic nephropathy.

In hyperglycemia-induced diabetic nephropathy, a high level of blood glucose increased ROS generation, producing a high level of oxidative stress, which causes renal failure [58–61]. In this study, the decrease in antioxidant potential due to hyperglycemia in STZ-induced diabetic rats was improved by the administration of the MR extract, with the marked restoration of the SOD and CAT levels. MDA and 8-OHdG levels were increased in the kidneys of STZ-treated rats, and the administration of the MR extract markedly reduced MDA and 8-OHdG. In diabetic nephritic disorders, 8-OHdG concentration is increased in the urine [50,62]. Hyperglycemia induces oxidative stress; hence, the GSH to GSSG ratio was decreased [63]. The concentration of GSH and GSSH was higher and lower, respectively, in the STZ-treated diabetic rats than in the normal control group, and the administration of MR restored this abnormality.

Cytokines play an important role in intercellular kidney injury and produce secondary messengers such as acute phase proteins, cell adhesion molecules, and transcription factors [64]. According to clinical data, the concentration of IL-1β and IL-6 was increased in patients with diabetic nephropathy [65]. In our study, STZ-treatment increased the concentrations of both IL-1β and IL-6. Although the exact molecular mechanism through which pro-inflammatory cytokines affect the regulation of diabetic nephropathy is not determined, it has been theorized that pro-inflammatory cytokines are closely associated with the thickening of the basement membrane and podocyte abnormalities [66]. According to Shahzad et al., 2015, IL-1β levels were elevated in the renal cortex and the serum of db/db mice in an age-dependent manner [67]. Here, the administration of the MR extract markedly reversed the increase in IL-1β levels. Our study supported other studies, which also reported that the IL-6 concentration was unusually increased in STZ-induced diabetic rats and was normalized after administration of the MR extract. Owing to the high concentration of IL-6, thickening of the glomerular membrane, mesangial cell proliferation, and alteration of endothelial permeability were detected in DN [68].

Renal fibrosis depends on the accumulation rate of ECM proteins, mesangial cell proliferation, and thickening of the glomerular basement membrane [69,70]. TGF-β1 has a critical role in the maintenance of diabetic-induced renal fibrosis, which is regulated by the α-SMA expression [71]. In this study, the expression of TGF-β, α-tubulin, vimentin, fibronectin, collagen-1, and α-SMA was upregulated markedly, but E-cadherin expression was downregulated after STZ treatment and was further decreased by MR administration. These data were further supported by the IHC staining for collagen-1, fibronectin, and α-SMA; STZ increased their expression level, which was restored by treatment with the MR extract. The administration of the MR extract decreased collagen deposition, which was confirmed by MT staining. Hydroxyproline is a well-known basic marker of kidney fibrosis [42]. Its levels were considerably restored after MR treatment in STZ-induced diabetic rats. Therefore, the MR extract possesses anti-fibrotic activity against diabetic-induced renal injury.

SIRT1 plays a vital role in the metabolism and inflammation of diabetic-induced renal injury [20,72] and reduced activity of SIRT1 in the renal cortex is common in diabetic disorders [21]. Hence, it is a possible target in the treatment of DN [73,74]. The expression of SIRT1 and claudin-1 were negatively correlated with each other both in the glomerular sections and proximal tubules [75]. An increase in SIRT1 expression level restored diabetic renal injury and impaired ROS-mediated apoptosis in mesangial cells [76,77]. In our study, the administration of the MR extracts upregulated SIRT1 and downregulated claudin-1 protein expression in the STZ-induced group. SIRT3 inhibits the inflammation associated with oxidative stress [78] and can scavenge ROS to protect kidney tissues against cell senescence and apoptosis [79,80]. It is an essential constituent of the acetyl-proteome in the mitochondria [81], and the upregulated expression of SIRT3 improved mitochondrial dysfunction and protected against renal injury. In this study, the MR extract upregulated SIRT3 expression in diabetic rats. SIRT4 also plays a role in the regulation of mitochondrial function and the pathogenesis of metabolic disorders including renal dysfunction, but its exact role in diabetic-induced kidney damage remains unclear. In the mitochondria, SIRT4 assists in the conversion of glutamate to α-ketoglutarate and controls insulin secretion from pancreatic β-cells in response to glucose and amino acids [82]. Our western blotting results showed that STZ-treatment markedly downregulated SIRT4 expression, which was consistent with the results of previous studies in a mouse model of type-2 diabetes [83] and insulin-resistant rats [84]. Our immunohistochemistry results also supported the western blotting data. The overexpression of SIRT4 improved renal injury in hyperglycemic conditions.

The pharmacological effect of our crude leaf extract on blood glucose level and it's possible mechanism depend on the modulation of sirtuin proteins (SIRT1, SIRT3, and SIRT4). SIRT1 possesses anti-inflammatory activity via deacetylation or inactivation of the p65 subunit by reducing ROS production. It also regulates apoptosis, metabolism, and mitochondrial biogenesis [85–87]. The downregulation of SIRT1 in proximal tubule upregulates the claudin-1 expression in podocyte that results in the interruption of glomerular filtration, podocyte malfunction, and albuminuria [75]. Our current study also revealed the lower expression of SIRT1 in DN model. However, after treatment with the MR extract, the expression of SIRT1 was restored. The anti-inflammatory role of SIRT3 is associated with oxidative stress in diabetes induced kidney injury. This protective function of SIRT3 is highly related to the reduction in ROS production and attenuation of inflammation formation [78–80]. SIRT4 overexpression inhibits apoptosis and increases proliferation along with reduced ROS production and increased mitochondrial membrane potential. Under hyperglycemic conditions, SIRT4 overexpression may decrease podocyte injury and inhibit podocyte apoptosis [88]. Therefore, the up regulation of SIRT1, SIRT3 and SIRT4 and downregulation of Claudin 1 by MR extract might prevent renal fibrosis by suppressing mitochondrial oxidative stress and the production of inflammatory cytokines. However, further experiments are necessary to identify the individual isolated compound and its mechanisms in our study.

In our study, the administration of the MR extract upregulated the expression of all SIRT proteins (SIRT-1, SIRT-3, and SIRT-4), which may be associated with the protective effect of the MR extract against DN. Our data suggested that the MR extract has an excellent therapeutic efficacy against DN. However, our extract has the potential to reduce the STZ-induced diabetic nephropathy, but isolation and characterization of the crude extract limits our present study. Further research is necessary to characterize the active constituent behind this protective effect of the MR extract against STZ-induced diabetic nephropathy to unfold the mechanistic approach.

4. Materials and Methods

4.1. Chemicals and Reagents

Streptozotocin (STZ) and metformin (Mef) were acquired from Sigma–Aldrich Biotechnology (St. Louis, MO, USA). The primary antibodies including kidney injury molecule-1 (KIM-1), neutrophil gelatinase-associated lipocalin (NGAL), selenium binding protein-1 (SBP1), pyruvate kinase muscle isozyme M2 (PKM2), collagen-1, E-cadherin, TGF-β1, vimentin, fibronectin, α-tubulin, claudin-1, α-SMA, SIRT1, SIRT3, SIRT4, and β-actin were obtained from Abcam (Cambridge, MA, USA). Immobilon Forte Western HRP substrate (cat. no. WBLUF0100) and polyvinylidene difluoride (PVDF) membrane were procured from Millipore (Burlington, MA, USA). Other chemicals were obtained from Sigma–Aldrich (St. Louis, MO, USA).

4.2. Preparation of the MR Extract

MR leaves were collected in February and March from the rural belt of Tripura (India). This plant was authenticated by the Dr. B. K. Datta (Department of Botany, Tripura University, India). The reference number was BOT/HEB/AC7228. The leaves were cleaned with distilled water, dried in the shade, powdered, and then extracted with methanol (1 L) at 90 °C for 6 h. The solvent was removed under reduced pressure in a rotary evaporator and MR extracts were stored in a sealed container at 4 °C until required [14,15].

4.3. *Design of Animals Experiment and Acute Toxicity Study*

Eight-week-old Sprague-Dawley male rats (body weight 250 ± 5 g) were purchased from Charles River Animal Laboratories (Orient, Seoul, Korea) and accommodated in a specific-pathogen-free (SPF) room with a 12 h light/dark cycle. The relative air temperature and humidity were set at 23 ± 0.5 °C and 55 ± 2%, respectively. Before the start of the experiment, the animals received a 2-week adjustment period. All rats were given ad libitum access to tap water and food (PMI, Brentwood, MO, USA). This study was approved by Sungkyunkwan University Laboratory Animal Care Service (A20180420-2865) following the regulations of the Korea Food and Drug Safety Administration (KFDA). The oral toxicity study of the MR extract was performed following OECD guidelines. MR extract (500 and 1000 mg/kg) was administered orally to Sprague-Dawley rats for 7 days, and the rats were observed for toxicity symptoms including diarrhea, changes in sleep, changes in skin color, convulsions, and respiratory symptoms.

4.4. *Experimental Design*

In all rats, diabetes was induced by a single intraperitoneal (i.p.) injection of STZ (45 mg/kg; dissolved in 0.1 M cold citrate buffer, pH 4.5). After 5 days, the fasting blood glucose level was measured, and the animals with blood glucose levels higher than 300 mg/dL were considered diabetic and selected for subsequent experiments. The experimental design is shown in Figure 6. The rats were divided into four groups (n = 6 animals per group): Group 1, the normal control group (NC), in which the animals received citrate buffer by oral gavage for 14 days; Group 2, the STZ group (STZ), in which the animals received a single STZ (45 mg/kg) injection i.p. [6]; Group 3, the STZ + Mef group (standard; STD), in which the animals received STZ injection and Mef (200 mg/kg) daily by oral gavage for 14 days; and Group 4, the STZ + MR extract group (test), in which the animals received STZ injection and MR extract (200 mg/kg) daily by oral gavage for 14 days. According to Figure 1, the MR extract and Mef were administered on day 6 after STZ injection. The Mef dose was selected as 200 mg/kg, based on the study by Niture et al., 2014 [89], which reported the dose to be effective at controlling the glucose level. Moreover, a dose of $\geq$ 600 mg/kg/day can result in toxic signs and symptoms [90]. Therefore, 200 mg/kg Mef selected for this study was appropriate for regulating the blood glucose level without producing any toxicity. All animals were observed for any clinical abnormality and dose-associated toxicities, and body weight was regularly measured. Blood from the animals was collected from the tail vein, and blood glucose level was measured on days 0, 7, and 14 days using a glucometer (ACCU-CHEK; Daeil Pharm. Co Ltd., Seoul, Korea). Urine samples were collected on days 7 and 14 using a metabolic cage. After sacrifice, major organs (kidney, liver, and pancreas) were perfused with saline to eliminate any trace of blood and then stored at −80 °C for biochemical analysis. Urine and serum samples were frozen at −20 °C for urinary and biochemical analysis.

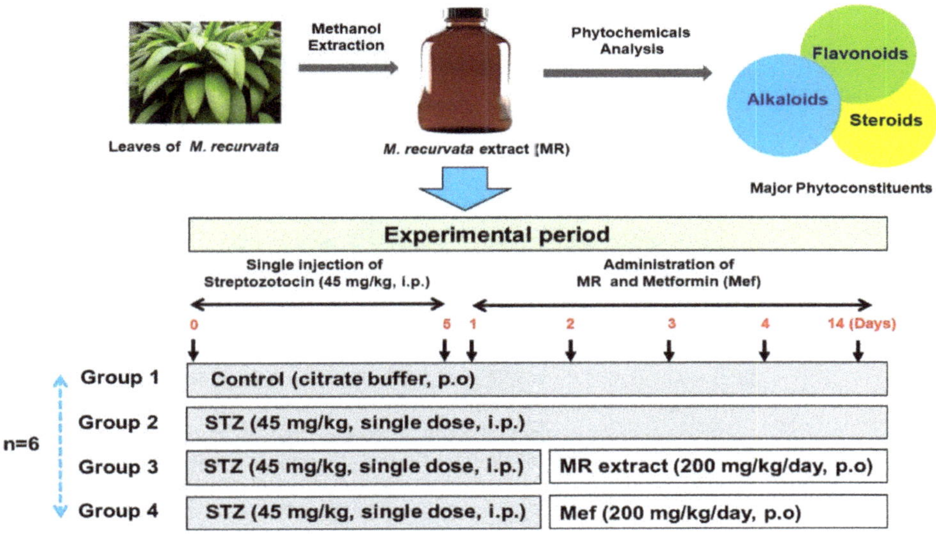

Figure 6. Experimental plan. Diabetes was induced by a single intraperitoneal (i.p.) injection of STZ (45 mg/kg). Citrate buffer was administered to the normal control group (n = 6). After 5 days of STZ treatment, the remaining rats were divided randomly into three groups: the STZ group (n = 6), STZ + MR (200 mg/kg of MR extract) group (n = 6), and the STZ + Mef (200 mg/kg) group (n = 6). The subsequent two groups (group 3 and 4) received MR extract and Mef (standard drug), respectively, for a total of 14 days. STZ: Streptozotocin, Mef: Metformin, and MR: *Molineria recurvate*. p.o.: mouth orally.

4.5. Histological Analysis

The right kidney of each animal was fixed in formaldehyde with 10% phosphate-buffered saline (PBS). The tissue sections (5-µm slices) were embedded in paraffin and were stained with hematoxylin and eosin (H&E) for studying histopathological alterations. All paraffin sections were de-paraffinized with xylene, rehydrated in a graded alcohol series, rinsed with deionized water, and stained with H&E for 1 min. After washing in tap water for 5 min, the stain was developed. The tissue sections were destained with acidified ethanol, washed in tap water, dehydrated, mounted, and stained with eosin for 30 s. Finally, the stained slides were examined for any histopathological changes using a Zeiss Axiphot light microscope (Zeiss, Carl Zeiss, Oberkochen, Germany).

4.6. Urine Analysis and Quantification of Biochemical Parameters

All the animals were shifted to metabolic cages overnight for 24 h before they were sacrificed on day 10. The total urine volume was collected and measured and then centrifuged at 879× g and 4 °C for 10 min. The supernatant was collected for the investigation of urinary biomarkers. Furthermore, the blood from each rat was collected from the hepatic vein and centrifuged at 2000× g and 4 °C for 10 min to separate the serum. The urine and serum of each rat were kept at −80 °C for subsequent experiments. The quantitative analysis of creatinine, microalbumin, and total protein (from the collected urine samples) was performed using the TBA-200FR NEO urine chemistry analyzer (Toshiba, Tochigi-Ken, Japan).

4.7. Detection of 3-Indoxyl Sulfate

The concentration of 3-indoxyl sulfate (3-IS) was quantified in the urine, plasma, and kidney tissue using high-performance liquid chromatography (HPLC), as described in our previous study [88]. The 3-IS was detected and quantified using a C18 column (5 µm; 250 mm × 4.6 mm) from Azilant (Torrance, CA, USA) at a flow rate of 0.9 mL/min at room temperature. First, kidney tissue (100 mg) was homogenized with phosphate buffer, centrifuged, and the supernatant was collected. The plasma, urine, and kidney samples were then extracted with acetonitrile (1:1 ratio; precipitation of protein occurs) and centrifuged (1000× g). The resulting supernatants were analyzed for 3-IS. A blank urine sample containing an internal standard (2-naphthalene sulfonic acid) and 10 different concentrations of 3-IS (1.56–800 μg/mL) was used to construct the calibration plots. After extraction, the collected samples were transferred into a sample tube and thoroughly mixed with acetonitrile containing an internal standard (diclofenac sodium salt, Tokyo Chemical Industry, Tokyo, Japan). The isocratic mobile phase (a mixture of 0.2% TFA in Milli-Q water and acetonitrile in a ratio of 80:20 v/v) was used at a flow rate of 1 mL/min at 25 °C. The wavelength of the IS was 280 nm, and the retention time was 7.5 min.

4.8. Examination of Advanced Glycation End Products

AGE assay was completed as per manufacturer's instructions, as previously described by Kaur et al. [91]. In this ELISA, the AGE samples (test and standard) were blocked in a microplate precoated with the AGE-specific antibody. The wells were washed, incubated, washed again to remove unbound conjugate, and incubated at 37 °C for 1 h. All the wells were again supplemented with an HRP-conjugated secondary antibody and further incubated at room temperature for 1 h. Subsequently, the substrate solution was added and incubated for 2–20 min to develop the color. The absorbance at 450 nm was recorded using a microplate reader.

4.9. Evaluation of Oxidative Stress

The kidney tissues (200 mg) from each rat were homogenized in ice-cold HEPES buffer (20 mM, pH 7.2), and then centrifuged (1500× g). The protein concentration of the supernatant was quantified using a Protein Assay Kit (500–0006, Bio-Rad, Hercules, CA, USA). Intercellular reactive oxygen species (ROS; Cell Biolabs, San Diego, CA, USA), malondialdehyde (MDA; MDA-586, OxisResearCh™, Portland, OR, USA), reduced forms of glutathione (GSH) and oxidized GSH (GSSG; Cayman Chemical Co., Ann Arbor, MI, USA), superoxide dismutase (SOD) activity, and catalase (CAT) peroxidation (Cayman Chemical Co., Ann Arbor, MI, USA) were quantified in accordance with the manufacturer's protocol. In contrast, 8-hydroxy-2′-deoxy-guanosine (8-OHdG) in urine samples was investigated using an ELISA assay kit (Cell Biolabs, San Diego, CA, USA) in accordance with the company's instructions.

4.10. Assessment of Inflammatory Cytokines

The inflammatory biomarkers, namely IL-1β, IL-6, IL-10, and TGF-β, of diabetic nephropathy, were investigated using ELISA Kit (Abcam, Cambridge, MA, USA) according to the manufacturer's protocol.

4.11. Investigation of Hydroxyproline

Hydroxyproline was quantified using a hydroxyproline assay kit (Cell BioLabs, San Diego, CA, USA) in accordance with the manufacturer's protocol, and the levels were reported as micromoles per milligram of protein.

4.12. Western Blot Analyses

Western blotting was performed as previously described [92–94]. For the urine samples, the urine was centrifuged (1000× g for 10 min), and the supernatant (considered to contain the total urinary proteins) was collected and diluted in a 1:2 ratio with double distilled water (DDW). Protein from the kidney tissue was collected using a PRO-PREP cell lysis buffer. The samples were electrophoresed on a 6–12% sodium dodecyl sulfate (SDS) polyacrylamide gel. After electrophoresis, the resolved protein bands were transferred onto PVDF membranes, blocked, incubated with primary and corresponding secondary antibodies, and developed using Immobilon Forte Western HRP substrate. All uncropped western blot images are mentioned in Supplementary Figure S5A–Q.

4.13. Immunohistochemical Evaluation

Immunohistochemical (IHC) analysis of rat kidney tissues was performed as previously described [95]. To examine the expression of α-SMA, fibronectin, collagen-1, SIRT1, and claudin-1 in tissues from all groups, the slides were exposed to the appropriate primary antibodies after de-paraffinization, hydration, and antigen retrieval. The slides were moved to a xylene chamber, dipped in alcohol and water, and then incubated with 3% hydrogen peroxide. Subsequently, non-specific binding to the slides was blocked, and the slides were incubated with primary antibodies and then with horseradish peroxidase (HRP)-conjugated secondary antibodies, in accordance with the manufacturer's instructions. Finally, all immunostained slides were visualized using diaminobenzidine tetrahydrochloride (DAB), counterstained with hematoxylin, and examined using a microscope.

4.14. Statistical Analysis

Data are presented as the mean ± standard deviation (S.D.; $n = 6$). One-way analysis of variance (ANOVA) was followed by Tukey's (honest significant difference) post hoc test for multiple comparisons. Analyses were performed using GraphPad Prism 5 software (Version 5.0, San Diego, CA, USA). Post hoc testing was performed for intergroup comparisons using the least significant difference test. The following levels of statistical significance were considered: *** $p < 0.001$, ** $p < 0.01$, and * $p < 0.05$.

5. Conclusions

In our study, methanolic extract of MR leaves markedly normalized STZ-induced increases in the levels of blood glucose, AGEs, oxidative stress, inflammatory cytokines, and kidney function biomarkers, and repaired the histopathological damage to the kidney and pancreas. The administration of the MR extract enhanced the antioxidant capability by increasing GSH, SOD, and CAT levels in the STZ-treated diabetic group. Hence, the MR extract can be utilized as a novel therapeutic agent for renal disorders including diabetic nephropathy. Future experiments will highlight the targets and molecular mechanism of the MR extract to aid the discovery and development of new drugs for diabetic nephropathy.

Supplementary Materials: The following supporting information can be downloaded at: https://www.mdpi.com/article/10.3390/molecules27154985/s1, Figure S1: Effect of MR extract on the (A) oral glucose tolerance and (B) insulin tolerance in STZ treated diabetic rats. Figure S2: Effect of MR extract on the (A) urinary volume and (B) urinary creatinine level in STZ treated diabetic rats. Figure S3: Effects of MR extract on expression of renal fibrosis biomarkers in STZ-treated rats. The band intensity was analyzed densitometrically using ImageJ software. (*** $p < 0.001$, ** $p < 0.01$, * $p < 0.05$). Figure S4: Effects of MR extract on SIRTs and claudin-1 expression in the kidneys of diabetic rats. The band intensity was analyzed densitometrically using ImageJ software. (*** $p < 0.001$, ** $p < 0.01$, * $p < 0.05$). Figure S5: Uncrop images of the western blots. (A) Kim-1, (B) SBP-1, (C) NGAL, (D) PKM2, (E) E-Cadherin, (F) TGF-b1, (G) 1-Tubulin, (H) Vimentin, (I) Fibronectin, (J) Collagen-I, (K) a-SMA, (L) b-actin, (M) SIRT1, (N) SIRT3, (O) SIRT4, (P) Claudin-1, (Q) b-actin.

Author Contributions: H.S.K., T.B. and P.D. conceived and designed the experiments. P.D. and A.K. performed all experiments. P.D., A.K., H.E.L., B.K., V.V., T.B. and H.S.K. analyzed the data. P.D. and H.S.K. wrote the manuscript. S.D. and I.S.K. critically revised the manuscript. All authors have read and agreed to the published version of the manuscript.

Funding: This work was supported by the National Research Foundation of Korea (NRF) grants funded by the Korean Government (NRF-2019R1A2C2002923 and NRF-2022R1A4A1018930).

Institutional Review Board Statement: This study was approved by Sungkyunkwan University Laboratory Animal Care Service (A20180420-2865) following the regulations of the Korea Food and Drug Safety Administration (KFDA).

Informed Consent Statement: Not applicable.

Data Availability Statement: Not applicable.

Acknowledgments: The authors are very grateful to B. K. Datta (Department of Botany, Tripura University, India) for the identification of the plant specimen. We would like to acknowledge technical support from Keon Wook Kang (College of Pharmacy, Seoul National University, Seoul, Korea). We also would like to thank Whan Sup Cho from Dong-a University (Injae, Korea) for the evaluation of histological examinations.

Conflicts of Interest: The authors declare no conflict of interest. The funders had no role in the design of the study; in the collection, analyses, or interpretation of data; in the writing of the manuscript; or in the decision to publish the results.

Sample Availability: Not applicable.

Abbreviations

AGEs	Advanced glycation end products
CAT	Catalase
DAB	Diaminobenzidine tetrahydrochloride
DN	Diabetic nephropathy
ECM	Extracellular matrix
EMT	Epithelial-mesenchymal transition
ESRD	End-stage renal disease
GSH	Glutathione
GPx	Glutathione peroxidase
H&E	Hematoxylin and eosin
HPLC	High-performance liquid chromatography
HRP	Horseradish peroxidase
3-IS	3-indoxyl sulfate
KIM-1	Kidney injury molecule-1
IL-1β	Interleukin-1β
MDA	Malondialdehyde
MR	Molineria recurvata
ROS	Reactive oxygen species
8-OHdG	8-Hydroxy-deoxyguanosine
PBS	Phospahte-buffered saline
PKM2	Pyruvate kinase M2
PVDF	Polyvinylidene difluoride
SBP-1	Selenium binding protein 1
SDS	Sodium dodecyl sulfate
SOD	Superoxide dismutase
STZ	Streptozotocin
α-SMA	α-Smooth muscle actin
TBA	Thiobarbituric acid
TGF-β	Transforming growth factor-β
TNF-α	Tumor necrosis factor-α

References

1. Kitada, M.; Zhang, Z.; Mima, A.; King, G.L. Molecular mechanisms of diabetic vascular complications. *J. Diabetes Investig.* **2010**, *1*, 77–89. [CrossRef] [PubMed]
2. Giacco, F.; Brownlee, M. Oxidative stress and diabetic complications. *Circ. Res.* **2010**, *107*, 1058–1070. [CrossRef] [PubMed]
3. Keane, W.F.; Brenner, B.M.; De Zeeuw, D.; Grunfeld, J.-P.; McGill, J.; Mitch, W.E.; Ribeiro, A.B.; Shahinfar, S.; Simpson, R.L.; Snapinn, S.M. The risk of developing end-stage renal disease in patients with type 2 diabetes and nephropathy: The RENAAL study. *Kidney Int.* **2003**, *63*, 1499–1507. [CrossRef] [PubMed]
4. Qian, X.; Li, X.; Ma, F.; Luo, S.; Ge, R.; Zhu, Y. Novel hydrogen sulfide-releasing compound, S-propargyl-cysteine, prevents STZ-induced diabetic nephropathy. *Biochem. Biophys. Res. Commun.* **2016**, *473*, 931–938. [CrossRef]
5. Wang, D.; Guan, M.-P.; Zheng, Z.-J.; Li, W.-Q.; Lyv, F.-P.; Pang, R.-Y.; Xue, Y.-M. Transcription factor Egr1 is involved in high glucose-induced proliferation and fibrosis in rat glomerular mesangial cells. *Cell. Physiol. Biochem.* **2015**, *36*, 2093–2107. [CrossRef]
6. Mestry, S.N.; Dhodi, J.B.; Kumbhar, S.B.; Juvekar, A.R. Attenuation of diabetic nephropathy in streptozotocin-induced diabetic rats by *Punica granatum* Linn. leaves extract. *J. Tradit. Complement. Med.* **2017**, *7*, 273–280. [CrossRef]
7. Brownlee, M. The pathobiology of diabetic complications: A unifying mechanism. *Diabetes* **2005**, *54*, 1615–1625. [CrossRef]

8. Steffes, M.W.; Østerby, R.; Chavers, B.; Mauer, S.M. Mesangial expansion as a central mechanism for loss of kidney function in diabetic patients. *Diabetes* **1989**, *38*, 1077–1081. [CrossRef]
9. Kolset, S.; Reinholt, F.; Jenssen, T. Diabetic nephropathy and extracellular matrix. *J. Histochem. Cytochem.* **2012**, *60*, 976–986. [CrossRef]
10. Mason, R.M.; Wahab, N.A. Extracellular matrix metabolism in diabetic nephropathy. *J. Am. Soc. Nephrol.* **2003**, *14*, 1358–1373. [CrossRef]
11. Badid, C.; Desmouliere, A.; Babici, D.; Hadj-Aissa, A.; McGregor, B.; Lefrancois, N.; Touraine, J.L.; Laville, M. Interstitial expression of α-SMA: An early marker of chronic renal allograft dysfunction. *Nephrol. Dial. Transplant.* **2002**, *17*, 1993–1998. [CrossRef]
12. Valcourt, U.; Kowanetz, M.; Niimi, H.; Heldin, C.-H.; Moustakas, A. TGF-β and the Smad signaling pathway support transcriptomic reprogramming during epithelial-mesenchymal cell transition. *Mol. Biol. Cell* **2005**, *16*, 1987–2002. [CrossRef] [PubMed]
13. Brosius, F.C.; Khoury, C.C.; Buller, C.L.; Chen, S. Abnormalities in signaling pathways in diabetic nephropathy. *Expert Rev. Endocrinol. Metab.* **2010**, *5*, 51–64. [CrossRef] [PubMed]
14. Dey, P.; Bhakta, T. Evaluation of in vitro anticoagulant activity of *Molineria recurpata* leaf extract. *J. Nat. Prod. Plant Resour.* **2012**, *2*, 685–688.
15. Dey, P.; Debnath, P.; Bhakta, T. Evaluation of anthelmintic activity of *Molineria Recurvata* leaf extracts. *Int. Res. J. Pharm. Appl. Sci.* **2012**, *2*, 16–20.
16. Dey, P.; Mukherjee, M.; Bhakta, T.; Ghosh, T.K. Preliminary phytochemical studies of leaf extracts of *Molineria recurvata*. *J. Chem. Pharm. Res.* **2012**, *4*, 3727–3730.
17. Hostetter, T.H. Hyperfiltration and glomerulosclerosis. In *Seminars in Nephrology*; Elsevier: Amsterdam, The Netherlands, 2003; pp. 194–199.
18. Won, A.J.; Kim, S.; Kim, Y.G.; Kim, K.-B.; Choi, W.S.; Kacew, S.; Kim, K.S.; Jung, J.H.; Lee, B.M.; Kim, S. Discovery of urinary metabolomic biomarkers for early detection of acute kidney injury. *Mol. BioSyst.* **2016**, *12*, 133–144. [CrossRef]
19. Forbes, J.M.; Thallas, V.; Thomas, M.C.; Founds, H.W.; Burns, W.C.; Jerums, G.; Cooper, M.E. The breakdown of pre-existing advanced glycation end products is associated with reduced renal fibrosis in experimental diabetes. *FASEB J.* **2003**, *17*, 1762–1764. [CrossRef]
20. Price, N.L.; Gomes, A.P.; Ling, A.J.; Duarte, F.V.; Martin-Montalvo, A.; North, B.J.; Agarwal, B.; Ye, L.; Ramadori, G.; Teodoro, J.S. SIRT1 is required for AMPK activation and the beneficial effects of resveratrol on mitochondrial function. *Cell Metab.* **2012**, *15*, 675–690. [CrossRef]
21. Xu, Y.; Nie, L.; Yin, Y.-G.; Tang, J.-L.; Zhou, J.-Y.; Li, D.-D.; Zhou, S.-W. Resveratrol protects against hyperglycemia-induced oxidative damage to mitochondria by activating SIRT1 in rat mesangial cells. *Toxicol. Appl. Pharmacol.* **2012**, *259*, 395–401. [CrossRef]
22. Dey, P.; Bardalai, D.; Kumar, N.R.; Subramani, C.; Mukherjee, M.; Bhakta, T. Ethnomedicinal knowledge about various medicinal plants used by the tribes of Tripura. *Res. J. Pharmacogn. Phytochem.* **2012**, *4*, 297–302.
23. Dey, P.; Kundu, A.; Chakraborty, H.J.; Kar, B.; Choi, W.S.; Lee, B.M.; Bhakta, T.; Atanasov, A.G.; Kim, H.S. Therapeutic value of steroidal alkaloids in cancer: Current trends and future perspectives. *Int. J. Cancer* **2019**, *145*, 1731–1744. [CrossRef] [PubMed]
24. Dey, P.; Karuna, D.; Bhakta, T. Medicinal plants used as anti-acne agents by tribal and non-tribal people of Tripura, India. *Am. J. Phytomed. Clin. Ther.* **2014**, *2*, 556–570.
25. Dey, P.; Kundu, A.; Kumar, A.; Gupta, M.; Lee, B.M.; Bhakta, T.; Dash, S.; Kim, H.S. Analysis of alkaloids (indole alkaloids, isoquinoline alkaloids, tropane alkaloids). In *Recent Advances in Natural Products Analysis*; Elsevier: Amsterdam, The Netherlands, 2020; pp. 505–567.
26. Garg, A.; Sharma, R.; Dey, P.; Kundu, A.; Kim, H.S.; Bhakta, T.; Kumar, A. Analysis of triterpenes and triterpenoids. In *Recent Advances in Natural Products Analysis*; Elsevier: Amsterdam, The Netherlands, 2020; pp. 393–426.
27. Dey, P.; Debnath, P.; Bhakta, T. Evaluation of anthelmintic activity of pineapple fruit extract using Indian earthworm (*Pheritima posthuma*). *Mintage J. Pharm. Med. Sci.* **2013**, *2*, 26–27.
28. Dey, P.; Kundu, A.; Kar, B.; Bhakta, A.; Vishal, V.; Keerthana, S.; Kumar, A.; Bhakta, T.; Dash, S.; Kim, H.S. Bioactive Natural Leads Targeting Cancer Cell Metabolism. In *Evidence Based Validation of Traditional Medicines: A Comprehensive Approach*; Springer: Singapore, 2021; pp. 29–75.
29. Karuna, D.; Dey, P.; Das, S.; Kundu, A.; Bhakta, T. In vitro antioxidant activities of root extract of *Asparagus racemosus* Linn. *J. Tradit. Complement. Med.* **2018**, *8*, 60–65. [CrossRef]
30. Karuna, D.; Dey, P.; Kundu, A.; Vishal, V.; Bhakta, T. Evaluation of in vitro antioxidant potential of *Aconitum napellus* Linn. root extract. *Int. J. Pharmacogn. Chin. Med.* **2018**, *2*, 125–129.
31. Debnath, P.; Dey, P.; Chanda, A.; Bhakta, T. A Survey on Pineapple and its medicinal value. *Sch. Acad. J. Pharm.* **2012**, *1*, 24–29.
32. Islam, M.T.; Ali, E.S.; Uddin, S.J.; Shaw, S.; Islam, M.A.; Ahmed, M.I.; Shill, M.C.; Karmakar, U.K.; Yarla, N.S.; Khan, I.N. Phytol: A review of biomedical activities. *Food Chem. Toxicol.* **2018**, *121*, 82–94. [CrossRef]
33. Lenzen, S. The mechanisms of alloxan-and streptozotocin-induced diabetes. *Diabetologia* **2008**, *51*, 216–226. [CrossRef]
34. Zafar, M.; Naqvi, S.N.-u.-H. Effects of STZ-Induced Diabetes on the Relative Weights of Kidney, Liver and Pancreas in Albino Rats: A Comparative Study. *Int. J. Morphol.* **2010**, *28*, 135–142. [CrossRef]

35. Kishore, L.; Kaur, N.; Singh, R. Nephroprotective effect of *Paeonia emodi* via inhibition of advanced glycation end products and oxidative stress in streptozotocin–nicotinamide induced diabetic nephropathy. *J. Food Drug Anal.* **2017**, *25*, 576–588. [CrossRef] [PubMed]
36. Saravanan, S.; Pari, L. Protective effect of thymol on high fat diet induced diabetic nephropathy in C57BL/6J mice. *Chem.-Biol. Interact.* **2016**, *245*, 1–11. [CrossRef] [PubMed]
37. Xue, W.; Lei, J.; Li, X.; Zhang, R. *Trigonella foenum* graecum seed extract protects kidney function and morphology in diabetic rats via its antioxidant activity. *Nutr. Res.* **2011**, *31*, 555–562. [CrossRef] [PubMed]
38. Cheon, J.H.; Kim, S.Y.; Son, J.Y.; Kang, Y.R.; An, J.H.; Kwon, J.H.; Song, H.S.; Moon, A.; Lee, B.M.; Kim, H.S. Pyruvate kinase M2: A novel biomarker for the early detection of acute kidney injury. *Toxicol. Res.* **2016**, *32*, 47–56. [CrossRef]
39. Fiseha, T. Urinary biomarkers for early diabetic nephropathy in type 2 diabetic patients. *Biomark. Res.* **2015**, *3*, 1–7. [CrossRef]
40. Kim, K.S.; Yang, H.Y.; Song, H.; Kang, Y.R.; Kwon, J.; An, J.; Son, J.Y.; Kwack, S.J.; Kim, Y.-M.; Bae, O.-N. Identification of a sensitive urinary biomarker, selenium-binding protein 1, for early detection of acute kidney injury. *J. Toxicol. Environ. Health Part A* **2017**, *80*, 453–464. [CrossRef] [PubMed]
41. Lee, E.K.; Shin, Y.-J.; Park, E.Y.; Kim, N.D.; Moon, A.; Kwack, S.J.; Son, J.Y.; Kacew, S.; Lee, B.M.; Bae, O.-N. Selenium-binding protein 1: A sensitive urinary biomarker to detect heavy metal-induced nephrotoxicity. *Arch. Toxicol.* **2017**, *91*, 1635–1648. [CrossRef]
42. Parikh, C.R.; Edelstein, C.L.; Devarajan, P.; Cantley, L. Biomarkers of acute kidney injury: Early diagnosis, pathogenesis, and recovery. *J. Investig. Med.* **2007**, *55*, 333–340. [CrossRef]
43. Forbes, J.M.; Coughlan, M.T.; Cooper, M.E. Oxidative stress as a major culprit in kidney disease in diabetes. *Diabetes* **2008**, *57*, 1446–1454. [CrossRef]
44. Ha, H.; Hwang, I.-A.; Park, J.H.; Lee, H.B. Role of reactive oxygen species in the pathogenesis of diabetic nephropathy. *Diabetes Res. Clin. Pract.* **2008**, *82*, S42–S45. [CrossRef]
45. Bolati, D.; Shimizu, H.; Yisireyili, M.; Nishijima, F.; Niwa, T. Indoxyl sulfate, a uremic toxin, downregulates renal expression of Nrf2 through activation of NF-κB. *BMC Nephrol.* **2013**, *14*, 56. [CrossRef] [PubMed]
46. Wu, V.-C.; Young, G.-H.; Huang, P.-H.; Lo, S.-C.; Wang, K.-C.; Sun, C.-Y.; Liang, C.-J.; Huang, T.-M.; Chen, J.-H.; Chang, F.-C. In acute kidney injury, indoxyl sulfate impairs human endothelial progenitor cells: Modulation by statin. *Angiogenesis* **2013**, *16*, 609–624. [CrossRef] [PubMed]
47. Yabuuchi, N.; Sagata, M.; Saigo, C.; Yoneda, G.; Yamamoto, Y.; Nomura, Y.; Nishi, K.; Fujino, R.; Jono, H.; Saito, H. Indoxyl sulfate as a mediator involved in dysregulation of pulmonary aquaporin-5 in acute lung injury caused by acute kidney injury. *Int. J. Mol. Sci.* **2017**, *18*, 11. [CrossRef] [PubMed]
48. Owada, S.; Goto, S.; Bannai, K.; Hayashi, H.; Nishijima, F.; Niwa, T. Indoxyl sulfate reduces superoxide scavenging activity in the kidneys of normal and uremic rats. *Am. J. Nephrol.* **2008**, *28*, 446–454. [CrossRef]
49. Sohn, E.; Kim, J.; Kim, C.-S.; Lee, Y.M.; Jo, K.; Shin, S.D.; Kim, J.H.; Kim, J.S. The extract of Litsea japonica reduced the development of diabetic nephropathy via the inhibition of advanced glycation end products accumulation in db/db mice. *Evid.-Based Complement. Altern. Med.* **2013**, *2013*, 769416. [CrossRef]
50. Niedowicz, D.M.; Daleke, D.L. The role of oxidative stress in diabetic complications. *Cell Biochem. Biophys.* **2005**, *43*, 289–330. [CrossRef]
51. Fakhruddin, S.; Alanazi, W.; Jackson, K.E. Diabetes-induced reactive oxygen species: Mechanism of their generation and role in renal injury. *J. Diabetes Res.* **2017**, *2017*, 8379327. [CrossRef]
52. Volpe, C.M.O.; Villar-Delfino, P.H.; dos Anjos, P.M.F.; Nogueira-Machado, J.A. Cellular death, reactive oxygen species (ROS) and diabetic complications. *Cell Death Dis.* **2018**, *9*, 119. [CrossRef]
53. Horie, K.; Miyata, T.; Maeda, K.; Miyata, S.; Sugiyama, S.; Sakai, H.; de Strihou, C.v.Y.; Monnier, V.M.; Witztum, J.L.; Kurokawa, K. Immunohistochemical colocalization of glycoxidation products and lipid peroxidation products in diabetic renal glomerular lesions. Implication for glycoxidative stress in the pathogenesis of diabetic nephropathy. *J. Clin. Investig.* **1997**, *100*, 2995–3004. [CrossRef]
54. Soetikno, V.; Suzuki, K.; Veeraveedu, P.T.; Arumugam, S.; Lakshmanan, A.P.; Sone, H.; Watanabe, K. Molecular understanding of curcumin in diabetic nephropathy. *Drug Discov. Today* **2013**, *18*, 756–763. [CrossRef]
55. Cheng, D.; Liang, B.; Li, Y. Antihyperglycemic effect of Ginkgo biloba extract in streptozotocin-induced diabetes in rats. *BioMed Res. Int.* **2013**, *2013*, 162724. [CrossRef] [PubMed]
56. Gujjala, S.; Putakala, M.; Nukala, S.; Bangeppagari, M.; Ramaswamy, R.; Desireddy, S. Renoprotective effect of *Caralluma fimbriata* against high-fat diet-induced oxidative stress in Wistar rats. *J. Food Drug Anal.* **2016**, *24*, 586–593. [CrossRef] [PubMed]
57. Nakai, K.; Fujii, H.; Kono, K.; Goto, S.; Kitazawa, R.; Kitazawa, S.; Hirata, M.; Shinohara, M.; Fukagawa, M.; Nishi, S. Vitamin D activates the Nrf2-Keap1 antioxidant pathway and ameliorates nephropathy in diabetic rats. *Am. J. Hypertens.* **2014**, *27*, 586–595. [CrossRef] [PubMed]
58. Đorđević, M.; Mihailović, M.; Jovanović, J.A.; Grdović, N.; Usković, A.; Tolić, A.; Sinadinović, M.; Rajić, J.; Mišić, D.; Šiler, B. *Centaurium erythraea* methanol extract protects red blood cells from oxidative damage in streptozotocin-induced diabetic rats. *J. Ethnopharmacol.* **2017**, *202*, 172–183. [CrossRef]

59. Giribabu, N.; Karim, K.; Kilari, E.K.; Salleh, N. *Phyllanthus niruri* leaves aqueous extract improves kidney functions, ameliorates kidney oxidative stress, inflammation, fibrosis and apoptosis and enhances kidney cell proliferation in adult male rats with diabetes mellitus. *J. Ethnopharmacol.* **2017**, *205*, 123–137. [CrossRef]
60. Achi, N.; Ohaeri, O.; Ijeh, I.; Eleazu, C. Modulation of the lipid profile and insulin levels of streptozotocin induced diabetic rats by ethanol extract of *Cnidoscolus aconitifolius* leaves and some fractions: Effect on the oral glucose tolerance of normoglycemic rats. *Biomed. Pharmacother.* **2017**, *86*, 562–569. [CrossRef]
61. Zhang, H.; Zhao, T.; Gong, Y.; Dong, X.; Zhang, W.; Sun, S.; Wang, H.; Gu, Y.; Lu, X.; Yan, M. Attenuation of diabetic nephropathy by Chaihuang-Yishen granule through anti-inflammatory mechanism in streptozotocin-induced rat model of diabetics. *J. Ethnopharmacol.* **2014**, *151*, 556–564. [CrossRef]
62. Hinokio, Y.; Suzuki, S.; Hirai, M.; Suzuki, C.; Suzuki, M.; Toyota, T. Urinary excretion of 8-oxo-7, 8-dihydro-2′-deoxyguanosine as a predictor of the development of diabetic nephropathy. *Diabetologia* **2002**, *45*, 877–882. [CrossRef]
63. Ying, C.; Mao, Y.; Chen, L.; Wang, S.; Ling, H.; Li, W.; Zhou, X. Bamboo leaf extract ameliorates diabetic nephropathy through activating the AKT signaling pathway in rats. *Int. J. Biol. Macromol.* **2017**, *105*, 1587–1594. [CrossRef]
64. Navarro-González, J.F.; Jarque, A.; Muros, M.; Mora, C.; García, J. Tumor necrosis factor-α as a therapeutic target for diabetic nephropathy. *Cytokine Growth Factor Rev.* **2009**, *20*, 165–173. [CrossRef]
65. Shikano, M.; Sobajima, H.; Yoshikawa, H.; Toba, T.; Kushimoto, H.; Katsumata, H.; Tomita, M.; Kawashima, S. Usefulness of a highly sensitive urinary and serum IL-6 assay in patients with diabetic nephropathy. *Nephron* **2000**, *85*, 81–85. [CrossRef] [PubMed]
66. Nakamura, A.; Shikata, K.; Hiramatsu, M.; Nakatou, T.; Kitamura, T.; Wada, J.; Itoshima, T.; Makino, H. Serum interleukin-18 levels are associated with nephropathy and atherosclerosis in Japanese patients with type 2 diabetes. *Diabetes Care* **2005**, *28*, 2890–2895. [CrossRef] [PubMed]
67. Shahzad, K.; Bock, F.; Dong, W.; Wang, H.; Kopf, S.; Kohli, S.; Ranjan, S.; Wolter, J.; Wacker, C.; Biemann, R. Nlrp3-inflammasome activation in non-myeloid-derived cells aggravates diabetic nephropathy. *Kidney Int.* **2015**, *87*, 74–84. [CrossRef] [PubMed]
68. Lim, A.K.; Tesch, G.H. Inflammation in diabetic nephropathy. *Mediat. Inflamm.* **2012**, *2012*, 146154. [CrossRef]
69. Lu, Q.; Zuo, W.-Z.; Ji, X.-J.; Zhou, Y.-X.; Liu, Y.-Q.; Yao, X.-Q.; Zhou, X.-Y.; Liu, Y.-W.; Zhang, F.; Yin, X.-X. Ethanolic Ginkgo biloba leaf extract prevents renal fibrosis through Akt/mTOR signaling in diabetic nephropathy. *Phytomedicine* **2015**, *22*, 1071–1078. [CrossRef] [PubMed]
70. Ponnusamy, M.; Zhou, X.; Yan, Y.; Tang, J.; Tolbert, E.; Zhao, T.C.; Gong, R.; Zhuang, S. Blocking sirtuin 1 and 2 inhibits renal interstitial fibroblast activation and attenuates renal interstitial fibrosis in obstructive nephropathy. *J. Pharmacol. Exp. Ther.* **2014**, *350*, 243–256. [CrossRef]
71. Hinz, B.; Celetta, G.; Tomasek, J.J.; Gabbiani, G.; Chaponnier, C. Alpha-smooth muscle actin expression upregulates fibroblast contractile activity. *Mol. Biol. Cell* **2001**, *12*, 2730–2741. [CrossRef]
72. Zhong, Y.; Lee, K.; He, J.C. SIRT1 is a potential drug target for treatment of diabetic kidney disease. *Front. Endocrinol.* **2018**, *9*, 624. [CrossRef]
73. Kitada, M.; Takeda, A.; Nagai, T.; Ito, H.; Kanasaki, K.; Koya, D. Dietary restriction ameliorates diabetic nephropathy through anti-inflammatory effects and regulation of the autophagy via restoration of Sirt1 in diabetic Wistar fatty (fa/fa) rats: A model of type 2 diabetes. *Exp. Diabetes Res.* **2011**, *2011*, 908185. [CrossRef]
74. Li, C.; Cai, F.; Yang, Y.; Zhao, X.; Wang, C.; Li, J.; Jia, Y.; Tang, J. Tetrahydroxystilbene glucoside ameliorates diabetic nephropathy in rats: Involvement of SIRT1 and TGF-β1 pathway. *Eur. J. Pharmacol.* **2010**, *649*, 382–389. [CrossRef]
75. Hasegawa, K.; Wakino, S.; Simic, P.; Sakamaki, Y.; Minakuchi, H.; Fujimura, K.; Hosoya, K.; Komatsu, M.; Kaneko, Y.; Kanda, T. Renal tubular Sirt1 attenuates diabetic albuminuria by epigenetically suppressing Claudin-1 overexpression in podocytes. *Nat. Med.* **2013**, *19*, 1496–1504. [CrossRef]
76. Kume, S.; Haneda, M.; Kanasaki, K.; Sugimoto, T.; Araki, S.-i.; Isono, M.; Isshiki, K.; Uzu, T.; Kashiwagi, A.; Koya, D. Silent information regulator 2 (SIRT1) attenuates oxidative stress-induced mesangial cell apoptosis via p53 deacetylation. *Free Radic. Biol. Med.* **2006**, *40*, 2175–2182. [CrossRef] [PubMed]
77. Sharma, S.; Anjaneyulu, M.; Kulkarni, S.; Chopra, K. Resveratrol, a polyphenolic phytoalexin, attenuates diabetic nephropathy in rats. *Pharmacology* **2006**, *76*, 69–75. [CrossRef] [PubMed]
78. Zhao, W.-Y.; Zhang, L.; Sui, M.-X.; Zhu, Y.-H.; Zeng, L. Protective effects of sirtuin 3 in a murine model of sepsis-induced acute kidney injury. *Sci. Rep.* **2016**, *6*, 33201. [CrossRef] [PubMed]
79. Salminen, A.; Kaarniranta, K.; Kauppinen, A. Crosstalk between oxidative stress and SIRT1: Impact on the aging process. *Int. J. Mol. Sci.* **2013**, *14*, 3834–3859. [CrossRef] [PubMed]
80. Lombard, D.B.; Alt, F.W.; Cheng, H.-L.; Bunkenborg, J.; Streeper, R.S.; Mostoslavsky, R.; Kim, J.; Yancopoulos, G.; Valenzuela, D.; Murphy, A. Mammalian Sir2 homolog SIRT3 regulates global mitochondrial lysine acetylation. *Mol. Cell. Biol.* **2007**, *27*, 8807–8814. [CrossRef]
81. Hebert, A.S.; Dittenhafer-Reed, K.E.; Yu, W.; Bailey, D.J.; Selen, E.S.; Boersma, M.D.; Carson, J.J.; Tonelli, M.; Balloon, A.J.; Higbee, A.J. Calorie restriction and SIRT3 trigger global reprogramming of the mitochondrial protein acetylome. *Mol. Cell* **2013**, *49*, 186–199. [CrossRef]

82. Haigis, M.C.; Mostoslavsky, R.; Haigis, K.M.; Fahie, K.; Christodoulou, D.C.; Murphy, A.J.; Valenzuela, D.M.; Yancopoulos, G.D.; Karow, M.; Blander, G. SIRT4 inhibits glutamate dehydrogenase and opposes the effects of calorie restriction in pancreatic β cells. *Cell* **2006**, *126*, 941–954. [CrossRef]
83. Mahlknecht, U.; Voelter-Mahlknecht, S. Fluorescence in situ hybridization and chromosomal organization of the sirtuin 4 gene (Sirt4) in the mouse. *Biochem. Biophys. Res. Commun.* **2009**, *382*, 685–690. [CrossRef]
84. Chen, Y.-R.; Fang, S.-R.; Fu, Y.-C.; Zhou, X.-H.; Xu, M.-Y.; Xu, W.-C. Calorie restriction on insulin resistance and expression of SIRT1 and SIRT4 in rats. *Biochem. Cell Biol.* **2010**, *88*, 715–722. [CrossRef]
85. Pacholec, M.; Bleasdale, J.E.; Chrunyk, B.; Cunningham, D.; Flynn, D.; Garofalo, R.S.; Griffith, D.; Griffor, M.; Loulakis, P.; Pabst, B. SRT1720, SRT2183, SRT1460, and Resveratrol Are Not Direct Activators of SIRT1. *J. Biol. Chem.* **2010**, *285*, 8340–8351. [CrossRef] [PubMed]
86. Yeung, F.; Hoberg, J.E.; Ramsey, C.S.; Keller, M.D.; Jones, D.R.; Frye, R.A.; Mayo, M.W. Modulation of NF-kappaB-dependent transcription and cell survival by the SIRT1 deacetylase. *EMBO J.* **2004**, *23*, 2369–2380. [CrossRef]
87. Lee, I.H.; Cao, L.; Mostoslavsky, R.; Lombard, D.B.; Liu, J.; Bruns, N.E.; Tsokos, M.; Alt, F.W.; Finkel, T. A role for the NAD-dependent deacetylase Sirt1 in the regulation of autophagy. *Proc. Natl. Acad. Sci. USA* **2008**, *105*, 3374–3379. [CrossRef]
88. Kundu, A.; Richa, S.; Dey, P.; Kim, K.S.; Son, J.Y.; Kim, H.R.; Lee, S.-Y.; Lee, B.-H.; Lee, K.Y.; Kacew, S. Protective effect of EX-527 against high-fat diet-induced diabetic nephropathy in Zucker rats. *Toxicol. Appl. Pharmacol.* **2020**, *390*, 114899. [CrossRef] [PubMed]
89. Niture, N.T.; Patil, D.G.; Somani, R.S.; Sahane, R.S. Effect of rutin on early diabetic neuropathy in experimental animals. *J. Nat. Prod. Plant Resour.* **2014**, *4*, 1–9.
90. Quaile, M.P.; Melich, D.H.; Jordan, H.L.; Nold, J.B.; Chism, J.P.; Polli, J.W.; Smith, G.A.; Rhodes, M.C. Toxicity and toxicokinetics of metformin in rats. *Toxicol. Appl. Pharmacol.* **2010**, *243*, 340–347. [CrossRef] [PubMed]
91. Kaur, N.; Kishore, L.; Singh, R. *Dillenia indica* L. attenuates diabetic nephropathy via inhibition of advanced glycation end products accumulation in STZ-nicotinamide induced diabetic rats. *J. Tradit. Complement. Med.* **2018**, *8*, 226–238. [CrossRef]
92. Dey, P.; Son, J.Y.; Kundu, A.; Kim, K.S.; Lee, Y.; Yoon, K.; Yoon, S.; Lee, B.M.; Nam, K.T.; Kim, H.S. Knockdown of Pyruvate Kinase M2 Inhibits Cell Proliferation, Metabolism, and Migration in Renal Cell Carcinoma. *Int. J. Mol. Sci.* **2019**, *20*, 5622. [CrossRef]
93. Dey, P.; Kundu, A.; Sachan, R.; Park, J.; Ahn, M.Y.; Yoon, K.; Lee, J.; Kim, N.D.; Kim, I.S.; Lee, B.M. PKM2 Knockdown Induces Autophagic Cell Death via the AKT/mTOR Pathway in Human Prostate Cancer Cells. *Cell. Physiol. Biochem.* **2019**, *52*, 1535–1552.
94. Kundu, A.; Dey, P.; Sarkar, P.; Karmakar, S.; Tae, I.H.; Kim, K.S.; Park, J.H.; Lee, S.H.; Lee, B.M.; Renthlei, L. Protective effects of *Croton hookeri* on streptozotocin-induced diabetic nephropathy. *Food Chem. Toxicol.* **2020**, *135*, 110873. [CrossRef]
95. Sachan, R.; Kundu, A.; Dey, P.; Son, J.Y.; Kim, K.S.; Lee, D.E.; Kim, H.R.; Park, J.H.; Lee, S.H.; Kim, J.-H. *Dendropanax morbifera* Protects against Renal Fibrosis in Streptozotocin-Induced Diabetic Rats. *Antioxidants* **2020**, *9*, 84. [CrossRef] [PubMed]

molecules

Article

Phytochemical Discrimination, Biological Activity and Molecular Docking of Water-Soluble Inhibitors from *Saussurea costus* Herb against Main Protease of SARS-CoV-2

Hajo Idriss [1,2], Babeker Siddig [3,4], Pamela González Maldonado [5], H. M. Elkhair [1], A. I. Alakhras [1,6], Emad M. Abdallah [7], Pablo Hernán Sotelo Torres [5] and Amin O. Elzupir [1,2,*]

1 Deanship of Scientific Research, Imam Mohammad Ibn Saud Islamic University (IMSIU), P.O. Box 90905, Riyadh 11623, Saudi Arabia; hiidriss@imamu.edu.sa (H.I.); hmdirar@imamu.edu.sa (H.M.E.); aakhrasi@imamu.edu.sa (A.I.A.)
2 Department of Physics, College of Science, Imam Mohammad Ibn Saud Islamic University (IMSIU), P.O. Box 90905, Riyadh 11623, Saudi Arabia
3 Alawia Imam Institute for Pharmaceutical Research and Development, University of Medical Science and Technology, Khartoum 11115, Sudan; gcfid@savola.com
4 Savola Edible Oils, Khartoum 11115, Sudan
5 Biotechnology Department, Facultad de Ciencias Químicas, Universidad Nacional de Asunción, San Lorenzo 111421, Paraguay; pame.54.pg12@gmail.com (P.G.M.); phsotelo@qui.una.py (P.H.S.T.)
6 Department of Chemistry, College of Science, Imam Mohammad Ibn Saud Islamic University (IMSIU), Riyadh 11623, Saudi Arabia
7 Department of Science Laboratories, College of Science and Arts, Qassim University, Ar Rass 51921, Saudi Arabia; emad100sdl@yahoo.com
* Correspondence: aoalamalhuda@imamu.edu.sa

Citation: Idriss, H.; Siddig, B.; Maldonado, P.G.; Elkhair, H.M.; Alakhras, A.I.; Abdallah, E.M.; Torres, P.H.S.; Elzupir, A.O. Phytochemical Discrimination, Biological Activity and Molecular Docking of Water-Soluble Inhibitors from *Saussurea costus* Herb against Main Protease of SARS-CoV-2. *Molecules* **2022**, *27*, 4908. https://doi.org/10.3390/molecules27154908

Academic Editors: Sokcheon Pak and Soo Liang Ooi

Received: 20 March 2022
Accepted: 12 July 2022
Published: 1 August 2022

Abstract: Siddha medicine is one of the oldest medical systems in the world and is believed to have originated more than 10,000 years ago and is prevalent across ancient Tamil land. It is undeniable that inhibitor preferences rise with increasing solubility in water due to the considerations pertaining to the bioavailability and the ease of which unabsorbed residues can be disposed of. In this study, we showed the phytochemical discrimination of *Saussurea costus* extracted with water at room temperature as a green extraction procedure. A total of 48 compounds were identified using gas chromatography-mass spectrometry (GC-MS). The fatty acids had a high phytochemical abundance at 73.8%, followed by tannins at 8.2%, carbohydrates at 6.9%, terpenoids at 4.3%, carboxylic acids at 2.5%, hydrocarbons at 2.4%, phenolic compounds at 0.2%, and sterols at 1.5%. Of these compounds, 22 were docked on the active side and on the catalytic dyad of His41 and Cys145 of the main protease of SARS-CoV-2 (M^{pro}). Eight active inhibitors were carbohydrates, five were fatty acids, three were terpenoids, two were carboxylic acids, one was a tannin, one was a phenolic compound, and one was a sterol. The best inhibitors were 4,8,13-Cyclotetradecatriene-1,3-diol, 1,5,9-trimethyl-12-(1-methylethyl), Andrographolide, and delta.4-Androstene-3.beta.,17.beta.-diol, with a binding affinity that ranged from −6.1 kcal/mol to −6.5 kcal/mol. The inhibitory effect of *Saussurea costus* of SARS-CoV-2 entry into the cell was studied using a pseudovirus with Spike proteins from the D614G variant and the VOC variants Gamma and Delta. Based on the viral cycle of SARS-CoV-2, our results suggest that the *Saussurea costus* aqueous extract has no virucidal effect and inhibits the virus in the events after cell entry. Furthermore, the biological activity of the aqueous extract was investigated against HSV-1 virus and two bacterial strains, namely *Staphylococcus aureus* ATCC BAA 1026 and *Escherichia coli* ATCC 9637. According to this study, an enormous number of water-soluble inhibitors were identified from *Saussurea costus* against the M^{pro}, and this is unprecedented as far as we know.

Keywords: coronavirus SARS-CoV-2; COVID-19; main protease; *Saussurea costus;* molecular docking; GC-MS profiling

1. Introduction

Coronaviruses are a large family of single-stranded, enveloped, zoonotic RNA viruses, and some cause fewer mild disease, such as fevers and the common cold, than others [1–4]. However, other viruses cause more severe illnesses such as Middle East Respiratory Syndrome (MERS) and severe acute respiratory syndrome coronavirus 2 (SARS-CoV-2), and some of these viruses are easily transmitted from person to person, unlike other viruses [5,6]. It has been found that civet cats can transmit SARS-CoV to humans; MERS-CoV is transmitted between dromedary camels and humans, but SARS-CoV-2 was first found in seafood markets [7]. However, the origin of the SARS-CoV-2 virus is not known, and the two opposing theories that have recently dominated discussion are the "laboratory escape" hypothesis and zoonotic evolution [8].

The severe acute respiratory syndrome coronavirus was isolated from the airway epithelial cells of infected humans, and a phylogenetic analysis of full-length genome sequences obtained from infected patients revealed that SARS-CoV-2 is similar to SARS-CoV and uses the same cell entrance receptor, angiotensin-converting enzyme 2 [9]. Additionally, many mutants of the SARS-CoV-2 virus have been discovered, including the Omicron variant (B. 1.1.529) and the Delta variant (AY.3), which have high transmissibility rates [10–12]. As a consequence, they pose a huge threat to public health around the world [13–15]. The World Health Organization (WHO) has declared a medical emergency; scientific research is tackling this condition worldwide [16].

In the laboratory, drug-testing techniques assist scientists in identifying new and suitable candidates versus different convergent diseases [17]. However, developing drugs consumes time, money, humans, and specialized equipment. Therefore, many developing countries have resorted to using herbs to treat many diseases [18]. In ancient times and throughout history, plants were the main source of medicine and pharmaceuticals. However, modern medicine has mostly ignored medicinal plants [19]. Medicinal plants are widely used to treat many common diseases such as malaria, cholera, and asthma. Herbal medicine is considered to be the principal healthcare system of choice in various developing countries for several factors, including cost and the lack of availability of other systems. Antiviral plant compounds may work by inactivating virus particles, reducing endocytic activity, inhibiting viral enzymes and molecular reproduction mechanisms, altering virus capsid properties, blocking virus adsorption and penetration into human cells, inhibiting reverse transcriptase, inhibiting translation, reducing the expression level, and apoptosis [20]. Investigating the possible applications of medicinal plants could be beneficial for the control of the pandemic. Several studies have found that medicinal herbs such as Vernonia amygdalina, *Nigella sativa*, *Eurycoma longifolia*, and *Azadirachta indica* can assist in the prevention of and hasten the recovery from COVID-19 disease [21].

One of the essential herbs is *Saussurea costus*, which is widely used in Asia and the Arab world [22]. The plant is a member of the Asteraceae family, which is located worldwide; however, its most prevalent regions are India, Pakistan, and the Himalayas [23,24]. *Saussurea costus* is an herb used in many traditional medicine systems to treat asthma, inflammatory diseases, ulcers, and stomach problems [25,26]. Furthermore, several laboratory and animal model studies have demonstrated that *Saussurea costus* shows anti-inflammatory, anti-trypanosomal, potent anticancer, antibacterial, antifungal, and antiviral activity, confirming its tradition of use in various forms of medicine [26–34].

In spite of vaccination efforts and distinguished therapies for some viral illnesses, humans continue to lose the battle against viruses [35]. Therefore, for infectious diseases in particular, people need to go back to medicinal plants, which are a huge and all-natural drug store [36]. This study put a special emphasis on finding inhibitors against the main protease (M^{pro}), as it is considered an ideal target for the treatment of COVID-19 [37–42]. The blind molecular docking approach was performed utilizing the elucidated compounds extracted in water at room temperature from *Saussurea costus*. It is indisputable that the inhibitor preference increases with increased solubility in water and answers why people in rural communities use this plant to prevent and treat COVID-19.

2. Materials and Methods

2.1. Cell Cultures and Plasmids

DMEM (Gibco, Grand Island, NY, USA) supplemented with 10% fetal bovine serum (Gibco), antibiotics (Sigma-Aldrich, Saint Louis, MO, USA), and non-essential amino acids (Gibco) was used to sustain HEK293T and stable HEK293T-ACE2-expressing cells at 37 °C and in a 5% CO_2 environment. Puromycin was also present in the HEK293-ACE2 cells at a final concentration of 1 µg/mL. the following plasmids were used: pNL4.3-Env-FLuc, spike-G614-19, and pCMV-VSV-G [43].

2.2. Extraction and Preparation for GC-MS Analysis

Saussurea costus root was purchased from an herbal store in Riyadh, Saudi Arabia. An amount of 50 gm of dry *Saussurea* roots was extracted with 250 mL of distilled water for 96 h at room temperature using a multifunctional Orbital Shaker (BioSan PSU-20i). The supernatant was then filtered through filter paper, and the extract was allowed to dry. To 0.1 g of the extract, 1.5 g of anhydrous sodium sulfate was added and then dissolved with 10 mL of analytical-grade ethanol. The solution was then passed through a 0.45 mm syringe filter and into a 1.5 mL vial, where it was prepared for injection into GC-MS.

2.3. GC-MS Analysis Conditions

This study was carried out on samples through the use of the GM-MS technique model (GC-MS-QP2010-Ultra) with serial number 020525101565SA (Shimadzu, Kyoto, Japan), and a capillary column (Rtx-5 ms-30 m 0.25 mm 0.25 mm) manufactured by Restek, D-81379 Munich, German. The sample was injected using the split mode with helium as the carrier gas passing through at a flow rate of 1.61 mL/min. The temperature program was started at 50 °C with a rate of 10 °C per minute and ended at 300 °C with a hold time of 10 min. The injection port temperature was 300 °C, the ion source temperature was 200 °C, and the interface temperature was 250 °C. It took 35 min to analyze the sample using the scan mode in the m/z 40–500 charges to ratio range, with a run time of 40 min. The sample's constituents were determined by evaluating their retention index and mass fragmentation patents to tose available in the National Institute of Standards and Technology library (NIST).

2.4. Molecular Docking

The crystal structure of the M^{pro} of SARS-CoV-2 (PDB ID: 6Y2E) was downloaded from the Protein Data Bank database. Then, water residues were removed before minimization for 1000 steepest descent steps at 20 conjugate gradient steps. The 3D structure of the screened compounds was generated as a PDB file utilizing SMILES strings and PubChem ID using the build structure function in UCSF Chimera which developed by Resource for Biocomputing, Visualization, and Informatics (RBVI) at the University of California, San Francisco, USA [44–47]. Their energy was minimized with an antechamber plugin in UCSF Chimera for 14,000 steepest descent steps at 8000 conjugate gradient steps. Molecular docking was accomplished using the AutoDock Vina bulging in UCSF Chimera [48]. A grid box of (−15 × −24 × 15) Å centered at (35, 65, 65) Å was used while maintaining the default parameter values. The predicted affinity score was explored through the UCSF Chimera View Dock tool. UCSF Chimera has been used to process and visualize images, hydrogen bonds, and van der Waals interactions [44,45,48–51].

2.5. Antibacterial Activity Evaluation

The disc diffusion assay was utilized to evaluate the extract's initial antimicrobial activity against the selected microbial strains, including a Gram-positive bacterium (*Staphylococcus aureus* ATCC BAA 1026) and a Gram-negative bacterium (*Escherichia coli* ATCC 9637). The Kirby–Bauer disc diffusion susceptibility protocol was used with minor modifications [52]. Briefly, the aqueous crude extract was reconstituted in sterile distilled water to obtain a concentration of 100 mg/mL. Microbial strains were adjusted to 0.5 Mc-

Farland (10^8 CFU/mL) suspension using normal saline (0.9% NaCl) and were directly swabbed onto nutrient agar plates (Oxoid, UK). An amount of 10 µL of the aqueous extract (100 mg/mL) was placed on sterile 6-millimeter paper discs and allowed to dry under aseptic conditions. A disc holding 10 µL of normal saline served as a negative control, while discs containing chloramphenicol (2.5 mg/mL) served as a reference drug (positive control), and the plates were incubated at ambient temperature for up to 18 h. Plates were then inspected for the inhibition zones around the discs and the test was repeated twice.

2.6. Cytotoxicity Assays

Cytotoxicity assays were performed as previously described [53]. Using the resazurin technique, the maximum non-toxic concentration was determined. In a 96-well plate, HEK293T-ACE2 cells were grown at a density of 1×10^4 cells per well while being exposed to various doses of the natural products. Cells treated with the vehicle (DMSO) were utilized as a standard and control. At 48 h after resazurin addition, cell viability was assessed after 3 h. At 570 and 630 nm, the Multiskan TM GO (Thermo Scientific, Waltham, MA, USA) was used to detect absorbance. The MNTC has the highest concentration and less than 10% cytotoxicity.

2.7. Antiviral Activity against SARS-CoV-2

The assay was performed in HEK293-ACE2 cells as previously described [54]. Briefly, 1×10^4 cells in suspension were added to each well of 96-well plates and infected in the presence of 2 mg/mL of the extract. Firefly luciferase activity was measured 48 h later using the Dual-Luciferase Reporter Assay System kit (Promega, Madison, WI, USA) and the Fluoroskan FL (Thermo Scientific). HEK293T-ACE2 cells transduced with the pseudotyped virus without the extract were used as untreated control cells. The following formula was used to calculate the percentage of inhibition: 100 − (RLUs of treated cells/RLUs of control untreated cells) × 100.

2.8. Time of Addition Assay of HSV-1

The time of addition assay was carried out in three different experimental conditions according to the time the extract was added. In the pre-infection condition, the extract was added two hours before viral adsorption. Subsequently, cells were washed with PBS, the viral inoculum was added, and the infection was performed as previously described. In the adsorption condition, the extract was added with the viral inoculum and was incubated for 1 h at 37 °C. After that, the viral inoculum was removed, cells were washed with PBS and replaced with DMEM 2%, and infection was performed as previously described. For the post-entry condition, the viral inoculum was added and incubated for 1 h at 37 °C, washed with PBS, and DMEM 2% with the extract was added. After 72 h, the virus genome in the supernatant was quantitated by qPCR, and the antiviral activity was quantified as previously described [54].

2.9. Statistical Analysis

The data of at least three separate experiments were provided as the mean and standard deviation (SD). Each group's differences were evaluated through a t-test. Lines between the groups being compared represent statistically significant differences (* $p = 0.05$, ** $p = 0.01$, *** $p = 0.001$, and **** $p = 0.001$).

3. Results and Discussion

Phytochemical analysis of an extract of the *Saussurea* root revealed the presence of forty-eight chemical substances, all of which exhibited specific mass spectral fragmentation characteristics similar to known compounds in the National Institute of Standards and Technology (NIST) library, as shown in Table S1. According to the data, fatty acids account for the highest percentage of phytochemicals at 73.8%, followed by tannins at 8.2%, carbohydrates at 6.9%, terpenoids at 4.3%, carboxylic acids at 2.5%, hydrocarbons at 2.4%, and sterols at 1.5%. Phenolic compounds account for the lowest percentage of phytochemicals, at 0.2%, as shown in Figure 1.

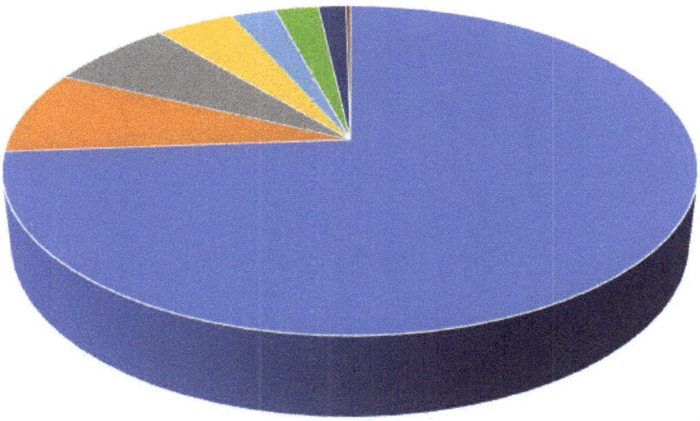

Figure 1. Phytochemicals in water-extracted compounds from *Saussurea costus* as identified by GC-MS.

The identified extracted compounds were used in blind docking experiments against the M^{pro}, with a particular focus on the interaction with the catalytic dyad of His41 and Cys145 in addition to Glu166 residue (Figure 2). In Table 1, we present the results of the docking experiment, which clearly show the uniqueness of the water extract of *Saussurea costus*, as 22 of the 48 compounds showed the ability to interact with the active sites of the M^{pro}. Of these compounds, eight were carbohydrates, five were fatty acids, three were terpenoids, two were carboxylic acids, one was a tannin, one was a phenolic compound, and one was a sterol. Figure 3 depicts the docked complexes of the top seven candidates. In Figure 3, the hydrogen bonds and van der Waals interactions that were formed between these screened compounds and the M^{pro} expected from the molecular docking were monitored. The compounds 4,8,13-Cyclotetradecatriene-1,3-diol-1,5,9-trimethyl-12-(1-methylethyl) (cmd34), Andrographolide (cmd35), and delta.4-Androstene-3.beta.,17.beta.-diol (cmd42) showed a higher binding affinity among the identified compounds, with the binding affinity ranging from −6.3 kcal/mol to −6.5 kcal/mol. They are a hydrocarbon, a terpenoid, and a sterol, respectively. These results indicate that *Saussurea costus*, which is used in traditional medicine as a COVID-19 remedy, produced an enormous number of inhibitors against the M^{pro} that were water soluble in green conditions. Therefore, the plant's mixture of active components may have a more effective therapeutic impact than a single isolated chemical.

The literature has shown that tannins, terpenoids, and phenolic substances can inhibit SARS-CoV-2 [55–58]. In addition, it has been reported that the quinolin-2-carboxylic acids found in *Ephedra sinica* are being suggested as possible treatment agents for COVID-19 [59]. Fatty acids are self-defense agents in organisms and have a variety of biological actions, particularly anti-inflammatory properties [60]. Carbohydrates have not been demonstrated to have a therapeutic impact on their own, but they may boost the efficacy of the therapeutically essential components and be used to produce polysaccharide immunomodulatory with potential medicinal and vaccine applications [61,62]. Sterol molecules have exhibited a wide range of biological actions, the majority of which act as cancer inhibitors, anti-inflammatory agents, and immunomodulatory and anti-viral agents [63]. Some of the compounds isolated from this plant, including costsunolide, dehydrocostus lactone, and cynaropicrin, appear to be capable of being developed as bioactive molecules [64–66].

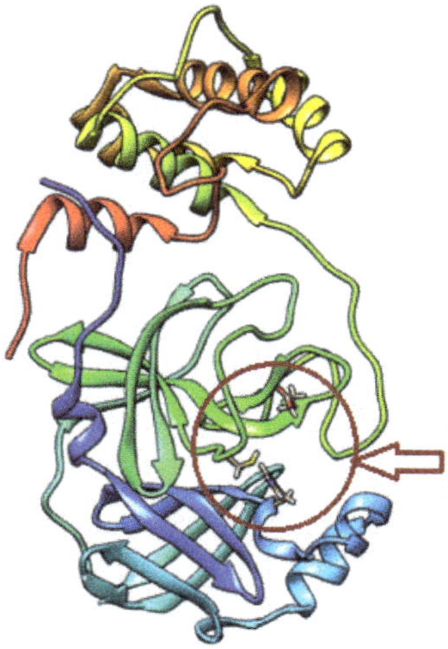

Figure 2. The crystal structure of the main protease of SARS-CoV-2 (PDB ID: 6Y2E).

Table 1. The binding affinity of the extracted compounds active against 3-chymotrypsin-like protease ($3CL^{pro}$).

Compound ID/Class	Structure	Binding Affinity (Kcal/Mol)/(RMSD)	Hydrogen Bond	van der Waals
Cmd1/carbohydrate	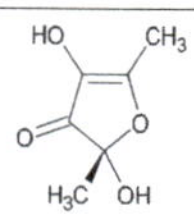	−4.7/ (0.00–3.2)	GLU 166, HIS 164, HIS 163	HIS 164, MET 165, HIS 163, LEU 141, ASN 142, GLU 166, PHE 140, SER 144, CYS 145
Cmd2/carbohydrate	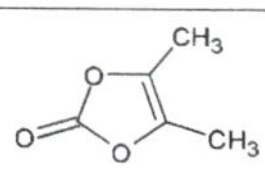	−3.9/ (40.82–41.72)	HIS 163	HIS 163, MET 165, GLU 166, SER 144, ASN 142, HIS 164, PHE 140
Cmd3/carbohydrate	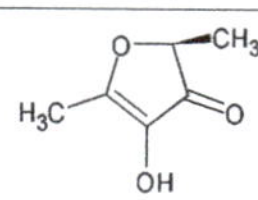	−4.3/ (40.10–41.20)	HIS 163	GLU 166, HIS 163, MET 165, HIS 164, ASN 142, LEU 141, PHE 140, CYS 145
Cmd5/carbohydrate	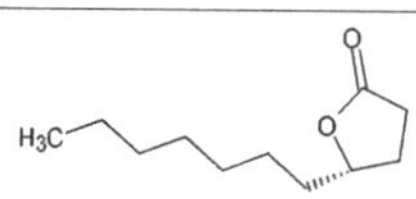	−4.4/ (27.24–29.57)	HIS 163, GLU 166	HIS 163, MET 49, GLU 166, MET 165, HIS 41, GLN 189
Cmd6/carboxylic acid	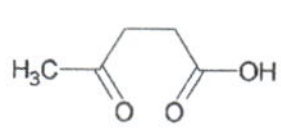	−3.8/ (25.23–26.15)	HIS 163, GLU 166, PHE 140	GLU 166, MET 165, CYS 145, HIS 163, ASN 142, SER 144, PHE 140, LEU 141
Cmd7/carbohydrate	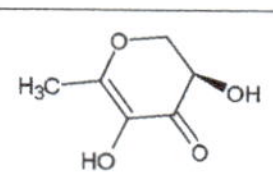	−5.0/ (0.00–0.00)	HIS 163	HIS 163, GLU 166, PHE 140, MET 165, HIS 164, CYS 145, ASN 142, SER 144, LEU 141
Cmd8/fatty acid	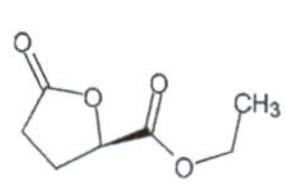	−4.3/ (40.97–42.42)	HIS 163, GLU 166	HIS 163, GLU 166, MET 165, HIS 164, PHE 140, MET 49, CYS 145, ASN 142, SER 144, LEU 141
Cmd11/carbohydrate	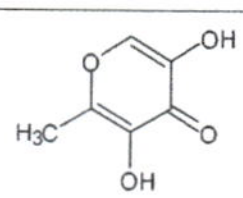	−4.9/ (44.57–45.67)	-	HIS 163, GLU 166, MET 165, HIS 164, PHE 140, LEU 141, CYS 145, ASN 142, SER 144

Table 1. *Cont.*

Compound ID/Class	Structure	Binding Affinity (Kcal/Mol)/(RMSD)	Hydrogen Bond	van der Waals
Cmd12/tannins	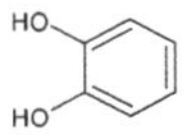	−4.6/ (44.33–45.19)	HIS 164	HIS 163, HIS 164, MET 165, GLU 166, PHE 140, LEU 141, ASN 142, SER 144, CYS 145
Cmd13/carbohydrate	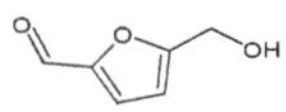	−4.3/ (24.50–25.23)	PHE 140, HIS 163	HIS 163, HIS 164, MET 165, GLU 166, PHE 140, LEU 141, ASN 142, SER 144, CYS 145, HIS 41
Cmd15/fatty acid	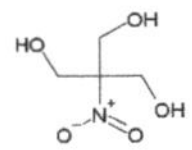	−4.2/ (25.99–27.47)	HIS 163	HIS 163, HIS 164, MET 165, GLU 166, PHE 140, LEU 141, GLY 143, ASN 142, SER 144, CYS 145
Cmd18/Phenolic Compound	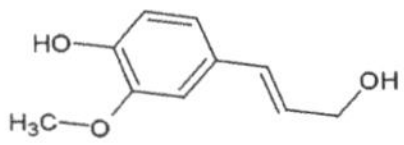	−4.9/ (42.17–45.62)	HIS 164, GLU 166	GLN 189, PRO 168, THR 190, HIS 164, MET 165, GLU 166, MET 49, HIS 41, CYS 145
Cmd20/carbohydrate	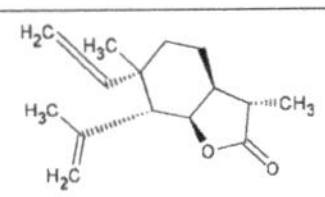	−5.1/ (29.29–33.07)	GLY 143	MET 49, GLU 166, ASN 142, GLY 143, GLN 189, MET 165, LEU 141, HIS 41, CYS 145
Cmd21/Carboxylic acid	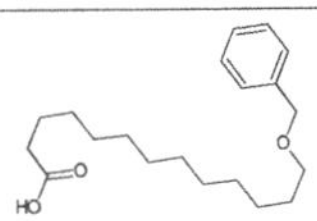	−4.3/ (28.11–30.02)	GLU 166, GLY 143	GLY 143, ASN 142, MET 49, THR 26, LEU 27, HIS 41, HIS 164, GLU 166, GLN 189, MET 165, CYS 145
Cmd26/Fatty acid	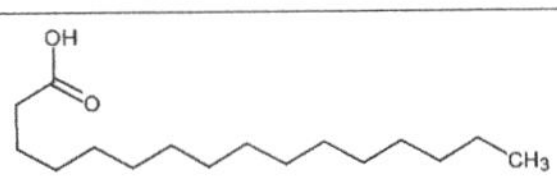	−4.2/ (27.28–29.77)	HIS 163, SER 144	SER 144, HIS 163, THR 25, LEU 27, GLU 166, GLN 189, ASN 142, MET 49, PHE 140, LEU 141
Cmd30/Terpenoid	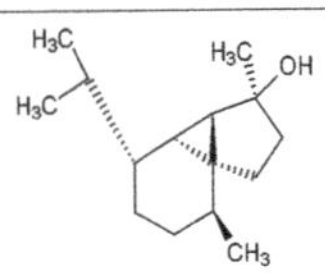	−5.2/ (29.65–31.24)	-	MET 49, THR 25, THR 26, LEU 27, HIS 41, ASN 142, CYS 145, HIS 164
Cmd34/Hydrocarbon	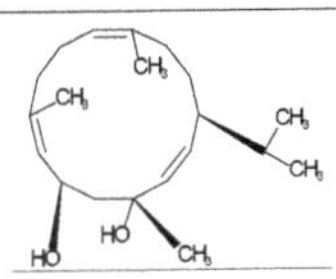	−6.3/ (42.35–46.32)	GLU 166	GLU 166, ASN 142, PRO 168, PHE 140, LEU 141, HIS 163, MET 165, ASN 142, CYS 145, LEU 167, HIS 164, GLN 189
Cmd35/Terpenoid	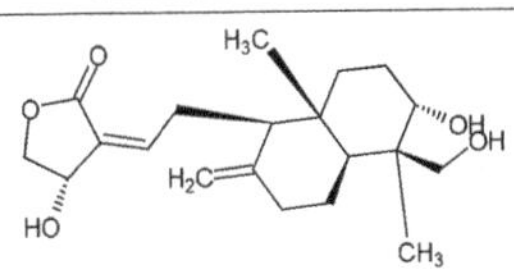	−6.3/ (30.17–32.16)	GLU 166	THR 25, MET 165, GLU 166, GLN 189, ASN 142, THR 45, SER 46, MET 49, CYS 44, HIS 41, CYS 145, THR 24
Cmd38/Fatty acid	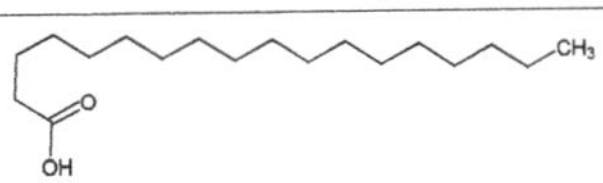	−3.7/ (26.63–30.65)	HIS 163	HIS 163, GLU 166, MET 49, SER 144, GLN 189, PHE 140, LEU 141, ASN 142, HIS 41, HIS 164, MET 165
Cmd40/Terpenoid	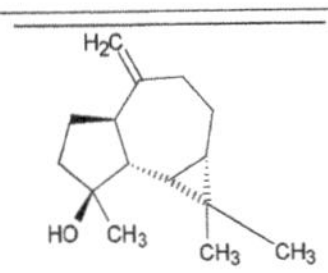	−5.4/ (43.11–45.74)	HIS 164	GLU 166, MET 49, GLN 189, ASN 142, MET 165, CYS 145, HIS 164

Table 1. *Cont.*

Compound ID/Class	Structure	Binding Affinity (Kcal/Mol)/(RMSD)	Hydrogen Bond	van der Waals
Cmd41/Fatty acid	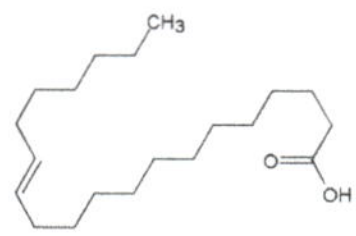	−4.3/ (28.45–31.03)	HIS 163, PHE 140	HIS 163, ASN 142, SER 144, LEU 141, PHE 140, LEU 27, GLN 189, MET 49, THR 25, THR 26, GLU 166, CYS 145, GLY 143, HIS 41, MET 165
Cmd42/Sterol	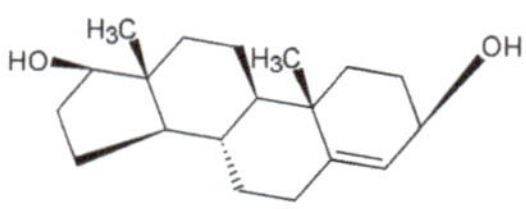	−6.5/ (30.43–33.11)	-	MET 49, HIS 163, ASN 142, SER 144, LEU 141, PHE 140, LEU 27, GLN 189, THR 25, THR 26, GLU 166, CYS 145, GLY 143, HIS 41, MET 165

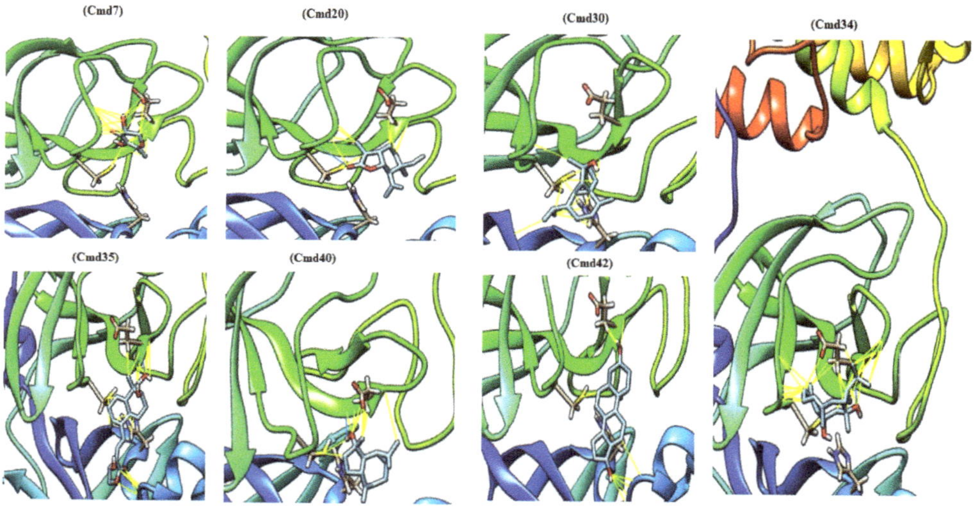

Figure 3. The active inhibiter docked with the active sites of the main protease of SARS-CoV-2.

As shown in Figure 4, the results of the antibacterial evaluation of the crude aqueous extract of *Saussurea costus* showed no activity in an inhibition zone of less than 7.0 mm (the diameter of the paper disc is 6.0 mm). Previous studies reported that methanol and ethanol extracts have remarkable antibacterial activities against various Gram-positive and Gram-negative bacteria [67,68]. Our results are also in agreement with a previous study that revealed that the water root extract of *Saussurea costus* did not show antibacterial activity against a Gram-positive bacterium (*Bacillus subtilis*) and a Gram-negative bacterium (*Escherichia coli*); however, these results showed higher activity with the non-polar extract (Chloroform extract) [69], indicating that the bioactive phytochemical molecules were not attracted by the aqueous extract and that the bioactive compounds might have a non-polar or a hydrophobic nature.

To determine the antiviral activity against SARS-CoV-2, the inhibitory activity was evaluated using the pseudovirus method. This system allows for the study of inhibitors of virus entry into the cell [54]. The antiviral activity was performed using a pseudovirus with Spike proteins from the D614G variant and the VOC variants Gamma and Delta. As shown in Figure 5, no antiviral effect of the extract from *Saussurea costus* was observed, and unexpectedly, an increase in pseudovirus infection by the extract was observed. This effect was similar for all three SARS-CoV-2 variants. These results suggest that the extract from *Saussurea costus* facilitates the cell entry of SARS-CoV-2.

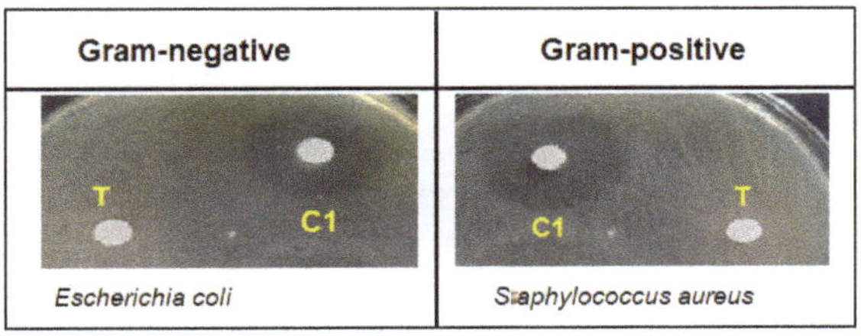

Figure 4. Negative zone of inhibition of *Saussurea costus* aqueous extract (100 mg/mL) (T) compared to chloramphenicol (2.5 mg/mL) (C1) against tested bacteria.

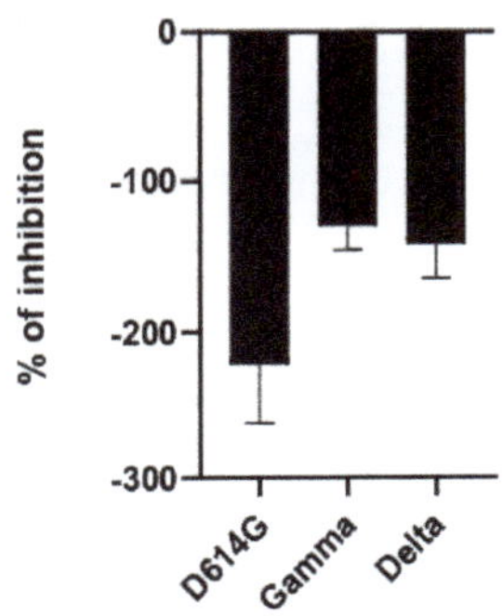

Figure 5. Antiviral activity of *Saussurea costus* aqueous extract against SARS-CoV-2. HEK−293T ACE2 cells were infected with the corresponding Spike-pseudotyped virus in the presence and absence of the extract. After 24 h, the luciferase activity was measured. The % of inhibition was determined as the ratio between treated and untreated cells.

Also, the antiviral activity of the *S. costus* aqueous extract against HSV-1 was evaluated. For this purpose, serial dilutions of the extract were performed, and the amount of virus produced was evaluated by qPCR. As shown in Figure 6, the *Saussurea costus* aqueous extract showed antiviral activity with an effective concentration 50 (EC_{50}) of 1.35 mg/mL. In addition, a cytotoxic concentration (CC_{50}) of 4.92 mg/mL and a selectivity index of 3.6 were observed.

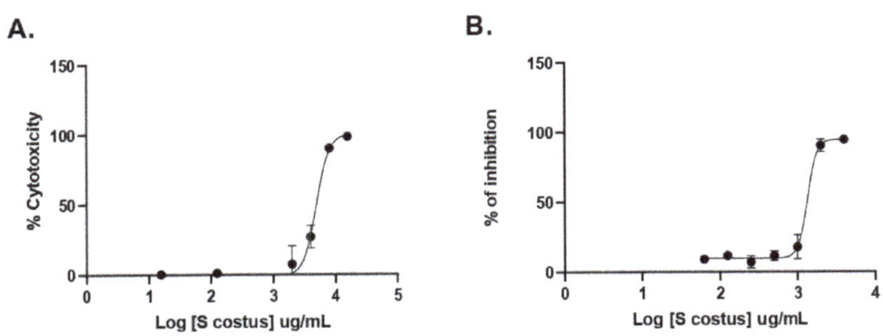

Figure 6. Antiherpetic activity and cytotoxicity of *Saussurea costus* aqueous extract. (**A**) Vero cells were incubated with increasing concentrations of the extract, and after 72 h, the cytotoxicity was measured as described in Section 2. (**B**) Vero cells were infected at MOI 1.5 with HSV-1 and incubated with increasing concentrations of the extract. After 72 h, the virus genome production was quantitated in the supernatant by qPCR. The % of inhibition was determined as the ratio between treated and untreated infected cells. Data are expressed as mean +/− SD for n = 2.

To obtain some information on the step of the viral cycle at which the extract exhibits antiviral activity, a time-of-infection study was performed. For this purpose, the extract was added at different times during infection. In the pre-infection condition, the extract was added 2 h prior to the addition of the virus and then removed to continue with the infection. In the adsorption condition, the virus was added together with the extract and incubated for one hour and then removed. In the post-entry condition, the extract was added after the virus entered the cell. As can be seen in Figure 7, the aqueous extract of *Saussurea costus* showed post-entry antiviral activity. This suggests that the *Saussurea costus* aqueous extract has no virucidal effect and inhibits the virus during the events after cell entry.

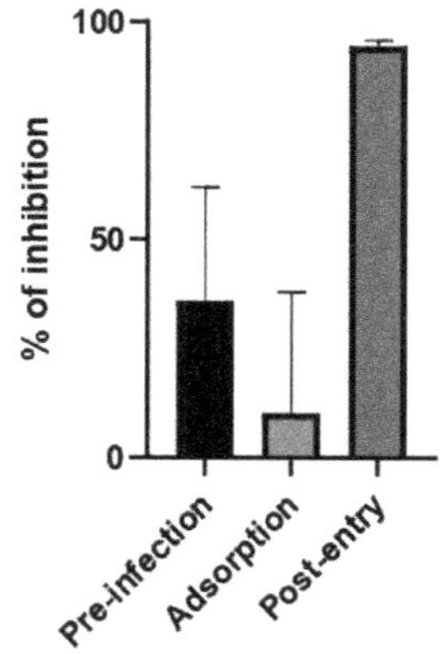

Figure 7. Time of addition assay of *Saussurea costus* aqueous extract. Vero cells were infected at MOI 1.5 and incubated with the extract at different times during the infection as is described in Section 2. After 24 h.p.i, the virus genome production was quantitated in the supernatant by qPCR. The % of inhibition was determined as the ratio between treated and untreated infected cells. Data are expressed as mean +/− SD for $n = 3$.

The information presented here describes the chemical profiles of *Saussurea costus* derived from a green extraction procedure using a water-based solvent at ambient temperature. Eighty-four compounds were identified, including fatty acids, tannins, carbohydrates, terpenoids, carboxylic acids, hydrocarbons, phenolic compounds, and sterols. The inhibitory effect of *Saussurea costus* on SARS-CoV-2 entry into the cell was studied using a pseudovirus with Spike proteins from the D614G variant and the VOC variants Gamma and Delta. Furthermore, the inhibition of the M^{pro} by these compounds was investigated using the molecular docking approach. Twenty-two candidates showed a good affinity to bind with the active site of the M^{pro}. According to this study, twenty-two candidates were identified from *Saussurea costus* against the M^{pro}, which has been used in traditional medicine as a COVID-19 remedy. The number of water-soluble inhibitors that were identified from *Saussurea costus* against the M^{pro} is unprecedented to our knowledge.

Supplementary Materials: The following supporting information can be downloaded at: https://www.mdpi.com/article/10.3390/molecules27154908/s1, Table S1: GC-MS discrimination of phytochemicals from the water extract of *Saussurea costus*.

Author Contributions: Conceptualization, H.I., E.M.A., P.H.S.T. and A.O.E.; methodology, H.I., B.S., H.M.E., A.I.A., E.M.A., P.G.M., P.H.S.T. and A.O.E.; GC-MS analysis, B.S.; antibacterial analysis, E.M.A.; antiviral analysis, P.G.M. and P.H.S.T.; molecular docking and software, H.M.E. and A.O.E.; formal analysis and investigation, H.I. and A.O.E.; writing—original draft preparation, H.I., B.S. and A.O.E.; writing—review and editing H.I., E.M.A., P.G.M., P.H.S.T. and A.O.E. All authors have read and agreed to the published version of the manuscript.

Funding: This research was supported by the Deanship of Scientific Research, Imam Mohammad Ibn Saud Islamic University (IMSIU), Saudi Arabia, Grant No. (21-13-18-031).

Informed Consent Statement: Not applicable.

Data Availability Statement: The data will be available upon suitable request.

Acknowledgments: The authors would like to thank, the laboratories of Ricardo Soto-Rifo and Fernando Valiente−Echeverría for providing the pseudoviral SARS-CoV-2 system.

Conflicts of Interest: The authors declare no conflict of interest.

Sample Availability: Samples of the compounds extracted from *Saussurea costus* are available from the authors.

References

1. Zimmermann, P.; Curtis, N. Coronavirus infections in children including COVID-19: An overview of the epidemiology, clinical features, diagnosis, treatment and prevention options in children. *Pediatric Infect. Dis. J.* **2020**, *39*, 355. [CrossRef] [PubMed]
2. Nadda, R.K.; Ali, A.; Goyal, R.C.; Khosla, P.K.; Goyal, R. Aucklandia costus (syn. *Saussurea costus*): Ethnopharmacology of an endangered medicinal plant of the Himalayan region. *J. Ethnopharmacol.* **2020**, *263*, 113199. [CrossRef] [PubMed]
3. Wald, E.R.; Schmit, K.M.; Gusland, D.Y. A pediatric infectious disease perspective on COVID-19. *Clin. Infect. Dis.* **2021**, *72*, 1660–1666. [CrossRef] [PubMed]
4. Saif, L.J.; Wang, Q.; Vlasova, A.N.; Jung, K.; Xiao, S. Coronaviruses. *Dis. Swine* **2019**, *3*, 488–523.
5. McIntosh, K.; Perlman, S. Coronaviruses, including severe acute respiratory syndrome (SARS) and Middle East respiratory syndrome (MERS). *Mandell Douglas Bennett's Princ. Pract. Infect. Dis.* **2015**, *2*, 1928–1936.
6. Sagar, M.; Reifler, K.; Rossi, M.; Miller, N.S.; Sinha, P.; White, L.F.; Mizgerd, J.P. Recent endemic coronavirus infection is associated with less-severe COVID-19. *J. Clin. Investig.* **2021**, *131*, 1–5. [CrossRef]
7. Fani, M.; Teimoori, A.; Ghafari, S. Comparison of the COVID-19 (SARS-CoV-2) pathogenesis with SARS-CoV and MERS-CoV infections. *Future Virol.* **2020**, *15*, 317–323. [CrossRef]
8. Holmes, E.C.; Goldstein, S.A.; Rasmussen, A.L.; Robertson, D.L.; Crits-Christoph, A.; Wertheim, J.O.; Anthony, S.J.; Barclay, W.S.; Boni, M.F.; Doherty, P.C. The origins of SARS-CoV-2: A critical review. *Cell* **2021**, *184*, 4848–4856. [CrossRef]
9. Wang, H.; Li, X.; Li, T.; Zhang, S.; Wang, L.; Wu, X.; Liu, J. The genetic sequence, origin, and diagnosis of SARS-CoV-2. *Eur. J. Clin. Microbiol. Infect. Dis.* **2020**, *39*, 1629–1635. [CrossRef]
10. Hu, J.; Peng, P.; Cao, X.; Wu, K.; Chen, J.; Wang, K.; Tang, N.; Huang, A.-L. Increased immune escape of the new SARS-CoV-2 variant of concern Omicron. *Cell. Mol. Immunol.* **2022**, *19*, 293–295. [CrossRef]
11. Glaß, M.; Misiak, D.; Misiak, C.; Müller, S.; Rausch, A.; Angermann, K.; Hoyer, M.; Zabel, R.; Kehlen, A.; Möbius, B. Fast forward evolution in real time: The rapid spread of SARS-CoV-2 variant of concern lineage B. 1.1. 7 in Saxony-Anhalt over a period of 5 months. *Lab. Med.* **2022**, *46*, 71–75. [CrossRef]
12. Grabowski, F.; Kochańczyk, M.; Lipniacki, T. The spread of SARS-CoV-2 variant Omicron with a doubling time of 2.0–3.3 days can be explained by immune evasion. *Viruses* **2022**, *14*, 294. [CrossRef] [PubMed]
13. Pollard, C.A.; Morran, M.P.; Nestor-Kalinoski, A.L. The COVID-19 pandemic: A global health crisis. *Physiol. Genom.* **2020**, *52*, 549–557. [CrossRef] [PubMed]
14. World Health Organization. *Coronavirus Disease 2019 (Covid-19): Situation Report, 73*; World Health Organization: New York, NY, USA, 2020.
15. World Health Organization. *Coronavirus Disease (Covid-19): Situation Report, 182*; World Health Organization: New York, NY, USA, 2020.
16. Kokudo, N.; Sugiyama, H. Call for international cooperation and collaboration to effectively tackle the COVID-19 pandemic. *Glob. Health Med.* **2020**, *2*, 60–62. [CrossRef] [PubMed]
17. Shrestha, J.; Razavi Bazaz, S.; Aboulkheyr Es, H.; Yaghobian Azari, D.; Thierry, B.; Ebrahimi Warkiani, M.; Ghadiri, M. Lung-on-a-chip: The future of respiratory disease models and pharmacological studies. *Crit. Rev. Biotechnol.* **2020**, *40*, 213–230. [CrossRef]
18. Ekor, M. The growing use of herbal medicines: Issues relating to adverse reactions and challenges in monitoring safety. *Front. Pharmacol.* **2014**, *4*, 177. [CrossRef]
19. Sadeek, A.; Abdallah, E. Medicinal Plants with Antiviral Properties to Tackle Covid-19 Pandemic: A Short-Review. *Antivirals* **2021**, *2*, 122–127.
20. Adhikari, B.; Marasini, B.P.; Rayamajhee, B.; Bhattarai, B.R.; Lamichhane, G.; Khadayat, K.; Adhikari, A.; Khanal, S.; Parajuli, N. Potential roles of medicinal plants for the treatment of viral diseases focusing on COVID-19: A review. *Phytother. Res.* **2021**, *35*, 1298–1312. [CrossRef]
21. Abdallah, E.M. Controlling Covid-19 Needs More Than Just a Vaccine; It Requires An Integrated Control Strategy. *Open Access J. Biol. Sci.* **2021**, *3*, 1395–1396. [CrossRef]
22. Hussain, M.; Khera, R.A.; Iqbal, J.; Khalid, M.; Hanif, M.A. Phytochemicals: Key to effective anticancer drugs. *Mini-Rev. Org. Chem.* **2019**, *16*, 141–158. [CrossRef]
23. Wang, K.; Zhang, J.; Ping, S.; Ma, Q.; Chen, X.; Xuan, H.; Shi, J.; Zhang, C.; Hu, F. Anti-inflammatory effects of ethanol extracts of Chinese propolis and buds from poplar (Populus × canadensis). *J. Ethnopharmacol.* **2014**, *155*, 300–311. [CrossRef]

24. Ali, S.I.; Venkatesalu, V. Botany, traditional uses, phytochemistry and pharmacological properties of *Saussurea costus*—An endangered plant from Himalaya—A review. *Phytochem. Lett.* **2022**, *47*, 140–155. [CrossRef]
25. Schippmann, U.; Leaman, D.J.; Cunningham, A. Impact of cultivation and gathering of medicinal plants on biodiversity: Global trends and issues. In *FAO. 2002. Biodiversity and the Ecosystem Approach in Agriculture, Forestry and Fisheries. Satellite Event on the occasion of the Ninth Regular Session of the Commission on Genetic Resources for Food and Agriculture, Rome, Italy, 12–13 October 2002*; Inter-Departmental Working Group on Biological Diversity for Food and Agriculture: Rome, Italy, 2002.
26. Pandey, M.M.; Rastogi, S.; Rawat, A.K.S. Saussurea costus: Botanical, chemical and pharmacological review of an ayurvedic medicinal plant. *J. Ethnopharmacol.* **2007**, *110*, 379–390. [CrossRef] [PubMed]
27. Cho, J.Y.; Baik, K.U.; Jung, J.H.; Park, M.H. In vitro anti-inflammatory effects of cynaropicrin, a sesquiterpene lactone, from *Saussurea lappa*. *Eur. J. Pharmacol.* **2000**, *398*, 399–407. [CrossRef]
28. Julianti, T.; Hata, Y.; Zimmermann, S.; Kaiser, M.; Hamburger, M.; Adams, M. Antitrypanosomal sesquiterpene lactones from *Saussurea costus*. *Fitoterapia* **2011**, *82*, 955–959. [CrossRef] [PubMed]
29. Lee, B.-K.; Park, S.-J.; Nam, S.-Y.; Kang, S.; Hwang, J.; Lee, S.-J.; Im, D.-S. Anti-allergic effects of sesquiterpene lactones from *Saussurea costus* (Falc.) Lipsch. determined using in vivo and in vitro experiments. *J. Ethnopharmacol.* **2018**, *213*, 256–261. [CrossRef] [PubMed]
30. Avdeeva, E.; Reshetov, Y.; Domrachev, D.; Gulina, E.; Krivoshchekov, S.; Shurupova, M.; Brazovskii, K.; Belousov, M. Constituent composition of the essential oils from some species of the genus Saussurea DC. *Nat. Prod. Res.* **2022**, *36*, 660–663. [CrossRef] [PubMed]
31. Soliman, M.F.; Shetaia, Y.M.; Tayel, A.A.; Munshi, A.M.; Alatawi, F.A.; Alsieni, M.A.; Al-Saman, M.A. Exploring the Antifungal Activity and Action of Saussurea costus Root Extracts against Candida albicans and Non-albicans Species. *Antibiotics* **2022**, *11*, 327. [CrossRef]
32. Abdel-Wahhab, K.G.; Mannaa, F.A.; El-Sahra, D.G.; Morsy, F.A.; Gomaa, H.F. Effect of oral administration of methanolic root extract of *Saussurea costus* to rats after propylthiouracil-induced hypothyroid obesity. *Comp. Clin. Pathol.* **2022**, *31*, 377–390. [CrossRef]
33. Ashry, M.; Galal ELSahra, D.; Abdel-Wahhab, K.G.; Abdelsalam, M.E.; Elmashad, W.; El-Bitar, A.M.; Gomaa, H.F. Saussurea Costus Extract Has Anti-İnflammatory, Antioxidant and Hormonal Effects Against Testicular Toxicity Induced by Oxaliplatin in Male Albino Rats. *Iran. J. Toxicol.* **2022**, *16*, 83–90. [CrossRef]
34. Mohsen, E.; El-Far, A.H.; Godugu, K.; Elsayed, F.; Mousa, S.A.; Younis, I.Y. SPME and solvent-based GC–MS metabolite profiling of Egyptian marketed *Saussurea costus* (Falc.) Lipsch. concerning its anticancer activity. *Phytomed. Plus* **2022**, *2*, 100209. [CrossRef]
35. Rumlová, M.; Ruml, T. In vitro methods for testing antiviral drugs. *Biotechnol. Adv.* **2018**, *36*, 557–576. [CrossRef] [PubMed]
36. Abdallah, E.M. Plants: An alternative source for antimicrobials. *J. Appl. Pharm. Sci.* **2011**, *1*, 16–20.
37. Elzupir, A.O. Inhibition of SARS-CoV-2 main protease 3CLpro by means of α-ketoamide and pyridone-containing pharmaceuticals using in silico molecular docking. *J. Mol. Struct.* **2020**, *1222*, 128878. [CrossRef] [PubMed]
38. Anju, A.; Chaturvedi, S.; Chaudhary, V.; Pant, P.; Hussain, F.; Mishra, A.K. Virtual screening of quinoline derived library for SARS-COV-2 targeting viral entry and replication. *J. Biomol. Struct. Dyn.* **2021**, 1–30. [CrossRef]
39. Hamed, M.I.; Darwish, K.M.; Soltane, R.; Chrouda, A.; Mostafa, A.; Shama, N.M.A.; Elhady, S.S.; Abulkhair, H.S.; Khodir, A.E.; Elmaaty, A.A. β-Blockers bearing hydroxyethylamine and hydroxyethylene as potential SARS-CoV-2 Mpro inhibitors: Rational based design, in silico, in vitro, and SAR studies for lead optimization. *RSC Adv.* **2021**, *11*, 35536–35558. [CrossRef]
40. Rana, S.; Kumar, P.; Sharma, A.; Sharma, S.; Giri, R.; Ghosh, K.S. Identification of Naturally Occurring Antiviral Molecules for SARS-CoV-2 Mitigation. *Open COVID J.* **2021**, *1*, 38–46. [CrossRef]
41. El-Shamy, N.T.; Alkaoud, A.M.; Hussein, R.K.; Ibrahim, M.A.; Alhamzani, A.G.; Abou-Krisha, M.M. DFT, ADMET and Molecular Docking Investigations for the Antimicrobial Activity of 6, 6′-Diamino-1, 1′, 3, 3′-Tetramethyl-5, 5′-(4-chlorobenzylidene) bis [pyrimidine-2, 4 (1H, 3H)-dione]. *Molecules* **2022**, *27*, 620. [CrossRef]
42. Wang, B.; Zhong, C.; Tieleman, D.P. Supramolecular Organization of SARS-CoV and SARS-CoV-2 Virions Revealed by Coarse-Grained Models of Intact Virus Envelopes. *J. Chem. Inf. Modeling* **2021**, *62*, 176–186. [CrossRef]
43. Beltrán-Pavez, C.; Riquelme-Barrios, S.; Oyarzún-Arrau, A.; Gaete-Argel, A.; González-Stegmaier, R.; Cereceda-Solis, K.; Aguirre, A.; Travisany, D.; Palma-Vejares, R.; Barriga, G.P. Insights into neutralizing antibody responses in individuals exposed to SARS-CoV-2 in Chile. *Sci. Adv.* **2021**, *7*, eabe6855. [CrossRef]
44. Pettersen, E.F.; Goddard, T.D.; Huang, C.C.; Couch, G.S.; Greenblatt, D.M.; Meng, E.C.; Ferrin, T.E. UCSF Chimera—A visualization system for exploratory research and analysis. *J. Comput. Chem.* **2004**, *25*, 1605–1612. [CrossRef] [PubMed]
45. Wang, J.; Wang, W.; Kollman, P.A.; Case, D.A. Automatic atom type and bond type perception in molecular mechanical calculations. *J. Mol. Graph. Model.* **2006**, *25*, 247–260. [CrossRef] [PubMed]
46. Elzupir, A.O. Caffeine and caffeine-containing pharmaceuticals as promising inhibitors for 3-chymotrypsin-like protease of SARS-CoV-2. *J. Biomol. Struct. Dyn.* **2022**, *40*, 2113–2120. [CrossRef] [PubMed]
47. Al-Janabi, A.S.; Elzupir, A.O.; Yousef, T.A. Synthesis, anti-bacterial evaluation, DFT study and molecular docking as a potential 3-chymotrypsin-like protease (3CLpro) of SARS-CoV-2 inhibitors of a novel Schiff bases. *J. Mol. Struct.* **2021**, *1228*, 129454. [CrossRef] [PubMed]

48. Elzupir, A.O. Molecular Docking and Dynamics Investigations for Identifying Potential Inhibitors of the 3-Chymotrypsin-like Protease of SARS-CoV-2: Repurposing of Approved Pyrimidonic Pharmaceuticals for COVID-19 Treatment. *Molecules* **2021**, *26*, 7458. [CrossRef]
49. O'Boyle, N.M.; Banck, M.; James, C.A.; Morley, C.; Vandermeersch, T.; Hutchison, G.R. Open Babel: An open chemical toolbox. *J. Cheminformatics* **2011**, *3*, 1–14. [CrossRef]
50. Shapovalov, M.V.; Dunbrack, R.L., Jr. A smoothed backbone-dependent rotamer library for proteins derived from adaptive kernel density estimates and regressions. *Structure* **2011**, *19*, 844–858. [CrossRef]
51. Trott, O.; Olson, A.J. AutoDock Vina: Improving the speed and accuracy of docking with a new scoring function, efficient optimization, and multithreading. *J. Comput. Chem.* **2010**, *31*, 455–461. [CrossRef]
52. Al-Mijalli, S.H.; Assaggaf, H.; Qasem, A.; El-Shemi, A.G.; Abdallah, E.M.; Mrabti, H.N.; Bouyahya, A. Antioxidant, Antidiabetic, and Antibacterial Potentials and Chemical Composition of *Salvia officinalis* and *Mentha suaveolens* Grown Wild in Morocco. *Adv. Pharmacol. Pharm. Sci.* **2022**, *2022*, 2844880. [CrossRef]
53. Gabaglio, S.; Alvarenga, N.; Cantero-González, G.; Degen, R.; Ferro, E.; Langjahr, P.; Chnaiderman, J.; Sotelo, P. A quantitative PCR assay for antiviral activity screening of medicinal plants against Herpes simplex 1. *Nat. Prod. Res.* **2021**, *35*, 2926–2930. [CrossRef]
54. González-Maldonado, P.; Alvarenga, N.; Burgos-Edwards, A.; Flores-Giubi, M.E.; Barúa, J.E.; Romero-Rodríguez, M.C.; Soto-Rifo, R.; Valiente-Echeverría, F.; Langjahr, P.; Cantero-González, G. Screening of Natural Products Inhibitors of SARS-CoV-2 Entry. *Molecules* **2022**, *27*, 1743. [CrossRef] [PubMed]
55. Zia, M.; Muhammad, S.; Bibi, S.; Abbasi, S.W.; Al-Sehemi, A.G.; Chaudhary, A.R.; Bai, F.Q. Exploring the potential of novel phenolic compounds as potential therapeutic candidates against SARS-CoV-2, using quantum chemistry, molecular docking and dynamic studies. *Bioorganic Med. Chem. Lett.* **2021**, *43*, 128079. [CrossRef] [PubMed]
56. Kuo, C.J.; Liang, P.H. Characterization and inhibition of the main protease of severe acute respiratory syndrome coronavirus. *ChemBioEng Rev.* **2015**, *2*, 118–132. [CrossRef]
57. Muhseen, Z.T.; Hameed, A.R.; Al-Hasani, H.M.; ul Qamar, M.T.; Li, G. Promising terpenes as SARS-CoV-2 spike receptor-binding domain (RBD) attachment inhibitors to the human ACE2 receptor: Integrated computational approach. *J. Mol. Liq.* **2020**, *320*, 114493. [CrossRef] [PubMed]
58. Giofrè, S.V.; Napoli, E.; Iraci, N.; Speciale, A.; Cimino, F.; Muscarà, C.; Molonia, M.S.; Ruberto, G.; Saija, A. Interaction of selected terpenoids with two SARS-CoV-2 key therapeutic targets: An in silico study through molecular docking and dynamics simulations. *Comput. Biol. Med.* **2021**, *134*, 104538. [CrossRef]
59. Mei, J.; Zhou, Y.; Yang, X.; Zhang, F.; Liu, X.; Yu, B. Active components in *Ephedra sinica* stapf disrupt the interaction between ACE2 and SARS-CoV-2 RBD: Potent COVID-19 therapeutic agents. *J. Ethnopharmacol.* **2021**, *278*, 114303. [CrossRef]
60. Hanh, T.T.H.; Hang, D.T.T.; Van Minh, C.; Dat, N.T. Anti-inflammatory effects of fatty acids isolated from *Chromolaena odorata*. *Asian Pac. J. Trop. Med.* **2011**, *4*, 760–763. [CrossRef]
61. Auwal, M.S.; Saka, S.; Mairiga, I.A.; Sanda, K.A.; Shuaibu, A.; Ibrahim, A. Preliminary Phytochemical and Elemental Analysis of Aqueous and Fractionated Pod Extracts of *Acacia nilotica* (Thorn Mimosa). *Vet. Res. Forum* **2014**, *5*, 95–100.
62. Villaseñor, I.M.; Cabrera, M.; Meneses, K.; Rivera, V.; Villasenor, R. Comparative antidiabetic activities of some medicinal plants. *Philipp. J. Sci.* **1998**, *127*, 261–266.
63. Loza-Mejía, M.A.; Salazar, J.R. Sterols and triterpenoids as potential anti-inflammatories: Molecular docking studies for binding to some enzymes involved in inflammatory pathways. *J. Mol. Graph. Model.* **2015**, *62*, 18–25. [CrossRef]
64. Kim, D.Y.; Choi, B.Y. Costunolide—A bioactive sesquiterpene lactone with diverse therapeutic potential. *Int. J. Mol. Sci.* **2019**, *20*, 2926. [CrossRef] [PubMed]
65. Gwari, G.; Bhandari, U.; Andola, H.C.; Lohani, H.; Chauhan, N. Volatile constituents of *Saussurea costus* roots cultivated in Uttarakhand Himalayas, India. *Pharmacogn. Res.* **2013**, *5*, 179.
66. Kim, E.J.; Lim, S.S.; Park, S.Y.; Shin, H.-K.; Kim, J.-S.; Park, J.H.Y. Apoptosis of DU145 human prostate cancer cells induced by dehydrocostus lactone isolated from the root of *Saussurea lappa*. *Food Chem. Toxicol.* **2008**, *46*, 3651–3658. [CrossRef] [PubMed]
67. Abdallah, E.M.; Qureshi, K.A.; Ali, A.M.; Elhassan, G.O. Evaluation of some biological properties of *Saussurea costus* crude root extract. *Biosci. Biotechnol. Res. Commun.* **2017**, *10*, 601–611. [CrossRef]
68. Sagar, A.; Chauhan, V.; Prakash, V. Studies on endophytes and antibacterial activity of *Saussurea costus* (falc.). *J. Drug Deliv. Ther.* **2017**, *7*, 5–10. [CrossRef]
69. Alaagib, R.M.O.; Ayoub, S.M.H. On the chemical composition and antibacterial activity of *Saussurea lappa* (Asteraceae). *Pharma Innov.* **2015**, *4*, 73.

molecules

Article

Steroidal Antimetabolites Protect Mice against *Trypanosoma brucei*

Minu Chaudhuri [1,*], Ujjal K. Singha [1,†], Boden H. Vanderloop [2], Anuj Tripathi [1] and W. David Nes [2,*]

1 Department of Microbiology, Immunology, and Physiology, Meharry Medical College, Nashville, TN 37208, USA; ujjal.singha@vumc.org (U.K.S.); anuj1tripathi@yahoo.com (A.T.)
2 Department of Chemistry & Biochemistry, Texas Tech University, Lubbock, TX 79409, USA; boden.h.vanderloop@vanderbilt.edu
* Correspondence: mchaudhuri@mmc.edu (M.C.); wdavid.nes@ttu.edu (W.D.N.)
† Present address: Department of Allergy, Pulmonary and Critical Care Medicine, Vanderbilt University Medical Center, Nashville, TN 37232, USA.

Abstract: *Trypanosoma brucei*, the causative agent for human African trypanosomiasis, is an emerging ergosterol-dependent parasite that produces chokepoint enzymes, sterol methyltransferases (SMT), not synthesized in their animal hosts that can regulate cell viability. Here, we report the lethal effects of two recently described natural product antimetabolites that disrupt *Acanthamoeba* sterol methylation and growth, cholesta-5,7,22,24-tetraenol (CHT) and ergosta-5,7,22,24(28)-tetraenol (ERGT) that can equally target *T. brucei*. We found that CHT/ERGT inhibited cell growth in vitro, yielding EC_{50} values in the low nanomolar range with washout experiments showing cidal activity against the bloodstream form, consistent with their predicted mode of suicide inhibition on SMT activity and ergosterol production. Antimetabolite treatment generated altered *T. brucei* cell morphology and death rapidly within hours. Notably, in vivo ERGT/CHT protected mice infected with *T. brucei*, doubling their survival time following daily treatment for 8–10 days at 50 mg/kg or 100 mg/kg. The current study demonstrates a new class of lead antibiotics, in the form of common fungal sterols, for antitrypanosomal drug development.

Keywords: ergosta-5,7,22,24(28)-tetraenol (ERGT); cholesta-5,7,22,24-tetraenol (CHT); ergosterol biosynthesis; antimetabolite; suicide substrate; *Trypanosoma brucei*

Citation: Chaudhuri, M.; Singha, U.K.; Vanderloop, B.H.; Tripathi, A.; Nes, W.D. Steroidal Antimetabolites Protect Mice against *Trypanosoma brucei*. *Molecules* **2022**, *27*, 4088. https://doi.org/10.3390/molecules27134088

Academic Editor: Lillian Barros

Received: 18 May 2022
Accepted: 21 June 2022
Published: 25 June 2022

1. Introduction

Trypanosoma brucei, a group of flagellated parasitic protozoa, infect both human and domestic animals and cause a fatal disease, African trypanosomiasis (AT) [1]. The disease is transmitted by an insect vector, the tsetse fly. During its digenetic life cycle, *T. brucei* undergoes a complex developmental process [2]. The procyclic and the bloodstream forms are the two major developmental forms of *T. brucei* that are found in the insect gut and the mammalian blood, respectively. World Health Organization (WHO) classified AT as a neglected tropical disease. The available drugs for AT originate as synthetic compounds, are mostly toxic, antiquated, develop resistance and difficult to administer [3]. Recent efforts of the Drugs for Neglected Diseases Initiative (DNDi) developed two new promising oral drugs, fexinidazole and oxabcroles, by screening small molecule chemical libraries against *T. brucei* [4,5]. Despite their promiscuity, fexinidazole has potential to develop cross-resistance to nifurtimox [6], a widely used current therapy for AT, and the actual target for oxaboroles is yet to be identified. Interestingly, an alternative approach that has recently advanced new trypanocides against AT derives from the isolation and testing natural products through traditional bioactivity-guided isolation steps or from the recently applied genome mining and engineering efforts [7,8]. Therefore, these successes involving synthetic compounds or natural products notwithstanding, there continues to be an urgent need to develop new anti-parasitic drugs that can prevent parasite growth through novel modes of action that might include blockage of crucial enzymes in essential metabolic pathways in African trypanosomes [9,10].

Antimetabolites in primary metabolism have emerged as effective chemotherapeutic agents against some forms of cancer by selectively disrupting a productive biosynthetic step [11,12]. Typically, these compounds are designed to show substrate mimicry and under physiological conditions they are expected to enter cells, then outcompete a relevant intermediate for a targeted enzyme that provides an essential end-product. Most of these antimetabolites are synthetic compounds with a functional warhead, making them suicide substrates. Consequently, the inhibitory influence of these drugs goes beyond simple competitive inhibition; following irreversible binding of inhibitor via covalent attachment to the active site of the enzyme causing its inactivation accompanied by muted biosynthesis of the essential end-product that, therefore, produces suppression of cell growth and development [13–16]. Covalent drugs have gained further notoriety [17,18] in their use to treat AT. Thus, a highly effective fluorinated substrate analog was developed in the later part of the 20th century against the ornithine decarboxylase in the polyamine biosynthesis pathway and appears under the brand name, Ornidyl, among others, and its chemical name, eflornithine [9].

Noting recent therapeutic successes involving other small molecule inhibitors, VNI targeting 14α-sterol demethylase enzyme in the trypanosome ergosterol biosynthesis pathway [19], we considered the sterol methyltransferase (SMT) enzyme as another druggable target for ergosterol biosynthesis in trypanosomes. SMTs have been shown to form covalent attachments to their substrate analogs, similar structurally to zymosterol (fungi/protozoa) or cycloartenol (plants), making these analogs suicide substrates (Figure 1) [20]. In contradistinction to the mechanism of action of typical ergosterol biosynthesis inhibitors, such as VNI, which bind through electrostatic interactions, the substrate mimics we designed produce their electrophilic handle only after the catalytic reaction cycle is turned on, yielding a sterol methyl intermediate positively charged in an unconventional side chain location (Supplementary Materials, Figure S1), otherwise these analogs are functionally inert upon binding. So far as is known, inhibitors coupled to cell death that interfere with ergosterol production and processing typically generate altered plasma-membrane structures, or in some cases altered mitochondrial membranes, as a consequence of changes to apoptosis-like and autophagic events [21,22]. Fascinated by the possibility that suicide inhibitors could interfere with sterol methylation in *Acanthamoeba* [22], it was thought that they could be developed elsewhere to control the ergosterol-dependent disease processes associated with parasites. Thus, we initiated a program several years ago to identify suicide inhibitors (synthetic) that specially block the sterol methylation pathways of *T. brucei* (*Tb*) or *Acanthamoeba castellanii* (*Ac*) and inhibit the ergosterol biosynthesis pathways in these protozoa without effect on cholesterogenesis in animals [23–26]. Our work is based on the condition that the major genetic difference in host C_{27}-cholesterol biosynthesis pathway and parasites *Tb* and *Ac* C_{28}-ergosterol and C_{29}-7-dehydroporifersterol biosynthesis pathways [22,24] (Figure 1) reside in the sterol methyltransferase (SMT) gene synthesized in *Ac* and *Tb* but not in animals [27]. Because the variant SMTs in these pathogens are product specific and mechanistically distinct in kinetoplastids and amoebae yielding $\Delta^{24(28)}$- or $\Delta^{25(27)}$-24-alkyl sterols (Figure S1), the differences in the reaction pathway of sterol methylation in these organisms and their absence in the host organism affords a path to drug selectivity in the form of our newly discovered steroidal chemotherapeutics (synthetic compound) and steroidal antibiotics (natural compound) that uniquely target SMT activities.

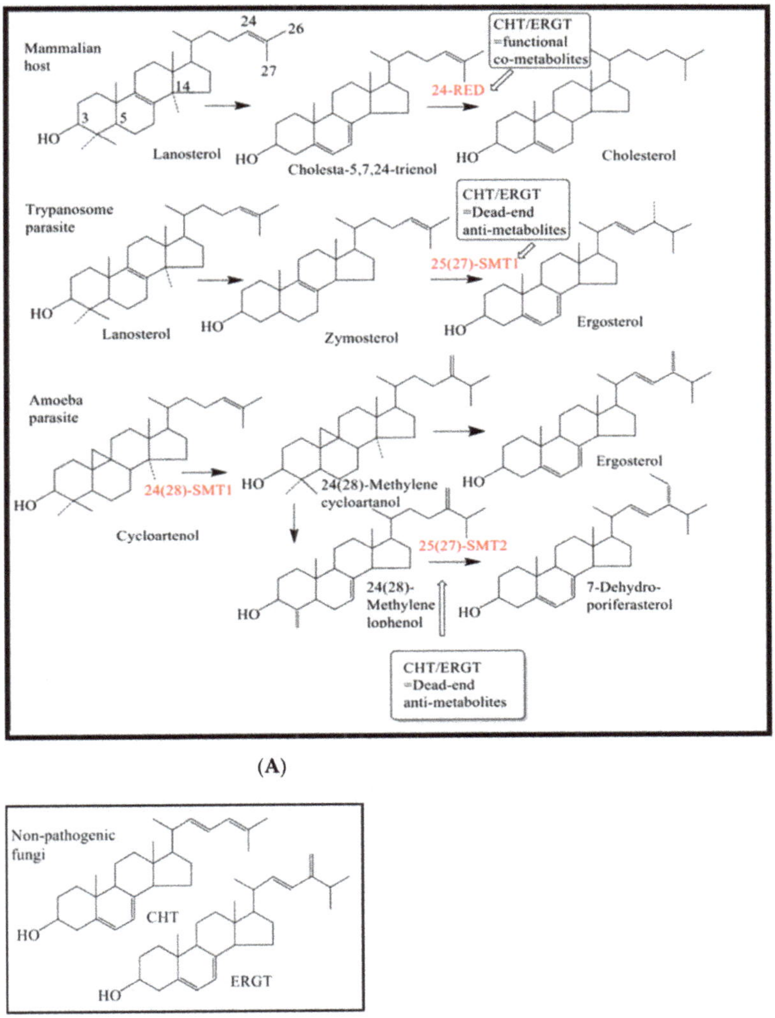

(A)

(B)

Figure 1. Differences in inhibition of steroidogenesis affected by steroidal antimetabolites. As shown, in ergosterol synthesis CHT and ERGT specifically inhibit the D25(27)-SMT in *Tb* and *Ac* via suicide (irreversible) inhibition. Alternatively, in cholesterogenesis neither CHT nor ERGT attach irreversibly to any enzyme in the pathway (**A**). Antimetabolites ERGT and CHT derived from yeast (**B**).

In our search for more effective synthetic mimics to complex SMT, we surreptitiously observed a pair of natural substrates, cholesta-5,7,22,24-teraenol (CHT) and ergosta-5,7,22,24(28)-tetraenol (ERGT), produced in the yeast ergosterol biosynthesis pathway that could bind and alter catalysis of either the *Ac* or *Tb* SMT [26–29] in similar fashion to that of 26,27-dehydrozymosterol tested against the yeast SMT [30,31]. For these reasons, we asked whether CHT or ERGT could bind *Tb* SMT in vivo and in so doing, play a protective role in animals infected with the parasite. We chose to study mice infected with *Tb* since several reports revealed the ergosterol biosynthesis pathway can be uncoupled, producing cell death, when the cells are subjected to inhibitors targeting crucial enzymes in the post-squalene pathways of trypanosome steroidogenesis [10,32]. Consistent with these reports, in our studies of the transition state inhibitor 25-azalanosterol against *Tb* SMT, we observed the analog prevented Tb cell proliferation in the low micromolar range [33] and quelled the parasite burden in vivo, effectively protecting the infected mice from the disease [25]. Because the recently discovered steroidal antimetabolites CHT and ERGT, distinct from the conventional anti-infectives of ergosterol synthesis [21,34], can covalently bind the *Tb* SMT during catalysis [30], this predicts their suicide inhibitor properties in *Tb* ergosterol biosynthesis and in so doing, provides a new armamentarium for therapeutic study. Here, mice infected with *Tb* parasites were treated with the suicide inhibitors CHT and ERGT or with no inhibitor. The effects of these short-term treatments upon cell proliferation and subcellular organization in vitro and in vivo were determined.

2. Results

2.1. ERGT/CHT Generates Marked Inhibition of T. brucei BF Growth in Cell Culture

From our recent observation that a sparking level of ergosterol, i.e., hormonal/trace amount of 24-alkyl sterol produced biosynthetically in bloodstream form (BF) [25] against bulk amounts of cellular cholesterol derived from the culture medium [35], is necessary for trypanosome growth and the report showing in *Trypanosoma cruzi* that sterol methylation is also essential in vivo and in vitro for trypanosome growth [36], we were encouraged to look for new leads, in the form of CHT and ERGT, that could be used to treat pathogens capable of catalyzing the sterol methylation route yielding a D25(27)-product. As observed previously, certain intermediates of the yeast sterol biosynthesis pathway, cholesta-5,7,22,24-tetraenol (CHT) and ergosta-5,7,22,24(28)-tetraenol (ERGT), while ineffective against the *Ac* SMT1 enzyme, which generates a $\Delta^{24(28)}$-product, can nonetheless inhibit sterol methylation catalyzed by other SMTs such as the *Ac* SMT2 isoform capable of generating an alternate $\Delta^{25(27)}$-product. Importantly, incubation of these fungal sterols at the low nanomolar range against *Ac* trophozoites cells led the substrate to distinctly bind *Ac* SMT2, followed by cellular loss of 24-alkyl sterol and cell death within hours of the initial cell treatment [28]. Strikingly, as determined for the related suicide inhibitors previously studied, the depleting effect of the foreign sterol containing a reactive conjugated diene on ergostanoid biosynthesis was specific to the pathogen while ineffective against the cholesterol biosynthesis pathway in cultured HEK cells, which did accumulate CHT and ERGT and metabolize them to $\Delta^{5,22,24}$-trienol products [28].

Noting the similarity in sterol methylation patterning catalyzed by *Ac* SMT2 and *Tb* SMT1 in generating a $\Delta^{25(27)}$-product (Figure S1) [28,29] and the understanding that *Tb* cell growth is vulnerable to a loss of sterol methylation synthesis, we proceeded to test CHT and ERGT on *Tb* cell growth in culture. Preliminary tests of these compounds yielded potent growth inhibition of bloodstream form (BF) cells in cell culture, relevant to any future studies of their efficacy in animal models of infection. Consistent with the *Ac* studies [28], the BF is highly sensitive to both CHT and ERGT (Figure 2), displaying calculated EC_{50} values of 2.9 $\pm$ 0.2 nM and 0.52 $\pm$ 0.05 μM for CHT and ERGT, respectively, while the EC_{90} for these compounds was 8.8 $\pm$ 0.5 nM and 2 $\pm$ 0.3 μM, respectively. We have shown previously that CHT and ERGT do not have any effect on growth of the human epithelial kidney cells (HEK) in culture when treated with up to 40 μM concentrations [28] of either CHT or ERGT. Consequently, the approximate specificity index of CHT and ERGT calculated against the cell growth of HEK and *Tb*-BF is ~10^3, showing targeted potency toward trypanosomes and not the human host.

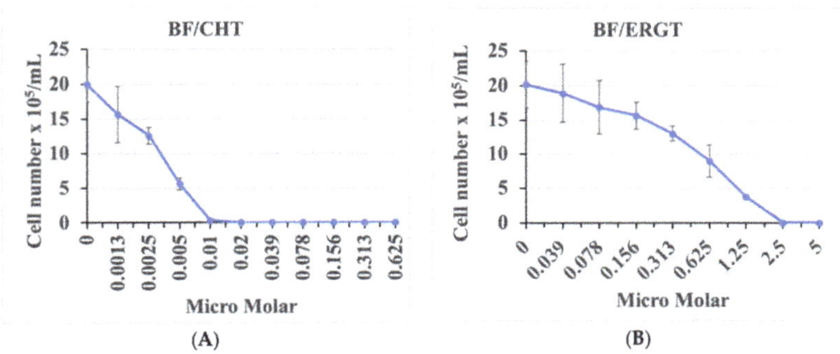

Figure 2. *T. brucei* cell growth at different concentrations of CHT and ERGT. *T. brucei* BF cells were inoculated at a cell density of 10^4/mL in HMI-9 medium containing different concentrations of CHT (0–0.62 μM) (**A**) and ERGT (0–5 μM) (**B**). Cells were counted after 2 days. Cell numbers were plotted against the concentrations of drugs to determine the EC_{50} of each compound for each type of cell. Experiments were done in triplicate to calculate the standard deviations.

2.2. CHT and ERGT Cause Cell Death within Hours of Treatment

Microscopically evident, altered morphology associated with death was detected when the bloodstream form (BF) cells were incubated at EC_{90} of either ERGT or CHT. To investigate this process further, we incubated cells (1×10^6/mL) at $5 \times EC_{50}$ concentrations of CHT and ERGT separately for a shorter period (2–4 h) and observed the cell morphology under the microscope (Figure 3A–C). During these conditions cell viability was minimally affected. Cells treated with vehicle control were compared in parallel. The control cells were intact, fully motile, and alive at both time points (2 and 4 h), as expected (Figure 3A). In contrast, treated cells acted differently. At the 2 h time point, BF cells treated with either CHT or ERGT were mostly intact and motile but some cells (1–2%) tended to circularize (Figure 3B,C, left panel). However, at the 4 h time point almost 90% of cells were circularized, flagella detached, and appeared dead (Figure 3B,C, right panel). Therefore, CHT/ERGT kills *Tb*-BF very rapidly, like *Ac*. The effect of CHT on *Tb*-BF cell morphology was more drastic in comparison to ERGT.

2.3. Antimetabolites Cause Mitochondrial Swelling and Mitophagy

To observe if there were any changes in the intracellular structure of *T. brucei* due to the treatment with antimetabolites, we harvested BF cells after 4 h treatment with a representative antimetabolite, ERGT, at $5 \times EC_{50}$ concentration and prepared them for electron microscopy (EM) (Figure 4A,B). Captured images revealed many vesicular structures inside the treated cells in comparison with control (Figure 4B). We found the nucleus (N) and kinetoplast (K) DNA, the mitochondrial DNA in trypanosomes [37] in the control cells as expected (Figure 4A). The presence of dividing nucleus and extended kinetoplast in the control indicated that the cells were replicating. In contrast, in the ERGT-treated cells we did not identify any dividing nucleus, instead the presence of autophagosomes (Figure 4B) was seen. In contrast to the tubular mitochondria in the control (Figure 4A), mitochondria in the ERGT-treated cells were enlarged without any cristae structures (Figure 4B). These results show a significant alteration in the intracellular structure, including a major change in the mitochondria in the BF that may lead to cell death upon treatment with these antimetabolites.

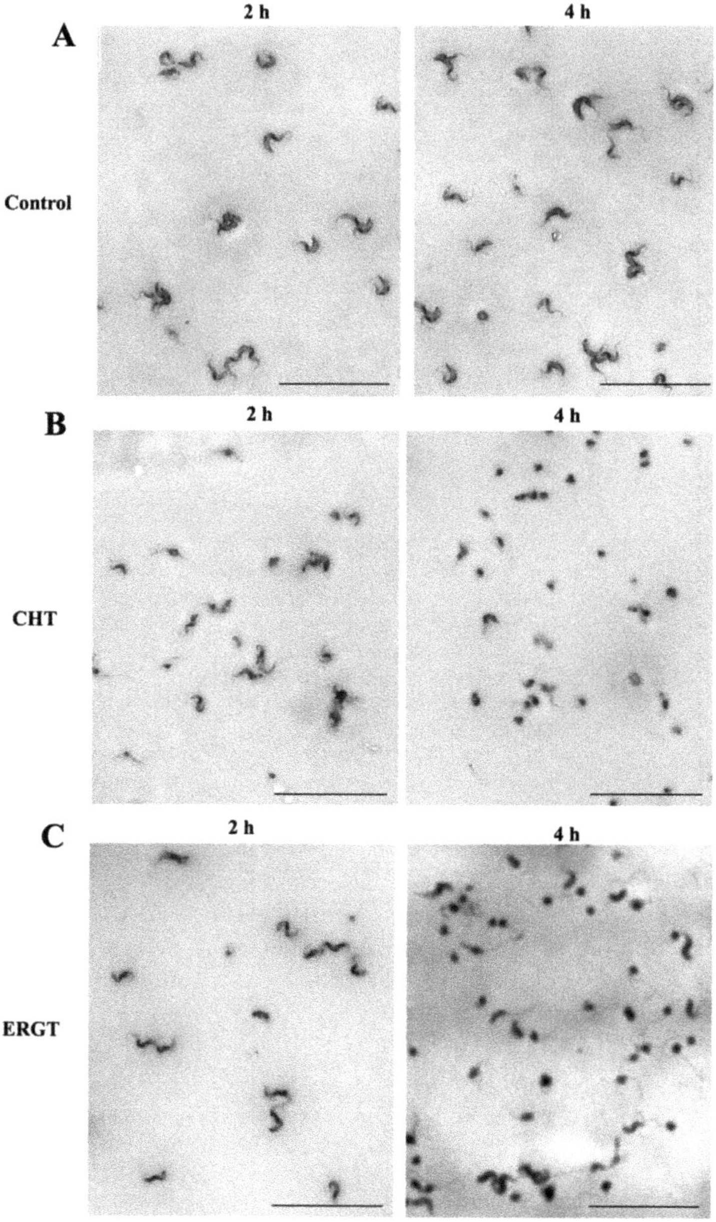

Figure 3. Effect of CHT/ERGT treatment on cell morphology. *T. brucei* BF cells (1×10^6/mL) were incubated with CHT and ERGT at their respective $5 \times EC_{50}$ concentrations. Untreated control cells were used in parallel (**A**). At different time points (2 and 4 h) cells (100 µL of the culture) were harvested, spread on microscope slides, and stained with Giemsa solution. Images were taken with a Keyence phase-contrast microscope using 40× objective. A major change in morphology was observed for cells treated with CHT (**B**) and ERGT (**C**) at 4 h time point. Experiments were repeated three times and multiple areas on the slides were observed for each sample. Scale bars represent 50 µm.

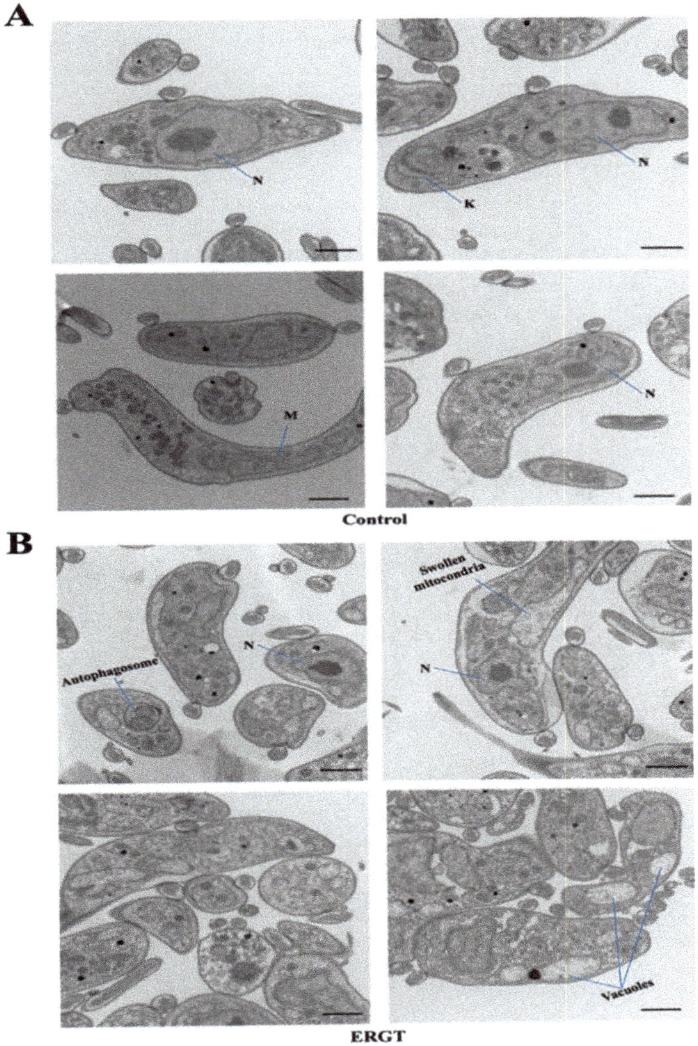

Figure 4. Electron microscopy of the ERGT-treated *T. brucei*. *T. brucei* BF cells were incubated with ERGT at 5 × EC_{50} for 4 h. Cells were harvested, treated with fixative solution, and prepared for electron microscopy as described in the materials and methods. Untreated control cells were run in parallel (**A**). Images from cells treated with ERGT are shown in (**B**). N and K represent nucleus and kinetoplast. M represents mitochondria. Presence of autophagosome, swollen mitochondria, and large vacuoles found in treated cells are indicated. Two individual biological replicates were used for EM and multiple sections from each sample were examined. Scale bars represent 1 μm.

2.4. Washout Experiments of CHT and ERGT Reveal Inhibition of BF Growth and by Analogy Ergosterol Production

We investigated to determine whether washout experiments in the form of an extended duration of action can be observed through growth response. Here, we assumed that covalent adduct of antimetabolite SMT should remain upon seeding treated cells into fresh medium that thereby prevents cell growth, otherwise the steroidal antimetabolite should undergo conversion to product, freeing up SMT for new catalysis that therefore permits a resumption in ergosterol production and cell growth to the control levels. For this purpose, the BF cells were treated with CHT and ERGT at respective 5 × EC_{50} concentrations for 2 h. Cell numbers were not decreased during this short-term treatment and cells were mostly intact and motile as shown in Figure 3B,C (left panels). Cells were harvested, washed to remove any residual compounds and reinoculated in fresh medium with different cell numbers (10^4/mL, 10^3/mL, 10^2/mL, and 10^1/mL) and allowed to grow in appropriate conditions for 4 days, the point when the BF cells reached the stationary phase when inoculated at 10^3/mL (Figure 5A). Once reaching the stationary phase, BF cells die very rapidly due to a quorum-sensing-like phenomenon [38]. Therefore, control cells inoculated at 10^4/mL reached at the maximum level before day 4 and some cells died thereafter (Figure 5A), as expected. In contrast, we found that BF cells treated with either CHT or ERGT had significant growth inhibition in comparison with control after removal of the antimetabolites. Cell growth was almost undetectable till day 4 after washout of the CHT (Figure 5B). This indicates that CHT treatment caused an inherent damage in the BF, which was not recovered even after removal of the antimetabolites. Cells treated with ERGT also showed a lingering growth inhibitory effect (Figure 5C). Cell numbers were below the detection levels up to 3 days, even when inoculated at 10^4/mL or 10^3/mL. After that, cells grew slowly and reached the level of 2 × 10^6/mL at day 4. Therefore, as observed for inhibitor sensitivity, treatment with CHT had longer effects on BF cell growth in comparison with ERGT. These results strongly support our previous observation that these antimetabolites covalently bound and inactivate the target enzyme SMT. Moreover, these different antimetabolite treatment outcomes are to be expected since the ERGT/CHT inhibited growth response is subject to endogenous *Tb* SMT1 acceptance of the fungal sterols. As reported, a first methylation substrate (e.g., CHT) is converted by the *Tb* enzyme about ten times more effectively than a second methylation substrate (e.g., ERGT) [23].

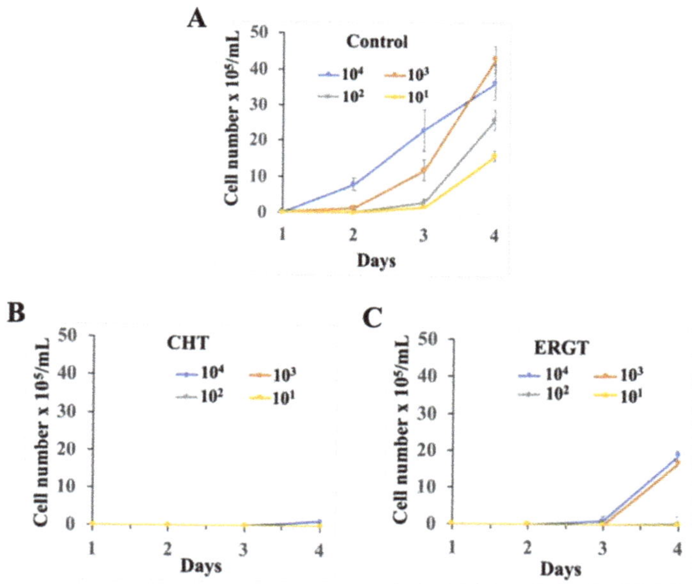

Figure 5. Extended effect of CHT and ERGT on *T. brucei* BF cell growth. The BF cells were pretreated with CHT and EREGT at 5 × EC_{50} concentrations for 2 h. Cells were harvested, washed to remove drugs and reinoculated in fresh medium at different cell numbers (10^4/mL, 10^3/mL, 10^2/mL and 10^1/mL). The untreated control cells were run in parallel starting with different inoculum sizes. Cell number was counted each day for 4 days and plotted against time in culture. (**A**) control, (**B**) and (**C**) are CHT and ERGT treated cells, respectively. Standard errors were calculated from four independent experiments.

2.5. Protective Role of CHT/ERGT on Mice Infected with *T. brucei*

To test the efficacy of CHT and ERGT in *T. brucei* infection, groups of mice were infected with the *T. brucei* 427 strain, which creates an acute form of the disease. Since we had some success with treating mice infected with *Tb* at doses of 5 mg/kg 25-azalanosterol [25], we tested CHT or ERGT at this dosage in the *Tb*-infected mouse with no beneficial results. Consequently, we then tested a higher level of steroidal antimetabolite to outcompete the circulating cholesterol for lipoprotein binding necessary for uptake by the *Tb* bloodstream form [35]. For these studies, mice were divided into six groups (five mice in each group). Starting at the same day of infection, group-I and -II mice were treated with ERGT (50 mg/kg and 100 mg/kg body weight, respectively) each day. The group-III and -IV infected mice were similarly treated with CHT (50 mg/kg and 100 mg/kg body weight, respectively) each day and the group-V and -VI were vehicle control for 50 mg/kg and 100 mg/kg dosages, respectively. As expected, the control infected mice had all died by 5–6 days post-infection, whereas the treated mice lived longer. In further detail, three of the five ERGT-treated (50 mg/kg body weight) mice died on day 8 and rest had died by 10 days post-infection (DPI) (Figure 6A). We didn't observe any better results when increasing the dose to 100 mg/kg/day (Figure 6B). The results show that like in in vitro data, CHT has a slightly higher trypanocidal efficiency than ERGT, consistent with their substrate recognition. However, doubling the dosages didn't show much additive effect. This is possibly because the compounds are metabolized at the sterol core structure [28] by the host. Counting the parasite numbers in the blood of the infected mice revealed that the parasite numbers were below the detection levels (~10^3/mL) in the first 2 days of post-infection, increased exponentially after that period, and reached maximum by day 5 in the control groups (Figure 6C,D), which are the characteristics of the acute form of the disease in laboratory animals [39,40]. In contrast to the control, we couldn't detect any parasites in the blood till 6 days post-infection, when treated with ERGT (50 mg/kg/day) (Figure 6C). The blood parasitemia reached maximum levels at different time points and most of the mice stayed alive till day 8. Blood parasite counts were below the detection level up to day 7 in the group of mice treated with CHT (50 mg/kg body weight/day); however, after that it increased exponentially (Figure 6C). Increasing the doses of ERGT or CHT (100 mg/kg/day) didn't show any further delay in the appearance of the parasite number in the blood above the detection limit (Figure 6D). Neither did it take any longer time to reach lethal levels. Overall, our studies showed that ERGT/CHT protected mice infected with *T. brucei* by doubling their survival time, which is equivalent to 100% increase in life expectancy, following daily treatment for 8–10 days at 50 mg/kg.

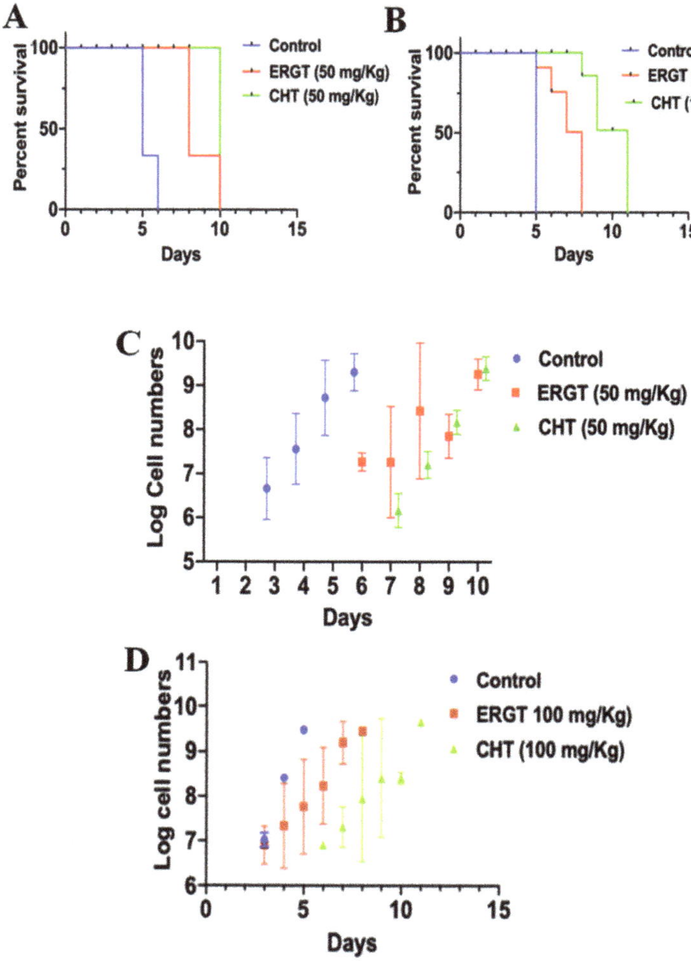

Figure 6. Protective role of CHT/ERGT on animal infection with *T. brucei*. Six groups of Balb/C mice (5 mice in each group) were injected intraperitoneally with *T. brucei* bloodstream form parasite (1×10^4 cells/mouse). Two groups were treated with ERGT (50 mg/kg) or CHT (50 mg/kg) via intraperitoneal injection once per day post-infection (**A**). Another two groups of mice were treated with ERGT 100 mg/kg) or CHT (100 mg/kg) (**B**) via the same route once per day post-infection. Controls for these groups were injected with appropriate amount of buffer used to prepare the antimetabolites. The blood parasitemia levels were monitored each day post-infection by counting the number of *T. brucei* in blood collected from the tail vein. Survival curves for the control (infected but not treated with any drug) and infected mice treated with ERGT were plotted using Graphpad Prism. To reduce the pain and distress mice were sacrificed when the parasitemia levels reached $>5 \times 10^8$ cells/mL of blood and death is considered one day after euthanasia. (**C**,**D**) Log of parasite numbers in the blood of the infected mice are plotted against days post infection. Standard deviations were calculated from 5 counts from each mouse on each day. Parasite numbers on day 1 and 2 post-infection were below the detection limit.

3. Discussion

Fungi, plants, and protozoan species use SMT for generation of a vast array of 24-alkyl sterol structures necessary for cell growth and development [41,42]. The primary sequences and tetrameric subunit organization of a wide range of these catalysts reflect a common active-site topography that can accept a wide range of structurally similar substrates. However, under selection pressure, this enzyme evolved to accept alternate substrate side chain groups enabling species-specific product formation and an SMT-specific dead-end complex [43–45]. As determined, it is straight-forward to detect the isolated CHT-SMT complex through characterization of the C_{28}-diol product by GC-MS/^{1}H-NMR analysis, as reported for studies on *Tb* SMT1 and *Ac* SMT2 (Figure S1), [20,28,29]. Importantly, the C_{28}-diol product is not a biosynthesis product but a saponification product; its identification serves as a chemical signature for the compound in its biosynthetic state as a C_{28}-intermediate-SMT adduct, reported first in studies on the yeast SMT incubated with the suicide substrate 26,27-dehydrozymosterol [30,31], then for the natural plant substrate 24-methyl cycloartenol tested against the soybean SMT1 [44,45]. An issue now is whether fungal sterols (such as CHT), which can bind irreversibly to *Tb* SMT [29] could behave as growth inhibitors against the kinetoplastid parasite in vitro and in vivo.

In our testing of CHT and ERGT against *T. brucei* we observed that they are trypanocidal at nanomolar concentrations, consistent with their predicted suicide inhibitor/antimetabolite effect on ergosterol biosynthesis. Consequently, we desired to explore their efficacy in a mouse model of infection. Previous efforts to identify more selective irreversible inhibitors for this enzyme continued and led to the discovery of pro-drugs and their associated zymosterol derivatives [27,34,44] that can serve as suicide inhibitors of ergosterol biosynthesis [33,46]. The EC_{50} of these compounds were in the range of 10–20 μM with specificity indexes 3–6 [34,46]. In contrast to synthetic 26,27-dehydrolanosterol or 26-fluorolanosterol [34,46], the EC_{50}s of natural CHT and ERGT against the *T. brucei* bloodstream form reported here are 2.9 ± 0.2 nM and 0.52 ± 0.05 μM, which are much lower than the previously identified suicide inhibitors. CHT is more potent than ERGT. This is possibly because of differences in productive binding to the *Tb* SMT, which could be due to the distinct binding sites for the two inhibitors and/or affected by the expression level of *Tb* SMT, sensitive to pressures of lipoprotein cholesterol. In any case, the selectivity index of these compounds for the BF was about three orders of magnitude as determined in vitro, showing that the compound is non-toxic to the parasite host. In addition, these antimetabolites showed a long-lasting effect for the bloodstream form, showing irreversible inactivation of the target enzyme, 24-SMT.

In animal models for *T. brucei* acute infection, CHT/ERGT showed promising results. CHT/ERGT-treated mice survived longer than the untreated control group. Although the parasite counts in the blood at the initial 5–6 days of infection were below the detection levels, once they cross this threshold they increase exponentially. There are multiple possibilities for these results. (1) *T. brucei* could be adapted by changing the means of drug uptake/efflux at the latter days of treatment or became resistant by other means. (2) It is also possible that the parasites were cleared from the bloodstream during initial treatment; however, a population that hides inside tissues (adipose tissue or brain) escaped, because drug concentration could be below the cidal levels in these tissues and that creates a space for their adaptation. These adapted parasites emerged in larger numbers at later time points. (3) The third possibility is that the antimetabolites are metabolized in the host and that metabolism was induced with a longer treatment period, thus the serum levels of these compounds dropped at the latter days and parasite population increased exponentially. We didn't find any toxicity of CHT/ERGT at the dosage of treatment. We also didn't observe any better effect by increasing the dose from 50 to 100 mg/kg/day via intraperitoneal injection. It was also not recommended for more frequent administration of these inhibitors due to the smaller size of the animals and volumes required to administer. However, these results warrant further trials with larger animals and more frequent administration of these effective compounds and perhaps in combination therapy with conventional antitrypanosomal drugs. Overall, we found that CHT and ERGT are promising candidates to develop drugs targeting 24-SMT in both *A. castelanii* and *T. brucei* that cause deadly diseases in humans.

4. Material and Methods

4.1. Tb Cell Culture

T. brucei strain Lister 427 BF cells (ATCC NR-42009) were cultured in HMI-9 medium supplemented with 10% heat-inactivated fetal bovine serum (Bio-Techne, Minneapolis, MN, USA) and 10% Serum Plus (Thermo Fisher Scientific, Waltham, MA, USA) in a CO_2 incubator (5% saturation) at 37 °C as described [47]. To measure cell growth, BF cells were inoculated at 10^4/mL or as indicated, the parasite numbers were counted in a hemocytometer chamber using a phase-contrast microscope and plotted versus time of incubation.

4.2. Preparation of ERGT and CHT for Treatment and Sterol Analytics

Cholesta-5,7,22,24-tetraenol (CHT) and ergosta-5,7,22,24(28)-tetraenol (ERGT) were purified from the yeast mutants ERG6 and ERG5, respectively [28]. Stock solutions of ERGT and CHT were made in DMSO at a concentration of 10 mM, diluted accordingly before addition to the culture medium. For in vivo experiments, the stock solution of ERGT and CHT was made in DMSO at 100 mg/mL. At the time of treatment stock solution was diluted 25-fold in 1X PBS and injected intraperitoneally at the dose of 50 mg/kg or 100 mg/kg body weight for each mouse.

4.3. Inhibition Studies in Culture

Determination of EC_{50}/EC_{90} and wash out experiments to assess the effect of sterol biosynthesis inhibitor on cell growth in vitro were performed against cultured *T. brucei* cells as described previously [24,28]. Briefly, for determination of EC_{50}/EC_{90}, BF cells from a logarithmic phase culture were inoculated at 10^4/mL in 24-well plates. CHT/ERGT were serially diluted in medium from the stock solution as described above. Triplicate wells were set up for each concentration. Cells were counted each day for 2 days from each well twice and plotted against time with calculated standard deviations. For washout studies, the BF cells were pretreated with CHT and ERGT at $5 \times EC_{50}$ concentrations for 2 h. Cells were harvested, washed to remove drugs, and reinoculated in fresh medium at different cell numbers (10^4/mL, 10^3/mL, 10^2/mL and 10^1/mL). The untreated control cells were run in parallel starting with different inoculum sizes. Cell numbers were counted each day for 4 days and plotted against time in culture.

4.4. Giemsa Staining

T. brucei BF was grown in appropriate medium in the presence or absence of the drugs. At different time points cells were harvested, resuspended in fresh medium at a concentration of 10^5–10^6/mL, spread on a slide and left to air dry. Slides were flooded with 10% Giemsa stain solution and kept at room temperature for 20–30 min. Slides were washed with water to remove the excess dye and allowed to dry. Image was taken with a Keyence phase-contrast microscope using 40× objective

4.5. Electron Microscopy

Cells were fixed in 2% (*v*/*v*) glutaraldehyde and 2% (*w*/*v*) paraformaldehyde in 0.1 M sodium cacodylate buffer (SCB), pH 7.2 [48]. Cells were then washed with SCB, post-fixed with 1% osmium tetroxide in SCB, stained with 0.5% aqueous magnesium uranyl acetate, dehydrated, and embedded in Spurr's resin. Blocks were sectioned at 50–70 nm thickness and stained with 5% (*w*/*v*) uranyl acetate in 1% acetic acid and 0.4% (*w*/*v*) lead citrate in 0.1 N NaOH. Section grids were inserted into FEI CM12 twin lens 420 transmission electron microscopes (FEI) to capture images.

4.6. Animal Experiments

In the model of acute infection (Haubrich et al., 2015), female Balb/C mice (6–8 weeks old) for *T. brucei* infection were purchased from the Envigo laboratory, weight 25 g on average: 5 mice/group. Animals were allowed to be acclimated for 3–5 days in the Animal Care facility in Meharry Medical College. For *T. brucei* infection, BF cells were harvested from exponentially growing cultures and resuspended in 1X PBSG (phosphate buffered saline containing 5 mM glucose) at a concentration of 2×10^4/mL. Each mouse was injected intraperitoneally with 0.5 mL of cell suspension (1×10^4 cells). In these conditions, parasitemia against Tb reaches maximum of 1×10^9 cells on day 6 after infection. Mice were treated separately with ERGT and CHT (50 or 100 mg/kg body weight/mice/day) via intraperitoneal injection starting at the day of infection. The control mouse group received only vehicle. Parasitemia was monitored daily by counting the parasite numbers in blood (3–5 µL) collected by tail snipping. To reduce the pain and distress and to set up a humane end-point, mice were sacrificed when the parasitemia levels reached >5×10^8 cells/mL of blood and death is considered one day after euthanasia. All mice were closely monitored for any distress or pain by periodical assessment of body weight, food intake, hair coat, and activities in accordance with the animal protocol guidelines of the Meharry Medical College. Mice under severe distress were sacrificed after consultation with the Animal Care Facility Veterinarians. All studies were conducted in accordance with National Institutes of Health guidelines for the use of experimental animals, and the protocols were approved by the Meharry Institutional Animal Care and Use Committee (protocol number: 141017MC172, Animal Welfare assurance number: A3420-01)

5. Conclusions

We identified two natural byproducts of yeast sterol metabolism with potent anti-trypanocidal property but non-toxic to human cells. We believe these studies bridged, for the first time, the natural products involving sterols other than cholesterol and their chemical biology with the health sciences. We showed that CHT/ERGT, which can act as substrates of *Tb* SMT, and form an adduct with the enzyme during catalysis, can carry out their inhibitory effects on ergosterol biosynthesis with lethal consequences. Importantly, these antimetabolites work at nanomolar concentrations in vitro to inhibit cell growth. These products are also effective in vivo in a mouse model of *T. brucei* acute infection, doubling the survival time of the animals. Therefore, these potent trypanocidals deserve further investigation to optimize their treatment doses and pharmacokinetics to develop novel chemotherapies for AT.

Supplementary Materials: The following supporting information can be downloaded at: https://www.mdpi.com/article/10.3390/molecules27134088/s1, Figure S1: Sterol methylation steps catalyzed by SMT1 (first C24-methylation step) and SMT2 (second C24-methylation step) enzymes that occur across kingdoms after recognition of their natural or substrate analogue.

Author Contributions: Conceptualization; W.D.N. and M.C.; Data curation; M.C., U.K.S., A.T. and B.H.V.; Formal analysis; M.C., U.K.S., A.T., B.H.V. and W.D.N.; Investigation; M.C., U.K.S., A.T., B.H.V. and W.D.N.; Methodology; M.C., U.K.S., A.T., B.H.V. and W.D.N.; Software; M.C., U.K.S., A.T., B.H.V. and W.D.N.; Supervision; M.C. and W.D.N.; Visualization; M.C., U.K.S., A.T., B.H.V. and W.D.N.; Writing—original draft preparation; M.C.; Writing—review and editing; M.C. and W.D.N. All authors have read and agreed to the published version of the manuscript.

Funding: This work was supported by grants from the National Institute of Health (R21/R33 AI 119782) to W.D.N. and M.C. The electron microscopy core at Vanderbilt University Medical Center is supported by NIH grants P30DK058404 and P30CA068485. The content is solely the responsibility of the authors and does not necessarily represent the official views of the National Institutes of Health.

Institutional Review Board Statement: Institutional Animal Care and Use Committee, Meharry Medical College. Protocol #141017MC172, 4/29/21.

Informed Consent Statement: Not applicable.

Data Availability Statement: Not applicable.

Acknowledgments: We acknowledge the veterinarian and technicians of the Animal Care Facility in Meharry Medical for their support in animal work and Tanusree Singha for laboratory assistance.

Conflicts of Interest: The authors declare no conflict of interest.

Sample Availability: Samples of the compounds are not available from the authors.

References

1. Sternberg, J.M.; MacLean, L. A spectrum of disease in Human African trypanosomiasis: The host and parasite genetics of virulence. *Parasitology* **2010**, *137*, 2007–2015. [CrossRef] [PubMed]
2. Silvester, E.; McWilliam, K.; Matthews, K. The Cytological Events and Molecular Control of Life Cycle Development of *Trypanosoma brucei* in the Mammalian Bloodstream. *Pathogens* **2017**, *6*, 29. [CrossRef] [PubMed]
3. Sekhar, G.N.; Watson, C.P.; Fidanboylu, M.; Sanderson, L.; Thomas, S.A. Delivery of antihuman African trypanosomiasis drugs across the blood-brain and blood-CSF barriers. *Adv. Pharmacol.* **2014**, *71*, 245–275. [PubMed]
4. Torreele, E.; Bourdin Trunz, B.; Tweats, D.; Kaiser, M.; Brun, R.; Mazué, G.; Bray, M.A.; Pécoul, B. Fexinidazole—A new oral nitroimidazole drug candidate entering clinical development for the treatment of sleeping sickness. *PLoS Negl. Trop. Dis.* **2010**, *4*, e923. [CrossRef]
5. Jacobs, R.T.; Nare, B.; Wring, S.A.; Orr, M.D.; Chen, D.; Sligar, J.M.; Jenks, M.X.; Noe, R.A.; Bowling, T.S.; Mercer, L.T.; et al. SCYX-7158, an orally-active benzoxaborole for the treatment of stage 2 human African trypanosomiasis. *PLoS Negl. Trop. Dis.* **2011**, *5*, e1151. [CrossRef]
6. Wyllie, S.; Bernardo, J.F.; Kelner, A.; Sokolova, A.Y.; Berriman, M.; Fairlamb, A.H. Nitroheterocyclic drug resistance mechanisms in *Trypanosoma brucei*. *J. Antimicrob. Chemother.* **2016**, *71*, 625–634. [CrossRef]
7. Atanasov, A.G.; Zotchev, S.B.; Dirsch, V.M.; The International Natural Products Sciences Taskforce; Supuran, C.T. Natural products in drug discovery: Advances and opportunities. *Nat. Rev. Drug Discov.* **2021**, *28*, 1–17. [CrossRef]
8. Cockram, P.E.; Smith, T.K. Active natural product scaffolds against trypanosomatid parasites: A review. *J. Nat. Prod.* **2018**, *81*, 2138–2154. [CrossRef]
9. Heby, O.; Roberts, S.C.; Ullman, B. Polyamine biosynthetic enzymes as drug targets in parasitic protozoa. *Biochem. Soc. Trans.* **2003**, *31*, 7719–7726. [CrossRef]
10. Roberts, C.W.R.; McCleod, R.; Rice, D.W.; Ginger, M.; Chance, M.L.; Goad, L.J. Fatty acid and sterol metabolism: Potential antimicrobial targets in apicomplexan and trypanosomatid parasitic protozoa. *Mol. Biochem. Pharmacol.* **2003**, *126*, 129–142. [CrossRef]
11. Schimmel, K.J.; Gelderblom, H.; Guchelaar, H.J. Cyclopentenyl cytosine (CPEC): An overview of its in vitro and in vivo activity. *Curr. Cancer Drug Targets* **2007**, *7*, 504–509. [CrossRef]
12. Peters, G.J. Novel developments in the use of antimetabolites. *Nucleosides Nucleotides Nucleic Acids* **2014**, *33*, 358–374. [CrossRef]
13. Ray, S.; Murkin, A.S. New electrophiles and strategies for mechanism-based and targeted covalent inhibitor design. *Biochemistry* **2019**, *58*, 5234–5244. [CrossRef]
14. Azijli, K.; van Roosmalen, I.A.; Smit, J.; Pillai, S.; Fukushima, M.; de Jong, S.; Peters, G.J.; Bijnsdorp, I.V.; Kruyt, F.A. The novel thymidylate synthase inhibitor trifluorothymidine (TFT) and TRAIL synergistically eradicate non-small cell lung cancer cells. *Cancer Chemother. Pharmacol.* **2014**, *73*, 1273–1283. [CrossRef]
15. Chavoushi, S.F.; Jharap, B.; Friedrich, P.; Smid, K.; Peters, G.J.; Malingré, M. Thiopurines with low-dose allopurinol (ThiLDA)- a prospective clinical one-way crossover trial. *Eur. J. Clin. Pharmacol.* **2019**, *75*, 1669–1674. [CrossRef]
16. Balboni, B.; El Hassouni, B.; Honeywell, R.J.; Sarkisjan, D.; Giovannetti, E.; Poore, J.; Heaton, C.; Peterson, C.; Benaim, E.; Lee, Y.B.; et al. RX-3117 (fluorocyclopentenyl cytosine): A novel specific antimetabolite for selective cancer treatment. *Expert Opin. Investig. Drugs* **2019**, *28*, 311–322. [CrossRef]
17. Halford, B. Covalent drugs go from fringe to fashionable endeavor. *Chem. Eng. News* **2020**, *98*, 28–32. [CrossRef]
18. World Health Organization. World Health Organization Model List of Essential Medicines: 21st List 2019. Geneva: World Health Organization. hdl: 10665/3225771. License: CC BY-NC-SA 3.0 IGO. Available online: https://hdl.handle.net/10665%2F325771 (accessed on 21 September 2020).
19. Villalta, F.; Dobish, M.C.; Nde, P.N.; Kleshchenko, Y.Y.; Hargrove, T.Y.; Johnson, C.A.; Waterman, M.R.; Johnston, J.N.; Lepesheva, G.I. VNI cures acute and chronic experimental Chagas disease. *J. Infect. Dis.* **2013**, *208*, 504–511. [CrossRef]
20. Liu, J.; Nes, W.D. Steroidal triterpene: Design of substrate-based inhibitors of ergosterol and sitosterol synthesis. *Molecules* **2009**, *14*, 4690–4706. [CrossRef]
21. De Macedo-Silva, S.T.; de Souza, W.; Rodrigues, J.C.F. Sterol biosynthesis pathway as an alternative for the anti-protozoan parasite chemotherapy. *Cur. Med. Chem.* **2015**, *22*, 2186–2198. [CrossRef]
22. Zhou, W.; Warrilow, A.G.S.; Thomas, C.D.; Ramos, E.; Parker, J.E.; Price, C.L.; Vanderloop, B.H.; Fisher, P.M.; Loftis, M.D.; Kelly, D.E.; et al. Functional importance for developmental regulation of sterol biosynthesis in *Acanthameoba castelanii*. *Biochim. Biophys. Acta. Mol. Cell. Biol. Lipids.* **2018**, *1863*, 1164–1178. [CrossRef]
23. Zhou, W.; Lepesheva, G.I.; Waterman, M.R.; Nes, W.D. Mechanistic analysis of a multiple product sterol methyltransferase implicated in ergosterol biosynthesis in *Trypanosoma brucei*. *J. Biol. Chem.* **2006**, *281*, 6290–6296. [CrossRef]
24. Nes, C.R.; Singha, U.K.; Liu, J.; Ganapathy, K.; Villalta, F.; Waterman, M.R.; Lepesheva, G.I.; Chaudhuri, M.; Nes, W.D. Novel sterol metabolic network of *Trypanosoma brucei* procyclic and bloodstream forms. *Biochem. J.* **2012**, *443*, 267–277. [CrossRef]
25. Haubrich, B.A.; Singha, U.K.; Miller, M.B.; Nes, C.R.; Anyatonwu, H.; Lecordier, L.; Patkar, P.; Leaver, D.J.; Villalta, F.; Vanhollebeke, B.; et al. Discovery of an ergosterol-signaling factor that regulates *Trypanosoma brucei* growth. *J. Lipid Res.* **2015**, *56*, 331–341. [CrossRef]

26. Kidane, M.E.; Vanderloop, B.H.; Zhou, W.; Thomas, C.D.; Ramos, E.; Singha, U.; Chaudhuri, M.; Nes, W.D. Sterol methyltransferase a target for anti-ameoba therapy: Towards transition state analog and suicide substrate drug design. *J. Lipid Res.* **2017**, *58*, 2310–2323. [CrossRef]
27. Nes, W.D. Biosynthesis of cholesterol and other sterols. *Chem. Rev.* **2011**, *111*, 6423–6451. [CrossRef]
28. Zhou, W.; Ramos, E.; Zhu, X.; Fisher, P.M.; Kidane, M.E.; Vanderloop, B.H.; Thomas, C.D.; Yan, J.; Singha, U.; Chaudhuri, M.; et al. Steroidal antibiotics are antimetabolites of *Acanthameoba* steroidogenesis with phylogenetic implications. *J. Lipid Res.* **2019**, *60*, 981–994. [CrossRef]
29. Liu, J.; Ganapathy, K.; Wyyial, E.; Buknicki, J.M.; Nwogwugwu, C.A.; Nes, W.D. Effect of substrate features and mutagenesis of the active site tyrosine residues on the reaction course catalyzed by *Trypanosma brucei* sterol C-24-methyltransferase. *Biochem. J.* **2011**, *439*, 413–422. [CrossRef]
30. Nes, W.D.; Marshall, J.A.; Jia, Z.; Jaradat, T.T.; Song, Z.; Jaysimha, P. Active site mapping and substrate channeling in the sterol methyltransferase pathway. *J. Biol. Chem.* **2002**, *277*, 42459–42565. [CrossRef]
31. Soape, M. Protein Chemistry, Peptide Mapping, and Preliminary Structural Characterization of *Saccharomyces cerevisiae* Sterol C-24-Methyltransferase Expressed in *Escherichia coli*. Master's Thesis, Texas Tech University, Lubbock, TX, USA, 2006; pp. 1–106.
32. Magaraci, F.; Jimenez, C.J.; Rodrigues, C.; Rodrigues, J.C.; Braga, M.V.; Yardley, V.; de Luca-Fradley, K.; Croft, S.L.; de Souza, W.; Ruiz-Perez, L.M.; et al. Azasterol as inhibitors of sterol 24-methyltransferase in Leishmania species and *Trypanosoma cruzi*. *J. Med. Chem.* **2003**, *46*, 4714–4727. [CrossRef]
33. Zhou, W.; Cross, A.M.; Nes, W.D. Cholesterol import fails to prevent catalyst-based inhibition of ergosterol synthesis and cell proliferation of *Trypanosoma brucei*. *J. Lipid Res.* **2007**, *48*, 665–673. [CrossRef] [PubMed]
34. Leaver, D.J.; Patkar, P.; Singha, U.K.; Miller, M.B.; Haubrich, B.A.; Chaudhuri, M.; Nes, W.D. Fluorinated sterols are suicide inhibitors of ergosterol biosynthesis and growth in *Trypanosoma brucei*. *Chem. Biol.* **2015**, *22*, 1374–1383. [CrossRef] [PubMed]
35. Coppens, I.; Courtoy, P.J. Host plasma low density lipoprotein particles as an essential source of lipids for the bloodstream forms of *Trypanosoma brucei*. *J. Biol. Chem.* **1995**, *270*, 5736–5741. [CrossRef] [PubMed]
36. Urbina, J.A.; Vivas, J.; Lazardi, K.; Molina, J.; Payares, G.; Piras, M.M.; Piras, R. Antiproliferative effects of delta24(25) sterol methyl transferase inhibitors on *Trypanosoma (Schizotrypanum) cruzi*: In vitro and in vivo studies. *Chemotherapy* **1996**, *42*, 294–307. [CrossRef]
37. Mensa-Wilmot, K.; Hoffman, B.; Wiedeman, J.; Sullenberger, C.; Sharma, A. Kinetoplast Division Factors in a Trypanosome. *Trends Parasitol.* **2019**, *35*, 119–128. [CrossRef]
38. Matthews, K.R. Trypanosome signaling-quorum sensing. *Ann. Rev. Microbiol.* **2021**, *75*, 495–514. [CrossRef]
39. Bitoni, A.J.; Dumont, J.A.; McCann, P.P. Characterization of *Trypanosoma brucei brucei* S-adenosyl-L-methionine decarboxylase and its inhibition by berenil, pentamidine and methylglyoxal (guanyhydrazone). *Biochem. J.* **1986**, *237*, 518–521.
40. Kipkorir, L.W.; John, T.K.; Owino, O.B.; John, O.; Robert, S.; Daniel, M.; Owino, A.V. Mouse experiments demonstrate differential pathogenicity and virulence of *Trypanosoma brucei rhodesiense* strains. *Exp. Parasitol.* **2021**, *228*, 108135. [CrossRef]
41. Darnet, S.; Blary, A.; Chevalier, Q.; Schaller, H. Phytosterol profiles, genomes and enzymes—An overview. *Front. Plant Sci.* **2021**, *12*, 1–31. [CrossRef]
42. Nes, W.D. Control of Sterol biosynthesis and its importance to developmental regulation. *Rec.Adv. Phytochem.* **1990**, *24*, 283–327.
43. Neelakandan, A.K.; Song, Z.; Wang, J.; Richards, M.H.; Wu, X.; Valliyodan, B.; Nguyen, H.T.; Nes, W.D. Cloning, functional expression and phylogenetic analysis of sterol C24-methyltransferases involved in sitosterol biosynthesis. *Phytochemistry* **2009**, *70*, 1982–1998. [CrossRef]
44. Nes, W.D.; Song, A.L.; Dennis, W.; Zhou, W.; Nam, J.; Miller, M.B. Biosynthesis of phytosterols: Kinetic mechanism for the enzymatic C-methylation of sterols. *J. Biol. Chem.* **2003**, *278*, 34505–34516. [CrossRef]
45. Wang, J.; Nes, W.D. Cyclobranol: A substrate for C25-methyl sterol side chains and potent mechanistic-based inactivator of plant sterol methyltransferase. *Bioorg. Med. Chem. Lett.* **2008**, *18*, 3878–3881. [CrossRef]
46. Miller, M.B.; Patkar, P.; Singha, U.K.; Chaudhuri, M.; Nes, W.D. 24-Methylenecyclopropane steroidal inhibitors: A Trojan horse in ergosterol biosynthesis that prevents growth of *Trypanosoma brucei*. *Biochim. Biophys. Acta. Mol. Cell. Biol. Lipids* **2017**, *1862*, 305–313. [CrossRef]
47. Hirumi, H.; Hirumi, K. In vitro cultivation of *Trypanosoma congolense* bloodstream forms in the absence of feeder cell layers. *Parasitology* **1991**, *112*, 225. [CrossRef]
48. Nebesarova, J.; Hozák, P.; Frank, L.; Štěpan, P.; Vancová, M. The cutting of ultrathin sections with the thickness less than 20 nm from biological specimens embedded in resin blocks. *Microsc. Res. Tech.* **2016**, *79*, 512–517. [CrossRef]

molecules

MDPI

Article

Anti-Obesity and Anti-Adipogenic Effects of Administration of Arginyl-Fructose-Enriched Jeju Barley (*Hordeum vulgare* L.) Extract in C57BL/6 Mice and in 3T3-L1 Preadipocytes Models

Soo-Young Lee [1], Tae-Yang Kim [1], Ji-Yoon Hong [1], Gi-Jung Kim [1], Jung-Bae Oh [2], Min-Joo Kim [1], Emmanouil Apostolidis [3], Jung-Yun Lee [3,*] and Young-In Kwon [1,*]

1 Department of Food and Nutrition, Hannam University, Daejeon 34054, Korea; dltndud1221@naver.com (S.-Y.L.); xodid5606@naver.com (T.-Y.K.); ysmieee@naver.com (J.-Y.H.); homina97@daum.net (G.-J.K.); minjookim@hnu.kr (M.-J.K.)
2 Institute of Functional Foods, Kunpoong Bio Co., Ltd., Jeju 63010, Korea; denisoh89@gmail.com
3 Department of Chemistry and Food Science, Framingham State University, Framingham, MA 01701, USA; eapostolidis@framingham.edu
* Correspondence: seembeeks@hanmail.net (J.-Y.L.); youngk@hnu.kr (Y.-I.K.); Tel.: +1-508-626-4777 (J.-Y.L.); +82-42-629-8790 (Y.-I.K.)

Abstract: In our previous study, we reported that arginyl-fructose (AF), one of the Amadori rearrangement compounds (ARCs) produced by the heat processing of Korean ginseng can reduce carbohydrate absorption by inhibiting intestinal carbohydrate hydrolyzing enzymes in both in vitro and in vivo animal models. This reduced absorption of carbohydrate might be helpful to control body weight gain due to excessive carbohydrate consumption and support induced calorie restriction. However, the weight management effect, except for the effect due to anti-hyperglycemic action, along with the potential mechanism of action have not yet been determined. Therefore, the efforts of this study are to investigate and understand the possible weight management effect and mechanism action of AF-enriched barley extracts (BEE). More specifically, the effect of BEE on lipid accumulation and adipogenic gene expression, body weight gain, body weight, plasma lipids, body fat mass, and lipid deposition were evaluated using C57BL/6 mice and 3T3-L1 preadipocytes models. The formation of lipid droplets in the 3T3-L1 treated with BEE (500 and 750 μg/mL) was significantly blocked ($p < 0.05$ and $p < 0.01$, respectively). Male C57BL/6 mice were fed a high-fat diet (30% fat) for 8 weeks with BEE (0.3 g/kg-body weight). Compared to the high fat diet control (HFD) group, the cells treated with BEE significantly decreased in intracellular lipid accumulation with concomitant decreases in the expression of key transcription factors, peroxisome proliferator-activated receptor gamma (PPARγ), CCAAT/enhancer-binding protein alpha (CEBP/α), the mRNA expression of downstream lipogenic target genes such as fatty acid binding protein 4 (FABP4), fatty acid synthase (FAS), and sterol regulatory element-binding protein 1c (SREBP-1c). Supplementation of BEE effectively lowered the body weight gain, visceral fat accumulation, and plasma lipid concentrations. Compared to the HFD group, BEE significantly suppressed body weight gain (16.06 ± 2.44 g vs. 9.40 ± 1.39 g, $p < 0.01$) and increased serum adiponectin levels, significantly, 1.6-folder higher than the control group. These results indicate that AF-enriched barley extracts may prevent diet-induced weight gain and the anti-obesity effect is mediated in part by inhibiting adipogenesis and increasing adiponectin level.

Keywords: Amadori rearrangement compounds; barley; anti-obesity; arginyl-fructose; adipogenic

Citation: Lee, S.-Y.; Kim, T.-Y.; Hong, J.-Y.; Kim, G.-J.; Oh, J.-B.; Kim, M.-J.; Apostolidis, E.; Lee, J.-Y.; Kwon, Y.-I. Anti-Obesity and Anti-Adipogenic Effects of Administration of Arginyl -Fructose-Enriched Jeju Barley (*Hordeum vulgare* L.) Extract in C57BL/6 Mice and in 3T3-L1 Preadipocytes Models. *Molecules* **2022**, 27, 3248. https://doi.org/10.3390/molecules27103248

Academic Editors: Sokcheon Pak and Soo Liang Ooi

Received: 13 April 2022
Accepted: 17 May 2022
Published: 19 May 2022

1. Introduction

The energy imbalance resulting from excessive food intake and increased caloric intake results in obesity, a serious health concern that can lead to increased incidence of chronic diseases such as coronary heart disease, diabetes, nonalcoholic fatty liver disease (NAFLD), and cancer [1]. Obesity can be characterized by weight gain resulting from excessive body fat accumulation. Genetic and environmental factors may induce weight gain, but it has been clearly defined that calorie intake is a predominant factor. Diet-induced obesity (DIO) animal models are commonly used to study obesity since these models reflect the state of human obesity much better than genetically modified models [2]. These models are based on the fact that a high fat diet increased weight gain and initiated the diabetic stage in mice and rats [3,4].

Adipogenesis is a process in which preadipocytes are hyperplastically transformed into adipocytes, resulting in hypertrophy and dysfunctional adipocytes due to the excessive storage of lipids as triglycerides through increased adipogenesis. It has been reported that adipogenesis and adipogenesis are initiated by the expression of differentiation-related transcription factors when preadipocytes are exposed to adipogenic inducers. Recent evidence suggests that dietary factors containing several bioactive compounds can effectively change adipocyte number as well as adipocyte size by regulating the expression of these adipocyte differentiation-related transcription factors [5,6], potentially resulting in fewer adipocytes. Therefore, the regulation of key transcription factors such as PPARγ and C/EBPα in the early stage of adipogenesis, and the control of expression of fatty acid synthase (FAS), lipoprotein lipase (LPL), fatty acid binding protein 4 (FABP), and hormone sensitive lipase (HSL) are being studied in various strategies using bioactive compounds derived from foods, especially in plant foods.

Barley is a widely consumed cereal among the most ancient cereal crops. Almost 80–90% of barley production is for animal feeds and malt, but now barley is gaining renewed interest as an ingredient for the production of functional foods due to their concentration of bioactive compounds, such as β-glucans and tocols [7]. Moreover, there are many classes of phenolic compounds in barley, such as benzoic and cinnamic acid derivatives, proanthocyanidins, quinines, flavonols, chalcones, flavones, flavanones, and amino phenolic compounds [8]. Barley is traditionally roasted in some Asian countries to produce a beverage (Japanese roasted barley tea, Mugicha) and during this roasting process various aroma compounds are formed [9]. The predominant amino acids present in barley have been reported to be glutamine and proline, but arginine is also found in significant levels [10], and barley is also an excellent source of glucose due to the presence of maltose.

During the heat process of food products that contain reducing sugars and amino acids, non-enzymatic browning (or Maillard reaction) takes place [11]. More specifically, food products containing arginine and glucose or maltose and undergoing heat processing, initially form arginyl-fructose (AF) and arginyl-fructosyl-glucose (AFG) through Amadori rearrangement (ARC) in products such as grains [12]. Previous reports have demonstrated that AF and AFG can modulate glucose uptake from dietary carbohydrates by exhibiting an inhibitory effect on α-glucosidases [12,13]. Furthermore, AF's anti-hyperglycemic effect has been demonstrated by using Korean Ginseng as a model food, in both animal and clinical models [14,15]. Through this mechanism, the inhibiting carbohydrate hydrolyzing enzymes, ARCs, additionally contribute towards the reduction in the caloric uptake, which can in turn result in weight and fat gain management.

The development of melanoidins and their antioxidant effects in roasted barley have been investigated in the past [16]. However, very limited information is available about the development of the ARC's content in roasted barley, other than aroma compounds, and their subsequent enhanced anti-obesity effect. Therefore, to evaluate the effect of ARC-rich barley extract on weight management and investigate the possible mechanism of action on improving weight gain, it is essential to measure the changes in body weight gain, body weight, plasma lipids, body fat mass, and lipid deposition. Consequently, the purpose of this study was (i) to investigate the changes in physical condition and (ii) fat metabolism-related biochemical markers such as transcriptional factors, gene expression, and (iii) to measure the changes in cell-based fat accumulation and its morphology on supplementation of heat-processed/AF-enriched barley extract (BEE) in C57BL/6 mice fed high-fat diets.

2. Results

2.1. BEE Inhibits Adipocyte Differentiation

Initially we evaluated the possible cytotoxicty of BEE on preadipocyte 3T3-L1 cells. As shown in Figure 1a, when 3T3-L1 cells were treated with 250 to 750 μg/mL of BEE for 48 h at 37 °C, our findings demonstrated no cytotoxicity at the tested doses (Figure 1a). Adipogenesis during the differentiation of 3T3-L1 preadipocytes into mature adipocytes was evaluated with the supplementation of BEE. These results suggest that higher concentrations of BEE administration (250 to 750 μg/mL) inhibit adipogenesis (Figure 1b). More specifically, we observed that BEE inhibited lipid accumulations in a dose-dependent manner. The formation of lipid droplets in the 3T3-L1 treated with BEE (500 and 750 μg/mL) was significantly blocked ($p < 0.05$ and $p < 0.01$, respectively, Figure 1b).

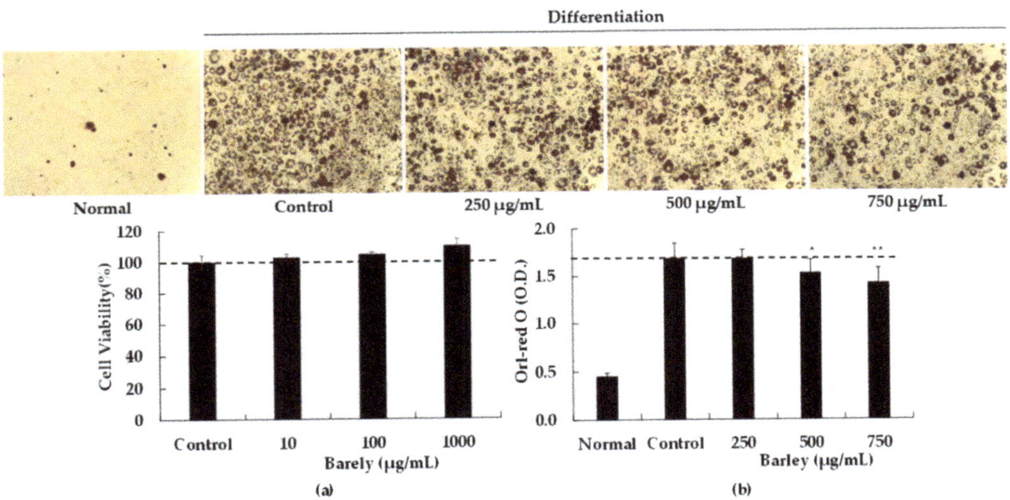

Figure 1. Effects of BEE on the lipid accumulation of 3T3-L1 cells. (**a**): 3T3-L1 preadipocytes cells were seeded at a density of 1 × 104 cells per well in 96-well plates and incubated in culture medium at 37 °C for 24 h to allow attachment. The attached cells were either untreated control or treated with 10, 100, or 1000 μg/mL of BEE at 37 °C for 48 h. After 48 h of incubation. The effects of BEE on cell viability were measured by MTT assay. The data are presented as relative cell viability values. Data are the means ± standard deviation (S.D.) values of at least 3 independent experiments. (**b**): 3T3-L1 preadipocytes were grown and differentiated with the differentiation cocktail in the absence and presence of varying concentrations (0, 250, 500, and 750 μg/mL) of BEE throughout the differentiation for 8 days. After 8 days of differentiation, these cells were subjected to Oil Red O staining for control and BEE to compare intracellular lipid accumulation (control: without BEE, normal: no differentiation). The results are expressed as the mean ± S.D ($n \geq 6$). Significantly different from control group (* $p < 0.05$, ** $p < 0.01$).

2.2. BEE Alleviates HFD-Induced Obesity in In Vivo Model

To prove the results from the in vitro experiment, the effects of BEE administration (0.3 g/kg-body weight) were evaluated in a high fat diet (HFD)-induced obesity C57BL/6 mice model for 54 days with dietary composition described in the section of material and methods. BEE supplementation, along with an HFD, resulted in significant changes in food intake and weight gain after day 24, compared to the control group

(Figure 2). More specifically, by the last day of the experiment, there was no significant difference in food consumption between the control and BEE group (Figure 2a). Additionally, BEE consumption significantly reduced weight gain compared to the control (Figure 2b).

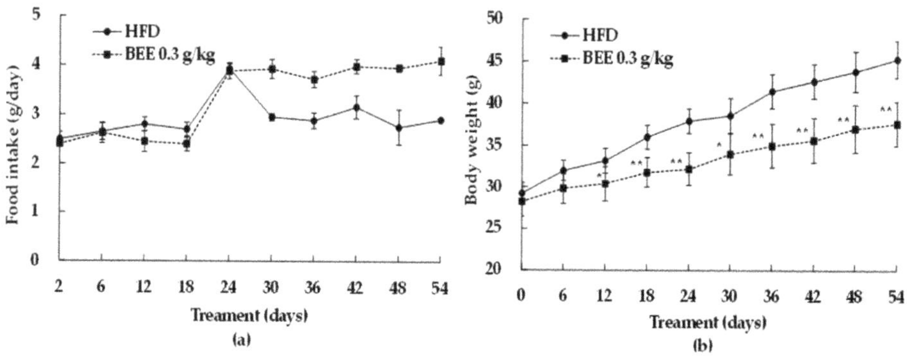

Figure 2. Changes in food intake (**a**) and body weight gains (**b**) before/after administration of BEE. Male C57BL/6 mice had free access to a high fat diet (HFD) (30% fat) to induce the weight gain for 54 days. During an HFD administration for 54 days, BEE was orally administrated (0.3 g/kg-body weight/day, peroral zonde injection) to BEE group C57BL/6 mice, 2 times per day (9–10 a.m. and 4–5 p.m.) with 0.15 g/kg-body weight each. Ten C57BL/6 mice were used for each group. Each point represents mean ± S.D. ($n = 10$). Food intake and body weight levels were compared between control (HFD) and treatment group (BEE) at each time point by unpaired Student's t-test (* $p < 0.05$; ** $p < 0.01$).

By looking at specific parameters, we observed that although the initial weight of the mice in the control and BEE groups were similar, a significant reduction was observed at the end of the experiment in the BEE-treated animals when compared to the control (HFD) ($p < 0.01$, Table 1). By estimating the changes from initial weight to final weight, BEE supplementation reduced weight gain by around 35% compared to the control.

Table 1. Effects of BEE treatment on various parameters in C57BL/6.

Parameters	C57BL/6	
	HFD	**BEE**
Initial body weight (g)	28.90 ± 1.66	28.90 ± 1.97
Final body weight (g)	45.26 ± 2.17	37.60 ± 2.60 **
Final body weight gain (g)	16.06 ± 2.44	9.40 ± 1.39 **
FER (%) †	8.70 ± 2.75	4.99 ± 1.63 **
Total cholesterol (mg/dL)	125.10 ± 11.42	112.26 ± 11.35
Triglyceride (mg/dL)	79.76 ± 5.60	52.36 ± 9.81 **
HDL Cholesterol (mg/dL)	38.34 ± 11.28	62.94 ± 13.06 *
LDL Cholesterol (mg/dL)	70.81 ± 8.62	38.85 ± 14.00 **

† FER, Feed efficiency ratio (%) = [Body weight gain (g)/Food intake (g)] × 100. Each experiment was compared between control (HFD) and BEE administration group (BEE) at each time point by unpaired Student's t-test (* $p < 0.05$; ** $p < 0.01$).

Interestingly, after the 24th day, in the case of the BEE-administered group, food intake increased, but compared to the HFD control group, the body weight decreased. These results are considered to be related to the postprandial blood glucose rise inhibitory effect of the arginyl-fructose (main key compound in BEE) reported in the previous study [12,13]. In addition to the regulation effect of BEE on adipogenesis/lipogenesis, it indicates that the effect of inhibiting the absorption of ingested carbohydrates into the small intestine also may act, giving a synergistic effect.

Additionally, we observed that BEE administration significantly reduced the LDL cholesterol ($p < 0.01$) and total triglyceride ($p < 0.01$) levels, and increased HDL cholesterol ($p < 0.05$) when compared to the control (HFD) (Table 1, Figure 3). As shown in Table 2, treatment with BEE also markedly decreased the weight of fat mass for each fat tissue location (Table 2).

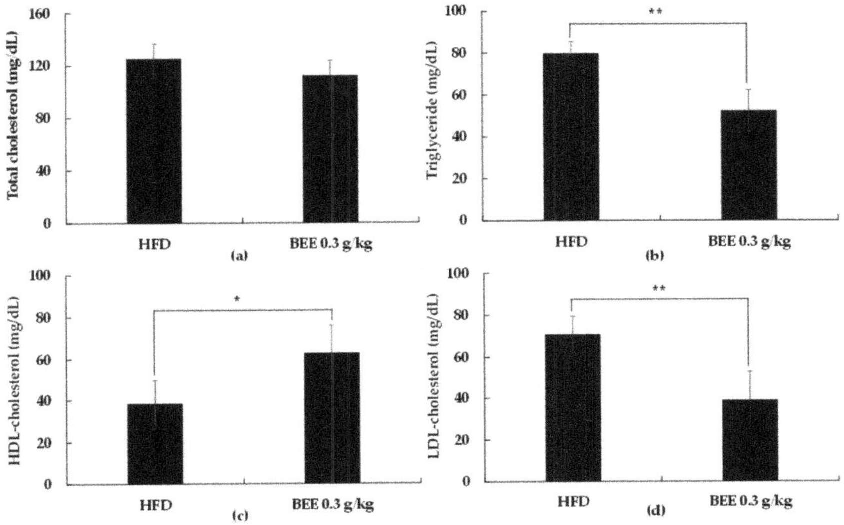

Figure 3. Comparison of total cholesterol (**a**), triglyceride (**b**), HDL (**c**), and LDL (**d**) contents with or without administration of BEE. Each point represents mean ± S.D. (n = 10). Triglyceride, total cholesterol, HDL, LDL contents were compared between control (HFD) and treatment group (BEE) at each time point by unpaired Student's t-test (* $p < 0.05$; ** $p < 0.01$).

Table 2. Effects of BEE treatment on various organs and adipose tissue weight in C57BL/6.

Parameters (mg/g)	C57BL/6	
	HFD	BEE
Liver	40.892 ± 4.691	35.059 ± 3.917
Kidney	7.215 ± 0.458	8.321 ± 0.564 **
Cecum	3.433 ± 0.619	4.589 ± 1.084
Mesenteric fat	20.805 ± 1.276	15.337 ± 1.791 **
Retroperitoneal fat	11.945 ± 1.801	11.301 ± 2.289 *
Kidney fat	4.027 ± 2.347	3.177 ± 0.699
Subcutaneous fat	54.673 ± 8.099	33.792 ± 8.907 **
Epididymal fat	49.196 ± 5.720	37.537 ± 6.018 *
Small intestine	17.553 ± 1.853	21.095 ± 4.944

Each experiment was compared between control (HFD) and BEE administration group (BEE) at each time point by unpaired Student's t-test (* $p < 0.05$; ** $p < 0.01$).

Additionally, BEE administration resulted in a significant increase in the levels of the hormone adiponectin ($p < 0.01$) (relevant to insulin sensitivity), TNF-α ($p < 0.05$), while reducing leptin ($p < 0.01$) (suggesting improvement of leptin resistance) and insulin levels ($p < 0.01$) when compared to the control (HFD) (Table 3).

Table 3. Effects of BEE treatment on various obesity-related hormone and cytokine parameters in C57BL/6.

Parameters	C57BL/6	
	HFD	BEE
Adiponectin (μg/mL)	34.93 ± 8.46	54.81 ± 10.81 *
Leptin (ng/mL)	41.26 ± 8.60	22.19 ± 8.24 **
TNF-α (pg/mL)	31.00 ± 2.58	42.03 ± 9.09 *
IL-6 (μg/mL)	67.55 ± 14.84	41.92 ± 11.82 *
Insulin (μg/mL)	0.77 ± 0.08	0.48 ± 0.06 **

Each experiment was compared between control (HFD) and BEE administration group (BEE) at each time point by unpaired Student's *t*-test (* $p < 0.05$; ** $p < 0.01$).

2.3. BEE Administration Decreases the Expression of Adipogenesis/Lipogenesis-Related Genes in Epididymal Fat Biopsy from C57BL/6 Mice

CEBPα, FAS, PPARγ, and SREBP-1c are the main key genes of fat metabolism and are involved in the regulation of fat metabolism. Increased CEBPα expression is related to adipocyte hypertrophy, impaired insulin signaling, and decreased glucose utilization [17]. It has been well reported that decreased FAS and PPARγ expressions are considered as regulation factors of adipogenesis [18]. SREBP-1c participated in adipocyte differentiation and adipogenesis is a major regulator of lipid homeostasis transcription [19]. Our observations demonstrated that PPARγ ($p < 0.05$) and C/EBPα ($p < 0.001$) mRNA levels were significantly decreased with the administration of BEE (0.3 g/kg-body weight) (Figure 4b,d). Additionally, we observed that BEE supplementation markedly decreased the expression levels of key target lipogenic genes, fatty acid synthase (FAS) ($p < 0.001$), lipoprotein lipase (LPL) ($p < 0.01$), and fatty acid binding protein 4 (FABP4) ($p < 0.001$), and sterol regulatory element-binding protein 1c (SREBP-1c) ($p < 0.01$) (Figure 4a,c,e,f), an observation that correlates with the reduced PPARγ and C/EBPα mRNA expression. The above lipid metabolism results indicated that BEE treatment affected the gene and/or protein expression of CEBPα, FAS, SREBP-1c, and PPARγ in C/57BL6 related to lipid metabolism.

2.4. BEE Administration Decreases the Fat Accumulation in the HFD-Induced C57BL/6 Mice

Our initial study with preadipocyte 3T3-L1 examined whether BEE treatment markedly alleviates hepatic fat accumulation in HFD-induced C57BL/6 mice. We investigated the liver weight of mice after 8 weeks of BEE administration. While the liver weight was found to be increased in HFD mice, the BEE treatment prevented this increase in the HFD mice (Figure 5b,c).

Diet-induced obesity through the administration of a western-type diet high in fat and sugar to mice can result in a higher weight of adipose and liver tissues. The increased adipose tissue weight can be due to hypertrophy (bigger cells) or hyperplasia (more cells) of adipocytes, or a combination of both. Hematoxylin and eosin (H&E) staining of adipose tissue sections was conducted to compare the sizes of the individual adipocytes.

For the BEE treatment group, the adipose tissue mass (subcutaneous and epididymal fat) was significantly lower than the HFD group (Table 2). The difference in adipocyte size was evident in H&E-stained samples of both normal control and HFD. Moreover, the BEE treatment group's adipose tissues showed smaller adipocytes than the HFD adipose tissue (Figure 6b,c).

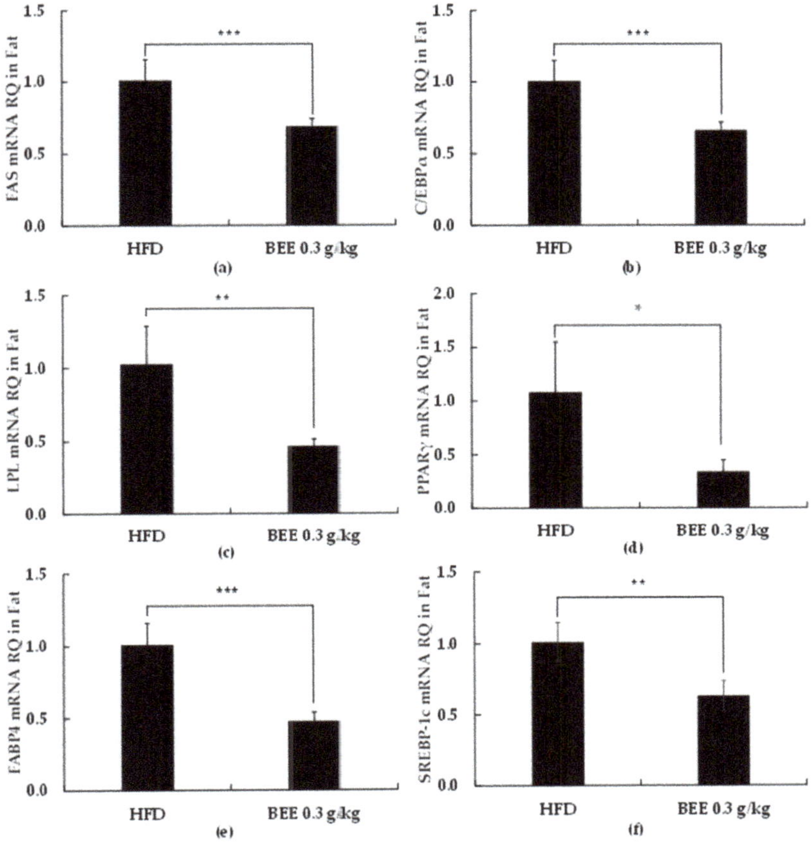

Figure 4. RT, real-time PCR quantitative analysis of adipocyte differentiation-related genes expression in the 3T3-L1. For real-time PCR, we used SYBR green mix with gene-specific primers ((**a**): fatty acid synthase (FAS), (**b**): CEBP/α, (**c**): lipoprotein lipase (LPL), (**d**): PPARγ, (**e**): FABP4, and (**f**): SREBP-1c). Each value is expressed as mean ± S.D. and is representative of at least three separate experiments. Different letters indicate statistically significant differences between groups with one-way ANOVA followed by Duncan's test of $p < 0.05$. The results are expressed as the mean ± S.D ($n \geq 6$). Statistical significances from control group were determined by Student's *t*-test (* $p < 0.05$, ** $p < 0.01$, *** $p < 0.001$).

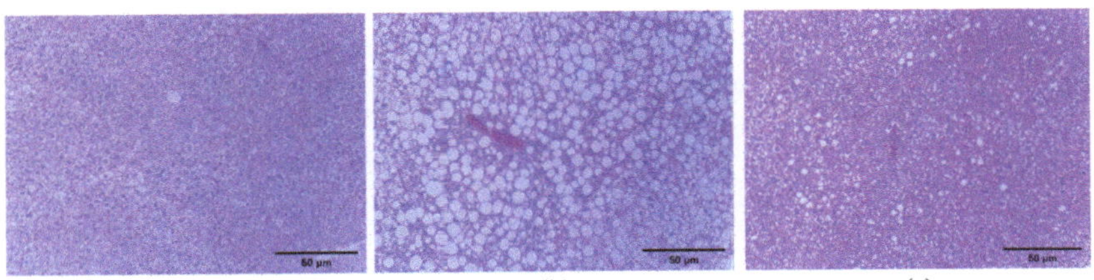

(a) (b) (c)

Figure 5. Effect of Jeju barley extract (BEE) on the histopathological change in liver tissues in the high fat diet-induced C57BL/6 mice. Liver tissues were stained with H&E (original magnification ×100). (**a**) Normal diet mice, (**b**) high fat diet mice, (**c**) high fat diet treated with BEE 0.3 g/kg-body weight.

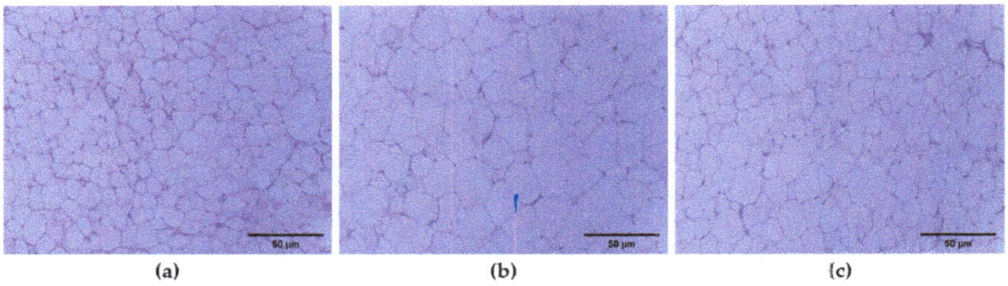

Figure 6. Effect of Jeju barley extract (BEE) on the histopathological change in epididymal fat tissues in the high fat diet-induced C57BL/6 mice. Epididymal fat tissues were stained with H&E (original magnification ×100). (**a**) Normal diet mice, (**b**) high fat diet mice, (**c**) high fat diet treated with BEE 0.3 g/kg-body weight.

3. Discussion

In this paper we present the anti-obesity effects of BEE in a rodent preadipocyte cell line, 3T3-L1 cells, and also in a C57BL/6 mice model of HFD-induced obesity. Our findings suggest the dose-dependent inhibition of lipid accumulation and 3T3-L1 preadipocyte differentiation with BEE supplementation. Additionally, BEE decreased the expression of PPARγ and C/EBPα, which are important transcription factors regulating adipocyte metabolism, and also decreased the expression of their downstream target genes including FABP4, LPL, and FAS. The latter genes are adipocyte-specific and are involved in maintaining the adipocyte phenotype. Additionally, we confirmed that the tested doses of BEE do not have any toxicity against 3T3-L1 cells.

The anti-obesity effect of BEE administration was evaluated in a mice model for 42 days and we observed that BEE supplementation decreased body weight compared to the control without a significant difference in food intake. In addition, BEE administration reduced the levels of triglycerides, total cholesterol, and LDL cholesterol levels while increasing the HDL cholesterol level.

Dietary fat intake and obesity have been strongly correlated through numerous epidemiological studies [20]. Hypertrophic adipocytes and the dysfunction of adipose tissue are the main obesity characteristics. Reducing the uptake of dietary lipids in adipocytes can result in elevated triglycerides and reduced HDL cholesterol levels. It has been well-defined that adipocyte differentiation is characterized by the increased expression of PPARγ and C/EBPα [21]. We observed that BEE treatment significantly reduced levels of PPARγ and C/EBPα. We also observed a significant reduction in FABP4, FAS, and LPL gene expression. The correlation between PPARγ and C/EBPα and lipogenesis has been well-defined, but our findings further confirm it. Our findings in the animal trial suggest that BEE can result in reduced body weight, triglyceride, and total cholesterol while increasing HDL cholesterol. Based on these observations, we suggest that BEE can possibly be used as a weight management strategy through adipogenesis inhibition.

Adiponectin is a hormone secreted by adipocytes that plays an important role in maintaining lipid homeostasis. Adiponectin is secreted from adipose tissue and binds to adiponectin receptors (adipoR1 and adipoR2) in the liver, enhancing insulin sensitivity while maintaining low levels of lipids in the liver through controlling lipogenesis, and increasing β-oxidation through the activation of AMP protein kinase. Obesity downregulates adiponectin secretion, leading to excessive lipid accumulation including triglyceride and LDL cholesterol. As a result, higher levels of adiponectin are important in the prevention of obesity. Our observations suggest that BEE administration increases the adiponectin levels, which can be an additional mechanism for the weight gain reduction effect observed in this study.

Leptin is a hormone produced by the adipose tissue and communicates the state of body energy repletion to the central nervous system (CNS) in order to suppress food intake and permit energy expenditure [22–24]. Most obese individuals exhibit elevated circulating leptin levels commensurate with their adipose mass, a condition defined as leptin resistance [25]. In our study we observed that leptin levels were higher in the HFD group when compared to the BEE group. This suggests that leptin resistance might have developed in the HFD group and not in the BEE-treated group, indicating that BEE prevents the development of leptin resistance. This could be another mechanism for the observed weight management effects of BEE supplementation.

In conclusion, BEE administration reduces weight gain and body fat accumulation in a mouse model. Our observations provide strong supporting evidence for additional studies to better define the potential weight management effect of BEE. Here we report a mechanism through the management of adipocyte differentiation and adiponectin secretion. Our previous research results related to arginyl-fructose, a major active ingredient in BEE, were that AF showed a calorie-restriction effect [12,13]. Considering this, AF, the main component of BEE, also inhibited the activity of α-glucosidase in the small intestine and blocked the absorption of disaccharides, which may have a synergistic effect on the anti-adipogenesis/obesity efficacy, resulting in a statistically significant decrease in weight gain. In the future, it is important to determine how BEE might affect glucose and lipid metabolism in various relevant tissues, such as skeletal muscle and the liver.

4. Materials and Methods

4.1. Materials

Arginyl-fructose (AF)-enriched barley powder (BEE) was purchased from Kunpoong Bio Co., Ltd. (Jeju, Korea). Corn starch, casein, vitamin mix, mineral mix, calcium phosphate, and sodium chloride were purchased from Raon Bio (Yonginsi, Korea). Total cholesterol and total glyceride kits were purchased from Stanbio laboratory (Boerne, TX, USA). Unless noted, all chemicals were purchased from Sigma-Aldrich Co. (St. Louis, MO, USA). The fast SYBR real-time PCR master mix, Dulbecco's modified Eagle's medium (DMEM), fetal bovine serum (FBS), bovine calf newborn serum (BCS), penicillin-streptomycin (*p*/S), and trypsin-EDTA were obtained from Life Technologies (Grand Island, NY, USA). Adiponectin ELISA kit was purchased from Thermo Fisher Scientific (Invitrogen, Carlsbad, CA, USA). 3T3-L1 cells (ATCCV®CL-173TM) were used below passage 12. 3T3-L1 preadipocytes were propagated and cultured in DMEM medium supplemented with 10% BCS and 1% P/S until confluent and maintained for additional 2 days and differentiated as reported previously [26] with or without BEE.

4.2. Sample Preparation

In order to obtain an arginyl-fructose (AF)-enriched barley powder (BEE) sample, gelatinization was performed at 80 °C for 1 h in 500 L distilled water after milling and pulverizing of 25 kg of barley. Then, 50 mL of a triple complex enzyme (AMG® glucoamlyase, Viscoflow® MG, Pectinex® (1:1:1 ratio), Bagsvaerd, Denmark) and 5.9 kg of citric acid were added and reaction was performed at 60 °C for 18 h. Arginine (14.9 kg) was added to adjust the neutral pH. To maximize the content of the Amadori derivative, heating was carried out at 95 °C for 4 h. Filtration of mixture was carried out using a filter paper, and 54.2 kg of dextrin was added. To obtain Amadori derivative containing barley powder, spray drying followed at 170 °C. AF content in BEE was 22%, approximately.

4.3. Determination of Cell Viability and Morphometric Analysis

The effects of BEE on 3T3-L1 cell viability were determined using an established MTT assay. Briefly, the 3T3-L1 preadipocytes cells were seeded at a density of 1 × 104 cells per well in a 96-well plate and incubated in culture medium at 37 °C for 24 h to allow at-tachment. The attached cells were either untreated control or treated with 250, 500, or 750 μg/mL of BEE at 37 °C for 48 h. After 48 h of incubation, the cells were washed with phosphate-buffered saline (PBS) prior to the addition of MTT (0.5 μg/mL PBS) and incubated at 37 °C for 2 h. Formazan crystals were dissolved with dimethyl sulfoxide

(100 µL/well) and detected at OD570 with a model Emax (Molecular Devices, Sunnyvale, CA, USA).

Fresh dissected mice epididymal adipose and liver tissue samples (n = 5) were fixed by immersion in 4% paraformaldehyde in 0.1 M phosphate buffer (pH 7.4) overnight at 4 °C. Samples were then washed in phosphate buffer and dehydrated in a graded series of ethanol, cleared in xylene, and embedded in paraffin blocks. Five-micrometer-thick sections of the tissues were stained with hematoxylin and eosin to assess morphology. Images from light microscopy were digitalized and the diameters of at least 100 adipocytes of each section were determined using Axio Vision software (Carl Zeiss Imaging Solutions).

4.4. Oil Red O (ORO) Staining

To determine the degree of differentiation as measured by intracellular lipid content, ORO was performed as previously described [27]. Briefly, 3T3-L1 preadipocytes were cultured in DMEM/high-glucose medium containing 10% calf serum until confluent (day −2) and maintained for an additional 2 days (until day 0). Differentiation was induced on day 0 by the addition of 0.5 mmol/L methylisobutylxanthine, 1 µmol/L dexamethasone, 1.0 µg/mL insulin, and 10% fetal bovine serum (FBS) in DMEM. After 48 h (day 2), the medium was replaced with DMEM containing 1.0 µg/mL insulin and 10% FBS. Medium was changed every 2 days thereafter until the cells were collected for analysis [18]. BEE was reconstituted as 1000 µg/mL stock solutions in DMSO (dimethyl sulfoxide) and added at the indicated concentrations on day 0. Cells were cultured with BEE until cells were collected for analysis. After 8 days of differentiation, 3T3-L1 adipocytes were washed with 4% paraformaldehyde once and fixed with 4% paraformaldehyde for 20 min at room temperature. Cells were then washed with 60% isopropanol once and stained with diluted Oil Red O solution for 30 min. After photographing the stained cells, the dye retained in 3T3-L1 cells was eluted with 100% isopropanol and the absorbance was measured by a microplate reader (SpectraMax M2, Molecular Devices, Sunnyvale, CA, USA) at 490 nm.

4.5. Quantitative Real-Time PCR

RNA was isolated with TRIzol® plus RNA purification kit according to manufacturer's protocol (Life Technologies, Grand Island, NY, USA). One microgram of total RNA was used to synthesize cDNA using Revert Aid First Strand cDNA Synthesis kit (Thermo Scientific, Waltham, MA, USA). The reaction was performed with Fast SYBR® Green Master Mix containing 1 µM of primer pair and 100 ng of cDNA under 40 cycles with each of 95 °C for 1 sec and 58 °C for 20 s. Relative levels of the target mRNA expression were determined by ViiA™ 7 real-time PCR system (Life technologies, Grand Island, NY, USA), normalized to GAPDH calculated with the $2-(^{\Delta\Delta}nd)$ method. The primer sequences are listed in Table 4. All the results were normalized to the housekeeping gene, glyceraldehyde 3-phosphate dehydrogenase (GAPDH), to control for variations in mRNA concentrations. Relative quantification was performed using the comparative delta-delta Ct method according to the manufacturer's instructions (Applied Biosystems).

4.6. In Vivo Experimental Design

Five-week-old male C57BL/6 mice were purchased from Joongang Experimental Animal Co. (Seoul, Korea) and fed a high fat diet (30% fat) (Table 5) for 54 days. After 3 days the normal diet (Pico 5053) was switched to a high fat diet (HFD) (Oriental Bio. Co., Seongnam, Korea) for 54 days. During the HFD administration for 8 weeks, C57BL/6 mice were divided into two groups, one group received HFD and the second group received HFD with BEE. BEE was orally administrated 2 times a day (9–10 a.m. and 4–5 p.m.). Distilled water (D.W.) was used as a vehicle for a solution of the experimental compound, BEE. In a control group, D.W. was used as a vehicle for oral administration without BEE using a zonde injection needle. The dose of each BEE administration was 0.15 g/kg-body weight, yielding a final dose of 0.3 g/kg-body weight/day. The animals were housed in individual cages in a room with a 12 h light/dark cycle (lights on from 06:00 h) with 50% ± 7% relative humidity. In this study, ten C57BL/6 mice were used for each group. The experimental protocols were approved by the Institutional Animal Care and Use Committee (IACUC) of the Hannam University (Approval number: HNU2020-0014). The mice had free access to water throughout the experimental period. The mice were anesthetized with pentobarbital and sacrificed, and blood was collected and serum was processed and stored at −80 °C until used. The retroperitoneal, mesenteric, kidney, subcutaneous, and epididymal fat tissue depots were rapidly removed and weighed. Samples for RNA and protein analysis, and for DNA quantification were immediately frozen in liquid nitrogen and stored at −70 °C until analysis.

Group I: Control.

Group II: BEE 0.3 g/kg-body weight/day.

Table 4. Primer for real-time quantitative PCR.

Genes	Primer Sequences	
Accession Number	**Forward (5′-3′)**	**Reverse (5′-3′)**
GAPDH NM_008084	CGTCCCGTAGACAAAATGGT	TTGATGGCAACAATCTCCAC
PPARγ NM_011146	GAAAGACAACGGACAAATCACC	GGGGGTGATATGTTTGAACTTG
C/EBPα NM_007678	TTGTTTGGCTTTATCTCGGC	CCAAGAAGTCGGTGGACAAG
FABP4 NM_024406	AGCCTTTCTCACCTGGAAGA	TTGTGGCAAAGCCCATC
FAS NM_007988	TGATGTGGAACACAGCAAGG	GGCTGTGGTGACTCTTAGTGATAA
SREBP-1c NM_011480	ACGGAGCCATGGATTGCACA	AAGGGTGCAGGTGTCACCTT
LPL NM_008509	GGACGGTAACGGGAATGTATGA	TGACATTGGAGTCAGGTTCTCTCT

PCR, polymerase chain reaction; PPARγ, peroxisome proliferator-activated receptor γ; C/EBPα, CCAAT/enhancer -binding protein α; FAS, fatty acid synthase; SREBP-1c, sterol regulatory element-binding protein-1c; LPL, lipoprotein lipase; GAPDH, glyceraldehyde 3-phosphate dehydrogenase.

Table 5. Composition of high fat diet (g/kg).

High Fat Diets (g/kg)	
Corn starch	321
Sucrose	100
Casein	200
Corn oil	100
Lard	200
Cellulose	30
DL-methionine	2
Vitamin mix (1)	10
Mineral mix (2)	35
Choline bitartrate	2

(1) Vitamin mixture: AIN-93VX; (2) Mineral mixture: AIN-93G.

4.7. Blood Analysis

The plasma total cholesterol and total glyceride concentration was measured using a kit (Stanbio lab., Boerne, TX, USA). Serum adiponectin levels in SD rats were detected by ELISA kit (Invitrogen, Carlsbad, CA, USA).

4.8. Statistical Analysis

Statistical analyses were carried out using the statistical package SPSS 10 (Statistical Package for Social Science, SPSS Inc., Chicago, IL, USA) program and significance of each group was verified with the analysis of One-way analysis of variance (ANOVA) followed by the Duncan's multiple range test of $p < 0.05$ and the Student's t-test for comparison of means.

Author Contributions: S.-Y.L., T.-Y.K., J.-Y.H. and G.-J.K. conducted the animal experiment and analyzed the data. J.-B.O. prepared the product to be tested. M.-J.K., E.A., J.-Y.L. and Y.-I.K. participated in the design of the study and preparation of the manuscript. All authors have read and agreed to the published version of the manuscript.

Funding: This research was financially supported by Hannam University.

Institutional Review Board Statement: The study was conducted according to the guidelines of the Declaration of Helsinki, and approved by the Institutional Animal Care and Use Committee (IACUC) of the Hannam University (Approval number: HNU2020-0014, October 2020).

Informed Consent Statement: Not applicable.

Data Availability Statement: The data are available on request.

Acknowledgments: This research was financially supported by the Ministry of Trade, Industry, and Energy (MOTIE), Korea, under the "Regional Specialized Industry Development Program" supervised by the Korea Institute for Advancement of Technology (KIAT) (S2891360).

Conflicts of Interest: The authors declare no conflict of interest.

Sample Availability: Samples of AF-enriched Jeju barley extracts are available from the authors.

References

1. Pigeyre, M.; Yazdi, F.T.; Kaur, Y.; Meyre, D. Recent progress in genetics, epigenetics and metagenomics unveils the pathophysiology of human obesity. *Clin. Sci.* **2016**, *130*, 943–986. [CrossRef] [PubMed]
2. Lutz, T.A.; Wood, S.C. Overview of animal models of obesity. *Curr. Protoc. Pharmacol.* **2012**, *58*, 5–61. [CrossRef] [PubMed]
3. Lamont, B.J.; Waters, M.F.; Andrikopoulos, S. A low-carbohydrate high-fat diet increases weight gain and does not improve glucose tolerance, insulin secretion or β-cell mass in NZO mice. *Nutr. Diabetes* **2016**, *6*, 194–200. [CrossRef] [PubMed]
4. Kaare, M.; Mikheim, K.; Lilleväli, K.; Kilk, K.; Jagomäe, T.; Leidmaa, E.; Piirsalu, M.; Porosk, R.; Singh, K.; Reimets, R.; et al. High-fat diet induces pre-diabetes and distinct sex-specific metabolic alterations in negr1-deficient mice. *Biomedicines* **2021**, *9*, 1148. [CrossRef] [PubMed]
5. Jeon, G.; Choi, Y.; Lee, S.M.; Kim, Y.; Jeong, H.S.; Lee, J. Anti-obesity activity of methanol extract from hot pepper (*Capsicum annuum* L.) seeds in 3T3-L1 adipocyte. *Food Sci. Biotechnol.* **2010**, *19*, 1123–1127. [CrossRef]
6. Klop, B.; Elte, J.W.F.; Cabezas, M.C. Dyslipidemia in obesity: Mechanisms and potential targets. *Nutrients* **2013**, *5*, 1218–1240. [CrossRef]
7. Jadhav, S.J.; Lutz, S.E.; Ghorpade, V.M.; Salunkhe, D.K. Barley: Chemistry and value-added processing. *Crit. Rev. Food Sci. Nutr.* **1998**, *38*, 123–171. [CrossRef]
8. Goupy, P.; Hugues, M.; Boivin, P.; Amiot, M.J. Antioxidant composition and activity of barley (*Hordeum vulgare*) and malt extracts and of isolated phenolic compounds. *J. Sci. Food Agric.* **1999**, *79*, 1625–1634. [CrossRef]
9. Mickowska, B.; Socha, P.; Urminská, D.; Cieślik, E. The comparison of prolamines extracted from different varieties of wheat, barley, rye triricale species; Amino acid composition, electrophoresis and immunodetection. *J. Microbiol. Biotechnol. Food Sci.* **2012**, *1*, 742–752.
10. Tatsu, S.; Matsuo, Y.; Nakahara, K.; Hofmann, T.; Steinhaus, M. Key Odorants in Japanese Roasted Barley Tea (Mugi-Cha)—Differences between Roasted Barley Tea Prepared from Naked Barley and Roasted Barley Tea Prepared from Hulled Barley. *J. Agric. Food Chem.* **2020**, *68*, 2728–2737. [CrossRef]
11. Lee, J.S.; Kim, G.N.; Lee, S.H.; Kim, E.S.; Ha, K.S.; Kwon, Y.I.; Jeong, H.S.; Jang, H.D. In vitro and cellular antioxidant activity of arginyl-fructose and arginyl-fructosyl-glucose. *Food Sci. Biotechnol.* **2009**, *18*, 1505–1510.
12. Ha, K.S.; Jo, S.H.; Kang, B.H.; Apostolidis, E.; Lee, M.S.; Jang, H.D.; Kwon, Y.I. In vitro and in vivo antihyperglycemic effect of 2 amadori rearrangement compounds, Arginyl-fructose and arginyl-fructosyl-glucose. *J. Food Sci.* **2011**, *76*, 188–193. [CrossRef] [PubMed]
13. Lee, K.-H.; Ha, K.S.; Jo, S.H.; Lee, C.-M.; Kim, Y.-C.; Chung, K.-H.; Kwon, Y.-I. Effect of long-term dietary arginyl-fructose (AF) on hyperglycemia and HbA1c in diabetic db/db mice. *Int. J. Mol. Sci.* **2014**, *15*, 8352–8359. [CrossRef] [PubMed]
14. Park, S.E.; Kin, O.-H.; Kwak, J.H.; Lee, K.H.; Kwon, Y.-I.; Chung, K.H.; Lee, J.H. Antihyperglycemic effect of short-term arginyl-fructose supplementation in subjects with prediabetes and newly diagnosed type 2 diabetes: Randomized, double-blinded, placebo-controlled trial. *Trials* **2015**, *16*, 521–528. [CrossRef] [PubMed]
15. Vuksan, V.; Sievenpipper, J.; Jovanovski1, E.; Jenkins, A.L. Current clinical evidence for Korean red ginseng in management of diabetes and vascular disease: A Toronto's ginseng clinical testing program. *J. Ginseng Res.* **2010**, *34*, 264–273. [CrossRef]
16. Papetti, G.; Daglia, M.; Aceti, C.; Quaglia, M.; Gregotti, C.; Gazzani, G. Isolation of an In Vitro and Ex Vivo Antiradical Melanoidin from Roasted Barley. *J. Agric. Food Chem.* **2006**, *54*, 1209–1216. [CrossRef]
17. Ariemma, F.; Desposito, V.; Liguoro, D. Low-dose bisphenol-A impairs adipogenesis and generates dysfunctional 3T3-L1 adipocytes. *PLoS ONE* **2016**, *11*, e0150762. [CrossRef]

18. Jang, B. Artesunate inhibits adipogeneis in 3T3-L1 preadipocytes by reducing the expression and/or phosphorylation levels of C/EBP-α, PPAR-γ, FAS, perilipin A, and STAT-3. *Biochem. Biophys. Res. Commun.* **2016**, *474*, 220–225. [CrossRef]
19. Lee, H.; Park, Y. Identifcation of metabolic pathways related to the bisphenol A-induced adipogenesis in diferentiated murine adipocytes by using RNA-sequencing. *Environ. Res.* **2019**, *171*, 161–169. [CrossRef]
20. Ali, A.T.; Penny, C.B.; Paiker, J.E.; Niekerk, C.V.; Smit, A.; Ferris, W.F.; Crowther, N.J. Alkaline phosphatase is involved in the control of adipogenesis in the murine preadipocyte cell line, 3T3-L1. *Clin. Chim. Acta.* **2005**, *354*, 101–109. [CrossRef]
21. Ntambi, J.M.; Kim, Y.C. Adipocyte differentiation and gene expression. *J. Nutr.* **2000**, *130*, 3122S–3126S. [CrossRef] [PubMed]
22. Friedman, J.M.; Halaas, J.L. Leptin and the regulation of body weight in mammals. *Nature* **1998**, *395*, 763–770. [CrossRef] [PubMed]
23. Elmquist, J.K.; Maratos-Flier, E.; Saper, C.B.; Flier, J.S. Unraveling the central nervous system pathways underlying responses to leptin. *Nat. Neurosci.* **1998**, *1*, 445–449. [CrossRef] [PubMed]
24. Bates, S.H.; Myers, M.G. The role of leptin receptor signaling in feeding and neuroendocrine function. *Trends Endocrinol. Metab.* **2003**, *14*, 447–452. [CrossRef]
25. Myers, M.G.; Cowley, M.A.; Munzberg, H. Mechanisms of leptin action and leptin resistance. *Ann. Rev. Physiol.* **2008**, *70*, 537–556. [CrossRef]
26. Lee, J.-Y.; Kim, T.Y.; Kang, H.; Oh, J.B.; Park, J.W.; Kim, S.-C.; Kim, M.; Apostolidis, E.; Kim, Y.-C.; Kwon, Y.I. Anti-Obesity and Anti-Adipogenic Effects of Chitosan Oligosaccharide (GO2KA1) in SD Rats and in 3T3-L1 Preadipocytes Models. *Molecules* **2021**, *26*, 331. [CrossRef]
27. Yu, S.Y.; Choi, Y.; Kwon, Y.I.; Lee, O.-H.; Kim, Y.C. Mechanism of Formononetin-Induced Stimulation of Adipocyte Fatty Acid Oxidation and Preadipocyte Differentiation. *J. Food. Nutr. Res.* **2021**, *9*, 90–99. [CrossRef]

molecules

Article

Antiproliferative Properties of Triterpenoids by ECIS Method—A New Promising Approach in Anticancer Studies?

Anna Hordyjewska [1], Monika Prendecka-Wróbel [2], Łukasz Kurach [3,*], Anna Horecka [1], Anna Olszewska [2], Dominika Pigoń-Zając [2], Teresa Małecka-Massalska [2] and Jacek Kurzepa [1]

1 Chair and Department of Medical Chemistry, Medical University of Lublin, 4A Chodzki Str., 20-093 Lublin, Poland; anna.hordyjewska@umlub.pl (A.H.); anna.horecka@umlub.pl (A.H.); jacek.kurzepa@umlub.pl (J.K.)
2 Chair and Department of Human Physiology, Medical University of Lublin, 11 Radziwiłłowska Str., 20-093 Lublin, Poland; monika.prendecka-wrobel@umlub.pl (M.P.-W.); anna.olszewska@umlub.pl (A.O.); dominika.pigon-zajac@umlub.pl (D.P.-Z.); teresa.malecka-massalska@umlub.pl (T.M.-M.)
3 Independent Laboratory of Behavioral Studies, Medical University of Lublin, 4A Chodzki Str., 20-093 Lublin, Poland
* Correspondence: lukasz.kurach@umlub.pl; Tel.: +48-814486196

Abstract: Electric cell–substrate impedance sensing is an advanced in vitro impedance measuring system which uses alternating current to determine behavior of cells in physiological conditions. In this study, we used the abovementioned method for checking the anticancer activities of betulin and betulinic acid, which are some of the most commonly found triterpenes in nature. In our experiment, the threshold concentrations of betulin required to elicit antiproliferative effects, verified by MTT and LDH release methods, were 7.8 μM for breast cancer (T47D), 9.5 μM for lung carcinoma (A549), and 21.3 μM for normal epithelial cells (Vero). The ECIS results revealed the great potential of betulin and betulinic acid's antitumor properties and their maintenance of cytotoxic substances to the breast cancer T47D line. Moreover, both substances showed a negligible toxic effect on healthy epithelial cells (Vero). Our investigation showed that the ECIS method is a proper alternative to the currently used assay for testing in vitro anticancer activity of compounds, and that it should thus be introduced in cellular routine research. It is also a valuable tool for live-monitoring changes in the morphology and physiology of cells, which translates into the accurate development of anticancer therapies.

Keywords: impedance; ECIS; antitumor activity; anticancer properties; betulin; betulinic acid

Citation: Hordyjewska, A.; Prendecka-Wróbel, M.; Kurach, Ł.; Horecka, A.; Olszewska, A.; Pigoń-Zając, D.; Małecka-Massalska, T.; Kurzepa, J. Antiproliferative Properties of Triterpenoids by ECIS Method—A New Promising Approach in Anticancer Studies? *Molecules* **2022**, *27*, 3150. https://doi.org/10.3390/molecules27103150

Academic Editors: Sokcheon Pak and Akihito Yokosuka

Received: 17 March 2022
Accepted: 12 May 2022
Published: 14 May 2022

1. Introduction

Routine analysis carried out on cell lines usually allows for the assessment of one parameter at the same time, such as the number, morphology, and phenotype of cells, as well as the viability and metabolic activity of the cell, the synthesis of intracellular transcription factors, or the release of many regulatory proteins. The impedance method seems to be a promising, non-invasive technique for studying cells and their morphological changes caused by various stimuli. Giaever and Keese were the pioneers of this technique. They developed electric cell–substrate impedance sensing (ECIS) as a continuous monitoring system to study cell behavior using a real-time and label-free method [1–3]. These measurements detect changes in the morphology of the cells, which, in turn, are derived from specific metabolic processes. In this system, electrodes are used to apply low alternating current. Those electrodes which measure the voltage change are mounted at the bottom of a standard matrix. When the cells attach to the matrix and electrodes, they act as insulators, increasing impedance. The flow of the current is hindered depending on the number of cells covering the electrode, cells' shape, and the type of cells attached to the electrode surface. The structure design of the ECIS measuring system includes (Figure 1): two electrodes (one is a small working electrode and the other a large counter electrode on the bottom of the culture plate) connected to the edge of a culture chip—the chip is connected to the lock-in amplifier. The whole setup is placed inside an incubator in steady conditions of 37 °C and under a 5% CO_2 atmosphere. After cell seeding, cells drift downward and attach to the stratum of the electrode, which then passes the impeded current directly into the bulk electrolyte as the result of anchored plasma membrane intrusion above the electrode surface [1,4,5].

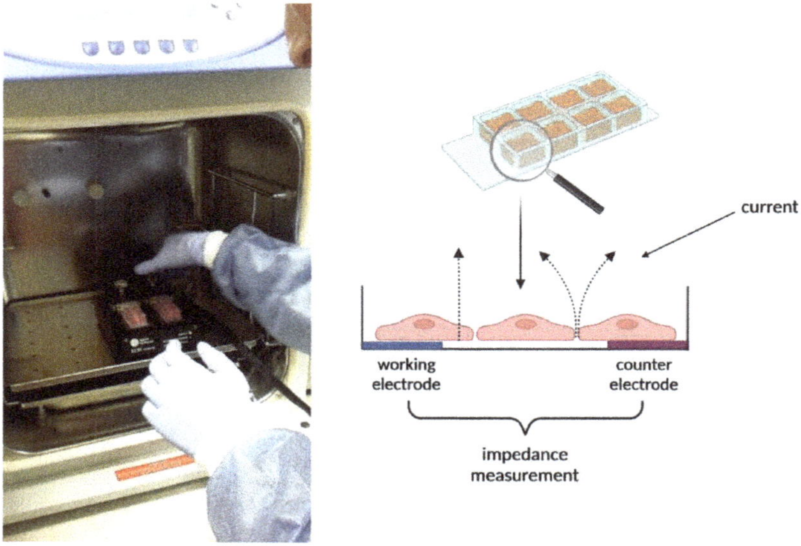

Figure 1. ECIS measuring station in the incubator (photo from the cell culture laboratory, Chair, and Department of Human Physiology, Medical University of Lublin). The scheme was created with BioRender.

In this system, frequency is a very important factor. At low frequencies (<2000 Hz), the majority of the current flows between the cells. At high frequencies (>40,000 Hz), most of the current flows directly through the isolated cell membranes. The impedance at high frequencies is affected more significantly by the cell membrane, while at low frequencies, the response is more dependent on the space below and between cells [2,4,6].

ECIS is capable of detecting morphology changes in the subnanometer to micrometer range. In ECIS, a small alternating current (I) is applied across the electrode pattern at the bottom of the ECIS arrays (a direct current (DC) cannot be used). This results in a potential (V) across the electrodes which is measured by the ECIS instrument. As cells grow and cover the electrodes, the current is impeded in a manner related to the number of cells covering the electrode, the morphology of the cells, and the nature of the cell attachment. When cells are stimulated to change their function, the accompanying changes in cell morphology alter the impedance. The data generated are impedance versus time [1,7,8].

Ohm's law is the basis for the measurement of the electrical impedance of biological objects (Formula (1)), which describes the relation between resistance (R), current (I), and voltage (U) in an electrical circuit at a given time (t):

$$R(t) = U(t)/I(t) \tag{1}$$

However, it should be noted that within the AC system, current and voltage not only differ in their amplitude but also their phase (φ). In this case, resistance alone is not sufficient to describe these relations. Instead, the complex impedance (Z) or, in most cases, the magnitude of the impedance (|Z|) is used, containing resistance plus reactance

(X), which results from AC flow through capacitors and inductors driving the phase shift between voltage and current [9,10].

Thus, in AC systems:

$$|Z(f)| = \sqrt{[R^2 + X(f)^2]}$$
$$\varphi = \arctan (X/R) \quad (2)$$

Due to the characteristics of cell membrane, when performing impedance measurements on intact cells, the cells form connections with the resistor and the capacitor, in parallel. Here, resistance represents the opposition to current flow, whereas capacitance (C) describes the separation of electric carriers at the insulating bilayer of the cell membrane that causes polarization of the cell. The direct parameters derived from impedance measurements are the resistance and capacitance of cells. The quality and function of the cell barrier are represented by the resistance and, therefore, the resistance towards para- and trans-cellular current flow should be considered. Capacitance provides an overall measure of electrode coverage [9,11,12].

Therefore, the different behavior of the cells after their seeding, adherence, proliferation, and reaction to the substances added to the substrate, as a result, produces a change in impedance. To check whether the ECIS method would be a good method for assessing antiproliferative and anticancer activity, we chose compounds belonging to the triterpene family: betulin and betulinic acid. Betulin (BE; lup-20(29)-ene-3β,28-diol) and betulinic acid (BA; 3β-hydroxy-lup-20(29)-en-28-oic acid) belong to triterpenes and are found in the outer layer of the bark of white birch species such as *Betula alba*, *Betula verrucasa*, or *Betula pendula* [13]. The most important pharmacological activity of these triterpenoids is the inhibition of the development of some chemo-resistant tumors, such as melanoma or gliomas [10,14–17]. Although BE and BA are very similar in structure, there is a major difference in their cytotoxic activity—BA is less cytotoxic to healthy cells than BE [18]. However, both of them are characterized by a lack of toxicity both in vitro and in vivo within normal cells, and that is the reason why BE and BA are considered potential precursors of many new medicinal preparations [14,19,20].

This study aimed to determine whether the anticancer effects of betulin and betulinic acid on human lung carcinoma (A549) and human breast carcinoma (T47D) lines can be determined by the ECIS method, or only by common tests such as MTT or LDH. We chose such a configuration of cancer cell lines because lung cancer usually metastasizes to the nipple first [21]. For the Vero cell line, non-pathological epithelial cells served as a control for the experiment.

2. Results

2.1. The MTT Results

To examine whether BE and BA have cytotoxic activity in chosen cell line types, we first assessed their effect on the viability of cell lines with an MTT assay during a 48 h culture period. As shown in Table 1, BE inhibited cell growth in T47D, A549, and Vero cells with an IC_{50} of 7.8, 9.5, and 21.3 μM, respectively. BA, a structure-related derivative of betulin, showed a potent antiproliferative effect with lower IC_{50} values in the same cell lines (5.4, 6.9, and 18.6 μM, respectively), indicating that betulin has a mechanism slightly distinct from BA in the inhibition of cell growth.

Table 1. IC_{50} values of BE and BA on T47D and A549 cell lines as determined during 48 h period of MTT assay, using non-linear, four-parameter regression analysis.

	IC_{50} (μM)		
Compound/Cell Line	**T47D**	**A549**	**Vero**
BE	7.8	9.5	21.3
BA	5.4	6.9	18.6

2.2. The LDH Results

To obtain a reliable IC_{50} value of BE and BA towards a Vero cell, we applied another cytotoxicity test but, this time, using a different mechanism of action to MTT. Vero cells were treated with increasing doses of BE and BA, ranging from 1 μM to 100 μM during a 48 h culture period. LDH release was observed after 48 h of BE and BA treatment, and this release was significantly increased with 20 μM of betulin and betulinic acid (Figure 2). The LDH assay revealed that both compounds caused modest cytotoxicity in normal Vero cell culture, consistent with the results obtained from the MTT assay. Accordingly, 20 μM of BA and BE were used in further studies concerning Vero cell lines.

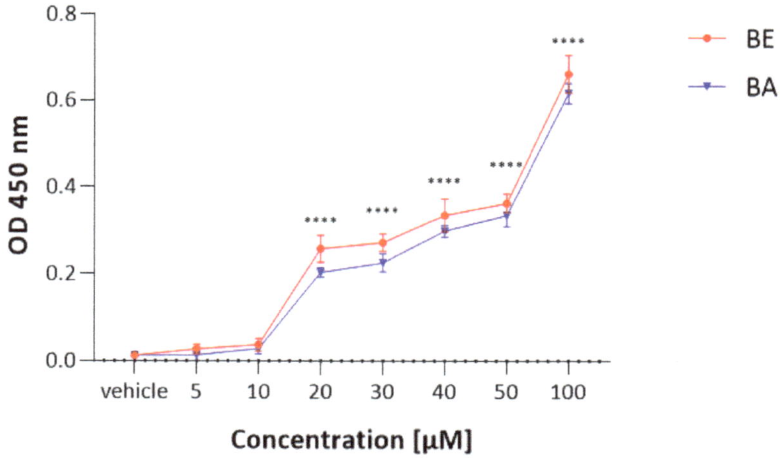

Figure 2. The influence of BE and BA on Vero cell viability measured by LDH assay. Data are shown as means ± SD; **** $p < 0.0001$ vs. vehicle; Dunnett's test.

2.3. The ECIS Results

Using the ECIS system, we monitored selected cell lines (T47D, A549, and Vero) continuously for up to 72 h and noted significant changes in impedance after the administration of appropriately selected concentrations of BE (8, 10, and 20 μM) and BA (5, 7, and 20 μM). Based on the obtained results, it can be noted that both substances affect electrical parameters during cultivation in a different mode (Figure 3). Every cell type has its characteristic adhesion and growth curve that can be manipulated by stimuli such as the chemical structure or concentration of substances in the medium. As made clear at the starting phase (cell attachment), the impedance increased in all cell cultures: A549 (Figure 3a), T47D (Figure 3b), and Vero (Figure 3c) cells, and it grew even further (negative control). After the addition of BE and BA, the impedance increase was insignificant for the A549 line, whereas recorded impedance values of the T47D line treated with BE showed a sharp decrease in impedance values and a slight decrease when treated with BA, followed by a linear increase in values for both cases. A similar trend of impedance value changes was observed for the Vero line.

Moreover, to present the influence of a well-known cytotoxic substance on the Vero cell line, rotenone was used. Cell viability was assessed after 48 h of incubation by the MTT assay (Figure 4a); then, two concentrations were used for ECIS monitoring (Figure 4b). The influence of a lower concentration of rotenone was characterized by a slow decrease in impedance, while treatment with a higher concentration resulted in an immediate decrease in impedance.

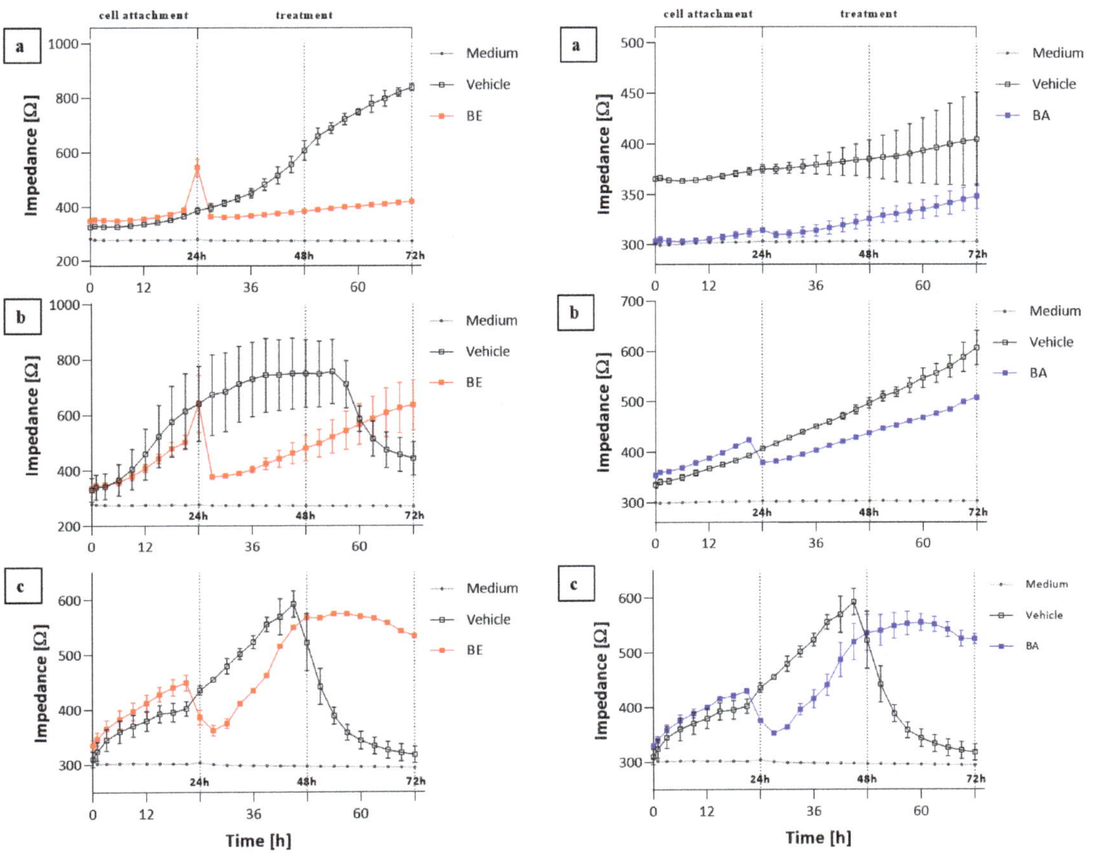

Figure 3. Impedance changes monitoring of the cell line (**a**) A549, (**b**) T47D, (**c**) Vero during 48 h treatment with BE (red line) and BA (blue line). Data are presented as mean value ± SEM.

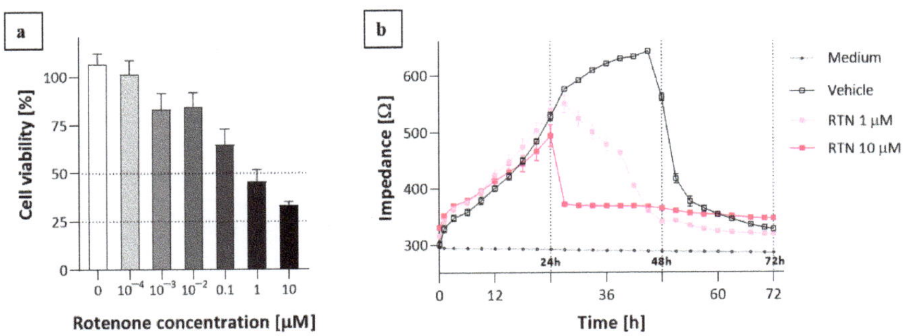

Figure 4. Influence of rotenone on Vero cell line during 48 h treatment assessed by (**a**) MTT assay and monitored by (**b**) ECIS method. Data are presented as mean value ± SEM.

3. Discussion

Explaining the relationship between changes in electrical parameters in cell cultures and the processes influencing its survival seems to be an interesting challenge. No matter how satisfactory this answer might be, in the face of new technical possibilities (such as ECIS) and assuming that changes in the electrical properties of cells precede changes on the biochemical level, it would be very interesting to examine the character and dynamics of these changes. It is possible only with the monitoring of selected electrical parameters, i.e., the impedance, resistance, and capacity of the cell membrane, in real time and after the application of chosen bioactive compounds to the examined cell lines [9–11,22].

Drug discovery and screening of bioactive compounds are often performed in cell-based test systems to reduce costs and save time. Traditionally, microscopy, spectrophotometry, and flow cytometry techniques are used to examine cell cultures. The abovementioned methods are considered standard in studies conducted on cell cultures. While these methods may provide insight into the physiological function of every single cell or into pathological changes that could have occurred, they usually require fluorescence, chemiluminescence, or radioactive ways of marking, which lead to cell destruction. The marking process causes the loss of important biological information about living cells [23,24]. ECIS is an innovative and non-invasive method used to monitor cell parameters such as cell membrane capacity, resistance, or impedance in real-time analysis. Additionally, ECIS measurements provide information about temporal changes in the cell–cell contacts which are not available for single-cell observations. Depending on the experimental setup, ECIS measurements show an excellent time resolution, ranging from seconds (for example, micromotion) to minutes. So far, this technique has been successfully used in the study of cell viability, the determination of IC_{50}, and the influence of various biological factors on cellular processes [25,26].

Previous studies have reported that BE and BA have an anticancer effect in human lung or breast cancer cell lines. In our experiment, we also confirmed the antiproliferative activity of these compounds on such lines, and our obtained IC_{50} values are in agreement with the literature [17,27,28] on the matter. Comparisons made using IC_{50} (Table 1) showed that the most sensitive cell line to BA and BE was T47D (breast carcinoma). This agrees with the results obtained by other authors [17,29]. Results regarding ECIS are rather similar, but their analysis allows us to observe the accurate influence of BE and BA on cells. Based on impedance measurements in the cultures of the examined cells exposed to BE, significant differences in the impedance values were observed. Likewise, the A549 line was the most sensitive to the action of BE. After BE addition, further cell growth was stopped rapidly compared to the negative control and lasted until the end of incubation, which indicates the antiproliferative effect of the test substance on these cells. If we compare the effect of BE on A549 cells, it can be seen that, unlike with betulin, the effect of betulinic acid on these cells cannot be demonstrated.

In the case of T47D cells (Figure 3b), the effect of BE was slightly different. After substance addition, a gradual but significant drop in impedance and, subsequently, a slow and linear growth was observed. Given that during the experiment the medium was not exchanged, the drop in impedance in the negative control cultures reflects the increase in cell deaths associated with the depletion of nutrients and the accumulation of metabolites. Therefore, the effect of betulin appears to be all the more interesting since, according to the results, it seems to prolong the life of T47D cells. On the other hand, BA had a pronounced and efficiently antiproliferative effect on T47D cells.

The effects of BE and BA on the cells of the Vero line are similar (Figure 3c) to those of T47D cells, although the increase in cell proliferation is greater, demonstrating the tolerance of these cells towards BE and BA, which was also confirmed with other normal lines [14,19,20]. Rotenone—an inhibitor of complex I of the electron transport chain (ETC)—was used as a positive control at two concentrations, which is a good example to illustrate the changes in impedance when a toxicant is used at concentrations higher than the IC50. Rotenone use at the 10 μM level caused an immediate decrease in the impedance value, proving a strong antiproliferative effect which was also confirmed by the MTT test.

The literature surrounding the study suggests that the selective cytotoxic activity of BE and BA may result from their direct influence on the mitochondrial bioenergetics and functioning of the lipid membrane. A well-known feature of neoplastic cells is their intensive proliferation, which means the intensification of their energy changes. Mitochondria lie at the centers of these changes, so the fate of the cell depends on their proper functioning. Although BE and BA are structurally similar, differing only in the substituent at C-17, they may act differently on the same cell type, for example, inducing the release of cytochrome C from the mitochondria in different ways [30]. Aside from this, possible triterpene–lipid interactions should also be considered. Rodrıguez et al. showed that pentacyclic monohydroxytriterpenes affect the dynamics and structural properties of the artificial lipid bilayer dipalmitoylphosphatidylcholine (DPPCB) and that some of the triterpenes, such as α-amyrin, are incorporated into the lipid bilayer at the same high concentrations as cholesterol [31]. The susceptibility of steroid-like compounds to incorporation into the artificial lipid membranes of multilamellar liposomes of DPPCB is strongly dependent on their structure [32,33]. Thus, the hypothesis for the existence of interactions between triterpenes and cellular membrane lipids (their having an ordering or destabilizing influence on the lipid components of the membrane) does not seem unfounded. Few studies focused on the structural similarity of BE and BA with cholesterol [33,34]. It is widely recognized that one of the most important structural functions of cholesterol is modulating the fluidity of lipid membranes. By blending in between the elastic membrane phospholipids, it modifies their interactions, consequently modulating the dynamics of the bilayer [35–37]. A similar effect is attributed to BE and BA: it is presumed that, due to their structural similarity to cholesterol, they may exhibit similar properties to the aforementioned cholesterol or to other steroids and thus show a very high affinity for lipid cell membranes. Nevertheless, the effect on the membranes may not be the same. Dubinin et al. reported that BE changes the surface properties of the lipid membrane, facilitating its aggregation or fusion. Furthermore, BE and BE can react with mitochondrial permeability transition (MPT) pores in the mitochondria [38]. Carvalho et al. came to similar conclusions on the effect of BE on the plasticity of lysosomal membranes [39]. Moreover, lipid membranes are mainly made of cardiolipin and have low cholesterol levels. Consequently, their activity may be disturbed by an excess of BE or BA, which could explain their cytotoxicity [33,34].

To sum up, it must be realized that a cytotoxic compound in one cell line may well be much less active or even inactive in other cell lines, even within the same cancer type. The antiproliferative effect of the compound may also depend on the particular phase of the cell cycle. Therefore, there may be some differences in its properties. In our opinion, for precise measurements of cell viability (e.g., the cytotoxicity of tested compounds), it is worth comparing two or more different vitality tests, e.g., MTT and LDH with the ECIS method. It should be emphasized that the widely used MTT test (which allows for measuring the activity of energy transformations in mitochondria) should be used with caution, since many living cells may not show oxidative activity in mitochondria. Additionally, MTT was reported to interact with thiol-containing antioxidants [40], plant extracts [41], and other biologically relevant substances [42], leading to false positive endpoint measurements. Therefore, it is very important to select appropriate research methods to best capture the changes taking place in cells as a result of the action of the tested compounds, as well as to develop and implement them.

4. Material and Methods

4.1. Preparation of Betulin and Betulinic Acid

Both betulin and betulinic acid with purity >99% and rotenone with purity >95% were purchased from Sigma-Aldrich (St. Louis, MO, USA). Stock solutions of betulin (100 mM), betulinic acid (100 mM), and rotenone (1 mM) were prepared in DMSO (Sigma-Aldrich, St. Louis, MO, USA) and stored at −20 °C.

4.2. Cell Lines Cultures

Human lung carcinoma (A549) and human breast carcinoma (T47D) were obtained from the European Collection of Cell Cultures and were cultured in RPMI-1640 media, supplemented with 10% heat-inactivated fetal bovine serum (FBS, Gibco, Waltham, MA, USA), penicillin G (100 U/mL) (Sigma-Aldrich, St. Louis, MO, USA), and streptomycin

(100 μg/mL) (Sigma-Aldrich, St. Louis, MO, USA). Cultures were kept at 37 °C in a humidified atmosphere of 95% air and 5% CO_2. The control group was a Vero (fibroblast-like kidney from African green monkey) cell line obtained from Sigma-Aldrich and cultured in DMEM (Sigma-Aldrich, St. Louis, MO, USA), supplemented with L-glutamine (Gibco, Waltham, MA, USA) and 10% heat-inactivated fetal bovine serum (FBS, Gibco). All examined cell lines were tested against mycoplasma contamination with microbiological assays.

4.3. The Cell Proliferation Assay—MTT Assay

The cell viability was assessed by employing a 3-(4,5-dimethylthiazol-2-yl)-2,5-diphenyltetrazolium bromide (MTT) assay (Roche, Basel, Switzerland) in which the yellow tetrazolium salt was metabolized by viable cells to purple formazan crystals. T47D, A549, and Vero cells were seeded on 96-well microplates (Nunc) at the density of 1×10^4 cells/well and left for 24 h. The next day, the culture medium was removed and the cells were exposed to serial dilutions (0, 1, 5, 10, 20, 40, 50, 75, and 100 μM) of BE and BA made in a serum-free medium, for 48 h. Additionally, the Vero cell line was exposed to rotenone in concentrations: 0.1, 1, 10, 100 nM, 1, and 10 μM. Each compound in each concentration was tested in triplicate. Next, the cells were incubated for 4 h with 20 μL of MTT solution (5 mg/mL). The formazan grains formed by viable cells were solubilized with 200 μL of DMSO, and the color intensity was measured at a 570 nm wavelength. The results were expressed as an IC_{50}—the concentration of compound (in μM) that inhibits the proliferation rate of the tumor cells by 50%, as compared to the untreated control cells. For further experiments, values close to the IC_{50} concentrations of BE and BA were chosen for each line. The experiment was performed in three independent repetitions.

4.4. Cytotoxicity Assay—LDH Assay

A cytotoxicity detection kit based on the measurement of lactate dehydrogenase activity was applied (Tox-7, Sigma). The assay is based on the reduction of NAD by the action of lactate dehydrogenase released from damaged cells. The resulting NADH is utilized in the stoichiometric conversion of a tetrazolium dye. To evaluate the effect on normal cell viability, Vero cells were treated with increasing doses of BE and BA, ranging from 1 μM to 100 μM during a 48 h culture period. Next, cells were collected and incubated with substrate mixture for 30 min at room temperature, in the dark. In the end, the reaction was terminated by the addition of 1 N HCl, and the color product was quantified spectrophotometrically at a 450 nm wavelength.

4.5. Impedance Sensing Assay—ECIS Assay

The ECIS system's Ztheta instrument (Applied Biophysics Ltd., Troy, NJ, USA) was used to measure the impedance. It contained two separate units: the station controller Zθ, located outside the incubator, and a docking station containing two 8-well plates, which were placed in the incubator space. The standard 8-well ECIS disposable arrays consist of gold film electrodes delineated with an insulating film and mounted on a 20 mil optically clear Lexan® polycarbonate substrate. The eight-well top assembly is made of polystyrene. The gold layer is sufficiently thin (approx. 50 nm) to allow microscopic observation of the cells using a standard inverted tissue culture microscope. Each well has a surface area for cell attachment and growth of ~0.8 cm^2 and holds a maximum volume of about 600 μL. Gold pads at the edge of the array connect electrodes to the ECIS electronics via contact with spring-loaded pins within the electrode array station. The electrodes used were 8W10E (Applied Biophysics Ltd., Troy, NJ, USA), which comprised 8 wells and 10 active electrodes in each well. ECIS electrodes were placed in a holder plate in a humid incubator at 37 °C and 5% CO_2. Prior to inoculation, the arrays were incubated for 24 h with DMEM (Vero cells) and RPMI (A549, T47D cells) in the incubator overnight. Following stabilization, the array was removed from the array station and inoculated with cells. Inoculation of arrays was carried out by 600 μL/well of cell suspension ~1.2×10^5 cells/mL. After 24 h, the examined compounds were added to inoculated wells in concentrations presented in Table 1, which were selected based on the MTT and LDH experiments' results.

After cell manipulation, the matrix holder was placed in an incubator, and real-time measurements were initiated. The maximum response for Z, R, and C occurred at different frequencies. In this study, the default optimal frequencies were used: resistance (R) 4000 Hz, impedance (Z) 32,000 Hz and capacitance (C) 64,000 Hz. The changes in cellular behavior in response to the compound were recorded as impedance signals, and the data obtained were processed through ECIS software (Applied Biophysics Ltd., Troy, NJ, USA). After cell stimulation with the tested substance, the morphological changes that followed were expressed in the impedance values measured with the ECIS system.

5. Conclusions

Our study, with the use of lung (A549) cancer, breast (T47D) cancer, and normal epithelial cells (Vero), confirmed the possibility and justification of using the ECIS technique in anticancer in vitro research. Due to its non-destructive measurements, ECIS allowed for further cell testing after the end of the experiment. An unquestionable advantage is the live-tracking of changes in cell morphology, which reflects the processes of proliferation or cytotoxicity. Therefore, the ECIS method can be successfully used as a replacement or in combination with the commonly used colorimetric assays for cell viability testing, which could be translated to develop better treatment protocols and obtain the best possible cytotoxicity effect against cancer cells in time.

Author Contributions: Investigation, A.H. (Anna Hordyjewska), M.P.-W., D.P.-Z. and A.O.; formal analysis, A.H. (Anna Hordyjewska), M.P.-W., D.P.-Z. and A.O.; methodology and supervision, A.H. (Anna Hordyjewska); writing—original draft, A.H. (Anna Hordyjewska), A.H. (Anna Horecka), M.P.-W. and Ł.K.; writing—review and editing, A.H. (Anna Hordyjewska), M.P.-W., Ł.K., T.M.-M. and J.K. All authors have read and agreed to the published version of the manuscript.

Funding: This research was funded by DS212 Medical University of Lublin.

Institutional Review Board Statement: Not applicable.

Informed Consent Statement: Not applicable.

Data Availability Statement: The data presented in this study are available on request from the corresponding author.

Conflicts of Interest: The authors declare no conflict of interest.

References

1. Giaever, I.; Keese, C.R. A morphological biosensor for mammalian cells. *Nature* **1993**, *366*, 591–592. [CrossRef] [PubMed]
2. Giaever, I.; Keese, C.R. Monitoring fibroblast behavior in tissue culture with an applied electric field. *Proc. Natl. Acad. Sci. USA* **1984**, *81*, 3761–3764. [CrossRef] [PubMed]
3. Pänke, O.; Balkenhohl, T.; Kafka, J.; Schäfer, D.; Lisdat, F. Impedance spectroscopy and biosensing. *Adv. Biochem. Eng. Biotechnol.* **2007**, *109*, 195–237.
4. Serrano, J.A.; Huertas, G.; Maldonado-Jacobi, A.; Olmo, A.; Pérez, P.; Martín, M.E.; Daza, P.; Yúfera, A. An Empirical-Mathematical Approach for Calibration and Fitting Cell-Electrode Electrical Models in Bioimpedance Tests. *Sensors* **2018**, *18*, 2354. [CrossRef] [PubMed]
5. Lo, C.M.; Keese, C.R.; Giaever, I. Impedance analysis of MDCK cells measured by electric cell-substrate impedance sensing. *Biophys. J.* **1995**, *69*, 2800–2807. [CrossRef]
6. Opp, D.; Wafula, B.; Lim, J.; Huang, E.; Lo, J.-C.; Lo, C.-M. Use of electric cell-substrate impedance sensing to assess in vitro cytotoxicity. *Biosens. Bioelectron.* **2009**, *24*, 2625. [CrossRef] [PubMed]
7. Robilliard, L.D.; Kho, D.T.; Johnson, R.H.; Anchan, A.; O'Carroll, S.J.; Graham, E.S. The importance of multifrequency impedance sensing of endothelial barrier formation using ECIS technology for the generation of a strong and durable paracellular barrier. *Biosensors* **2018**, *8*, 64. [CrossRef]
8. Szulcek, R.; Bogaard, H.J.; van Nieuw Amerongen, G.P. Electric cell-substrate impedance sensing for the quantification of endothelial proliferation, barrier function, and motility. *J. Vis. Exp.* **2014**, *28*, e51300. [CrossRef]
9. Bennet, D.; Kim, S. Impedance-based cell culture platform to assess light-induced stress changes with antagonist drugs using retinal cells. *Anal. Chem.* **2013**, *85*, 4902–4911. [CrossRef]

10. Dehelean, C.A.; Feflea, S.; Molnár, J.; Zupko, I.; Soica, C. Betulin as an Antitumor Agent Tested in vitro on A431, HeLa and MCF7, and as an Angiogenic Inhibitor in vivo in the CAM Assay. *Nat. Prod. Commun.* **2012**, *7*, 1934578X1200700. [CrossRef]
11. Xie, F.; Xu, Y.; Wang, L.; Mitchelson, K.; Xing, W.; Cheng, J. Use of cellular electrical impedance sensing to assess in vitro cytotoxicity of anticancer drugs in a human kidney cell nephrotoxicity model. *Analyst* **2012**, *137*, 1343–1350. [CrossRef] [PubMed]
12. Xu, Y.; Xie, X.; Duan, Y.; Wang, L.; Cheng, Z.; Cheng, J. A review of impedance measurements of whole cells. *Biosens. Bioelectron.* **2016**, *77*, 824–836. [CrossRef] [PubMed]
13. Ostapiuk, A.; Kurach, Ł.; Strzemski, M.; Kurzepa, J.; Hordyjewska, A. Evaluation of Antioxidative Mechanisms In Vitro and Triterpenes Composition of Extracts from Silver Birch (*Betula pendula* Roth) and Black Birch (*Betula obscura* Kotula) Barks by FT-IR and HPLC-PDA. *Molecules* **2021**, *26*, 4633. [CrossRef] [PubMed]
14. Fulda, S. Betulinic Acid for Cancer Treatment and Prevention. *Int. J. Mol. Sci.* **2008**, *9*, 1096. [CrossRef]
15. Kvasnica, M.; Sarek, J.; Klinotova, E.; Dzubak, P.; Hajduch, M. Synthesis of phthalates of betulinic acid and betulin with cytotoxic activity. *Bioorganic Med. Chem.* **2005**, *13*, 3447–3454. [CrossRef]
16. Oh, S.H.; Choi, J.E.; Lim, S.C. Protection of betulin against cadmium-induced apoptosis in hepatoma cells. *Toxicology* **2006**, *220*, 1–12. [CrossRef]
17. Rzeski, W.; Stepulak, A.; Szymański, M.; Juszczak, M.; Grabarska, A.; Sifringer, M.; Kaczor, J.; Kandefer-Szerszeń, M. Betulin elicits anti-cancer effects in tumour primary cultures and cell lines in vitro. *Basic Clin. Pharmacol. Toxicol.* **2009**, *105*, 425–432. [CrossRef]
18. Zuco, V.; Supino, R.; Righetti, S.C.; Cleris, L.; Marchesi, E.; Gambacorti-Passerini, C.; Formelli, F. Selective cytotoxicity of betulinic acid on tumor cell lines, but not on normal cells. *Cancer Lett.* **2002**, *175*, 17–25. [CrossRef]
19. Fulda, S.; Kroemer, G. Targeting mitochondrial apoptosis by betulinic acid in human cancers. *Drug Discov. Today* **2009**, *14*, 885–890. [CrossRef]
20. Hordyjewska, A.; Ostapiuk, A.; Horecka, A. Betulin and betulinic acid in cancer research. *J. Pre-Clin. Clin. Res.* **2018**, *12*, 72–75. [CrossRef]
21. Bhattarai, B.; Schmidt, M.F.; Ghosh, M.; Sinha Ray, A.; Manhas, S.; Oke, V.; Agu, C.C.; Basunia, M.R.; Enriquez, D.; Quist, J.; et al. Lung Cancer with Skin and Breast Metastasis: A Case Report and Literature Review. *Case Rep. Pulmonol.* **2015**, *2015*, 136970. [CrossRef] [PubMed]
22. McCoy, M.H.; Wang, E. Use of electric cell-substrate impedance sensing as a tool for quantifying cytopathic effect in influenza A virus infected MDCK cells in real-time. *J. Virol. Methods* **2005**, *130*, 157–161. [CrossRef] [PubMed]
23. Pennington, M.R.; Van de Walle, G.R. Electric Cell-Substrate Impedance Sensing To Monitor Viral Growth and Study Cellular Responses to Infection with Alphaherpesviruses in Real Time. *mSphere* **2017**, *2*, e00039-17. [CrossRef] [PubMed]
24. Wegener, J.; Keese, C.R.; Giaever, I. Electric cell-substrate impedance sensing (ECIS) as a noninvasive means to monitor the kinetics of cell spreading to artificial surfaces. *Exp. Cell Res.* **2000**, *259*, 158–166. [CrossRef] [PubMed]
25. Nahid, M.; Campbell, C.; Ong, K.; Barnhill, J.; Washington, M. An evaluation of the impact of clinical bacterial isolates on epithelial cell monolayer integrity by the electric Cell-Substrate Impedance Sensing (ECIS) method. *J. Microbiol. Methods* **2020**, *169*, 105833. [CrossRef] [PubMed]
26. An, Y.; Jin, T.; Zhang, F.; He, P. Electric cell-substrate impedance sensing (ECIS) for profiling cytotoxicity of cigarette smoke. *J. Electroanal. Chem.* **2019**, *834*, 180–186. [CrossRef]
27. Kutkowska, J.; Strzadala, L.; Rapak, A. Synergistic activity of sorafenib and betulinic acid against clonogenic activity of non-small cell lung cancer cells. *Cancer Sci.* **2017**, *108*, 2265–2272. [CrossRef]
28. Tiwari, R.; Puthli, A.; Balakrishnan, S.; Sapra, B.K.; Mishra, K.P. Betulinic Acid-Induced Cytotoxicity in Human Breast Tumor Cell Lines MCF-7 and T47D and its Modification by Tocopherol. *Case Rep. Pulmonol.* **2014**, *32*, 402–408. [CrossRef]
29. Amiri, S.; Dastghaib, S.; Ahmadi, M.; Mehrbod, P.; Khadem, F.; Behrouj, H.; Aghanoori, M.R.; Machaj, F.; Ghamsari, M.; Rosik, J.; et al. Betulin and its derivatives as novel compounds with different pharmacological effects. *Biotechnol. Adv.* **2020**, *38*, 107409. [CrossRef]
30. Fulda, S.; Scaffidi, G.; Susin, S.A.; Krammer, P.H.; Kroemer, G.; Peter, M.E.; Debatin, K.M. Activation of mitochondria and release of mitochondrial apoptogenic factors by betulinic acid. *J. Biol. Chem.* **1998**, *273*, 33942–33948. [CrossRef]
31. Rodríguez, S.; Garda, H.A.; Heinzen, H.; Moyna, P. Effect of plant monofunctional pentacyclic triterpenes on the dynamic and structural properties of dipalmitoylphosphatidylcholine bilayers. *Chem. Phys. Lipids* **1997**, *89*, 119–130. [CrossRef]
32. Schroeder, F.; Atshaves, B.P.; McIntosh, A.L.; Gallegos, A.M.; Storey, S.M.; Parr, R.D.; Jefferson, J.R.; Ball, J.M.; Kier, A.B. Sterol carrier protein-2: New roles in regulating lipid rafts and signaling. *Biochim. Biophys. Acta* **2007**, *1771*, 700–718. [CrossRef] [PubMed]
33. Flasiński, M.; Hąc-Wydro, K.; Broniatowski, M. Incorporation of Pentacyclic Triterpenes into Mitochondrial Membrane—Studies on the Interactions in Model 2D Lipid Systems. *J. Phys. Chem. B* **2014**, *118*, 12927–12937. [CrossRef] [PubMed]
34. Broniatowski, M.; Flasiński, M.; Zięba, K.; Miśkowiec, P. Interactions of pentacyclic triterpene acids with cardiolipins and related phosphatidylglycerols in model systems. *Biochim. Biophys. Acta Biomembr.* **2014**, *1838*, 2530–2538. [CrossRef]
35. Kimura, Y.; Kioka, N.; Kato, H.; Matsuo, M.; Ueda, K. Modulation of drug-stimulated ATPase activity of human MDR1/P-glycoprotein by cholesterol. *Biochem. J.* **2007**, *401*, 597–605. [CrossRef]
36. Ohvo-Rekilä, H.; Ramstedt, B.; Leppimäki, P.; Peter Slotte, J. Cholesterol interactions with phospholipids in membranes. *Prog. Lipid Res.* **2002**, *41*, 66–97. [CrossRef]

37. Prinz, W.A. Non-vesicular sterol transport in cells. *Prog. Lipid Res.* **2007**, *46*, 297–314. [CrossRef]
38. Dubinin, M.V.; Semenova, A.A.; Ilzorkina, A.I.; Mikheeva, I.B.; Yashin, V.A.; Penkov, N.V.; Vydrina, V.A.; Ishmuratov, G.Y.; Sharapov, V.A.; Khoroshavina, E.I.; et al. Effect of betulin and betulonic acid on isolated rat liver mitochondria and liposomes. *Biochim. Biophys. Acta Biomembr.* **2020**, *1862*, 183383. [CrossRef]
39. Carvalho, F.S.; Morais, C.M.; Holy, J.; Krasutsky, D.; Yemets, S.V.; Krasutsky, P.A.; Jurado, A.S.; Oliveira, P.J.; Serafim, T.L. Toxicity of lupane derivatives on anionic membrane models, isolated rat mitochondria and selected human cell lines: Role of terminal alkyl chains. *Chem. Biol. Interact.* **2018**, *296*, 198–210. [CrossRef]
40. Natarajan, M.; Mohan, S.; Martinez, B.; Meltz, M.; Herman, T. Antioxidant compounds interfere with the 3-[4,5-dimethylthiazol-2-yl]-2,5-diphenyltetrazolium bromide cytotoxicity assay. *Cancer Detect. Prev.* **2000**, *24*, 405–414.
41. Karakaş, D.; Ari, F.; Ulukaya, E. The MTT viability assay yields strikingly false-positive viabilities although the cells are killed by some plant extracts. *Turk. J. Biol.* **2017**, *41*, 919–925. [CrossRef] [PubMed]
42. Bruggisser, R.; von Daeniken, K.; Jundt, G.; Schaffner, W.; Tullberg-Reinert, H. Interference of plant extracts, phytoestrogens and antioxidants with the MTT tetrazolium assay. *Planta Med.* **2002**, *68*, 445–448. [CrossRef] [PubMed]

molecules

Review

Health-Promoting Properties of Medicinal Mushrooms and Their Bioactive Compounds for the COVID-19 Era—An Appraisal: Do the Pro-Health Claims Measure Up?

Jennifer Mary Phillips [1,2], Soo Liang Ooi [1] and Sok Cheon Pak [1,*]

1 School of Dentistry and Medical Sciences, Charles Sturt University, Bathurst, NSW 2795, Australia; jennifer-phillips@live.com.au (J.M.P.); sooi@csu.edu.au (S.L.O.)
2 LAGOM NutriHealing, 16 Gentile Court, Hobart, TAS 7010, Australia
* Correspondence: spak@csu.edu.au; Tel.: +61-2-6338-4952; Fax: +61-2-6338-4993

Abstract: Many mushroom species are consumed as food, while significant numbers are also utilised medicinally. Mushrooms are rich in nutrients and bioactive compounds. A growing body of in vitro, in vivo, and human research has revealed their therapeutic potentials, which include such properties as anti-pathogenic, antioxidant, anti-inflammatory, immunomodulatory, gut microbiota enhancement, and angiotensin-converting enzyme 2 specificity. The uses of medicinal mushrooms (MMs) as extracts in nutraceuticals and other functional food and health products are burgeoning. COVID-19 presents an opportunity to consider how, and if, specific MM compounds might be utilised therapeutically to mitigate associated risk factors, reduce disease severity, and support recovery. As vaccines become a mainstay, MMs may have the potential as an adjunct therapy to enhance immunity. In the context of COVID-19, this review explores current research about MMs to identify the key properties claimed to confer health benefits. Considered also are barriers or limitations that may impact general recommendations on MMs as therapy. It is contended that the extraction method used to isolate bioactive compounds must be a primary consideration for efficacious targeting of physiological endpoints. Mushrooms commonly available for culinary use and obtainable as a dietary supplement for medicinal purposes are included in this review. Specific properties related to these mushrooms have been considered due to their potential protective and mediating effects on human exposure to the SARS CoV-2 virus and the ensuing COVID-19 disease processes.

Keywords: COVID-19; β-glucans; immunomodulation; anti-inflammation; anti-oxidant; ACE2 regulation

Citation: Phillips, J.M.; Ooi, S.L.; Pak, S.C. Health-Promoting Properties of Medicinal Mushrooms and Their Bioactive Compounds for the COVID-19 Era—An Appraisal: Do the Pro-Health Claims Measure Up? *Molecules* **2022**, *27*, 2302. https://doi.org/10.3390/molecules27072302

Academic Editor: Ricardo Calhelha

Received: 25 February 2022
Accepted: 30 March 2022
Published: 1 April 2022

1. Introduction

Mushrooms have long been regarded as healthful and widely consumed for their culinary and nutritional values. Some species have an ancient tradition as medicinal therapies, and increasingly, this is being realised in a contemporary context [1,2]. However, until recently, the scientific understanding of mushrooms' application as a medicinal agent has been chiefly empiric [3]. Interest in advancing the health properties and pharmacological activities means that clinical research is growing, and much of the traditional knowledge is being documented and validated [4]. Indeed, there is an interdisciplinary field of science studying medicinal mushrooms (MMs), with an increasing number of human studies emerging. This, along with industry technological developments, means that some mushrooms are now regarded as a class of drugs called "mushroom pharmaceuticals" [1]. Physiological activities revealed from numerous studies include anti-pathogenic, antioxidant, anti-inflammatory, immunomodulatory, and anticoagulation effects, plus gut microbiota enhancement, pulmonary cytoprotection, and angiotensin-converting enzyme (ACE) 2 specificity [4,5]. These actions are mainly attributable to the bioactive compounds present in the fruiting bodies and the mycelium, depending on the species [2]. For example, *Ganoderma lucidum* contains more than 120 different triterpenes plus polysaccharides, proteins, and other bioactive compounds [6].

Coronavirus disease 2019 (COVID-19), caused by the severe acute respiratory syndrome coronavirus 2 (SARS CoV-2), presents an opportunity to consider how macro-fungi or their derivatives might be harnessed and utilised as therapeutic agents. In particular, this pertains to their use in optimising health to avoid or mitigate risk factors, prevent severe disease outcomes, and improve recovery prospects as commonly noted or experienced. For example, the immune response is crucial in COVID-19 pathogenesis, and its modulation to control the hyperinflammatory response would be advantageous. Additionally, the virus appears to have a higher prevalence for and more severe outcomes in populations with particular risk factors and comorbidities associated with age, obesity, metabolic, cardiovascular, and inflammation-mediated conditions [7,8]. Many of these factors are known to respond to, and are modifiable through, dietary or lifestyle interventions. Therefore, with their potential and specific therapeutic effects, MMs seem well placed to be considered as nutraceutical options, or at the very least, as health-promoting food sources. Moreover, as vaccines become the mainstay for preventing severe outcomes, finding effective immune-enhancing adjuncts will be beneficial.

MMs are increasingly being utilised as a nutritional food source in the "food as medicine" health-promoting dietary approach [9]. They are also used as dietary supplements, biocontrol agents, and cosmetics. Most pertinent is their use as nutraceuticals and natural products for pharmacological therapy, which are valued for their immunological, anti-inflammatory, and health-promoting effects [4]. However, of prime consideration is that mushrooms' properties, mechanisms of action, and potential efficacies can be influenced by many variables, including climate, location, cultivation, processing, and extraction techniques [4,10]. The nomenclature may also be a problem as it affects accurate identification and, thus, attribution. As more research is achieved, and commercialisation and application continue to increase, these variables must be addressed. However, more work may yet be required. Particular aspects such as extraction methods may be very pertinent in this quest.

In the broader context of COVID-19, this narrative review investigates the use of mushrooms as medicine, exploring the bioactive compounds they contain and the associated pro-health claims. Specific properties have been considered due to their potential health enhancement or mediating effects on human exposure to the SARS CoV-2 virus or the ensuing COVID-19 disease processes. Mechanisms of action, physiological effects, and potential capabilities such as antiviral, antioxidant, immunomodulatory, cardiovascular regulation, and health-promoting factors will be highlighted.

The research question has two parts; (1) does the research support the potential of MMs to enhance health and protect against or ameliorate symptoms associated with COVID-19, and (2) what limitations currently impact utilising mushrooms as a reliable medicinal treatment? Research findings associated with specific mushrooms such as *Agaricus* spp. (e.g., *bisporus*, (common white button/brown)), *Cordyceps militaris*, *Flammulina velutipes/enokitake* (Enoki), *G. lucidum* (Reishi), *Grifola frondosa* (Maitake), *Hericium erinaceus* (Lion's Mane), *Lentinus edodes* (Shiitake), *Pleurotus* spp. (e.g., *ostreatus*, (Oyster)), and *Trametes versicolor* (Turkey Tail) are considered.

2. Research Methods

Major databases were initially searched via the Primo search tool of Charles Sturt University Library using the keywords "mushrooms" and "health" and "COVID-19" or SARS CoV-2". Further searches included specific mushroom species such as "*Ganoderma* spp." and compounds such as "β-glucans" and health effects such as "inflammation" or "immunomodulation". Reference lists associated with pertinent papers were examined extensively for additional source material.

3. SARS CoV-2 Virus and COVID-19 Infection

Human coronaviruses such as the common cold generally cause mild illnesses but can mutate over time [11]. SARS CoV-2, a novel species of the Coronaviridae family, is the infective agent that causes COVID-19. Structural elements of the virus critically enable attachment, cell activity, multiplication, and infectiousness [12], with ACE2 being the primary entry receptor [13]. SARS CoV-2 has become a global concern due to its mutative nature, transmissibility, virulence factors, and the severity of the disease processes [12]. SARS CoV-2 enters the lungs through the respiratory tract, directly infecting upper and lower tract cells. Infection typically involves the upper respiratory area and nasal ciliated epithelial cells, the lung, and alveolar epithelial cells [7,13]. Additionally involved are other types of endothelial cells of the arteries, immune, smooth muscle, and intestine [7]. This presents implications for systemic organ involvement, particularly through the pro-inflammatory immune response that characterises the disease sequelae.

Presentation of the disease can range from no symptoms (asymptomatic) to severe illness with potentially life-threatening complications or fatal outcomes [11]. For the vast majority, symptoms are similar to those of mild to moderate influenza. Present may be fever, dry cough, sore throat, muscle pain, and fatigue in the initial phase, and it can extend to headache, anorexia, malaise, dyspnoea, nasal congestion, haemoptysis, diarrhoea, lymphopaenia, and difficulty with reading and distinguishing smells in the ensuing phases [7,12]. Continuing respiratory distress can lead to acute respiratory distress syndrome (ARDS), requiring intensive medical intervention [12]. While most people recover, at increased risk for severity are the immunocompromised, elderly, male gender, and people with comorbid conditions such as cardiovascular diseases (CVD), diabetes, hypertension, and poor nutritional status [7,14,15].

4. Immune and Inflammatory Responses to COVID-19

Host innate and adaptive immune cells, particularly lymphocytes (T cells, B cells, and natural killer (NK) cells), are called upon to defend against SARS CoV-2 viral invasion and tissue damage [12,15]. The optimal immune response in humans involves the coordination of cytokine and chemokine activation, recruitment of defence cells, and secretion of antibodies in a timely, localised, and functional way. The combined immune and antiviral activation is tightly regulated and balanced to eliminate and resolve the viral invasion and promote tissue repair [15].

However, interference with, and aberration of, the immune responses along with associated hyperinflammation is a key characteristic of the increased disease severity of COVID-19 [12,13]. In the progression from a normal response to one that may lead to death, total T cells along with CD8+ and CD4+ required for clearing viral invasion are known to decrease markedly, leading to cell exhaustion and dysfunction [14–16]. Triggered, recruited, and increased in a highly organised cellular and molecular cascade are pro-inflammatory type 1 and type 2 cells, particularly interleukins of IL-6, IL-10, IL-2, and IL-7, as well as granulocyte, colony-stimulating factor, and tumour necrosis factor (TNF) [12,13]. There is also interference in the native immune response generated via Toll-like and other receptors. This triggers the expression of interferons and activates antiviral effectors such as NK cells and macrophages, and DNA replication and transcription anomalies triggering apoptotic pathways [7].

Systemic manifestations involve severe inflammation, respiratory complications, alterations of the circulatory system through endothelial cell interactions and damage to the vascular barrier, capillaries, and organs, dysregulation of the ACE2 receptor gateway, and disruption of the renin–angiotensin–aldosterone system (RAAS) [14,15,17]. Furthermore, cardiometabolic disease risk factors are implicated in the increased severity, morbidity, and mortality associated with COVID-19. Of note is that cardiovascular mortality is greater in all influenza pandemics than in all other causes, and acute respiratory viral infections are triggering factors for cardiovascular disease. Events such as myocardial infarction and inflammation, thromboembolism, and vasculitis are evident, and high blood pressure, obesity, and diabetes are known comorbidities [7,18,19].

5. Mushrooms as Prevention or Treatment for COVID-19

5.1. General Features

Mushrooms are macro-fungi with a distinctive fruiting body, either hypogeous (underground) or epigeous (aboveground). There are estimated to be around 140,000 known species of macro-fungi belonging mainly to the phyla Basidiomycetes and some to Ascomycetes. Many more hundreds of thousands of fungi are assumed yet to be identified and classified [4,6,20]. Some 2000 are edible, and a few hundred wild and cultivated mushrooms have long been utilised as MMs [5,20]. Nutritionally, mushrooms are low in energy and are generally a good source of macro and micronutrients and trace elements, although there is variability [21–24]. As functional foods, they are also anti-inflammatory and known to modulate gut bacteria [25]. Species-specific structural and maturation elements of fruiting bodies and mycelia, and potentially even the fermented substrate from which the prepared product may have been produced, impact effects. Variances involve the chemistry, bioactive fractions derived from them, and biological activity [2,17,26].

5.2. Structural Elements and Bioactive Compounds

MMs produce bioactive primary and secondary metabolites and specific molecular weight compounds such as polysaccharides, polysaccharide–protein complexes, polyphenols, terpenoids, lectins, coumarins, ribosomal and non-ribosomal peptides, peptidoglycans, alkaloids, fatty acids, sterols, and antioxidants [1,2,5,6,20,27–29]. The most studied health-promoting properties and effects of MMs appear to be related to polysaccharides, lectins, protein complexes, sterols, and polyphenols such as terpenoids [2,17]. For example, lectins are non-immunoglobulin binding storage proteins that play a crucial role in such biological processes as cell signalling, cell–cell interactions in the immune system, and host defence mechanisms [20]. Mushroom polysaccharides have a strong ability to carry biological information, and via such function, they have antitumour, antioxidant, immunomodulatory, anti-inflammatory, antimicrobial, anti-obesity, and anti-diabetic effects [18]. To some extent, all of the bioactive compounds mentioned above have a body of knowledge that could be examined in respect of SARS CoV-2 virus protection and COVID-19 symptom reduction, as these compounds demonstrated beneficial effects that may be very helpful. Notably, these effects include inflammation regulation, immune and reactive oxygen species (ROS) modulation, host defence mechanisms involvement, microbial activity, pulmonary cytoprotection, and ACE2 specificity or inhibition [4,17,30].

However, β-glucans (D fraction) appear to have wide-ranging effects and benefits that deserves further examination. Additionally, mushroom-derived β-glucans are becoming more common in nutraceutical products and are promoted in a functional culinary sense. The limitations that may impact the appropriate application must also be considered. For example, their efficacy is very dependent on species, extraction techniques, and methods.

5.3. β-Glucans

Glucans are heterogeneous polysaccharides, with short or long-chain glucose polymers linked by large numbers of glucose sub-units, different branching, branch linkage, and backbone structures. They are a natural component of the cell walls. In addition to mushrooms, glucans are also a constituent of foods such as cereals. Thus, to be clear, the fungi glucans, in particular, β-glucans, are discussed and referred to as edible mushroom polysaccharides (EMPs).

The molecular weight, chain length, and side-branching structures of mushroom β-glucans can be species and cultivar specific, influencing variance, complexity, and biological activity [5,17,31,32]. The purity and purification processes are essential in characterising the structure [25]. Growing environments, drying conditions, and isolation/extraction methods are also crucial for the intensity of β-glucans' activities [18]. The solubility and particulate size of β-glucans are important physical features as many receptors involved can activate different immune responses [31]. For example, particulate β-glucans are known to directly stimulate immune cell activation through a dectin-1 pathway, while soluble β-glucans require complement receptor-3 dependent pathway activation.

The immunomodulatory properties of mushroom-derived β-glucans exert more potent immunoinflammatory effects than other types and have long been recognised for this [31,33]. Immunomodulation is characterised by the ability to correct deviated immune functions. This may be through supporting declined or suppressed parameters or normalising overactive or increased functions [34]. Notably, due to their confirmed complex mode of action, β-glucans are recognised as biological response modifiers (BRMs). They induce epigenetic programming in innate immune cells to produce a more robust immune response and act as pathogen-associated molecular patterns, binding to specific pathogen recognition receptors, inducing innate and adaptive immune responses [31,33]. They can also stimulate the activity of macrophages [35] and neutrophils, support NK cell activity, influence the production of cytokines and chemokines, and modulate antibody production, amongst many other functions [34].

Several authors have described the concept of trained immunity [14,33,36] as being a modified and epigenetic innate immune response capable of producing antibody-free memory to a secondary heterologous stimulus that is more robust. However, Geller and Yan [36] cautioned that due to the development of hyperinflammatory symptoms, such as those that may occur in COVID-19, the use of a therapy that could induce trained immunity effects warrants particular consideration. Notwithstanding, due to their immune enhancement benefits, studies examined the use of biomacromolecule compounds, including polymers such as β-glucans and chitosan as vaccine adjuvants [37–39]. Moreover, the potential of using β-glucans as a wide-spectrum immune-balancing food-supplement-based enteric vaccine adjuvant for COVID-19 was explored by Ikewaki et al. [33].

6. Systemic Pro-Health Responses and Activities Associated with Specific MMs

The following synopsis examines the various pro-health effects and activities of specific macro-fungi elucidated in research that may have applications to the pathological sequelae and outcomes associated with COVID-19. The effects and potential benefits of MMs are illustrated in Figure 1.

6.1. Anti-Pathogenic

The basic viral cycle, also associated with SARS CoV-2, involves attachment, penetration, uncoating, replication, assembly, and release [40]. Mushroom extracts and bioactive compounds impede viral entry into host cells and multiplication, inhibit virus adsorption, replication, nucleic acid synthesis, and disrupt other pathogens [3,6,14]. It is known that proteolytic enzymes facilitate the cleavage of S glycoprotein, which is a critical step in SARS CoV-2 viral attachment [12]. Since protease inhibitors have been isolated from *G. lucidum*, *C. militaris*, and *A. bisporus* [14], MMs may have therapeutic utility.

Cordycepin isolated from *C. militaris* exerted an antiviral effect through a protein kinase inhibitory mechanism and an inhibitory role towards ribonucleic acid (RNA) synthesis and Epstein–Barr virus (EBV) replication [41]. Ganoderma compounds isolated from *G. lucidum* effectively inhibited human immunodeficiency virus (HIV)-1 and HIV-1 protease [42]. Additionally, various triterpenoids (isolated from *G. lucidum* and other Ganoderma species) were active against HIV-1, influenza type A, and herpes simplex type 1 [6]. In vitro and in vivo studies on a range of common viral agents from *Agaricus* spp. including *Agaricus blazei* Murril (AbM), *H. erinaceus*, and *G. frondosa* demonstrated antiviral properties [3].

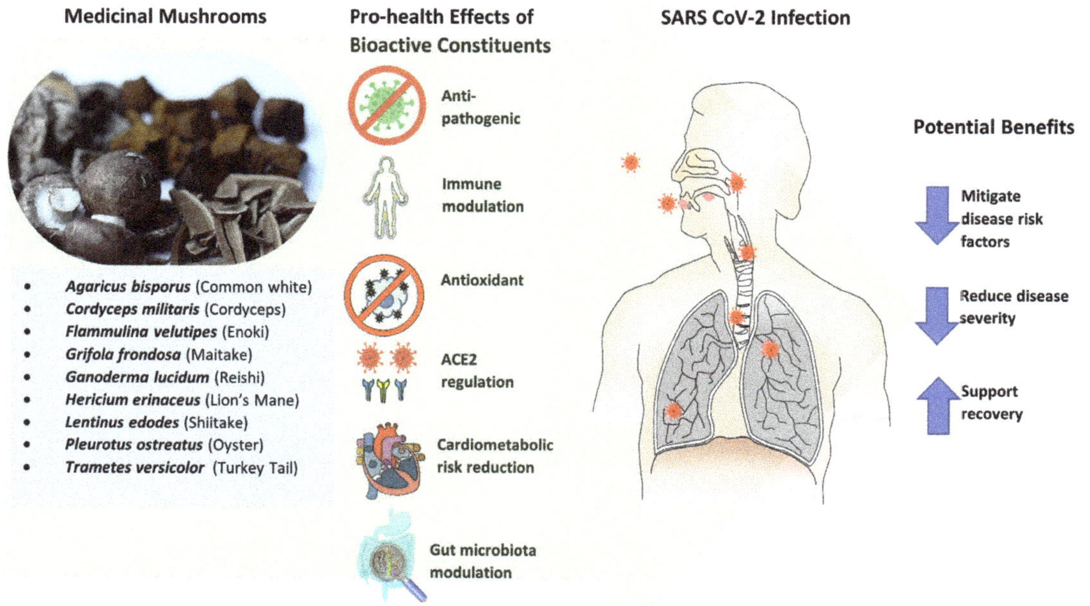

Figure 1. Medicinal mushrooms demonstrated modulatory and regulatory effects via the actions of bioactive compounds. These effects may apply in the pathophysiology and sequelae of SARS CoV-2 infection in humans. Potential benefits to consider and investigate specifically in this context involve mitigating disease risk factors, reducing disease severity, and supporting recovery.

MMs are also purported to have anti-bacterial functions [1]. To confirm this potential, Hearst et al. [43] conducted an in vitro microbiological assessment using aqueous extracts of *L. edodes* and *P. ostreatus*. The aqueous extract of *L. edodes* demonstrated potent activity when tested in culture against 29 bacterial isolates (Gram-positive and Gram-negative) and 10 fungal/yeast agents. Here, 85% of the bacterial and 50% of the fungal organisms were inhibited by the *L. edodes* extract. The results compared favourably against Ciproflaxin, which is a broad-spectrum antibiotic that was deployed as the control. In contrast, *P. ostreatus* aqueous extract showed minimal activity on the same range of pathogens, with only three out of 39 samples inhibited, while none of the yeast and mould species was affected. Additionally, a purified source of lentinan, a specific class of β-glucan, reduced populations of multiple antibiotic-resistant clinical isolate *Klebsiella pneumoniae* in an in vivo lung infection model and showed potential for treating sepsis-induced lung injury and boosting type 1 interferon response to RNA viruses such as influenza and coronavirus [17].

6.2. Immune Modulation

A response elicited from *L. edodes* named "the lentinan antiviral effect" has been attributed to innate immune responses and specific immunity regulation. Acting as a BRM, lentinan can promote T helper cell (Th) type 1 response and improve Th1/Th2 balance. It may also activate inflammasomes, enhance immune cells, activate the complement system, and promote cytotoxicity and phagocytosis [2,44]. An in-house hot water extract of *L. edodes* was compared to a commercially sourced lentinan extract (Carbosynth–Lentinan (CL)) to investigate if isolates could alleviate the immune cascade in conditions experienced by COVID-19 patients, such as ARDS. β-glucans from *L. edodes* reduced IL-1β and IL-6 in lung injury and activated macrophages in vitro [17]. β-glucans were also used to investigate oxidative stress alleviation in H_2O_2-treated THP-1 cells. Viability, apoptosis and necrosis were assessed. CL extract attenuated oxidative stress-induced early apoptosis, and the in-house lentinan extract attenuated late apoptosis [17].

Lectin derived from *P. ostreatus* has been studied as a hepatitis B virus DNA vaccine adjuvant and demonstrated effectiveness in enhancing surface protein antibodies [45]. Studies utilising pleuran (insoluble β-glucans derived from *P. ostreatus*) administered in oral liquid syrup form have suggested numerous positive immunomodulatory effects in recurrent upper and lower respiratory tract infection (RRTI). Demonstrated effects of pleuran, particularly in studies with children, include reduced incidences of RRTI, otitis media, tonsillopharyngitis, bronchitis, laryngitis, and other flu and cold-like symptoms, plus fewer days off school [46–48].

AbM extract is another rich source of BRMs. Via the actions of, for example, proteoglycans, β-glucans, and ergosterol, anti-inflammatory, anti-pathogenic, and immunomodulatory cytokine effects were stimulated, vaccine efficacy was improved, and cytotoxic effects were induced [49–52]. Andosan™, a product primarily manufactured from AbM extract, combined with *H. erinaceus* and *G. frondosa*, has been investigated in clinical studies [3]. Independently, these three mushrooms have demonstrated efficacy for their immunomodulatory, anti-infective, antitumour, and anti-inflammatory effects with reduced pro-inflammatory cytokines and oxidative stress, and beneficial gut microbiota responses [52]. *H. erinaceus* contains aromatic compounds such as hericerins and erinacines that appear to function as a nerve growth factor as well as the beneficial immunomodulating and antitumour properties derived from the glycoproteins and polysaccharides [52]. Further highlighting this, polysaccharides extracted from liquid-cultured mycelia and fruiting bodies of *G. frondosa* demonstrated antioxidant, antitumour, anti-inflammatory, hepatoprotection, and immunostimulatory activity [53]. Grifolan, a β-glucan isolated from *G. frondosa*, showed enhanced cellular immunity and modulation activities evidenced by increasing IL-2 and IL-10 production and augmentation of IL-6, IL-1, and TNF-α expression [54].

Water extract of four different MMs, including *G. lucidum*, caused NK cell-induced cytotoxicity against cancer cells, but an ethanol extract did the opposite by reducing intracellular pathway activation [55]. Various triterpene acids and sterols isolated from *G. lucidum* fruiting bodies revealed antitumour and anti-inflammatory effects as demonstrated via induction of EBV early antigen by 12-O-tetradecanoylphorbol-13-acetate [56].

T. versicolor has a long traditional history of use to promote health, strength, and longevity. More recently, numerous studies, including clinical trials, suggest properties and effects that include antimicrobial, antiviral, antitumour, anti-inflammatory, antioxidant, hepatoprotective, bone protective, and notably immunopotentiation [57,58]. Two bioactive mycelia extracts of protein bound polysaccharides from *T. versicolor*, namely polysaccharopeptide (PSP) and polysaccharide krestin (PSK), are currently utilised medicinally in some countries as integrated cancer therapy and adjuncts for chemotherapy and radiotherapy [2,57,59]. From a range of randomised and non-randomised controlled trials, both PSK and PSP promoted positive impacts on anticancer effects [60]. Deemed resulting from the immunomodulation and potentiation of immune surveillance, PSK and PSP positively affect immune parameters, haematological function, performance status, quality of life, body weight, fatigue, pain, nausea, anorexia, and median survival [59–61]. Additionally, antitumour and antimetastatic effects were noted through direct tumour-inhibiting experiments in vivo [60]. Of interest in the context of COVID-19 application is the mechanisms of *T. versicolor*. This appears to be through the inducement of predominantly pro-inflammatory cytokines: not only those associated with TNF-α and NK cells but also pleiotropic cytokines such as IL-1α and 1β and IL-6, plus prostaglandin E2, histamine, activation of complement-3, and T cell proliferation [57,59,61]. While this may be desired to improve cancer outcomes, such as enhancing the immunosuppressive status, a cautionary approach in applying *T. versicolor* due to the hyperinflammatory response associated with COVID-19 progression should be taken. However, perhaps, there may be a place for consideration in the context of long COVID or playing a role as a vaccine adjuvant.

6.3. Antioxidant

The antioxidant/ROS system plays a significant role in pathogenic protection, regulation, and homeostasis in the human body. For example, the increased activity of ROS is a key feature in the pathogenesis and progression of many disease states such as atherosclerosis, arterial thrombosis, hyperlipidaemia, hypertension, cancer, obesity, insulin resistance, diabetes mellitus, hepatic and renal conditions, amongst many others [62]. These disease states are representative comorbidities associated with SARS CoV-2 and COVID-19 sequalae and experience. The antioxidant capacity of MMs has been demonstrated in various studies through radical scavenging, lipid peroxidation inhibition, and increasing antioxidant enzyme activities [30,54,63,64]. Bioactive compounds such as phenolics, indoles, flavonoids, glycosides, polysaccharides, tocopherols, glutathione and ergothioneine, ascorbic acid, carotenoids, vitamin D, copper, manganese, zinc, and selenium in MMs all participate in reducing oxidative stress [62,65]. Ergothioneine deserves special mention, as it has a vast array of unique cytoprotective properties pertinent to COVID-19 pathologies, including scavenging reactive oxygen and nitrogen species. It is able to modulate inflammation, inhibit the expression of vascular adhesion proteins, and protect against respiratory burst, amongst many other antioxidant activities [63]. Notable amounts of bioavailable ergothioneine were demonstrated in the fruiting bodies of *A. bisporus* [66], *L. edodes*, *P. ostreatus*, and mycelia of *C. militaris* (strain cm5), *H. erinaceus*, and *P. eryngii* [67,68].

Liquid–liquid partitioned fractions of *H. erinaceus* were evaluated for their anti-atherosclerotic potential through evaluation of in vitro inhibitory effect on low-density lipoprotein (LDL) oxidation and 3-hydroxy-2methylglutaryl coenzyme A (HMG-CoA) reductase activity [69]. Several bioactive compounds with antioxidant activity were isolated, in particular ergosterol. Hexane solvent fraction demonstrated the most potent inhibiting oxidisation of LDL and HMG-CoA reductase activity. This indicates a possible role in preventing oxidative stress-mediated vascular disease processes [69].

Radical scavenging properties associated with catalase activity, glutathione reductase, and glutathione peroxidase activities were demonstrated in varying degrees from methanol and water extracts isolated from the gills, stipe, and caps of two wild strains and one cultivated strain of *A. bisporus* [70]. Fourteen selected culinary MMs were evaluated for in vitro antioxidant and ACE inhibitory activities [30]. The mushrooms were extracted by boiling water for 30 min. The total phenolic content was determined with *G. lucidum* demonstrating the highest phenolic content and the most potent ACE inhibitor. Antioxidant capacity was carried out via measuring the free radical scavenging effect, β-carotene, lipid peroxidation, reducing power ability, cupric-ion-reducing antioxidant capacity, and ACE inhibition. An antioxidant index was determined based on the average percentage relative to quercetin. *G. lucidum* and *H. erinaceus* were shown to be relatively high compared to the other mushrooms [30].

6.4. ACE2 Regulation

The deleterious effects of COVID-19, such as those associated with cardiometabolic and other hallmark disorders, demonstrate dysregulation of the homeostatic function within the RAAS [7,13,71,72] . RAAS maintains dynamic control of vascular function. ACE2 is an integral membrane protein present in the lungs, liver, heart, kidney, and endothelium. ACE2 dysregulation appears to strongly impact the RAAS, manifesting effects involving hyperinflammation and oxidative stress. MMs have been investigated for ACE inhibitory, antiplatelet, anti-inflammatory, and antioxidant activity [30,73,74].

In MMs, bioactive compounds such as triterpenes, sterols, phenolic compounds, and polysaccharide fractions possess metabolic-modulating capabilities [54]. These include blood pressure, glycaemia, cholesterol, triglyceride, and weight-lowering activities. The ACE inhibitory activity of several mushroom species was assessed via hot water and alcohol extracts [30]. *G. lucidum*, particularly as a hot water extract, and *Pleurotus* spp. demonstrated potent ACE inhibitory activity, which is assumed to be due to the phenolic content and antioxidant capacity. However, variations existed between species and depended on the extraction method [30]. In vitro digestion of *P. ostreatus* identified several peptides known to be ACE inhibitors [75]. A randomised, double-blind prevention trial is underway in the Democratic Republic of the Congo involving Tomeka®, a herbal mixture containing *A. bisporus* and other food-based nutrients such as soy, which is regarded for its potent ACE2 inhibition. The study aims to assess the intervention effect on COVID-19 markers of the RAAS, such as angiotensin-II and angiotensin-(1-7) [71]. Nutritional elements may support ACE inhibition indirectly by intercepting viral entry or via regulation and improvements in biomarkers associated with the involvement of the various systems [71]. For example, excessive sodium ions can impair the endothelial vasculature and risk hypertension, but manifestations may be ameliorated with higher potassium ion levels. Hence, mushrooms, which generally contain high potassium and low sodium may be a good nutritional source for ACE inhibition as well [1,24].

7. Comorbidities and Mortality Risk Reduction or Mitigation

7.1. Cardiometabolic Disorders Associated with COVID-19

Compounds of MMs have demonstrated biological activity with the potential to reduce the risk of cardiometabolic disease and comorbidity effects associated with COVID-19. One-third of patients with COVID-19, aged 40–60 years, have been identified as being afflicted with comorbidities such as CVD and hypertension [7]. Metabolic disorders, including obesity, diabetes, and hyperlipidaemia, are also featured in disease severity [7,8]. Additionally, meta-analyses have identified increased mortality risk and the need for intensive care for older patients with cardiovascular morbidities [7].

Fruiting bodies and mycelium extracts of edible mushrooms (some more than others) can be a valuable source of lovastatin, which is a statin group compound [76]. This compound inhibits HMG-CoA reductase, which is the rate-limiting step in cholesterol biosynthesis. Thus, MMs with activity associated with lovastatin's mechanisms of action may be a promising source of anti-hypercholesterolaemic agents. Most notably, such cholesterol-lowering potential has been observed in the fruiting bodies of *Pleurotus* spp. and others, including *A. bisporus* and *H. erinaceus* [66,69,75–77]. This potential, along with the absorption and stimulatory effect of dietary fibres and gut effects on faecal excretion, makes for a promising functional food application.

The effects of various MMs have also been studied with other cardiometabolic parameters. Dicks and Ellinger [75] undertook a systematic review of eight clinical studies of subjects with and without type 2 diabetes mellitus (T2DM) using fresh, cooked, or dry powder *P. ostreatus*. These studies seemed to reveal beneficial effects, although the risk of bias was high or unclear due to methodological weaknesses and/or inadequate reporting in most studies. Nevertheless, observed were effects in glycaemic control (reduction in fasting and/or 2 h postprandial glucose), lipids metabolism (decrease in total cholesterol (TC), LDL cholesterol, and/or triglycerides), some reduction in blood pressure, antioxidant effects, and a decrease in food intake with no weight change [75]. The administration of total polysaccharides extracted from *P. ostreatus* was given for four weeks in a high-fat diet and streptozotocin (STZ)-induced type 2 diabetic rats [78]. Elevated blood glucose levels were reduced, insulin resistance improved, and glycogen increased. The mechanisms occurred through the activation of GSK-2 phosphorylation in the liver and GLUT4 translocation in muscle tissue. In high-fat-fed rats, *A. bisporus* demonstrated increased high-density lipoprotein (HDL) along with reductions in TC and LDL, and additionally, in type 2 diabetic rats induced by STZ, glucose levels decreased [79]. The anti-hyperglycaemic effect, along with antioxidant protection on the pancreas, kidney, and liver, were also demonstrated with *H. erinaceus* polysaccharides on STZ-induced rats [80].

The pharmacological effects of *C. militaris* SU-12 residue polysaccharide were investigated in a study utilising high-fat emulsion-induced hyperlipidaemic mice models.

Identified were the characteristics of the polysaccharides, concluding these to have demonstrable antihyperlipidaemic, hepatoprotective activities, and an increase in antioxidant activity when serum and liver sections were analysed [81]. STZ-induced type 1 diabetic rats were used in a study that compared *G. frondosa* (water-soluble powdered whole Maitake fraction SX), two anti-diabetic drugs, and a control to assess circulating glucose levels and blood pressure (BP) [82]. All treatments generally decreased circulating glucose levels compared to control. However, only the Maitake group consistently demonstrated enhanced insulin sensitivity, significantly lowered systolic BP plus a decrease in the RAAS, and increased nitric oxide system activity. *F. velutipes* powder and extract were shown to have good antioxidant activity. They were a rich source of dietary fibre and mycosterol capable of impacting and reducing cardiometabolic disease parameters such TC, LDL cholesterol, and triglycerides (TGs) [83]. However, *G. lucidum* has produced less consistent results. In a study with high-fat-fed rabbits, Li et al. [84] assessed that atherosclerotic plaques were attenuated along with a reduced generation of ROS and malondialdehyde.

In contrast, no positive results were revealed in a prospective double-blind randomised, placebo-controlled trial of 84 subjects with type 2 diabetes mellitus and metabolic syndrome over 16 weeks using *G. lucidum*, *G. lucidum* with *C. sinesis*, or a placebo [85]. The primary outcome measures were blood glucose biomarkers (glycosylated haemoglobin (HBA1c) and fasting plasma glucose), and secondary outcome measures were HDL and LDL, TGs, BP, C-reactive protein, and apolipoproteins A and B markers. When the two intervention groups were combined due to sample size inadequacy, there was no effect on either the primary or secondary outcomes [85]. In a secondary analysis to investigate the benefits of daily intake of *A. bisporus* on cardiometabolic risk, the stored serum of prediabetic patients with features of metabolic syndrome were analysed. No significant changes to body weight, cardiovascular or metabolic parameters were observed, and plasma leptin did not change. Nevertheless, ergothioneine concentrations, oxygen radical absorbance capacity, and adiponectin were increased, with a reduction of advanced glycation end products (AGEs) [86].

7.2. Gut Microbiota Modulation as a COVID-19 Risk Reduction Consideration

It is recognised that the intestinal flora structure, the abundance and diversity of microbiota, or activity can influence positive or negative health outcomes. For example, aberrances of gut microbiota may contribute to chronic inflammatory diseases such as atherosclerosis, thrombosis, diabetes, and asthma, which may be in part due to oxidative stress [87]. On the other hand, symbiosis reduces cardiovascular and other metabolic diseases' risks, lowers postprandial blood glucose, increases satiety for weight management, and improves laxation [88]. These are all factors associated with COVID-19 disease severity and are also lifestyle associated; therefore, they are modifiable and applicable to lowering risk. Fortunately, the functions and diversity of gut microbiota are influenced by the non-digestible and digestible fibres and prebiotic properties of mushrooms. Ma et al. [25] reviewed collated studies from various disease models that demonstrated health improvement from EMPs. Correlations between EMPs and beneficial host microbiota suggest regulatory effects [25]. Similar effects may also be induced via mannitol, raffinose, resistant starch, and chitin found in mushrooms [88]. Benefits of EMPs highlighted by Ma et al. were metabolic improvements such as reduction in TC and other lipid markers, anti-obesity activities, inflammation and insulin resistance, improved gut mucosa integrity and intestinal morphology, signalling pathways, and pro-inflammatory cytokines inhibition [25].

Short-chain fatty acids (SCFAs), including acetic acid, propionic acid, butyric acid, and valeric acid, are known to provide beneficial health effects such as nutrient supply to the colonic epithelium, oxidative stress reduction, immune stimulation, and colonic pH modulation [87]. However, the proportion of SCFAs varies depending on the EMPs from different mushrooms [25]. *G. lucidum* appears most significant for microbiota-derived health effects. Specifically, *G. lucidum* polysaccharide (GLP) appeared to have notable anti-diabetic and anti-obesity effects [89,90]. For example, GLP restored the gut microbiota of T2DM rats to a normal level and modified metabolites [91], and GLPS3 (*G. lucidum* mycelium polysaccharide strain S3) increased the relative abundance of beneficial bacteria Lactobacillus, Roseburia, and Lachnospiracea in mice induced with repetitively intraperitoneal injection of diethyldithiocarbamate [87]. An increase in Roseburia after dietary administration with GLPS3, was considered an indicator of the health-promoting activities of the SCFAs and possibly also of immune stimulation.

Contrasting findings were reported by Lee et al. [92] when examining results from the Nurses' Health Study (for women) and the Health Professionals Follow Up Study (for men). In these studies, respondents were asked how often they consumed fresh, cooked, or canned mushrooms (species unidentified). Data collected were short and long-term changes in biomarkers such as lipids, insulin, inflammation, or cardiovascular risks such as sex, lifestyle factors, and certain medical conditions. No benefit of mushroom consumption on cardiovascular disease and T2DM was elucidated. However, a potential inverse relationship to T2DM was suggested.

Table 1 summarises the research findings exhibiting pro-health effects and benefits resulting from bioactive compounds or secondary metabolites of selected MMs, which are deemed applicable for potential COVID-19 therapy consideration.

Table 1. Research findings for specific mushroom species demonstrating beneficial effects and activities for potential pro-health therapeutic application.

Mushroom	Bioactive Compound	Pro-Health Effects	References
A. bisporus (Common white)	Phenolics e.g., Ergothioneine	Antioxidant; increased ORAC activity; increased adiponectin; reduced AGEs; increased glutathione reductase and catalase activities	[63,70,86]
	Polysaccharides	Gut microbiota regulation; intestinal barrier integrity	[25]
	Secondary metabolites	Anti-hyperglycaemic; inhibitory effects on LDL oxidation; reduced HMG-CoA reductase activity; LPS reduction; ACE2 inhibition; cardiometabolic parameters improvement	[30,66,79,80]
Andosan™	Ergosterol	Cytotoxic	[51]
	Commercial extract AbM + *H. erinaceus* and *G. frondosa*	Immunomodulatory; anti-inflammatory; anti-tumour	[49]
Tomeka™	Commercial extract *A. bisporus* + soy	Anti-pathogenic	[3,29]
C. militaris (Cordyceps)	Cordycepin	Antiviral; RNA synthesis inhibition; suppressed EB viral replication; cytotoxic	[14,41]
		Antioxidant; anti-hyperlipidaemic; hepatoprotective	[81]
	SCFAs	Immune regulation and health promoting	[25]
F. velutipes (Enoki)	Polysaccharides/dietary fibre	Reduced cardiometabolic parameters	[83]
	Antioxidant	Mycosterol	[83]
G. frondosa (Maitake)	Polysaccharides	Anti-inflammatory; antioxidant; immunomodulatory	[31,53]
		Increased insulin sensitivity; decreased systolic BP; decreased RAAS; increased NO	[82]

Table 1. *Cont.*

Mushroom	Bioactive Compound	Pro-Health Effects	References
G. lucidum (Reishi)	Polysaccharides	SCFAs production; gut microbiota regulation; anti-obesity; anti-inflammation; reduced metabolic endotoxaemia; decreased FBG and insulin levels	[87,89–91]
	Ganodermic compounds—triterpenoids, other phenolics	Antioxidant; atherosclerotic plaque attenuation; anti-tumour; anti-inflammation; antiviral HIV-1 and HIV-1 protease inhibition	[30,42,56,64,84]
	Mycelia fractions	ACE inhibition	[73]
H. erinaceus (Lion's Mane)	Mycelia polysaccharide fractions	Anti-hyperglycaemic; improved antioxidant enzymatic activities	[69,80]
	Fruiting body solvent fractions	LDL oxidation inhibition; HMG-CoA reductase inhibition	
L. edodes (Shiitake)	Polysaccharides; *β*-glucans	Antiviral; antioxidant; immunomodulatory; cytotoxic; anti-inflammatory; microbiome regulation	[1,2,6,17,25,44]
	Aqueous extract	Anti-bacterial; anti-fungal	[43]
	Lovastatin	Hypolipidaemic	[76]
P. ostreatus (Oyster)	*β*-glucans-pleuran	Immunomodulatory	[36,46–48]
	Lectins	Vaccine adjuvant	[20,45]
	Phenolics, peptides	ACE2 inhibition	[30,75]
	Lovastatin	HMG-CoA reductase inhibition	[75–77]
	Polysaccharides	Cardiometabolic parameter improvements	[75,78]
T. versicolor (Turkey Tail)	PSP, PSK	Cancer therapy/adjuvants; Immunomodulatory	[57,59–61]
		Gut microbiota modulation	[25]
	Glucans, phenolics	Antioxidant	[62]

8. Considering Limitations or Barriers in the Application of Mushrooms as a Medicine

Many factors may impact nutrient characteristics and potentially optimal and efficacious bioactivity of any particular plant compound, including MMs. These involve environmental influences such as plant provenance (e.g., geographic region, climate, and temperature); cultivation practices and production methods; plant structure; and isolation and extraction techniques and methods [4,29,55,70,93,94]. For example, in a study by Cohen et al. [95], fifteen dried and crushed Basidiomycetes MM strains, fruiting bodies, and mycelia were analysed. The protein and carbohydrate content ranges were 8.6–42.5% and 42.9–83.6%, respectively. Varying results for macro and micronutrients, and of concern, possible toxic elements, were reported as well. These results seem consistent with previous findings [24]. Therefore, consuming mushrooms with the intention of deriving some health or medicinal benefit may need to consider these influences to enhance or mitigate inhibitory effects.

The extraction methods and the part of the plant utilised appear to significantly impact bioactive potential. As an example, Pop et al. [96] demonstrated the isolation of a range of biocompounds from *Trametes* spp. using three different extraction solvents of water, ethanol, and methanol. Even within the same species, the bioactivity might differ depending on the type of extraction (water or ethanol/methanol) and the structure/section of the fruiting body [5,29,55,69].

Friedman's review [29] on *H. erinaceus* isolated specific bioactive secondary metabolites using various solvents and according to the different plant structures such as mycelia and the polysaccharide fractions. In cultivated *H. erinaceus* mycelia, ergosterol content and new sterols were isolated. Antioxidant properties in a water-soluble polysaccharide were identified but not in an ethanol extraction of another polysaccharide within the same species. Ma et al. [25] comprehensively examined the role EMPs have in their activities against obesity, inflammatory bowel disease, and cancer. They characterised the type of polysaccharide along with the extraction method and demonstrated that the carbohydrate structure and chemistry, and thus the methods to isolate, are vital information for assigning effects. Martel et al. [55] surveyed studies on cells, animals, and humans utilising various constituents of mushrooms and plants. Depending on the extraction method, the effects on a wide variety of immune cell types were either stimulatory or inhibitory. The dichotomy appeared to be due to the differential solubility and potency of the main constituents in the extracts. Within the same species, water-soluble polysaccharide extracts activated immune responses, but ethanol extracts inhibited them, although there were a few exceptions.

It is recognised that improved techniques and technologies may remedy disadvantages associated with the extraction of bioactive compounds, such as long extraction times, low selectivity, and solubility [53,97]. As an example of the potential associated with improved technologies, Wu et al. [53] undertook a study to evaluate the anti-inflammatory effects of total polysaccharides and β-glucans extracted from *G. frondosa* mycelia. The potential mechanisms were evaluated by examining effects on nitrite, prostaglandins, pro-inflammatory cytokines (TNF-α, IL-6 and IL-1β), and intracellular reactive oxygen species in lipopolysaccharide-induced macrophage cells. They compared a conventional extract method using ethanol with three different high pressure-assisted extractions (PE-200, PE-400, PE-600). The PE method in each case yielded greater extraction and content of polysaccharides and β-glucans and exerted stronger anti-inflammatory activities than the conventional method. Benson et al. [26] demonstrated that three components of *T. versicolor* (mycelium, initial substratem and fermented rice flour substrate), plus different extractions and combinations, could produce varied immune responses and activities when tested on human peripheral blood mononuclear cell cultures.

These highlights demonstrate that there may be variations in the isolates depending on the methods utilised. Additionally, more modern and improved techniques may produce a superior raw product, potentially creating differences in the efficacies associated with mechanisms of action and health benefit claims. Therefore, it seems essential to their application to expect mushroom therapeutic treatments to be produced from regions and in conditions that result in the most efficacious response, according to the most current research. As more and more commercialised products are becoming available for general purchase, as well as in therapeutic settings, it is vital to ascertain pharmacokinetics, timing of administration, dosage, mode, and formulation variances [27]. These are all factors to be considered in the application of bioactive compounds. This can only be accurate and informed if the whole plant or extract characteristics have been clearly identified and elucidated. Figure 2 summarises important considerations and some potential limitations as presented in this appraisal. Key inputs into isolating bioactive compounds are influenced by many factors. Extraction methods, in particular, appear to be very important. Effects from the extracts from the same and different species can vary and even produce an opposite effect in activity. Beyond this, other factors need consideration if MMs are to be utilised for therapeutic and treatment purposes.

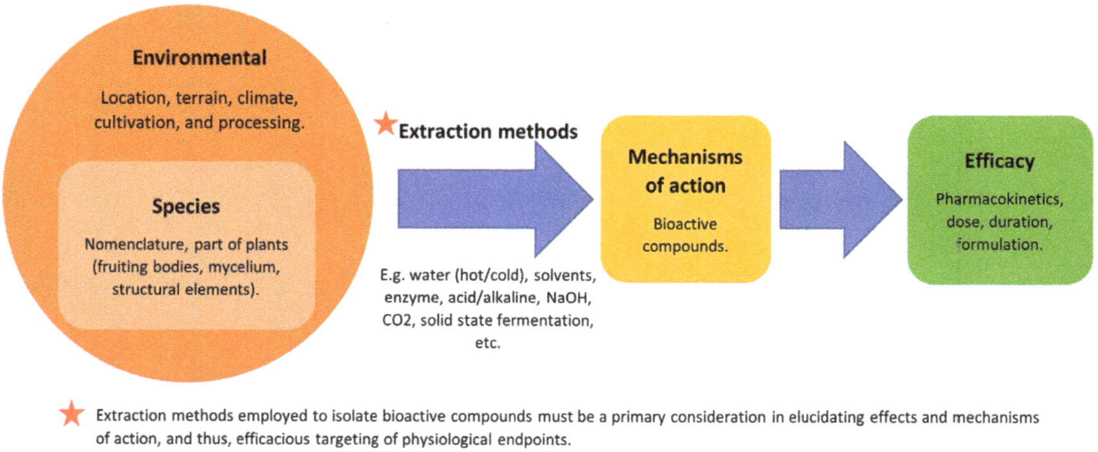

Figure 2. Factors to consider if medicinal mushrooms are to be utilised for therapeutic and treatment purposes.

9. Conclusions

This review identified major findings of in vitro, in vivo, and human studies for mushrooms utilised in the medicinal context. Particular species were used as examples and examined for their health effects due to their ready availability and/or use in nutraceutical products. COVID-19 pathogenesis presents a spectrum of systemic health impacts for consideration. In particular, this relates to the aberrant immune and inflammatory responses that can lead to severe and detrimental consequences. It was demonstrated that specific bioactive compounds derived from mushrooms are capable of inducing a physiological effect that could be considered applicable in preventing COVID-19, mitigating symptoms or reducing disease severity. These effects include anti-pathogenic, anti-inflammatory, immune-modulatory, antioxidant, and ACE inhibitory activity. Thus, MMs are well placed as a possible therapeutic option due to these functions and properties.

Increasingly, MMs are being commercialised and promoted for their health-promoting benefits. However, potentially, many of the health claims may be ahead of the actual research validation, particularly concerning human translatability in specific contexts—in this case, COVID-19. Many factors such as extraction method, growing conditions, standardisation practices, mechanisms of action, and formulation synergies require greater understanding and examination. Hence, continued efforts must be applied to realising more research and improving data collection and modelling, particularly in humans, with all these variables and factors in mind.

Author Contributions: J.M.P. conceptualisation, methodology, investigation, writing—original draft, visualization, S.L.O. writing—review and editing, validation, visualization, S.C.P. supervision, project administration, writing—review and editing, validation. All authors have read and agreed to the published version of the manuscript.

Funding: This research received no external funding.

Institutional Review Board Statement: Not applicable.

Informed Consent Statement: Not applicable.

Data Availability Statement: Not applicable.

Acknowledgments: In the spirit of reconciliation, the authors acknowledge the Traditional Custodians of the country throughout Australia and their connections to land, sea, and community. We pay respect to their Elders past and present and extend that respect to all Aboriginal and Torres Strait Islander peoples today.

Conflicts of Interest: The authors declare no conflict of interest.

Abbreviations

The following abbreviations are used in this manuscript:

AbM	*Agaricus blazei* Murril
ACE	angiotensin-converting enzyme
AGEs	advanced glycation end products
ARDS	acute respiratory distress syndrome
BP	blood pressure
BRMs	biological response modifiers
CD	cluster of differentiation
COVID-19	corona virus disease 2019
CVD	cardiovascular disease
DNA	deoxyribonucleic acid
EBV	Epstein–Barr virus
EMPs	edible mushroom polysaccharides
FS	fermented substrate
GLP	*Ganoderma lucidum* polysaccharide
GLUT4	glucose transporter type 4
GSK	glycogen synthase kinase
HbA1c	glycosylated haemoglobin A1c
HDL	high-density lipoprotein
HIV	human immunodeficiency virus
HMG-CoA	3-hydroxy-3-methyl-glutaryl-coenzyme A
IL	interleukin
LDL	low-density lipoprotein
MMs	medicinal mushrooms
NK	natural killer
NO	nitric oxide
PE	pressure-assisted extractions
PPR	pattern recognition receptor
PSK	polysaccharide Krestin
PSP	polysaccharopeptides
RAAS	renin–angiotensin–aldosterone-system
RNA	ribonucleic acid
ROS	reactive oxygen species
RRTI	recurrent respiratory tract infection
SARS-CoV-2	severe acute respiratory distress syndrome coronavirus-2
SCFAs	short-chain fatty acids
STZ	streptozotocin
T2DM	type 2 diabetes mellitus
TC	total cholesterol
TG	triglycerides
Th	helper T cell
TNF	tumour necrosis factor
TvM	*Trametes versicolor* mycelium

References

1. Gargano, M.L.; van Griensven, L.J.L.D.; Isikhuemhen, O.S.; Lindequist, U.; Venturella, G.; Wasser, S.P.; Zervakis, G.I. Medicinal mushrooms: Valuable biological resources of high exploitation potential. *Plant Biosyst.* **2017**, *151*, 548–565. [CrossRef]
2. Venturella, G.; Ferraro, V.; Cirlincione, F.; Gargano, M.L. Medicinal mushrooms: Bioactive compounds, use, and clinical trials. *Int. J. Mol. Sci.* **2021**, *22*, 634. [CrossRef] [PubMed]
3. Hetland, G.; Johnson, E.; Bernardshaw, S.V.; Grinde, B. Can medicinal mushrooms have prophylactic or therapeutic effect against COVID-19 and its pneumonic superinfection and complicating inflammation? *Scand. J. Immunol.* **2021**, *93*, e12937. [CrossRef] [PubMed]
4. Wasser, S.P. Current findings, future trends, and unsolved problems in studies of medicinal mushrooms. *Appl. Microbiol. Biotechnol.* **2011**, *89*, 1323–1332. [CrossRef] [PubMed]
5. Patel, D.K.; Dutta, S.D.; Ganguly, K.; Cho, S.J.; Lim, K.T. Mushroom-derived bioactive molecules as immunotherapeutic agents: A review. *Molecules* **2021**, *26*, 1359. [CrossRef] [PubMed]
6. Lindequist, U.; Niedermeyer, T.H.J.; Jülich, W.D. The pharmacological potential of mushrooms. *Evid.-Based Complement. Altern. Med.* **2005**, *2*, 285–299. [CrossRef] [PubMed]
7. Mansueto, G.; Niola, M.; Napoli, C. Can COVID 2019 induce a specific cardiovascular damage or it exacerbates pre-existing cardiovascular diseases? *Pathol.-Res. Pract.* **2020**, *216*, 153086. [CrossRef]
8. Schieffer, E.; Schieffer, B.; Hilfiker-Kleiner, D. Cardiovascular diseases and COVID-19: Pathophysiology, complications and treatment. *Herz* **2021**, *46*, 107–114. [CrossRef]
9. Downer, S.; Berkowitz, S.A.; Harlan, T.S.; Olstad, D.L.; Mozaffarian, D. Food is medicine: Actions to integrate food and nutrition into healthcare. *BMJ* **2020**, *369*, m2482. [CrossRef]
10. Rózsa, S.; Gocan, T.M.; Lazăr, V.; Andreica, I.; Rózsa, M.; Măniuțiu, D.N.; Sima, R. The effect of processing on chemical constituents of Agaricus spp. mushrooms. *Not. Bot. Horti Agrobot. Cluj-Napoca* **2017**, *45*, 507–516. [CrossRef]
11. Communicable Diseases Network Australia (CDNA). Coronavirus disease 2019 (COVID-19). *CDNA Natl. Guidel. Public Health Units* **2021**, *4*, 1–65.
12. Raj, C.T.D.; Kandaswamy, D.K.; Danduga, R.C.S.R.; Rajasabapathy, R.; James, R.A. COVID-19: Molecular pathophysiology, genetic evolution and prospective therapeutics—A review. *Arch. Microbiol.* **2021**, *203*, 2043–2057. [CrossRef] [PubMed]
13. Bohn, M.K.; Hall, A.; Sepiashvili, L.; Jung, B.; Steele, S.; Adeli, K. Pathophysiology of COVID-19: Mechanisms underlying disease severity and progression. *Physiology* **2020**, *35*, 288–301. [CrossRef] [PubMed]
14. Rahman, M.A.; Rahman, M.S.; Bashir, N.M.B.; Mia, R.; Hossain, A.; Saha, S.K.; Kakon, A.J.; Sarker, N.C. Rationalization of mushroom-based preventive and therapeutic approaches to COVID-19: Review. *Int. J. Med. Mushrooms* **2021**, *23*, 1–11. [CrossRef] [PubMed]
15. Zhou, X.; Ye, Q. Cellular immune response to COVID-19 and potential immune modulators. *Front. Immuncl.* **2021**, *12*, 646333. [CrossRef]
16. Shahzad, F.; Anderson, D.; Najafzadeh, M. The antiviral, anti-inflammatory effects of natural medicinal herbs and mushrooms and SARS-COV-2 infection. *Nutrients* **2020**, *12*, 2573. [CrossRef]
17. Murphy, E.J.; Masterson, C.; Rezoagli, E.; O'toole, D.; Major, I.; Stack, G.D.; Lynch, M.; Laffey, J.G.; Rowan, N.J. β-Glucan extracts from the same edible shiitake mushroom *Lentinus edodes* produce differential in-vitro immunomodulatory and pulmonary cytoprotective effects-Implications for coronavirus disease (COVID-19) immunotherapies. *Sci. Total Environ.* **2020**, *732*, 139330. [CrossRef]
18. Cerletti, C.; Esposito, S.; Iacoviello, L. Edible mushrooms and beta-glucans: Impact on human health. *Nutrients* **2021**, *13*, 2195. [CrossRef]
19. Costela-Ruiz, V.J.; Illescas-Montes, R.; Puerta-Puerta, J.M.; Ruiz, C.; Melguizo-Rodríguez, L. SARS-CoV-2 infection: The role of cytokines in COVID-19 disease. *Cytokine Growth Factor Rev.* **2020**, *54*, 62–75. [CrossRef]
20. Hassan, M.A.A.; Rouf, R.; Tiralongo, E.; May, T.W.; Tiralongo, J. Mushroom lectins: Specificity, structure and bioactivity relevant to human disease. *Int. J. Mol. Sci.* **2015**, *16*, 7802–7838. [CrossRef]
21. Barroetaveña, C.; Toledo, C.V. The nutritional benefits of mushrooms. In *Wild Plants, Mushrooms Nuts*; Ferreira, I.C.F.R., Morales, P., Barros, L., Eds.; John Wiley & Sons, Ltd.: Chichester, UK, 2016; pp. 65–81. [CrossRef]
22. Fulgoni, V.L.; Agarwal, S. Nutritional impact of adding a serving of mushrooms on usual intakes and nutrient adequacy using National Health and Nutrition Examination Survey 2011–2016 data. *Food Sci. Nutr.* **2021**, *9*, 1504–1511. [CrossRef]
23. Jayachandran, M.; Xiao, J.; Xu, B. A critical review on health promoting benefits of edible mushrooms through gut microbiota. *Int. J. Mol. Sci.* **2017**, *18*, 1934. [CrossRef] [PubMed]
24. Kalač, P. A review of chemical composition and nutritional value of wild-growing and cultivated mushrooms. *J. Sci. Food Agric.* **2013**, *93*, 209–218. [CrossRef] [PubMed]
25. Ma, G.; Du, H.; Hu, Q.; Yang, W.; Pei, F.; Xiao, H. Health benefits of edible mushroom polysaccharides and associated gut microbiota regulation. *Crit. Rev. Food Sci. Nutr.* **2021**, *2021*, 1–18. [CrossRef]
26. Benson, K.F.; Stamets, P.; Davis, R.; Nally, R.; Taylor, A.; Slater, S.; Jensen, G.S. The mycelium of the Trametes versicolor (Turkey tail) mushroom and its fermented substrate each show potent and complementary immune activating properties in vitro. *BMC Complement. Altern. Med.* **2019**, *19*, 342. [CrossRef] [PubMed]

27. Badalyan, S.M.; Barkhudaryan, A.; Rapior, S. Recent progress in research on the pharmacological potential of mushrooms and prospects for their clinical application. In *Medicinal Mushrooms*; Agrawal, D., Dhanasekaran, M., Eds.; Springer: Singapore, 2019; pp. 1–70. [CrossRef]
28. Cheung, P.C.K. The nutritional and health benefits of mushrooms. *Nutr. Bull.* **2010**, *35*, 292–299. [CrossRef]
29. Friedman, M. Chemistry, nutrition, and health-promoting properties of Hericium erinaceus (Lion's Mane) mushroom fruiting bodies and mycelia and their bioactive compounds. *J. Agric. Food Chem.* **2015**, *64*, 7108–7123. [CrossRef]
30. Abdullah, N.; Ismail, S.M.; Aminudin, N.; Shuib, A.S.; Lau, B.F. Evaluation of selected culinary-medicinal mushrooms for antioxidant and ACE inhibitory activities. *Evid.-Based Complement. Altern. Med.* **2012**, *2012*, 464238. [CrossRef]
31. Han, B.; Baruah, K.; Cox, E.; Vanrompay, D.; Bossier, P. Structure-functional activity relationship of β-glucans from the perspective of immunomodulation: A mini-review. *Front. Immunol.* **2020**, *11*, 658. [CrossRef]
32. Zhu, F.; Du, B.; Bian, Z.; Xu, B. Beta-glucans from edible and medicinal mushrooms: Characteristics, physicochemical and biological activities. *J. Food Compos. Anal.* **2015**, *41*, 165–173. [CrossRef]
33. Ikewaki, N.; Iwasaki, M.; Kurosawa, G.; Rao, K.S.; Lakey-Beitia, J.; Preethy, S.; Abraham, S.J. β-glucans: Wide-spectrum immune-balancing food-supplement-based enteric (β-WIFE) vaccine adjuvant approach to COVID-19. *Hum. Vaccines Immunother.* **2021**, *17*, 2808–2813. [CrossRef] [PubMed]
34. Jesenak, M.; Hrubisko, M.; Majtan, J.; Rennerova, Z.; Banovcin, P. Anti-allergic effect of pleuran (β-glucan from pleurotus ostreatus) in children with recurrent respiratory tract infections. *Phytother. Res.* **2014**, *28*, 471–474. [CrossRef] [PubMed]
35. Ren, Z.; Qin, T.; Qiu, F.; Song, Y.; Lin, D.; Ma, Y.; Li, J.; Huang, Y. Immunomodulatory effects of hydroxyethylated Hericium erinaceus polysaccharide on macrophages RAW264.7. *Int. J. Biol. Macromol.* **2017**, *105*, 879–885. [CrossRef] [PubMed]
36. Geller, A.; Yan, J. Could the induction of trained immunity by β-glucan serve as a defense against COVID-19? *Front. Immunol.* **2020**, *11*, 1782. [CrossRef]
37. Liu, H.; Meng, Z.; Wang, H.; Zhang, S.; Huang, Z.; Geng, X.; Guo, R.; Wu, Z.; Hong, Z. Robust immune responses elicited by a hybrid adjuvant based on β-glucan particles from yeast for the hepatitis B vaccine. *ACS Appl. Bio Mater.* **2021**, *4*, 3614–3622. [CrossRef]
38. Mallakpour, S.; Azadi, E.; Hussain, C.M. Chitosan, alginate, hyaluronic acid, gums, and β-glucan as potent adjuvants and vaccine delivery systems for viral threats including SARS-CoV-2: A review. *Int. J. Biol. Macromol.* **2021**, *182*, 1931–1940. [CrossRef]
39. Soares, E.; Groothuismink, Z.M.A.; Boonstra, A.; Borges, O. Glucan particles are a powerful adjuvant for the HBsAg, favoring antiviral immunity. *Mol. Pharm.* **2019**, *16*, 1971–1981. [CrossRef]
40. Seo, D.J.; Choi, C. Antiviral bioactive compounds of mushrooms and their antiviral mechanisms: A review. *Viruses* **2021**, *13*, 350. [CrossRef]
41. Ryu, E.; Son, M.; Lee, M.; Lee, K.; Cho, J.Y.; Cho, S.; Lee, S.K.; Lee, Y.M.; Cho, H.; Sung, G.H.; et al. Cordycepin is a novel chemical suppressor of Epstein-Barr virus replication. *Oncoscience* **2014**, *1*, 866–881. [CrossRef]
42. El-Mekkawy, S.; Meselhy, M.R.; Nakamura, N.; Tezuka, Y.; Hattori, M.; Kakiuchi, N.; Shimotohno, K.; Kawahata, T.; Otake, T. Anti-HIV-1 and anti-HIV-1-protease substances from Ganoderma lucidum. *Phytochemistry* **1998**, *49*, 1651–1657. [CrossRef]
43. Hearst, R.; Nelson, D.; Mccollum, G.; Millar, B.C.; Maeda, Y.; Goldsmith, C.E.; Rooney, P.J.; Loughrey, A.; Rao, J.R.; Moore, J.E. An examination of antibacterial and antifungal properties of constituents of Shiitake (*Lentinula edodes*) and Oyster (*Pleurotus ostreatus*) mushrooms. *Complement. Ther. Clin. Pract.* **2008**, *15*, 5–7. [CrossRef] [PubMed]
44. Ren, G.; Xu, L.; Lu, T.; Yin, J. Structural characterization and antiviral activity of lentinan from Lentinus edodes mycelia against infectious hematopoietic necrosis virus. *Int. J. Biol. Macromol.* **2018**, *115*, 1202–1210. [CrossRef] [PubMed]
45. Gao, W.; Sun, Y.; Chen, S.; Zhang, J.; Kang, J.; Wang, Y.; Wang, H.; Xia, G.; Liu, Q.; Kang, Y. Mushroom lectin enhanced immunogenicity of HBV DNA vaccine in C57BL/6 and HBsAg-transgenic mice. *Vaccine* **2013**, *31*, 2273–2280. [CrossRef] [PubMed]
46. Jesenak, M.; Majtan, J.; Rennerova, Z.; Kyselovic, J.; Banovcin, P.; Hrubisko, M. Immunomodulatory effect of pleuran (β-glucan from *Pleurotus ostreatus*) in children with recurrent respiratory tract infections. *Int. Immunopharmacol.* **2013**, *15*, 395–399. [CrossRef]
47. Jesenak, M.; Urbancikova, I.; Banovcin, P. Respiratory tract infections and the role of biologically active polysaccharides in their management and prevention. *Nutrients* **2017**, *9*, 779. [CrossRef] [PubMed]
48. Pasnik, J.; Slemp, A.; Cyswinska-Bernas, A.; Zeman, K.; Jezenak, M. Preventative effect of pleuran (β-glucan isolated from *Pleurotus ostreatus*) in children with recurrent respiratory tract infections-open-label prospective study. *Curr. Paediatr. Res.* **2017**, *21*, 99–104.
49. Johnson, E.; Førland, D.T.; Saetre, L.; Bernardshaw, S.V.; Lyberg, T.; Hetland, G. Effect of an extract based on the medicinal mushroom Agaricus blazei murill on release of cytokines, chemokines and leukocyte growth factors in human blood ex vivo and in vivo. *Scand. J. Immunol.* **2009**, *69*, 242–250. [CrossRef]
50. Hetland, G.; Johnson, E.; Lybergà, T.; Bernardshaw, S.; Tryggestad, A.M.A.; Grinde, B. Effects of the medicinal mushroom Agaricus blazei Murill on immunity, infection and cancer. *Scand. J. Immunol.* **2008**, *68*, 363–370. [CrossRef]
51. Bernardshaw, S.; Johnson, E.; Hetland, G. An extract of the mushroom Agaricus blazei Murill administered orally protects against systemic *Streptococcus pneumoniae* infection in mice. *Scand. J. Immunol.* **2005**, *62*, 393–398. [CrossRef]

52. Hetland, G.; Tangen, J.M.; Mahmood, F.; Mirlashari, M.R.; Nissen-Meyer, L.S.H.; Nentwich, I.; Therkelsen, S.P.; Tjønnfjord, G.E.; Johnson, E. Antitumor, anti-inflammatory and antiallergic effects of Agaricus blazei mushroom extract and the related medicinal basidiomycetes mushrooms, *Hericium erinaceus* and *Grifola frondosa*: A review of preclinical and clinical studies. *Nutrients* **2020**, *12*, 1339. [CrossRef]
53. Wu, S.J.; Chen, Y.W.; Wang, C.Y.; Shyu, Y.T. Anti-inflammatory properties of high pressure-assisted extracts of Grifola frondosa in lipopolysaccharide-activated RAW 264.7 macrophages. *Int. J. Food Sci. Technol.* **2017**, *52*, 671–678. [CrossRef]
54. Roncero-Ramos, I.; Delgado-Andrade, C. The beneficial role of edible mushrooms in human health. *Curr. Opin. Food Sci.* **2017**, *14*, 122–128. [CrossRef]
55. Martel, J.; Ko, Y.F.; Ojcius, D.M.; Lu, C.C.; Chang, C.J.; Lin, C.S.; Lai, H.C.; Young, J.D. Immunomodulatory properties of plants and mushrooms. *Trends Pharmacol. Sci.* **2017**, *38*, 967–981. [CrossRef] [PubMed]
56. Akihisa, T.; Nakamura, Y.; Tagata, M.; Tokuda, H.; Yasukawa, K.; Uchiyama, E.; Suzuki, T.; Kimura, Y. Anti-inflammatory and anti-tumor-promoting effects of triterpene acids and sterols from the fungus Ganoderma lucidum. *Chem. Biodivers.* **2007**, *4*, 224–231. [CrossRef]
57. Saleh, M.H.; Rashedi, I.; Keating, A. Immunomodulatory properties of Coriolus versicolor: The role of polysaccharopeptide. *Front. Immunol.* **2017**, *8*, 1087. [CrossRef]
58. Teng, J.F.; Lee, C.H.; Hsu, T.H.; Lo, H.C. Potential activities and mechanisms of extracellular polysaccharopeptides from fermented Trametes versicolor on regulating glucose homeostasis in insulin-resistant HepG2 cells. *PLoS ONE* **2018**, *13*, e0201131. [CrossRef]
59. Chang, Y.; Zhang, M.; Jiang, Y.; Liu, Y.; Luo, H.; Hao, C.; Zeng, P.; Zhang, L. Preclinical and clinical studies of *Coriolus versicolor* polysaccharopeptide as an immunotherapeutic in China. *Discov. Med.* **2017**, *23*, 207–219.
60. Fritz, H.; Kennedy, D.A.; Ishii, M.; Fergusson, D.; Fernandes, R.; Cooley, K.; Seely, D. Polysaccharide K and Coriolus versicolor extracts for lung cancer: A systematic review. *Integr. Cancer Ther.* **2015**, *14*, 201–211. [CrossRef]
61. Jedrzejewski, T.; Piotrowski, J.; Kowalczewska, M.; Wrotek, S.; Kozak, W. Polysaccharide peptide from Coriolus versicolor induces interleukin 6-related extension of endotoxin fever in rats. *Int. J. Hyperth.* **2015**, *31*, 626–634. [CrossRef]
62. Kozarski, M.; Klaus, A.; Jakovljevic, D.; Todorovic, N.; Vunduk, J.; Petrović, P.; Niksic, M.; Vrvic, M.; van Griensven, L. Antioxidants of edible mushrooms. *Molecules* **2015**, *20*, 19489–19525. [CrossRef]
63. Cheah, I.K.; Halliwell, B. Could ergothioneine aid in the treatment of coronavirus patients? *Antioxidants* **2020**, *9*, 595. [CrossRef] [PubMed]
64. Kozarski, M.; Klaus, A.; Nikšić, M.; Vrvić, M.M.; Todorović, N.; Jakovljević, D.; Van Griensven, L.J. Antioxidative activities and chemical characterization of polysaccharide extracts from the widely used mushrooms Ganoderma applanatum, *Ganoderma lucidum*, *Lentinus edodes* and *Trametes versicolor*. *J. Food Compos. Anal.* **2012**, *26*, 144–153. [CrossRef]
65. Podkowa, A.; Kryczyk-Poprawa, A.; Opoka, W.; Muszyńska, B. Culinary–medicinal mushrooms: A review of organic compounds and bioelements with antioxidant activity. *Eur. Food Res. Technol.* **2021**, *247*, 513–533. [CrossRef]
66. Muszyńska, B.; Kała, K.; Rojowski, J.; Grzywacz, A.; Opoka, W. Composition and biological properties of *Agaricus bisporus* fruiting bodies—A Review. *Pol. J. Food Nutr. Sci.* **2017**, *67*, 173–181. [CrossRef]
67. Chen, S.Y.; Ho, K.J.; Hsieh, Y.J.; Wang, L.T.; Mau, J.L. Contents of lovastatin, γ-aminobutyric acid and ergothioneine in mushroom fruiting bodies and mycelia. *LWT* **2012**, *47*, 274–278. [CrossRef]
68. Weigand-Heller, A.J.; Kris-Etherton, P.M.; Beelman, R.B. The bioavailability of ergothioneine from mushrooms (*Agaricus bisporus*) and the acute effects on antioxidant capacity and biomarkers of inflammation. *Prev. Med.* **2012**, *54*, S75–S78. [CrossRef]
69. Rahman, M.A.; Abdullah, N.; Aminudin, N. Inhibitory effect on in vitro LDL oxidation and HMG Co-A reductase activity of the liquid-liquid partitioned fractions of *Hericium erinaceus* (Bull.) persoon (Lion's *Mane mushroom*). *BioMed. Res. Int.* **2014**, *2014*, 828149. [CrossRef]
70. Savoie, J.M.; Minivielle, N.; Largeteau, M. Radical-scavenging properties of extracts from the white button mushroom, *Agaricus bisporus*. *J. Sci. Food Agric.* **2007**, *88*, 970–975. [CrossRef]
71. Abubakar, M.B.; Usman, D.; Batiha, G.E.S.; Cruz-Martins, N.; Malami, I.; Ibrahim, K.G.; Abubakar, B.; Bello, M.B.; Muhammad, A.; Gan, S.H.; et al. Natural products modulating angiotensin converting enzyme 2 (ACE2) as potential COVID-19 therapies. *Front. Pharmacol.* **2021**, *12*, 629935. [CrossRef]
72. Lichtenberger, L.M.; Vijayan, K.V. Is COVID-19–induced platelet activation a cause of concern for patients with cancer? *Cancer Res.* **2021**, *81*, 1209–1211. [CrossRef]
73. Ansor, N.M.; Abdullah, N.; Aminudin, N. Anti-angiotensin converting enzyme (ACE) proteins from mycelia of *Ganoderma lucidum* (Curtis) P. Karst. *Complement. Altern. Med.* **2013**, *13*, 256. [CrossRef] [PubMed]
74. Mohamed Yahaya, N.F.; Rahman, M.A.; Abdullah, N. Therapeutic potential of mushrooms in preventing and ameliorating hypertension. *Trends Food Sci. Technol.* **2014**, *39*, 104–115. [CrossRef]
75. Dicks, L.; Ellinger, S. Effect of the intake of oyster mushrooms (*Pleurotus ostreatus*) on cardiometabolic parameters—A systematic review of clinical trials. *Nutrients* **2020**, *12*, 1134. [CrossRef] [PubMed]
76. Kała, K.; Kryczyk-Poprawa, A.; Rzewińska, A.; Muszyńska, B. Fruiting bodies of selected edible mushrooms as a potential source of lovastatin. *Eur. Food Res. Technol.* **2020**, *246*, 713–722. [CrossRef]
77. Atlý, B.; Yamaç, M.; Yýldýz, Z.; Sölener, M. Solid state fermentation optimization of *Pleurotus ostreatus* for lovastatin production. *Pharm. Chem. J.* **2019**, *53*, 858–864. [CrossRef]

78. Aramabašić Jovanović, J.; Mihailović, M.; Usković, A.; Grdović, N.; Dinić, S.; Vidaković, M. The effects of major mushroom bioactive compounds on mechanisms that control blood glucose level. *J. Fungi* **2021**, *7*, 58. [CrossRef]
79. Jeong, S.C.; Jeong, Y.T.; Yang, B.K.; Islam, R.; Koyyalamudi, S.R.; Pang, G.; Cho, K.Y.; Song, C.H. White button mushroom (*Agaricus bisporus*) lowers blood glucose and cholesterol levels in diabetic and hypercholesterolemic rats. *Nutr. Res.* **2010**, *30*, 49–56. [CrossRef]
80. Zhang, C.; Li, J.; Hu, C.; Wang, J.; Zhang, J.; Ren, Z.; Song, X.; Jia, L. Antihyperglycaemic and organic protective effects on pancreas, liver and kidney by polysaccharides from *Hericium erinaceus* SG-02 in streptozotocin-induced diabetic mice. *Sci. Rep.* **2017**, *7*, 10847. [CrossRef]
81. Wang, L.; Xu, N.; Zhang, J.; Zhao, H.; Lin, L.; Jia, S.; Jia, L. Antihyperlipidemic and hepatoprotective activities of residue polysaccharide from *Cordyceps militaris* SU-12. *Carbohydr. Polym.* **2015**, *131*, 355–362. [CrossRef]
82. Preuss, H.G.; Echard, B.; Fu, J.; Perricone, N.V.; Bagchi, D.; Kaylor, M.; Zhuang, C. Fraction SX of maitake mushroom favorably influences blood glucose levels and blood pressure in streptozotocin-induced diabetic rats. *J. Med. Food* **2012**, *15*, 901–908. [CrossRef]
83. Yeh, M.Y.; Ko, W.C.; Lin, L.Y. Hypolipidemic and antioxidant activity of enoki mushrooms (*Flammulina velutipes*). *BioMed Res. Int.* **2014**, *2014*, 352385. [CrossRef] [PubMed]
84. Li, Y.; Tang, J.; Gao, H.; Xu, Y.; Han, Y.; Shang, H.; Lu, Y.; Qin, C. *Ganoderma lucidum* triterpenoids and polysaccharides attenuate atherosclerotic plaque in high-fat diet rabbits. *Nutr. Metab. Cardiovasc. Dis.* **2021**, *31*, 1929–1938. [CrossRef] [PubMed]
85. Klupp, N.L.; Kiat, H.; Bensoussan, A.; Steiner, G.Z.; Chang, D.H. A double-blind, randomised, placebo-controlled trial of *Ganoderma lucidum* for the treatment of cardiovascular risk factors of metabolic syndrome. *Sci. Rep.* **2016**, *6*, 29540. [CrossRef] [PubMed]
86. Calvo, M.S.; Mehrotra, A.; Beelman, R.B.; Nadkarni, G.; Wang, L.; Cai, W.; Boon, C.; Goh, C.; Kalaras, M.D.; Uribarri, J. A retrospective study in adults with metabolic syndrome: Diabetic risk factor response to daily consumption of *Agaricus bisporus* (white button mushrooms). *Plant Foods Hum. Nutr.* **2016**, *71*, 245–251. [CrossRef]
87. Li, K.; Zhuo, C.; Teng, C.; Yu, S.; Wang, X.; Hu, Y.; Ren, G.; Yu, M.; Qu, J. Effects of *Ganoderma lucidum* polysaccharides on chronic pancreatitis and intestinal microbiota in mice. *Int. J. Biol. Macromol.* **2016**, *93*, 904–912. [CrossRef]
88. Hess, J.; Wang, Q.; Gould, T.; Slavin, J. Impact of *Agaricus bisporus* mushroom consumption on gut health markers in healthy adults. *Nutrients* **2018**, *10*, 1402. [CrossRef]
89. Chang, C.J.; Lin, C.S.; Lu, C.C.; Martel, J.; Ko, Y.F.; Ojcius, D.M.; Tseng, S.F.; Wu, T.R.; Chen, Y.Y.M.; Young, J.D.; et al. *Ganoderma lucidum* reduces obesity in mice by modulating the composition of the gut microbiota. *Nat. Commun.* **2015**, *6*, 7489. [CrossRef]
90. Ganesan, K.; Xu, B. Anti-obesity effects of medicinal and edible mushrooms. *Molecules* **2018**, *23*, 2880. [CrossRef]
91. Chen, M.; Xiao, D.; Liu, W.; Song, Y.; Zou, B.; Li, L.; Li, P.; Cai, Y.; Liu, D.; Liao, Q.; et al. Intake of *Ganoderma lucidum* polysaccharides reverses the disturbed gut microbiota and metabolism in type 2 diabetic rats. *Int. J. Biol. Macromol.* **2020**, *155*, 890–902. [CrossRef]
92. Lee, D.H.; Yang, M.; Giovannucci, E.L.; Sun, Q.; Chavarro, J.E. Mushroom consumption, biomarkers, and risk of cardiovascular disease and type 2 diabetes: A prospective cohort study of US women and men. *Am. J. Clin. Nutr.* **2019**, *110*, 666. [CrossRef]
93. Atila, F.; Tüzel, Y.; Pekşen, A.; Cano, A.F.; Fernández, J.A. The effect of different fruiting temperatures on the yield and nutritional parameters of some wild and hybrid *Hericium isolates*. *Sci. Hortic.* **2021**, *280*, 109915. [CrossRef]
94. Feeney, M.J.; Miller, A.M.; Roupas, P. Mushrooms—Biologically distinct and nutritionally unique. *Nutr. Today* **2014**, *49*, 301–307. [CrossRef] [PubMed]
95. Cohen, N.; Cohen, J.; Asatiani, M.D.; Varshney, V.K.; Yu, H.T.; Yang, Y.C.; Li, Y.H.; Mau, J.L.; Wasser, S.P. *Chemical composition* and nutritional and medicinal value of fruit bodies and submerged cultured mycelia of culinary-medicinal higher *Basidiomycetes mushrooms*. *Int. J. Med. Mushrooms* **2014**, *16*, 273–291. [CrossRef] [PubMed]
96. Pop, R.M.; Puia, I.C.; Puia, A.; Chedea, V.S.; Leopold, N.; Bocsan, I.C.; Buzoianu, A.D. Characterization of Trametes versicolor: Medicinal mushroom with important health benefits. *Not. Bot. Horti Agrobot. Cluj-Napoca* **2018**, *46*, 343–349. [CrossRef]
97. Rodríguez-Seoane, P.; Díaz-Reinoso, B.; González-Muñoz, M.J.; Fernández de Ana Portela, C.; Domínguez, H. Innovative technologies for the extraction of saccharidic and phenolic fractions from *Pleurotus eryngii*. *LWT* **2019**, *101*, 774–782. [CrossRef]

molecules

MDPI

Article

Immunomodulatory Properties of Polysaccharide-Rich Young Green Barley (*Hordeum vulgare*) Extract and Its Structural Characterization

Marta Kinga Lemieszek [1,*], Iwona Komaniecka [2], Michał Chojnacki [1], Adam Choma [2] and Wojciech Rzeski [1,3]

1 Department of Medical Biology, Institute of Rural Health, Jaczewskiego 2, 20-090 Lublin, Poland; chojnacki.michal@imw.lublin.pl (M.C.); rzeskiw@hektor.umcs.lublin.pl (W.R.)
2 Department of Genetics and Microbiology, Institute of Biological Sciences, Maria Curie-Skłodowska University, Akademicka 19, 20-033 Lublin, Poland; iwona.komaniecka@poczta.umcs.lublin.pl (I.K.); adam.choma@mail.umcs.pl (A.C.)
3 Department of Functional Anatomy and Cytobiology, Institute of Biological Sciences, Maria Curie-Skłodowska University, Akademicka 19, 20-033 Lublin, Poland
* Correspondence: lemieszek.marta@imw.lublin.pl; Tel.: +48-81-71-84-513

Abstract: Young green barley (YGB) water extract has revealed a beneficial impact on natural killer (NK) cells' ability to recognize and eliminate human colon cancer cells, without any side effects for normal colon epithelial cells. The direct anticancer effect of the tested compounds has been also shown. The mixture of oligosaccharides found in this extract was characterized by chemical analyses and via FT-IR spectroscopy and MALDI-TOF MS techniques. The YGB preparation contained 26.9% of proteins and 64.2% of sugars, mostly glucose (54.7%) and fructose (42.7%), with a small amount of mannose (2.6%) and galactose (less than 0.5%). Mass spectrometry analysis of YGB has shown that fructose oligomers contained from 3 to 19 sugar units. The number of fructans was estimated to be about 10.2% of the dry weight basis of YGB. The presented results suggest the beneficial effect of the consumption of preparations based on young barley on the human body, in the field of colon cancer prevention.

Keywords: cancer immunotherapy; fructan; FT-IR; mass spectrometry; young green barley

Citation: Lemieszek, M.K.; Komaniecka, I.; Chojnacki, M.; Choma, A.; Rzeski, W. Immunomodulatory Properties of Polysaccharide-Rich Young Green Barley (*Hordeum vulgare*) Extract and Its Structural Characterization. *Molecules* **2022**, 27, 1742. https://doi.org/10.3390/molecules27051742

Academic Editors: Sokcheon Pak and Soo Liang Ooi

Received: 14 February 2022
Accepted: 4 March 2022
Published: 7 March 2022

1. Introduction

Oncological immunotherapy, also known as immuno-oncology, is a modern method of cancer treatment consisting of administering non-toxic preparations, modulating the immune system's response by increasing and/or improving its ability to prevent, control, and fight cancer. Immuno-oncology can educate the immune system to better recognize, more effectively attack, and successfully eliminate cancer cells. At the beginning of the 21st century, the American Society of Clinical Oncology announced that oncological immunotherapy will dominate the treatment of neoplastic diseases and, after 2020, it could be used in the treatment of even every second type of cancer. In the beginning, immuno-oncology mainly focused on immunostimulatory cytokine treatments. Recently, natural killer cell-based immunotherapy has emerged as the most promising therapeutic approach for both hematological malignancies and solid tumors [1]. The natural killer (NK) cells were first identified as unique large granular lymphocytes able to detect and rapidly kill abnormal cells without prior sensitization; thus, they are the first line of defense for the immune system against cells infected with bacteria and viruses, as well as malignant cells [2]. The importance of NK cells for cancer treatment has been confirmed in epidemiological studies, which revealed the negative correlation between NK cells activity and the risk of cancer development and metastasis [1]. The main considerations relating to NK cell-based immunotherapy are the NK cell source and the method of enhancement of

NK cell anticancer activity. The presented studies address the second issue, in particular, being focused on barley (*Hordeum vulgare*), which is the fourth most important grain crop in the world and the richest source of dietary fiber among the cereals. For many years, barley was valued solely for its nutritional properties; recently, more and more scientific reports have indicated its health benefits. In the early 1990s, Japanese scientists showed that the most valuable source of both nutrients and bioactive substances is young green barley, defined as seedlings up to 200 h after sprouting, with a height of between 20 and 30 cm [3–5]. More recent studies have also indicated barley grass as a functional food with many health-promoting effects (e.g., anti-aging, antioxidant, anti-inflammatory, immunity enhancement, detoxification, hypolipidemic effects, improvement of gastrointestinal activity, blood pressure regulation, anti-hypoxia, antifatigue) as well as a potential agent against several human chronic diseases, including cancer, cardiovascular disease, obesity, diabetes, and stroke [6]. The health benefits of young green barley are associated with many of its different elements; however, immunomodulatory properties are typically considered to be due to providing a large amount of dietary fiber, especially in the form of polysaccharides [6–8]. The immunoenhancement effects of polysaccharides isolated from young barley leaves have been demonstrated by Kim et al. and Han et al., in both in vitro and in vivo studies [9,10]. Kim et al. have reported the stimulation of proliferation and the production of cytokines (IL-6, GM-CSF) in cells derived from Peyer's patches, in response to arabinoxylan- and rhamnogalacturonan I-rich polysaccharides that were isolated from barley grass (BLE0) [9]. Furthermore, the intestinal immunostimulatory activity observed in vitro has been also proved in mice, wherein the oral administration of BLE0 augmented IgA production and increased the levels of IgA-related cytokines (IL-10, TGF-β) [9]. Further studies on the abovementioned crude barley polysaccharides (BLE0) confirmed their immunostimulatory effects. They have been shown to elicit an increase in the proliferation and secretion of cytokines (IL-2, IL-4, IL-10, IFN-γ) in CD3/CD28-activated mice splenocytes. Additionally, oral administration of barley polysaccharides enhanced the immune system of mice with cyclophosphamide-induced immunosuppression, via the stimulation of splenocyte proliferation and acceleration of NK cell cytotoxic activity [10]. Han et al. have shown that water-soluble barley polysaccharide (BP-1) significantly improved the immune ability of immunosuppressive mice in several different ways, including: increasing the number of bone marrow cells as well as white blood cells in peripheral blood; promoting the proliferation of spleen cells, NK cells and macrophages; stimulating NK cell cytotoxic activity; improving the phagocytosis activity of macrophages; increasing the level of IL-2, TNF-α and IFN-γ in serum and the spleen; and enhancing the production of IgG and IgM in the spleen [11]. Ryu et al. have shown that water-soluble barley β-glucan suppressed the proliferation of mice splenocytes, accelerated by concanavalin A, IL-2, or alloantigen. Additionally, the tested β-glucan suppressed the death-induced activity of allogenic lymphocytes Tc and allogenic lymphokine-activated killer cells [12]. These immunomodulatory properties were also reported in the case of commercially available barley β-glucan. Misra et al. demonstrated that β-glucan derived from barley improved immune system activity in fish through an increase in leukocyte number, complemented activation, and stimulated phagocytic activity as well as bactericidal activity [13]. Aoe et al., who investigated the impact of the high-fructan barley product named BARLEYmax (BM; containing fructan 9.0%, β-glucan 6.3%, and resistant starch 3%) on the growth of microbiota that are present in the rat gut, revealed a beneficial effect of the investigated carbohydrates on health-promoting species belonging to the *Bifidobacterium*, *Oscillospira*, *Parabacteroides* and *Sutterella* species, which may prevent colonic inflammation and enhance intestinal immunological functions [14]. Furthermore, Aoe et al. have shown that BM possesses stronger prebiotic properties than the high-β-glucan barley line (BG; containing β-glucan 9.0%, fructan, 2.2%, and resistant starch, 0.5%) [14]. The beneficial effect of BARLEYmax on the gut microbiome was also proven in clinical studies, wherein 4 weeks' consumption at 12 g/day of this high-fructan barley product increased the abundance of *Bacteroides* and decreased the abundance of *Clostridium subcluster* [15]. Similarly, Sasaki et al. revealed in an in vitro model of the human colonic microbiota that polysaccharide-rich young green barley leaf extract (YBL; containing carbohydrates at 46.5%, including fructan at 9.4%, proteins at 29.0%, and minerals at 17.5%) increased the abundance of bacteria related to the genus *Bifidobacterium* and stimulated the growth of bacteria related to the genera *Faecalibacterium*, *Lachnospira*, *Roseburia*, and unclassified *Ruminococcaceae*. The discovered bifidogenic and butyrogenic effects of YBL have been associated with health-promoting properties as a result of gut immune system stimulation, as well as via inflammatory inhibition [16].

Despite the fact that several scientific reports have revealed the immunoenhancement properties of polysaccharides derived from barley, due to the high variability of polysaccharides and the associated diversity of biological activities, it is worth continuing research into this group of compounds. The aim of the current study was to evaluate the immunomodulatory properties of the water extract of young green barley (*Hordeum vulgare*) reached with polysaccharides by evaluating its influence on the viability and proliferation of NK-92 cells and above all their ability to kill human colon cancer LS180 cells representative for the common cancers worldwide. At the same time, the selectivity of the anticancer effect of polysaccharide-activated NK cells was also evaluated on the human normal colon epithelial CCD841 CoN cells. Furthermore, to investigate the major compounds related to the immunomodulatory effect, the structural characterization of extract polysaccharides was conducted.

2. Results

2.1. Young Green Barley Extract Composition

In the first stage, the chemical composition of young green barley extract was investigated. The total sugar and protein contents were examined using colorimetric methods (phenol-sulphuric acid assay, Pierce BCA protein assay). These studies revealed that the tested extract consisted mainly of sugars (64.2 ± 1.14%), while the protein amount (26.9 ± 4.09%) was also significant. Electrophoretic separation (SDS-PAGE) of young green barley extract (YGB) confirmed the presence of proteins with a molecular weight of around 55 kDa and 35 kDa, as well as proteins weighing less than or equal to 15 kDa. The obtained data could indicate that YGB is a carbohydrate-protein complex (Figure 1).

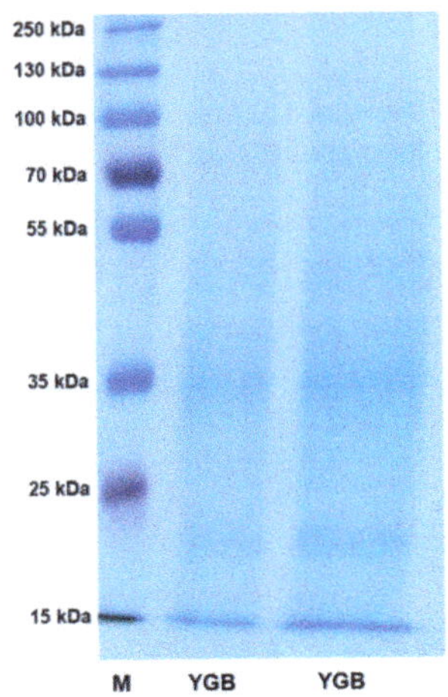

Figure 1. Electrophoretic separation (SDS-PAGE) of young green barley extract (YGB). The 12% polyacrylamide gel used after protein separation was stained with Coomassie Brilliant Blue. As a standard, a molecular protein mass marker (from 15 to 250 kDa) was used.

The FT-IR spectrum of YGB preparation ranged from 400 cm^{-1} to 4000 cm^{-1}, as shown in Figure 2. The results showed an intense and broad absorption peak at 3269 cm^{-1} for O-H and N-H stretching vibrations, a peak at 2933 cm^{-1} for C-H stretching vibrations, and a broad absorption band in the region of 950–1200 cm^{-1}, with a maximum at 1053 for coupled C-O and C-C stretching and C-OH bending vibrations. All these absorption bands are characteristic of polysaccharides [17]. The peaks at around 2933 cm^{-1} and 2880 cm^{-1} indicate the presence of C-H bonds (C-H stretching vibration of CH_3 and CH_2 groups from carbohydrate molecules, respectively) [18]. The broad symmetrical signal (ranging from 1740 to 1500 cm^{-1}), with a maximum at 1593 cm^{-1}, might be indicative of proteins as well as adsorbed water (bending vibrations from amide I and amide II; absorbing groups: C=O, C-N, N-H, and H-O-H) [19,20]. The next intense broad signal at 1397 cm^{-1} corresponded with the absorbance of the CH_2 and CH_3 functional groups (deformational vibrations) from both sugars and proteins. The prominent and overlapping absorption bands between 1150 and 970 cm^{-1} were assigned to the C-O-C glycosidic bond vibrations and ring vibrations, as well as the C-O-H stretching vibrations of side-group bounds. These are commonly present in sugars [21]. The YGB preparation also showed absorption peaks at 940 and 860 cm^{-1}, which are especially characteristic of α-d-glucose [22]. A broad signal at 618 cm^{-1} could be indicative of pyranose rings [23]. A very intense band with a maximum at ~1050 cm^{-1}, and two sharp lines of similar intensity in the regions of 820 and 780 cm^{-1} indicate the presence of fructose from fructans [18].

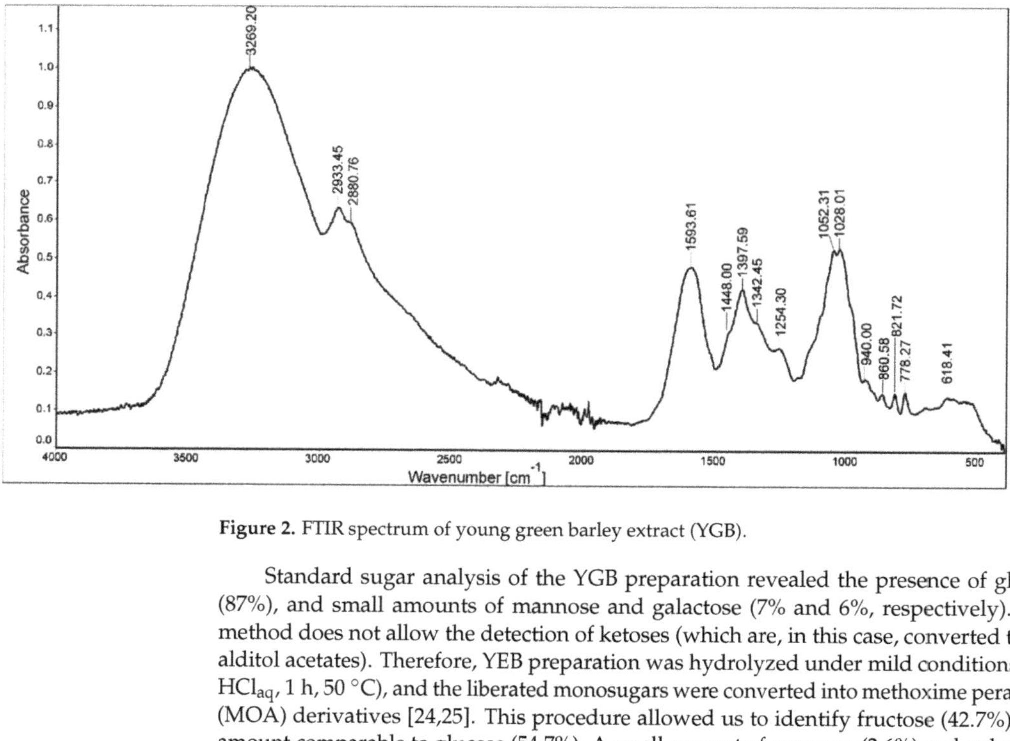

Figure 2. FTIR spectrum of young green barley extract (YGB).

Standard sugar analysis of the YGB preparation revealed the presence of glucose (87%), and small amounts of mannose and galactose (7% and 6%, respectively). This method does not allow the detection of ketoses (which are, in this case, converted to two alditol acetates). Therefore, YEB preparation was hydrolyzed under mild conditions (1 M HCl_{aq}, 1 h, 50 °C), and the liberated monosugars were converted into methoxime peracetate (MOA) derivatives [24,25]. This procedure allowed us to identify fructose (42.7%) in an amount comparable to glucose (54.7%). A small amount of mannose (2.6%) and only traces of galactose (less than 0.5%) were also detected.

Table 1 contains the results of the methylation analysis and shows all derivatized compounds that were found. Fructose derivatives were found as terminal, →6)-linked and branched, →1,6)-linked forms, whereas glucose was terminal, →3)-linked, →4)-linked, and →6)-linked, as well as branched, →3,6)-linked. Mannose was exclusively terminal, and galactose—terminal, →3)-linked and →6)-linked. Among all the above-mentioned derivatives, terminal sugars prevailed, giving about 75% of all methylated end-products.

Table 1. Linkage analysis of young green barley extract (YGB). The sample was subjected to methylation, hydrolysis, reduction, and acetylation. The obtained permethylated alditol acetates were identified by GLC–MS, based on their mass spectra and retention times. The main permethylated alditol acetates are shown in bold type.

Component	Amount (%)
***t*-Fru*f*-(2→**	**22.3**
t-Man*p*-(1→	1.2
***t*-Glc*p*-(1→**	**47.2**
t-Gal*p*-(1→	5
→6)-Fru*f*-(2→	**9.7**
→3)-Glc*p*-(1→	0.9
→4)-Glc*p*-(1→	3.1
→3)-Gal*p*-(1→	0.5
→6)-Glc*p*-(1→	3
→6)-Gal*p*-(1→	2
→1,6)-Fru*f*-(2→	3.5
→3,6)-Glc*p*-(1→	1.6

For the determination of fructan content, the PAHBAH reducing sugar method was used. The fructan content of young barley was calculated from the absorbance of the color complex and was then converted to a dry weight basis per 100 g of product. The average fructan content of the tested samples was 10.23 ± 0.22 g per 100 g dry weight basis (i.e., 10.2% of dry weight YEB extract).

The size of the oligomer/polymer of carbohydrates present in YEB was estimated using the mass spectrometry technique. The MALDI-TOF mass spectrum, in positive ion mode, is shown in Figure 3. A series of intensive signals appeared in the same intervals of 162.05 u, pointing to hexose as a repeating unit. In the MALDI-TOF mass spectrum of YEB (Figure 3A), only those polymers containing up to 10 hexose units (DP = 10) were shown, but, when tracing the spectrum at low intensities in the mass range from 1700 to 3000 *m/z*, it was possible to find oligomers up to DP = 19 (data not shown). The smallest oligomer visible in Figure 3A had *m/z* at 543.138. Its fragmentation pattern has been shown in Figure 3B. This MALDI MS-MS spectrum revealed the only fragment ion *m/z* at 381.062, which corresponded to a disaccharide potassium adduct, probably composed of glucose and fructose residues (sucrose, $[G+F+K]^+$). This disaccharide may be here a kind of backbone to which the next fructose units are linked during the biosynthesis process, forming the fructan detected in the YGB preparation.

The complexity of the MALDI-TOF mass spectrum is displayed in Figure 3C, showing that the main signal (oligomer potassium adduct, $[M+K]^+$) is accompanied by the sodium adduct ($[M+Na]^+$), as well as these molecules that are deprived of water ($[M\text{-}H_2O+K]^+$ and $[M\text{-}H_2O+Na]^+$, respectively). It should be noted that the protonated ($[M+H]^+$) ions were not detected at the mass spectrum; thus, it can be concluded that the YGB preparation is a rich source of potassium and sodium salts.

2.2. *Young Green Barley Extract Impacts NK-92 Cell Viability and Proliferation*

In order to evaluate the immunomodulatory potential of the YGB, its influence on NK-92 cell membrane integrity, metabolic activity, and proliferation was assessed via the LDH (lactate dehydrogenase), MTT (thiazolyl blue tetrazolium bromide), and BrdU (bromodeoxyuridine) tests, respectively (Figure 4). The LDH test revealed that the tested extract was not cytotoxic against NK-92 cells. Furthermore, the extract, when in concentrations from 25 µg/mL to 500 µg/mL, significantly decreased the lactate dehydrogenase release from treated cells. The MTT test has shown that YGB, when in concentrations from 25 µg/mL to 250 µg/mL, did not impact NK-92 viability, while at concentrations of 500 and 1000 µg/mL, it decreased the lymphocytes' metabolic activity by 4.4% and 5.1%, respectively. The BrdU test revealed no changes in the proliferation of NK-92 cells treated with YGB when used in concentrations equal to or less than 250 µg/mL. On the contrary, the tested extract, when used at concentrations of 500 and 1000 μg/mL, inhibited DNA synthesis in NK-92 cells, reducing cell proliferation by 9.8% and 11.4%, respectively.

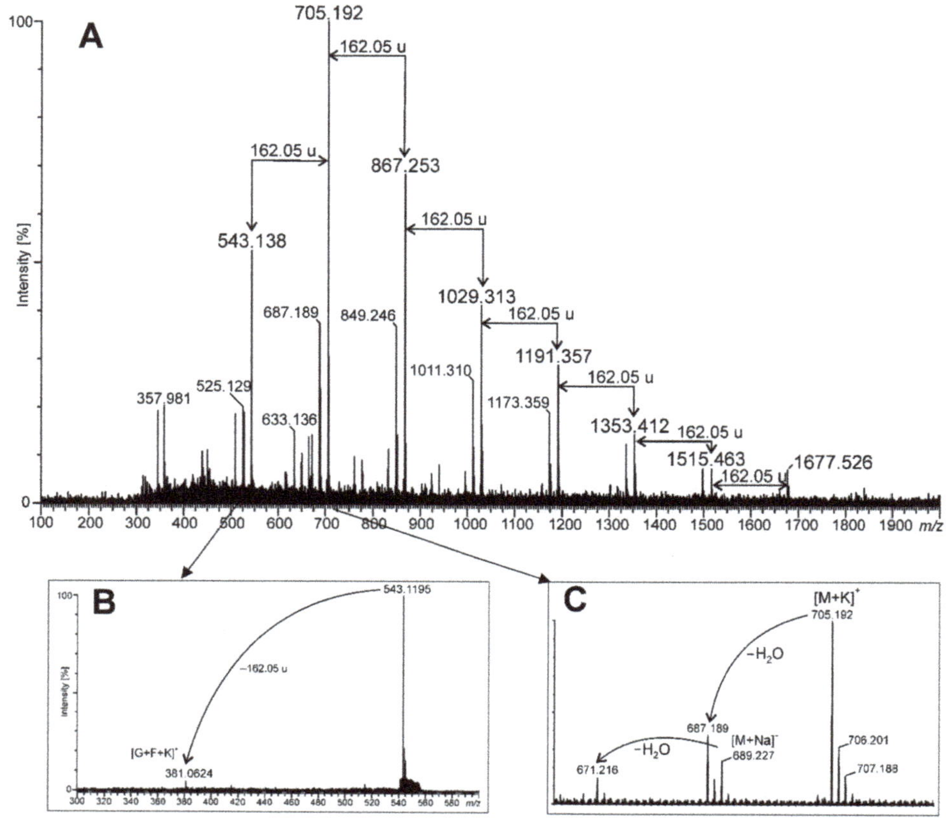

Figure 3. MALDI-TOF mass spectrum of young green barley extract (YGB) obtained in positive ion mode (**A**); MS-MS of the ion at *m/z* 543.1 (**B**); details of the spectrum A in the range of 670–710 *m/z* (**C**).

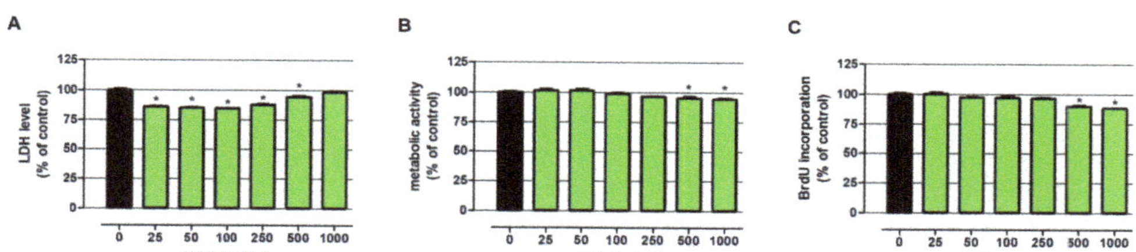

Figure 4. The influence of young green barley extract (YGB) on NK-92 cell viability and proliferation. Cells were exposed to culture medium alone (control) or with YGB at the following concentrations of 25, 50, 100, 250, 500, and 1000 μg/mL for 48 h. Cell viability was determined by the examination of both cell membrane integrity, using an LDH assay (**A**), and metabolic activity using an MTT assay (**B**), while cell proliferation was measured via an immunoassay based on BrdU incorporation (**C**). The results are presented as a mean ± SEM of at least 4 measurements. * $p < 0.05$ vs. control, one-way ANOVA test; post hoc test: Dunnett.

2.3. Young Green Barley Extract Enhances NK-92 Cells' Cytotoxicity in Human Colon Cancer LS180 Cells without any Side Effects on Human Colon Epithelial CCD841 CoN Cells

Before studying the impact of YGB on NK cells' cytotoxic activity, the extract's influence on normal and cancer colon cells was examined via LDH, MTT, and BrdU assays. As presented in Figure 5 (left panel), YGB in the whole range of analyzed concentrations (from 50 to 1000 µg/mL) did not affect the viability of the human colon epithelial cells (CCD841 CoN), which was seen in all the performed assays. On the contrary, when tested at the highest concentration, the extract decreased the metabolic activity of human colon adenocarcinoma LS180 cells by 21.3% (Figure 5D), and simultaneously inhibited DNA synthesis in the investigated cell line by 10.7% (Figure 5F). At the same time, when tested at concentrations of 50, 250, and 1000 µg/mL YGB, the extract was cytotoxic against colon cancer cells, as was shown by the increase in LDH release to the culture medium by 26.5%, 31.8%, and 36.4%, respectively (Figure 5D).

In the next step, YGB's influence on the NK-92 cells' cytotoxicity was determined under the same conditions. First, NK-92 cells, when used alone, did not cause any side effects in the investigated human colon epithelial cells. Similarly, the NK-92 cells' impact on CCD841 CoN viability did not change in the presence of tested extract (Figure 5; left panel). On the contrary, NK-92 cells distinctly inhibited the metabolic activity and DNA synthesis of human colon cancer cells and disrupted the integrity of their cell membranes. Furthermore, the anticancer properties of NK-92 cells were enhanced by YGB in a dose-dependent manner. As presented in Figure 5B, the cytotoxic effect of NK-92 cells in response to treatment with YGB increased from 133.9% to 143.5% (50 µg/mL), 147.3% (250 µg/mL), and 162.5% (1000 µg/mL). The beneficial effect of tested compounds was also observed in another assay. The MTT test (Figure 5D) revealed that the metabolic activity of colon cancer cells treated with both NK-92 cells and the tested extract decreased from 74.7% to 54.9% (50 µg/mL), 51.0% (250 µg/mL), and 47.5% (1000 µg/mL), while the BrdU assay (Figure 5F) showed the significant antiproliferative properties of NK cells against LS180 cells, which intensified in the presence of YGB, as reflected in the decrease in DNA synthesis from 85.5% to 66.5% (50 µg/mL), 62.8% (250 µg/mL), and 59.5% (1000 µg/mL). It should be highlighted that the inhibition of the LS180 cells' metabolic activity, as well as the DNA synthesis caused by NK-92 in the presence of YGB in the whole range of tested concentrations, was significantly stronger than the decrease in cancer cell viability induced by the extract when used alone (comparison of corresponding YGB concentrations). A similar pattern of changes was observed in the results of the LDH test. Nevertheless, a comparison of results obtained from LS180 cells treated with only NK-92 cells, vs. LS180 exposed simultaneously to both NK cells and YGB, revealed statistically significant differences in cancer cell viability and proliferation at the whole range of tested extract concentrations, while a significant distortion of cell membrane integrity was only noted at 1000 µg/mL YGB.

2.4. Young Green Barley Extract Enhances Colon Cancer Cell Death in Response to NK-92 Cells

To confirm the observed cytotoxic effect in LS180, cells of NK-92, used separately and together, a YGB cell death detection ELISA assay (Figure 6A) and nuclear double staining were performed (Figure 6B). The results of the ELISA assay revealed that the investigated extract in the tested concentrations was not able to induce apoptosis in LS180 cells at the same time as a number of nucleosomes were released into the cell culture medium (a marker of necrosis) and increased by 69.9% (50 µg/mL), 86.5% (250 µg/mL), and 103.2% (1000 µg/mL). NK-92 cells, when used alone, killed LS180 cells and induced both programmed cell death and necrosis, which was demonstrated by the increase in the number of DNA fragments recorded in the cytoplasmic fraction of colon cancer cells (126.5% of control) and the medium collected from above the cell culture (264.2% of control). Similar to the MTT and LDH results, this research revealed that the anticancer properties of NK-92 cells were enhanced by YGB in a dose-dependent manner. The proapoptotic abilities of NK-92 cells in LS180 cells increased in response to YGB, from 126.5% (LS180 cells treated with only NK-92 cells) to 131.8% (50 µg/mL), 136.4% (250 µg/mL), and 143.5% (1000 μg/mL); however, the statistically significant enhancement of NK-92 cells' ability to induce programmed cell death was observed only after NK cells were co-incubated with 1000 μg/mL YGB. Conversely, the number of necrotic colon cancer cells that were observed in response to NK-92 when co-incubated with the tested extract increased from 264.2% (LS180 cells treated with only NK-92 cells) to 302.9% (50 μg/mL), 337.6% (250 μg/mL), 388.8% (1000 μg/mL), and the reported YGB intensification of NK cells' cytotoxicity was statistically significant in the whole range of the investigated extract concentrations. Furthermore, LS180 cell-killing by NK-92 in the presence of YGB, in the whole range of tested concentrations, was significantly improved compared to the direct anticancer effect of the extract (comparison of corresponding YGB concentrations).

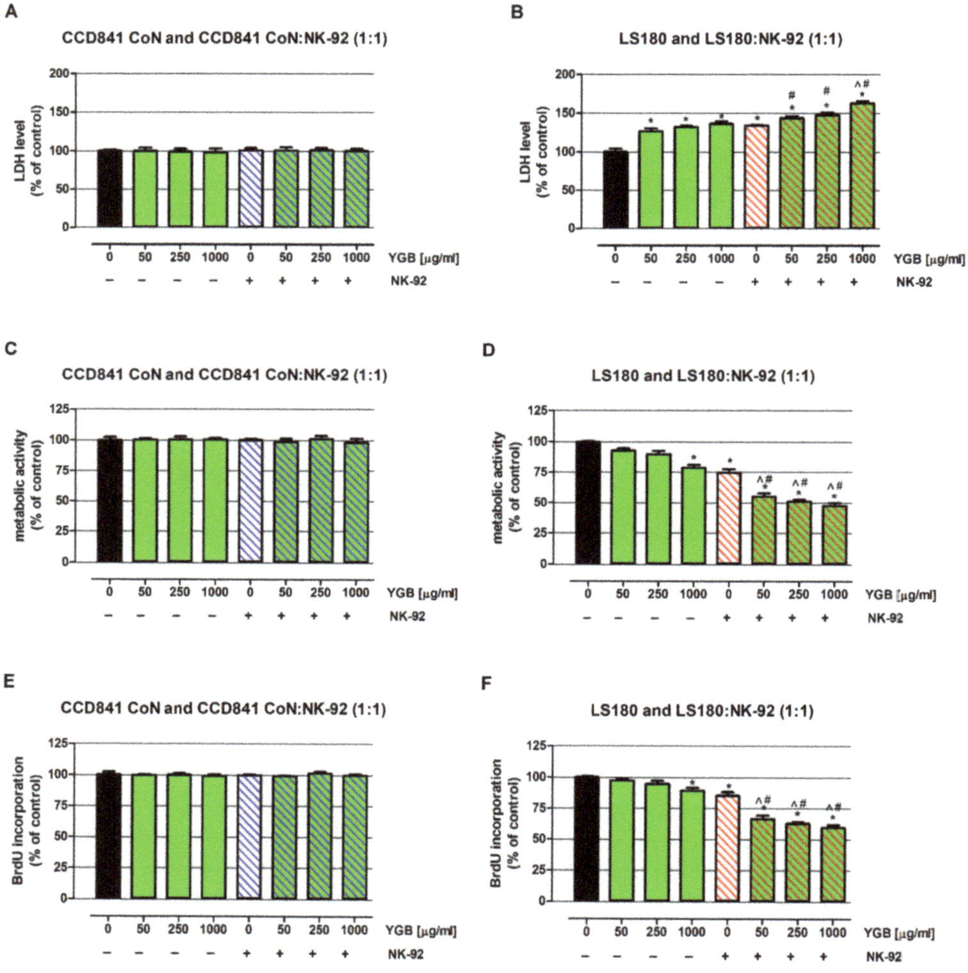

Figure 5. Influence of young green barley extract (YGB) on NK-92 cell cytotoxicity against normal and cancer colon cells. The human colon adenocarcinoma cell line LS180 and human colon epithelial cell line CCD841 CoN, and were exposed to culture medium alone (control) or the YGB (50, 250, and 1000 μg/mL) used alone, or to NK-92 cells in the absence or presence of YGB (50, 250 and 1000 μg/mL) for 48 h. NK-92 cells were used in the same concentration as treated cells. Changes in

investigated cells in response to YGB, used alone or together with NK cells, were determined by examining cell membrane integrity using an LDH assay (**A**,**B**), metabolic activity using an MTT assay (**C**,**D**), and DNA synthesis using the BrdU assay (**E**,**F**). Results are presented as mean ± SEM of at least 4 measurements. * $p < 0.05$ vs. control; ^ $p < 0.05$ colon cells treated with both NK-92 cells and YGB extract vs. colon cells treated with only NK-92 cells; # $p < 0.05$ colon cells treated with both NK-92 cells and YGB extract vs. colon cells exposed to YGB at the same concentration as those being tested in a one-way ANOVA test; Tukey post hoc test.

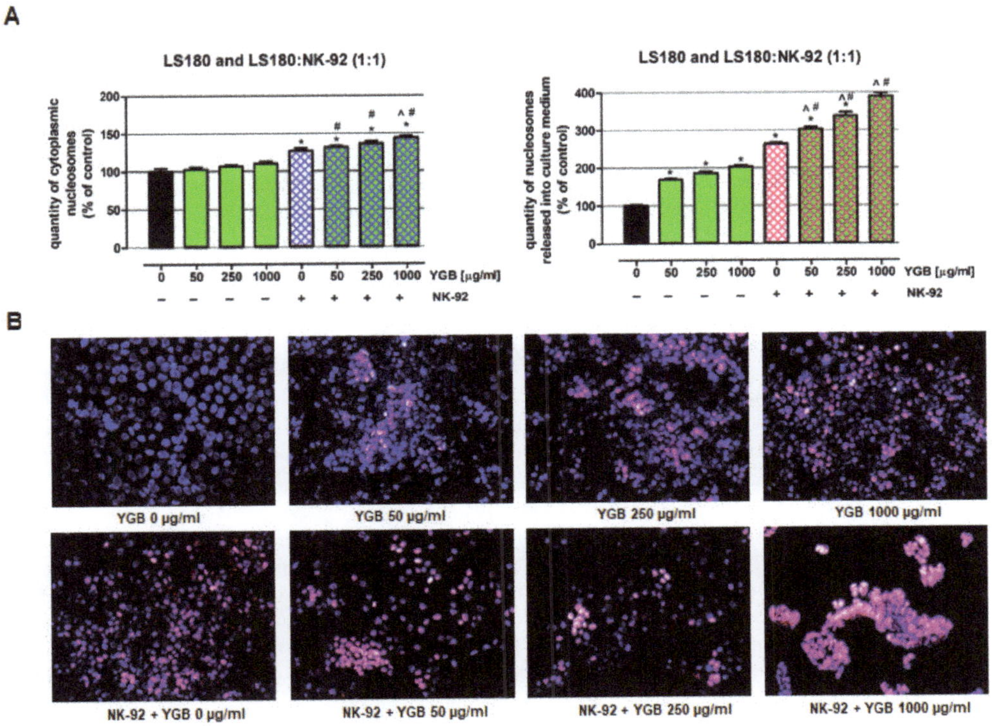

Figure 6. Cell death induction in human colon cancer cells, in response to NK cells co-incubated with young green barley extract (YGB). The impact of YGB (50, 250, and 1000 μg/mL) or NK-92 cells, used together or separately, on the viability of the human colon adenocarcinoma cell line LS180 was investigated after 48 h of incubation. (**A**) Enrichment of nucleosomes in the cytoplasmic fraction (an apoptosis marker) or cell culture medium (a necrosis marker) was determined by a cell-death detection ELISA assay. The results are presented as the mean ± SEM of at least 4 measurements. * $p < 0.05$ vs. control; ^ $p < 0.05$ cancer cells treated with both NK-92 cells and YGB extract vs. cancer cells treated with only NK-92 cells; # $p < 0.05$ cancer cells treated with both NK-92 cells and YGB extract vs. cancer cells exposed to YGB at the same concentration as tested. One-way ANOVA test; post hoc test: Tukey. (**B**) Untreated LS180 cells (control), as well as LS180 cells exposed to the extract with or without NK-92 cells, were double-stained (Hoechst and propidium iodide) and examined under fluorescence microscopy. Representative pictures are presented; the magnification is 400×.

The results described above were reflected in the microscopic analysis of LS180 cells that were double-stained with Hoechst and propidium iodide (Figure 6B). The conducted studies have shown single colon cancer cells undergo programmed cell death in both untreated and YGB-treated LS180 cells. Nevertheless, YGB, when used alone, significantly damaged colon cancer cell membranes, and the observed cytotoxicity was dose-dependent.

On the contrary, NK-92 cells induced both necrosis and apoptosis in the investigated colon cancer cells; however, the cytotoxic effect was dominant. The anticancer effect of the NK cells distinctly increased in the presence of the tested extract, which was associated with elevated levels of both necrotic and apoptotic LS180 cells; however, the disintegration of cell membranes was the dominant mode of cancer cell elimination. The enhancement of colon cancer cell death by YGB-treated NK-92 cells was directly dependent on extract concentration.

3. Discussion

Young green barley water extract has been shown to contain more than 64% of saccharides, composed mainly of glucose and fructose. Methylation analysis shows that the majority of them are in the terminal position, which might suggest that the preparation contains short oligosaccharides. Interestingly, the presence of saccharose—the metabolic precursor of fructans—could not be observed in the MALDI-TOF mass spectrum. This means that saccharose can be quickly metabolized or exported outside of the leaves. Fructans are considered to be the main saccharide reserves in the vegetative tissues of grasses, including barley [26]. The oligosaccharides observed in the analyzed extract can be recognized as plant fructan. Mixed-type fructan (containing both (1→2) and (6→2)-linked fructosyl units were found in the YGB preparation, based on methylation analysis and the mass spectrometry technique. The fructan content in YGB was also determined and represented approximately 10.2% of the dry matter basis of the analyzed samples. This type of sugar polymer is typical for cereals, such as barley [27,28]. Plant fructans are usually shorter than bacterial ones and, in general, contain less than 50 fructosyl units. Nemeth and co-workers showed changes in the distribution of different chain lengths and the pattern of fructan structures in the barleys derived from different cultivars or breeding lines [29]. In the case of YGB extract, the fructose oligomers had fewer than 20 units, and the most frequent fraction that was found contained 4 sugars—this is probably nystose. Nystose was described earlier as a component of barley grains [30].

Despite the fact that the immunomodulatory properties of plant fructans are becoming better and better known, most of the studies focused only on their indirect influence on the immune system components, which is a consequence of their prebiotic properties [31,32]. Nevertheless, the results from recent studies dedicated to fructans isolated from a wide range of plants revealed that the mentioned carbohydrates may also directly impact immune cells and, consequently, modulate the immune system responses [31]. Unfortunately, there is a lack of evidence presenting the direct impact of barley fructans on the components of the immune system. Nevertheless, considering the fact that the main component of the investigated extract was also fructooligosaccharide, the discovery of a beneficial impact on immune cells was predicted.

It should be emphasized that the chemical examination of YGB revealed that, next to carbohydrates, the second main component of the investigated extract was proteins, the total amount of which was around 27%. The electrophoretic separation of young green barley extract revealed the presence of proteins with a molecular weight of around 55 kDa and 35 kDa, as well as proteins weighing less than or equal to 15 kDa. The obtained data indicated that YGB was a carbohydrate–protein complex. Considering the proven enhancement impact of protein ligands on both the anticancer and immunomodulatory properties of polysaccharides [33], discovered that the presence of proteins in YGB seems to be beneficial. It should be emphasized that, according to our best knowledge, the data presented herein is the first report published about the immunomodulatory properties of the barley carbohydrate-protein complex.

The immunomodulatory properties of young green barley extract were examined using NK-92 cells, the best-described human NK cell line, the therapeutic utility of which has been confirmed in several clinical trials [34,35]. NK-92 cells characterized the ability to produce cytokines, as well as high cytotoxic activity. Although they are highly cancer-specific, the cells can be further modified to express different cancer receptors to increase the targeting and killing of cancer cells, as well as to express therapeutic antibody-binding receptors, which also enhances their anticancer activity. Furthermore, NK-92 cells are able to resist the suppression induced by many agents released by cancer cells, thanks to the absence of killer inhibitory receptors (KIR) [35,36]. The above-mentioned features, as well as their long-term growth potential, have made NK-92 cells an attractive model by which to study the function of human NK cells, and have become an excellent tool for evaluating the immunoenhancement properties of compounds of various origins; thus, this cell line was selected to examine the possibility of using polysaccharide-rich young green barley extract in oncological immunotherapy.

Evaluation of the direct impact of young green barley extract on NK cells revealed that the tested compounds are not cytotoxic in the whole range of investigated concentrations, and did not affect the viability and proliferation of NK-92 cells in a wide range of tested concentrations (from 25 to 250 μg/mL); however, at higher doses (500 and 1000 μg/mL) the extract decreased both lymphocyte metabolic activity and DNA synthesis. Several studies have reported the stimulation of leucocyte proliferation by barley polysaccharides; however, the beneficial effect on NK cell numbers was reported only by Han et al., who revealed the increase of NK cell proliferation in immunosuppressive BALB/c mice in response to water-soluble barley polysaccharide (BP-1; molecular weight 67 kDa) [11]. Nevertheless, it needs to be highlighted that NK-92 cells are human lymphoma-derived cells and, despite the fact that they are phenotypically similar to activated normal NK cells [36], the results obtained from studies conducted on this cell line should be interpreted in light of this data. Because of the reaction observed in response to higher doses of young green barley extract, the slight decrease of NK-92 cell viability and proliferation with a simultaneous lack of YGB cytotoxicity did not unequivocally determine the existence or absence of the immunomodulatory properties of the test substance.

In order to verify the possibility of using young green barley extract-derived carbohydrate-protein complex in oncological immunotherapy based on NK cells, YGB's influence on NK-92 anticancer activity was evaluated. Because immunotherapy should be both effective and safe, studies have been conducted on both normal and cancer cell lines (human colon epithelial CCD841 CoN cells and human colon cancer LS180 cells). An in vitro model of colon cancer was chosen because of the fact that this type of cancer is one of the most common cancers in terms of incidence and mortality rate in the world.

In the first step of these studies, the direct anticancer effect of YGB as well as NK cells used separately has been analyzed. The obtained results revealed that the examined compounds significantly decreased the viability and proliferation of colon cancer LS180 cells, as well as effectively damaging their cell membranes and causing necrosis induction. At the same time, YGB did not affect human colon epithelial CCD841 CoN cells. The observed high selectivity of the examined compounds can be understood as a lack of negative influence on normal colon cells, with an evident anti-cancer effect against colon cancer cells, which corresponds with the previous results of our research group [37,38].

Similar to YGB, NK-92 cells also revealed a high selectivity of action in relation to colon cancer cells. The recorded anticancer effect of NK cells was mostly associated with cytotoxicity; however, significant apoptosis induction was also noted. In terms of oncological immunotherapy, the observed ability of NK-92 cells to induce different cancer cell death would create the possibility of modulating the immunogenicity of the tumor microenvironment, which is important in the case of therapy efficiency and safety.

The discovered anticancer effect of NK-92 cells and YGB, when used alone, intensified when the lymphocytes and the extract were used together. The beneficial impact of NK cells co-incubated with the examined polysaccharide-rich young green barley extract was observed on metabolic activity, DNA synthesis, cell membrane integrity, and cell death induction in LS180 cells. The enhancement of NK cells' anticancer activity by YGB was dose-dependent; nevertheless, a significant improvement in the NK-92 cells' proapoptotic and cytotoxic abilities was noted only in the presence of tested compounds at the highest tested concentration (1000 μg/mL), while other investigated phenomena were evidently improved by YGB in the whole range of tested concentrations. It should be highlighted that cancer cell elimination by NK-92 in the presence of YGB in the whole range of tested concentrations was significantly stronger than the decrease in cancer cell viability and proliferation, as induced by the extract when used alone at the corresponding concentrations. Nevertheless, the synergy of the anticancer actions of NK-92 cells and the tested compounds was observed in all performed studies except the LDH assay. The YGB in the whole range of investigated concentrations that were given to cancer cells together with NK cells caused a stronger inhibition of metabolic activity and DNA synthesis, as well as improved apoptosis induction, than would have been anticipated from the mere summation of the direct effects of the lymphocytes and tested compounds. In the case of necrosis, a synergy of action was observed in an NK cell co-incubated with YGB at a concentration of 1000 μg/mL. The observed enhancement of the NK cells' cytotoxic abilities in response to barley polysaccharides corresponds with previously published data [10]. Ex vivo studies conducted by Han et al. revealed that the oral administration of crude barley polysaccharides (BLE0; mainly composed of neutral sugars and uronic acid and containing minor levels of 2-keto-3-deoxy-mannooctanoic acid-like materials) significantly recovered the cyclophosphamide-induced reduction of NK cell activity against YAC-1 mouse T cell lymphoma cells [10].

In summary, the results of the present studies indicate that polysaccharide-rich young green barley extract may have immunomodulatory properties associated with the enhancement of NK cells' ability to recognize and eliminate human colon cancer cells without any side effects on normal colon epithelial cells. Furthermore, the obtained data also revealed the direct anticancer effect of the tested compounds, as well as high selectivity against neoplastic cells. Although the presented studies were performed only on cell lines, their importance is supported by the fact that they focused on NK92 cells with proven therapeutic utility in clinical studies. The discovered beneficial impact of the investigated extract on NK92 cells' anticancer properties brings hope for its therapeutic use, especially as an adjuvant in the currently applied immunotherapies for colon cancers. Despite the fact that over 90% of the extract components have been determined, further study is necessary to recognize other functional ingredients, next to the discovered carbohydrate-protein complexes, that may also influence the discovered immunoenhancement properties of YGB.

4. Materials and Methods

4.1. Reagents

Unless indicated otherwise, the chemicals used in this study were purchased from Sigma-Aldrich Co. LLC, Saint Louis, MO, USA.

4.2. Preparation of Young Green Barley Extract

The powdered young green barley juice (*Hordeum vulgare*) was purchased from Green Ways (Prague, Czech Republic). First, 5 g of the product was dissolved in 150 mL sterile water. Extraction was then carried out for 24 h on a rotator at room temperature. The resulting mixture was centrifuged (4075× *g*, 10 min, 20 °C) and the collected supernatant was filtered through a microbiological filter. The resulting filtrates were then subjected to a freeze-drying process. The lyophilizate thus obtained was stored at −20 °C. The stock solution of young green barley (25 mg/mL) was prepared by dissolving the lyophilizate in PBS (phosphate-buffered saline). The stock solution was stored at −20 °C. Working solutions of the extract were prepared by dissolving a stock solution in a culture medium. In this manuscript, young green barley extract has been expressed as YGB.

4.3. Sugar Content Determination

The total sugar content in YGB was determined using the phenol-sulphuric acid method, as previously described [39], using glucose as a standard. The tested extract (200 μL) was mixed with 5% aqueous phenol solution (200 μL) and concentrated H_2SO_4 (1 mL). After 10 min of incubation at room temperature, absorbance was recorded on a microplate reader, the BioTek ELx800 (Highland Park, Winooski, VT, USA) at 490 nm wavelength. The results were expressed as the percentage of YGB.

4.4. Protein Content Determination

The protein concentration in YGB was determined using a Pierce BCA Protein Assay Kit (Thermo Fisher Scientific, Waltham, MA, USA) following the manufacturer's instructions. Bovine serum albumin was used as a standard. Absorbance was recorded on a microplate reader, a BioTek ELx800, at 570 nm wavelength. The results were expressed as the percentage of YGB.

4.5. Proteins Separation—SDS-PAGE

YGB at a concentration of 25 mg/mL was solubilized in sample buffer (30% glycerol, 10% SDS, 0.5 M Tris-HCl, pH 6.8, 0.012% bromophenol blue, 5% β-mercaptoethanol) and boiled for 5 min. Then, 10 μL of the YGB sample, as well as 5 μL of Page Ruler Plus prestained protein ladder (Thermo Fisher Scientific, Waltham, MA, USA), were loaded onto the 12% polyacrylamide gel. Protein separation was carried out at 80 V for 30 min, followed by 200 V for 60 min for the resolving gel, using a Power Pac (Bio-Rad Laboratories, Hercules, CA, USA). The gel was stained for 30 min with Coomassie Brilliant Blue, dissolved in water/methanol/acetic acid (5/4/1, *v/v/v*), and the destaining of the gel was achieved using Coomassie solvent for 3 h. The estimation of the molecular mass of proteins was performed using the abovementioned molecular protein mass marker (from 15 to 250 kDa).

4.6. Chemical Analyses of Sugars

Monosugars were liberated from 1 mg of the YGB sample by hydrolysis, using 2 M trifluoroacetic acid (100 °C, 4 h), and were converted into alditol acetates via reduction with $NaBD_4$ and acetylation, as described elsewhere [40]. The obtained alditol acetates were analyzed via gas-liquid chromatography coupled with mass spectrometry (GLC-MS) technique.

To obtain the acetylated methyl glycosides, the YGB sample was subjected to methanolysis (2 M HCl/methanol, 85 °C, 2 h) and acetylation (pyridine/acetic anhydride, 85 °C, 30 min).

To obtain the peracetylated derivatives of methyl oximes, the YGB preparation was mildly hydrolyzed (1 M HCl_{aq} 50 °C, 1 h) and dried under a vacuum. Methoxylamine hydrochloride (100 μL, 0.18 M in pyridine) was added and the sample was incubated at 70 °C for 1 h. Acetic anhydride was added (100 μL), then a sample was incubated at 45 °C for 1 h. The sample was dried, dissolved in ethyl acetate, and analyzed by GLC-MS. The standard aldoses (mannose, glucose, galactose) and ketoses (fructose, tagatose, sorbose) were prepared in the same way.

The linkage position between sugar components in the sample was established by methylation analysis [41] and the extraction of permethylated products into chloroform, followed by acid hydrolysis (2 M TFA, 100 °C, 4 h), reduction with $NaBD_4$, and acetylation. The obtained partly methylated alditol acetates were analyzed by GLC–MS.

4.7. Fructan Content Determination

The fructan content was determined using the PAHBAH reducing sugar method, with the help of the Fructan Assay Procecedure Kit (Megazyme, Bray, Ireland), following the manufacturer's instructions. First, 100 mg increments of three independent lyophilized samples were used in triplicate. The PAHBAH color complex absorbance was recorded on a microplate reader, BioTek ELx800, at a 405 nm wavelength. Fructan content was calculated using the Megazyme Mega-Calc spreadsheet (Megazyme, Ireland).

4.8. FTIR Spectroscopy

An infrared absorption spectrum (FTIR-ATR) of the YGB was recorded using an FTIR spectrometer, the Nicolet 8700A (Thermo Scientific). The spectral range was from 400 to 4000 cm^{-1}.

4.9. GLC-MS Analysis

The sugar derivatives were analyzed by GLC-MS using a gas chromatograph 7890A (Agilent Technologies, Inc., Wilmington, DE, USA) connected to a mass selective detector (MSD 5975C, inert XL EI/CI (Agilent Technologies, Inc., Wilmington, DE, USA)). The chromatograph was equipped with an HP-5MS column (30 m × 0.25 mm) and used helium as a carrier gas. The temperature program was as follows: 150 °C for 5 min, increased to 310 °C (5 °C min^{-1}); the final temperature was maintained for 10 min. The components were identified based on their mass spectra and retention time and then compared with the known standards.

4.10. Mass Spectrometry

The MALDI-TOF MS (Matrix-Assisted Laser Desorption/Ionization—Time of Flight Mass Spectrometry) and MS-MS (tandem mass spectrometry) of YGB were performed with a SYNAPT G2-Si HDMS instrument (Waters Corporation, Milford, MA, USA) equipped with a 1 kHz Nd: YAG laser system (355 nm wavelength). Acquisition of the data was performed using the MassLynx software, version 4.1 SCN916 (Waters Corporation, Wilmslow, UK). To act as a matrix, 2,5-dihydroxybenzoic acid solution (20 µg/µL) in 50% acetonitrile was used. The sample was dissolved in 50% methanol to a concentration of 20 µg/µL and then mixed with matrix solution in the proportion of 1:1 (*v*/*v*) [42]. Spectra were recorded in a positive ion mode. For MS/MS experiments, isolated precursor ions were fragmented using a collision voltage of 15 V. MS data were collected for 120 s. Mass spectra were assigned with multi-point external calibration using red phosphorous (Sigma Aldrich, St. Louis, MO, USA).

4.11. Cell Lines

Human natural killer cells (NK-92) and human colon epithelial cells (CCD841 CoN) were obtained from the American Type Culture Collection (ATCC, Manassas, VA, USA). Human colon adenocarcinoma cells (LS180) were obtained from the European Collection of Cell Cultures (ECACC, Center for Applied Microbiology and Research, Salisbury, UK).

NK-92 cells were grown in Alpha Minimum Essential Medium without ribonucleosides and deoxyribonucleosides, and with 2 mM L-glutamine and 1.5 g/L sodium bicarbonate, supplemented with 0.2 mM inositol, 0.1 mM 2-mercaptoethanol, 0.02 mM folic acid, 200 U/mL recombinant IL-2, 12.5% horse serum, and 12.5% fetal bovine serum (FBS). CCD841 CoN cells were grown in Dulbecco's Modified Eagle's Medium (DMEM) supplemented with 10% FBS. LS180 cells were grown in a 1:1 mixture of DMEM and nutrient mixture, Ham F-12, supplemented with 10% FBS. All media were supplemented with penicillin (100 U/mL) and streptomycin (100 µg/mL). Cells were maintained in a humidified atmosphere of 95% air and 5% CO_2 at 37 °C.

4.12. Cell Treatment

In the case of the MTT, LDH, and BrdU assays, cells were seeded on 96-well microplates at a density of 5×10^4 cells/mL. In the case of microscope evaluation, cells were seeded on Lab-Tek Chambers Slides (Nunc) at a density of 5×10^4 cells/mL.

Basic variant: Immediately after seeding on cell culture vessels, the NK-92 cells were exposed to a series of YGB dilutions. Conversely, adherent cells (CCD841 CoN, LS180) were treated with a series of YGB dilutions 24 h after cell seeding into cell culture vessels. YGB dilutions were prepared in a culture medium suitable for individual cell lines.

The variant in co-cultures: Adherent cells were seeded on cell culture vessels. The following day, the culture medium was removed, and the cells were exposed to NK92 cells in the absence or presence of a series of YGB dilutions. The proportion of adherent cells to NK cells was 1:1. The YGB dilutions were prepared in a culture medium suitable for NK-92 cells.

In both variants of the experiments, cells were treated with YGB, or NK-92 cells, or YGB + NK-92 cells for 48 h, and afterward, proper investigations were performed.

4.13. Cytotoxicity Assessment—LDH Assay

After treatment, the cells growing on microplates were centrifuged at 300× *g* for 10 min, at room temperature. Then, the culture supernatants were collected in new 96-well microplates, which were used to perform an LDH assay following the manufacturer's instructions (In Vitro Toxicology Assay Kit—Lactate Dehydrogenase-Based, Sigma Aldrich, St. Louis, MO, USA). Absorbance was recorded on a microplate reader (BioTek ELx800), at a wavelength of 450 nm.

4.14. Assessment of Metabolic Activity—MTT Assay

After cell treatment, the metabolic activity of cells was investigated using the MTT test. In brief, cells were incubated with MTT solution (5 mg/mL in PBS) for 6 h. Then, formazan crystals were solubilized overnight in SDS buffer (pH 7.4) (10% SDS in 0.01 N HCl), and the product was quantified spectrophotometrically by measuring the absorbance at 570 nm wavelength using the microplate reader (BioTek ELx800).

4.15. Assessment of DNA Synthesis—BrdU Assay

After cell treatment, DNA synthesis was measured with a colorimetric immunoassay, Cell Proliferation ELISA BrdU (Roche Diagnostics GmbH, Penzberg, Germany), according to the manufacturer's instructions. Absorbance was recorded on a microplate reader (BioTek ELx800) at a wavelength of 450 nm.

4.16. Cell Death Detection—ELISA

After cell treatment, cell death (both apoptosis and necrosis) was assessed using the Cell Death Detection ELISA PLUS kit (Roche Diagnostics, Mannheim, Germany), according to the manufacturer's instructions. Studies were conducted in cytoplasmic fractions as well as the cell medium collected from above the cell cultures, which allowed the determining of nucleosome quantity in apoptotic and necrotic cells, respectively. Absorbance was measured at a wavelength of 405 nm, using a microplate reader (BioTek ELx800).

4.17. Cell Death Detection—Nuclear Double Staining

After treatment, cell death was visualized using nuclear double-staining. Cells were incubated for 5 min with a fluorochrome mixture: Hoechst 33342 (0.24 mg/mL) and propidium iodide (PI) (0.15 mg/mL). The stained cells were observed under an Olympus BX51 System microscope and the micrographs were prepared using the CellFamily AnalySIS software.

4.18. Statistical Analysis

The obtained data were developed in the following programs: Microsoft Excel 2010 and GraphPad Prism 5.0. The results are presented as the mean value and standard error of the mean (SEM). The data were analyzed using a one-way ANOVA test with Dunnett's or Tukey's post hoc tests, and column statistics were used for comparisons. Significance was accepted at $p < 0.05$.

Author Contributions: Conceptualization, M.K.L., I.K., A.C. and W.R.; formal analysis, M.K.L., I.K. and A.C.; funding acquisition, M.K.L. and W.R.; investigation, M.K.L., I.K. and M.C.; methodology, M.K.L., I.K., M.C. and A.C.; visualization, M.K.L., I.K. and M.C.; writing—original draft, M.K.L., I.K. and A.C.; writing—review and editing, M.K.L., I.K., M.C., A.C. and W.R. All authors have read and agreed to the published version of the manuscript.

Funding: This research was funded by Green Ways International, Prague, Czech Republic (Project Agreement No. 713/11/2019, 2019).

Institutional Review Board Statement: Not applicable.

Informed Consent Statement: Not applicable.

Data Availability Statement: The data presented in this study are available on request from the corresponding author. The data are not publicly available due to privacy.

Conflicts of Interest: The authors declare no conflict of interest. The funders had no role in the design of the study; in the collection, analyses, or interpretation of data; in the writing of the manuscript, or in the decision to publish the results.

Sample Availability: Samples of the investigated extract are available from the authors.

References

1. Miller, J.S.; Lanier, L.L. Natural killer cells in cancer immunotherapy. *Annu. Rev. Cancer Biol.* **2019**, *3*, 77–103. [CrossRef]
2. Miller, J.S. Biology of natural killer cells in cancer and infection. *Cancer Investig.* **2002**, *20*, 405–419. [CrossRef] [PubMed]
3. Kubota, K.; Matsuoka, Y.; Seki, H. Isolation of potent anti-inflammatory protein from barley leaves. *Jpn. J. Inflamm.* **1983**, *3*, 4.
4. Ohtake, H.; Nonaka, S.; Sawada, Y.; Hagiwara, Y.; Hagiwara, H.; Kubota, K. Studies on the constituents of green juice from young barley leaves. Effect on dietarily induced hypercholesterolemia in rats. *J. Pharm. Soc. Jpn.* **1985**, *105*, 1052–1057. [CrossRef]
5. Ohtake, H.; Yuasa, H.; Komura, C.; Miyauchi, T.; Hagiwara, Y.; Kubota, K. Studies on the constituents of green juice from young barley leaves. Antiulcer activity of fractions from barley juice. *J. Pharm. Soc. Jpn.* **1985**, *105*, 1046–1051. [CrossRef]
6. Lahouar, L.; El-Bok, S.; Achour, L. Therapeutic potential of young green barley leaves in prevention and treatment of chronic diseases: An overview. *Am. J. Chin. Med.* **2015**, *43*, 1311–1329. [CrossRef]
7. Zeng, Y.; Pu, X.; Yang, J.; Du, J.; Yang, X.; Li, X.; Li, L.; Zhou, Y.; Yang, T. Preventive and therapeutic role of functional ingredients of barley grass for chronic diseases in human beings. *Oxid. Med. Cell. Longev.* **2018**, *2018*, 3232080. [CrossRef]
8. Zeng, Y.; Pu, X.; Du, J.; Yang, X.; Li, X.; Mandal, M.S.N.; Yang, T.; Yang, J. Molecular mechanism of functional ingredients in barley to combat human chronic diseases. *Oxid. Med. Cell. Longev.* **2020**, *2020*, 3836172. [CrossRef]
9. Kim, H.; Yu, K.W.; Hong, H.D.; Shin, K.S. Effect of arabinoxylan- and rhamnogalacturonan I-rich polysaccharides isolated from young barley leaf on intestinal immunostimulatory activity. *J. Funct. Foods* **2017**, *35*, 384–390. [CrossRef]
10. Han, H.S.; Shin, J.S.; Song, Y.R.; Rhee, Y.K.; Cho, C.W.; Ryu, J.H.; Inn, K.S.; Hong, H.D.; Lee, K.T. Immunostimulatory effects of polysaccharides isolated from young barley leaves (*Hordeum vulgare* L.) with dual activation of Th1 and Th2 in splenic T cells and cyclophosphamide-induced immunosuppressed mice. *Int. J. Biol. Macromol.* **2020**, *147*, 954–964. [CrossRef]
11. Han, L.; Meng, M.; Guo, M.; Cheng, D.; Shi, L.; Wang, X.; Wang, C. Immunomodulatory activity of a water-soluble polysaccharide obtained from highland barley on immunosuppressive mice models. *Food Funct.* **2019**, *10*, 304–314. [CrossRef] [PubMed]
12. Ryu, D.S.; Kim, S.H.; Oh, S.M.; Lee, S.M.; Jeong, S.M.; Kim, S.H.; Lee, D.S. Immunosuppressive activities of water-soluble barley β-glucan on alloantigen reactive cell proliferation and cytotoxicity. *Food Sci. Biotechnol.* **2011**, *20*, 267–271. [CrossRef]
13. Misra, C.K.; Das, B.K.; Mukherjee, S.C.; Pattnaik, P. Effect of multiple injections of β-glucan on non-specific immune response and disease resistance in Labeo rohita fingerlings. *Fish Shellfish Immunol.* **2006**, *20*, 305–319. [CrossRef] [PubMed]
14. Aoe, S.; Yamanaka, C.; Fuwa, M.; Tamiya, T.; Nakayama, Y.; Miyoshi, T.; Kitazono, E. Effects of BARLEYmax and high-β-glucan barley line on short-chain fatty acids production and microbiota from the cecum to the distal colon in rats. *PLoS ONE* **2019**, *14*, e0218118. [CrossRef] [PubMed]
15. Nishimura, A.; Kitazono, E.; Imose, K.; Urita, S.; Matsui, T. Effect of functional barley BARLEYmax (Tantangara) on intestinal regulation: A double-blind, randomized, placebo-controlled parallel group comparison clinical study. *Jpn. Pharmacol. Ther.* **2017**, *45*, 1047–1055.
16. Sasaki, D.; Sasaki, K.; Kadowaki, Y.; Aotsuka, Y.; Kondo, A. Bifidogenic and butyrogenic effects of young barley leaf extract in an in vitro human colonic microbiota model. *AMB Express* **2019**, *9*, 182. [CrossRef]
17. Liu, C.; Li, X.; Li, Y.; Feng, Y.; Zhou, S.; Wang, F. Structural characterization and antimutagenic activity of a novel polysaccharide isolated from Sepiella maindroni ink. *Food Chem.* **2008**, *110*, 807–813. [CrossRef]
18. Grube, M.; Bekers, M.; Upite, D.; Kaminska, E. Infrared spectra of some fructans. *Spectroscopy* **2002**, *16*, 289–296. [CrossRef]
19. Yang, W.; Pei, F.; Shi, Y.; Zhao, L.; Fang, Y.; Hu, Q. Purification, characterization and anti-proliferation activity of polysaccharides from Flammulina velutipes. *Carbohydr. Polym.* **2012**, *88*, 474–480. [CrossRef]
20. Gieroba, B.; Krysa, M.; Wojtowicz, K.; Wiater, A.; Pleszczyńska, M.; Tomczyk, M.; Sroka-Bartnicka, A. The FT-IR and Raman spectroscopies as tools for biofilm characterization created by cariogenic Streptococci. *Int. J. Mol. Sci.* **2020**, *21*, 3811. [CrossRef]
21. Galinari, É.; Sabry, D.A.; Sassaki, G.L.; Macedo, G.R.; Passos, F.; Mantovani, H.C.; Rocha, H. Chemical structure, antiproliferative and antioxidant activities of a cell wall α-d-mannan from yeast Kluyveromyces marxianus. *Carbohydr. Polym.* **2017**, *157*, 1298–1305. [CrossRef] [PubMed]
22. Seymour, F.R.; Julian, R.L.; Jeanes, A.; Lamberts, B.L. Structural analysis of insoluble d-glucans by Fourier-transform, infrared difference-spectrometry: Correlation between structures of dextrans from strains of leuconostoc mesenteroides and of d-glucans from strains of streptococcus mutans. *Carbohydr. Res.* **1980**, *86*, 227–246. [CrossRef]
23. Zhang, J.; Li, Z.; Zhou, L.; Bao, J.; Xu, J. The modifications of a fructan from Anemarrhena asphodeloides Bunge and their antioxidant activities. *Int. J. Biol. Macromol.* **2020**, *164*, 4435–4443. [CrossRef] [PubMed]
24. Guerrant, G.O.; Moss, C.W. Determination of monosaccharides as aldononitrile, O-methyloxime, alditol, and cyclitol acetate derivatives by gas chromatography. *Anal. Chem.* **1984**, *56*, 633–638. [CrossRef]

25. Wahjudi, P.N.; Patterson, M.E.; Lim, S.; Yee, J.K.; Mao, C.S.; Lee, W.N.P. Measurement of glucose and fructose in clinical samples using gas chromatography/mass spectrometry. *Clin. Biochem.* **2010**, *43*, 198–207. [CrossRef]
26. Müller, J.; Aeschbacher, R.A.; Sprenger, N.; Boller, T.; Wiemken, A. Disaccharide-mediated regulation of sucrose: Fructan-6-fructosyltransferase, a key enzyme of fructan synthesis in barley leaves. *Plant Physiol.* **2000**, *123*, 265–274. [CrossRef]
27. Carpita, N.C.; Kanabus, J.; Housley, T.L. Linkage structure of fructans and fructan oligomers from Triticum aestivum and Festuca arundinacea leaves. *J. Plant. Physiol.* **1989**, *134*, 162–168. [CrossRef]
28. Vijn, I.; Smeekens, S. Fructan: More than a reserve carbohydrate? *Plant Physiol.* **1999**, *120*, 351–360. [CrossRef]
29. Nemeth, C.; Andersson, A.A.M.; Andersson, R.; Mangelsen, E.; Sun, C.; Åman, P. Relationship of grain fructan content to degree of polymerisation in different barleys. *Food Nutr. Sci.* **2014**, *5*, 581–589. [CrossRef]
30. Peukert, M.; Thiel, J.; Peshev, D.; Weschke, W.; van den Ende, W.; Mock, H.P.; Matros, A. Spatio-temporal dynamics of fructan metabolism in developing barley grains. *Plant Cell.* **2014**, *26*, 3728–3744. [CrossRef]
31. Franco-Robles, E.; López, M.G. Implication of fructans in health: Immunomodulatory and antioxidant mechanisms. *Sci. World J.* **2015**, *2015*, 289267. [CrossRef]
32. Young, I.D.; Latousakis, D.; Juge, N. The immunomodulatory properties of β-2,6 fructans: A comprehensive review. *Nutrients* **2021**, *13*, 1309. [CrossRef]
33. Meng, X.; Liang, H.; Luo, L. Antitumor polysaccharides from mushrooms: A review on the structural characteristics, antitumor mechanisms and immunomodulating activities. *Carbohydr. Res.* **2016**, *424*, 30–41. [CrossRef]
34. Arai, S.; Meagher, R.; Swearingen, M.; Myint, H.; Rich, E.; Martinson, J.; Klingemann, H. Infusion of the allogeneic cell line NK-92 in patients with advanced renal cell cancer or melanoma: A phase I trial. *Cytotherapy* **2008**, *10*, 625–632. [CrossRef]
35. Cheng, M.; Zhang, J.; Jiang, W.; Chen, Y.; Tian, Z. Natural killer cell lines in tumor immunotherapy. *Front. Med.* **2012**, *6*, 56–66. [CrossRef] [PubMed]
36. Gong, J.H.; Maki, G.; Klingemann, H.G. Characterization of a human cell line (NK-92) with phenotypical and functional characteristics of activated natural killer cells. *Leukemia* **1994**, *8*, 652–658. [PubMed]
37. Kawka, K.; Lemieszek, M.K.; Rzeski, W. Chemopreventive properties of young green barley extracts in in vitro model of colon cancer. *Ann. Agric. Environ. Med.* **2019**, *26*, 174–181. [CrossRef] [PubMed]
38. Lemieszek, M.K.; Rzeski, W. Enhancement of chemopreventive properties of young green barley and chlorella extracts used together against colon cancer cells. *Ann. Agric. Environ. Med.* **2020**, *27*, 591–598. [CrossRef] [PubMed]
39. Dubois, M.; Gilles, K.A.; Hamilton, J.K.; Rebers, P.A.; Smith, F. Colorimetric method for determination of sugars and related substances. *Anal. Chem.* **1956**, *28*, 350–356. [CrossRef]
40. Sawardeker, J.S.; Sloneker, J.H.; Jeanes, A. Quantitative determination of monosaccharides as their alditol acetates by gas liquid chromatography. *Anal. Chem.* **1965**, *37*, 1602–1604. [CrossRef]
41. Ciucanu, J.; Kerek, F. A simple and rapid method for the permethylation of carbohydrates. *Carbohydr. Res.* **1984**, *131*, 209–217. [CrossRef]
42. Choma, A.; Komaniecka, I. Characterization of cyclic β-glucans of Bradyrhizobium by MALDI-TOF mass spectrometry. *Carbohydr. Res.* **2011**, *346*, 1945–1950. [CrossRef] [PubMed]

MDPI
St. Alban-Anlage 66
4052 Basel
Switzerland
Tel. +41 61 683 77 34
Fax +41 61 302 89 18
www.mdpi.com

Molecules Editorial Office
E-mail: molecules@mdpi.com
www.mdpi.com/journal/molecules

www.ingramcontent.com/pod-product-compliance
Lightning Source LLC
Chambersburg PA
CBHW061847210926
51850CB00006B/992
9783036578736